Radiation Damage in Biomolecules and Cells

Radiation Damage in Biomolecules and Cells

Editors

Francesca Ballarini
Mario Pietro Carante

MDPI • Basel • Beijing • Wuhan • Barcelona • Belgrade • Manchester • Tokyo • Cluj • Tianjin

Editors

Francesca Ballarini
Physics Department
University of Pavia
Pavia
Italy

Mario Pietro Carante
Physics Department
University of Pavia
Pavia
Italy

Editorial Office
MDPI
St. Alban-Anlage 66
4052 Basel, Switzerland

This is a reprint of articles from the Special Issue published online in the open access journal *International Journal of Molecular Sciences* (ISSN 1422-0067) (available at: www.mdpi.com/journal/ijms/special_issues/Radiation_Damage_DNA_Cell_death).

For citation purposes, cite each article independently as indicated on the article page online and as indicated below:

LastName, A.A.; LastName, B.B.; LastName, C.C. Article Title. *Journal Name* **Year**, *Volume Number*, Page Range.

ISBN 978-3-0365-7487-5 (Hbk)
ISBN 978-3-0365-7486-8 (PDF)

Contents

About the Editors

Francesca Ballarini

Following her graduation in Biomedical Physics (1997) and several post-doc positions, including a fellowship provided by the University of Houston (TX, USA), Francesca Ballarini became permanent researcher (2005) and then Associate Professor (2017) at the University of Pavia (Italy), where she teaches basic and applied Physics and leads the Computational Radiobiology group, working on modelling/simulating the effects of ionizing radiation with applications in cancer therapy and space radiation protection. This research activity has been carried out in the framework of many research projects, mainly funded by the European Community, the Italian Ministry of University and Research, the Italian Space Agency and the Italian National Institute for Nuclear Physics.

Until now, FB published 110 articles in peer-reviewed international journals (Scopus h-index=31), delivered more than 60 conference presentations (39 of which by invitation), organized several conferences and supervised 40 thesis works. She is Editor or Guest Editor of 7 peer-reviewed international journals and she was peer-reviewer for several agencies, including the Swedish Space Board and the French Institute for Health. Furthermore, she was President of the Radiation Research Italian Society (2021–22) and member of the Board of Councillors of IARR-International Association for Radiation Research, and she is currently a member of the Technical Scientific Committee of the Applied Nuclear Energy Laboratory of the University of Pavia.

Mario Pietro Carante

Mario Carante is a researcher in Applied Physics at the University of Pavia, Italy. He obtained his M.Sc. title in Physics in 2013 and his PhD in Applied Physics in 2017. His research activity is focused on the theoretical study and modelization of the biological effects of radiation, in view of applications in the field of innovative cancer radiotherapy and space radiation protection. He has authored about 40 papers in peer-reviewed scientific journals, he has delivered more than 20 conference presentations and he is currently professor for several courses at the University of Pavia.

Preface to "Radiation Damage in Biomolecules and Cells"

Considering that ionizing radiation is widely used in medicine, and that occupational and environmental exposure raise radiation protection issues, it is mandatory for the scientific community to continuously update and improve the knowledge about the biophysical mechanisms governing the induction of radiation damage in biological targets, as well as to use such knowledge to optimize the use of ionizing radiation and the corresponding radiation protection strategies.

In this framework, the papers collected in this Special Issue deal with the effects of ionizing radiation at molecular and cellular levels, as well as possible applications in medicine and radiation protection. In particular, some works analysed the early stages of the interaction between ionizing radiation and DNA focusing on DNA cluster damage, which is widely considered as an important step towards subsequent endpoints including chromosome aberrations and cell death. Other studies investigated the consequences of hypoxia on the evolution of the initial DNA damage, and one work analysed genome-wide DNA alterations in irradiated human cells. Concerning radiation therapy, the response of cancer cells to different radiation types was characterized both in vitro and in vivo, and possible strategies to enhance such responses were investigated; furthermore, FLASH radiotherapy was reviewed. In the framework of cancer therapy side effects, the consequences of high doses on gene expression in healthy cells were analyzed, and the RBE (Relative Biological Effectiveness) for late effects in the central nervous system was predicted by the BIANCA biophysical model. Finally, the protective role of melatonin and Vitamin D, as well as hypothermia, was reviewed, valine radiolysis was investigated, the changes in radio-sensitivity of embrios during the early developmental stages of the preimplantation were analyzed, and epigenetic modifications induced by ionizing radiation were critically reviewed.

Although mainly directed to scientists working in the field of radiobiology, we believe that the multi-disciplinary features of the works reported in this Special Issue make it interesting also for physicists, chemists and medical doctors. We are therefore very grateful to all the colleagues who accepted our invitation and provided their valuable contributions.

Francesca Ballarini and Mario Pietro Carante
Editors

International Journal of
Molecular Sciences

MDPI

Editorial

Radiation Damage in Biomolecules and Cells

Mario P. Carante [1] and Francesca Ballarini [1,2,*]

1 INFN (Italian National Institute for Nuclear Physics), Sezione di Pavia, via Bassi 6, I-27100 Pavia, Italy; mario.carante@pv.infn.it
2 Physics Department, University of Pavia, via Bassi 6, I-27100 Pavia, Italy
* Correspondence: francesca.ballarini@unipv.it

Received: 12 October 2020; Accepted: 28 October 2020; Published: 1 November 2020

Ionizing radiation is widely used in medicine, both as a diagnostic tool and as a therapeutic agent. Furthermore, several exposure scenarios, like occupational and environmental exposure (e.g., radon), raise radiation protection issues, since the initial damage to DNA and chromosomes may lead either to cell death [1,2], which in turn can cause early tissue damage, or to cell conversion to malignancy and, eventually, cancer [3]. It is therefore mandatory for the scientific community to continuously update and improve the knowledge about the mechanisms governing the induction of radiation effects in biological targets, as well as to apply the acquired information to optimize the use of ionizing radiation and the corresponding radiation protection strategies.

Although the DNA double helix is widely recognized as the main radiation target, the features of the critical DNA damage type(s) leading to cell death or cell conversion to malignancy are still unclear. Moreover, the role played by other potential targets, which may be involved in bystander effects and other phenomena that may occur at low doses, deserves further investigation. Among the many possible medical applications, several aspects of cancer hadron therapy should be further addressed, including a more and more accurate RBE (Relative Biological Effectiveness) evaluation [4,5]. Such investigations can be carried out both experimentally, by means of in vitro and/or in vivo studies, and theoretically, by biophysical models and simulation codes.

In this framework, this Special Issue reports studies on the effects of ionizing radiation at molecular and cellular level, as well as possible applications in medicine and radiation protection. More specifically, some works [6–8] analyzed the early stages of the interaction between ionizing radiation and DNA, with particular attention to the role played by radiation track-structure in modulating the induction of DNA cluster damage, which is thought to lead to important endpoints including chromosome aberrations and cell death [9–11]. Another work [12] studied the impact of such damage on nuclear architecture, which is known to play a relevant role in DNA damage induction and repair [13]; other studies [14,15] investigated the consequences of hypoxia on the evolution of the initial damage, and one work analyzed genome-wide DNA alterations in X-irradiated human fibroblasts [16].

Among the studies aimed at better understanding the mechanisms involved in cancer radiation therapy, as well as improving its effectiveness, the metabolic response of glioblastoma multiforme (GBM) cells to photons and C-ions was characterized [17], the therapeutic potential of a new targeted c-SRC inhibitor molecule on X- or proton-irradiated GBM cells was investigated [18,19], the sensitization of breast cancer cells by checkpoint kinase 1 inhibition was studied [20], and the in vivo molecular response to proton therapy and its efficacy in a xenograft model was analyzed [21]. Furthermore, potential differences in the effects induced by passive or active clinical proton beams were explored [22], and FLASH radiotherapy was reviewed [23]. Within the field of cancer therapy side effects, the consequences of high radiation doses on gene expression in healthy tissues were analyzed [24] and the RBE for late effects in the central nervous system following proton or C-ion therapy was predicted by a novel biophysical model called BIANCA [25]. Since ionizing radiation can also be used to treat diseases other than cancer, radiation-stimulated translocation of CD166 and

CRYAB proteins to the endothelial surface was found to provide potential vascular targets on irradiated brain Arteriovenous Malformations (bAVMs) [26].

Finally, in the framework of radiation protection studies, the protective role of melatonin and Vitamin D [27], as well as hypothermia [28], was reviewed, and valine radiolysis by fast ions was investigated [29]. Furthermore, the changes in radio-sensitivity and responses to ionizing radiation of embryos during the early developmental stages of the preimplantation were analyzed [30], and epigenetic modifications induced by ionizing radiation were critically reviewed [31].

Funding: This research was funded by the Italian National Institute for Nuclear Physics, project MC-INFN/FLUKA.

Conflicts of Interest: The authors declare no conflict of interest.

References

1. Ballarini, F.; Altieri, S.; Bortolussi, S.; Carante, M.P.; Giroletti, E.; Protti, N. The BIANCA model/code of radiation induced cell death: Application to human cells exposed to different radiation types. *Radiat. Environ. Biophys.* **2014**, *53*, 525–533. [CrossRef]
2. Carante, M.P.; Altieri, S.; Bortolussi, S.; Postuma, I.; Protti, N.; Ballarini, F. Modeling radiation-induced cell death: Role of different levels of DNA damage clustering. *Radiat. Environ. Biophys.* **2015**, *54*, 305–316. [CrossRef] [PubMed]
3. Ballarini, F.; Ottolenghi, A. A model of chromosome aberration induction and CML incidence at low doses. *Radiat. Environ. Biophys.* **2004**, *43*, 165–171. [CrossRef] [PubMed]
4. Karger, C.P.; Peschke, P. RBE and related modeling in carbon-ion therapy. *Phys. Med. Biol.* **2017**, *63*, 01TR02. [CrossRef] [PubMed]
5. Carante, M.P.; Aricò, G.; Ferrari, A.; Kozlowska, K.; Mairani, A.; Ballarini, F. First benchmarking of the BIANCA model for cell survival predictions in a clinical hadron therapy scenario. *Phys. Med. Biol.* **2019**, *64*, 215008. [CrossRef] [PubMed]
6. Matsuya, Y.; Nakano, T.; Kai, T.; Shikazono, N.; Akamatsu, K.; Yoshii, Y.; Sato, T. A Simplified Cluster Analysis of Electron Track Structure for Estimating Complex DNA Damage Yields. *Int. J. Mol. Sci.* **2020**, *21*, 1701. [CrossRef]
7. Boscolo, D.; Krämer, M.; Fuss, M.C.; Durante, M.; Scifoni, E. Impact of Target Oxygenation on the Chemical Track Evolution of Ion and Electron Radiation. *Int. J. Mol. Sci.* **2020**, *21*, 424. [CrossRef]
8. Tang, N.; Bueno, M.; Meylan, S.; Perrot, Y.; Tran, H.N.; Freneau, A.; Dos Santos, M.; Vaurijoux, A.; Gruel, G.; Bernal, M.A.; et al. Assessment of Radio-Induced Damage in Endothelial Cells Irradiated with 40 kVp, 220 kVp, and 4 MV X-rays by Means of Micro and Nanodosimetric Calculations. *Int. J. Mol. Sci.* **2019**, *20*, 6204. [CrossRef]
9. Georgakilas, A.G.; O'Neill, P.; Stewart, R.D. Induction and repair of clustered DNA lesions: What do we know so far? *Radiat. Res.* **2013**, *180*, 100–109. [CrossRef]
10. Testa, A.; Ballarini, F.; Giesen, U.; Gil, O.M.; Carante, M.P.; Tello, J.J.; Langner, F.; Rabus, H.; Palma, V.; Pinto, M.; et al. Analysis of radiation-induced chromosomal aberrations on cell-by-cell basis after 4He-ion microbeam irradiation: Experimental data and simulations. *Radiat. Res.* **2018**, *189*, 597–604. [CrossRef]
11. Ballarini, F.; Altieri, S.; Bortolussi, S.; Carante, M.; Giroletti, E.; Protti, N. The role of DNA cluster damage and chromosome aberrations in radiation-induced cell killing: A theoretical approach. *Radiat. Prot. Dosim.* **2015**, *166*, 75–79. [CrossRef] [PubMed]
12. Tonnemacher, S.; Eltsov, M.; Jakob, B. Correlative Light and Electron Microscopy (CLEM) Analysis of Nuclear Reorganization Induced by Clustered DNA Damage Upon Charged Particle Irradiation. *Int. J. Mol. Sci.* **2020**, *21*, 1911. [CrossRef]
13. Ottolenghi, A.; Ballarini, F.; Biaggi, M. Modelling chromosomal aberration induction by ionising radiation: The influence of interphase chromosome architecture. *Adv. Space Res.* **2001**, *27*, 369–382. [CrossRef]
14. Liew, H.; Klein, C.; Zenke, F.T.; Abdollahi, A.; Debus, J.; Dokic, I.; Mairani, A. Modeling the Effect of Hypoxia and DNA Repair Inhibition on Cell Survival after Photon Irradiation. *Int. J. Mol. Sci.* **2019**, *20*, 6054. [CrossRef] [PubMed]
15. Liew, H.; Mein, S.; Debus, J.; Dokic, I.; Mairani, A. Modeling Direct and Indirect Action on Cell Survival After Photon Irradiation under Normoxia and Hypoxia. *Int. J. Mol. Sci.* **2020**, *21*, 3471. [CrossRef]

16. Nath, N.; Hagenau, L.; Weiss, S.; Tzvetkova, A.; Jensen, L.R.; Kaderali, L.; Port, M.; Scherthan, H.; Kuss, A.W. Genome-Wide DNA Alterations in X-Irradiated Human Gingiva Fibroblasts. *Int. J. Mol. Sci.* **2020**, *21*, 5778. [CrossRef]
17. Palma, A.; Grande, S.; Ricci-Vitiani, L.; Luciani, A.M.; Buccarelli, M.; Biffoni, M.; Dini, V.; Cirrone, G.A.P.; Ciocca, M.; Guidoni, L.; et al. Different mechanisms underlie the metabolic response of GBM stem like cells to ionizing radiation: Biological and MRS studies on effects of photons and carbon ions. *Int. J. Mol. Sci.* **2020**, *21*, 5167. [CrossRef]
18. Torrisi, F.; Minafra, L.; Cammarata, F.P.; Savoca, G.; Calvaruso, M.; Vicario, N.; Maccari, L.; Pérès, E.A.; Özçelik, H.; Bernaudin, M.; et al. SRC Tyrosine Kinase Inhibitor and X-rays Combined Effect on Glioblastoma Cell Lines. *Int. J. Mol. Sci.* **2020**, *21*, 3917. [CrossRef]
19. Cammarata, F.P.; Torrisi, F.; Forte, G.I.; Minafra, L.; Bravatà, V.; Pisciotta, P.; Savoca, G.; Calvaruso, M.; Petringa, G.; Cirrone, G.A.P.; et al. Proton Therapy and Src Family Kinase Inhibitor Combined Treatments on U87 Human Glioblastoma Multiforme Cell Line. *Int. J. Mol. Sci.* **2019**, *20*, 4745. [CrossRef]
20. Choi, C.; Cho, W.K.; Park, S.; Shin, S.-W.; Park, W.; Kim, H.; Choi, D.H. Checkpoint Kinase 1 (CHK1) Inhibition Enhances the Sensitivity of Triple-Negative Breast Cancer Cells to Proton Irradiation via Rad51 Downregulation. *Int. J. Mol. Sci.* **2020**, *21*, 2691. [CrossRef]
21. Cammarata, F.P.; Forte, G.I.; Broggi, G.; Bravatà, V.; Minafra, L.; Pisciotta, P.; Calvaruso, M.; Tringali, R.; Tomasello, B.; Torrisi, F.; et al. Molecular Investigation on a Triple Negative Breast Cancer Xenograft Model Exposed to Proton Beams. *Int. J. Mol. Sci.* **2020**, *21*, 6337. [CrossRef]
22. Michaelidesová, A.; Vachelová, J.; Klementová, J.; Urban, T.; Brabcová, K.P.; Kaczor, S.; Falk, M.; Falková, I.; Depeš, D.; Vondráček, V.; et al. In Vitro comparison of passive and active clinical proton beams. *Int. J. Mol. Sci.* **2020**, *21*, 5650. [CrossRef]
23. Hughes, J.R.; Parsons, J.L. FLASH Radiotherapy: Current Knowledge and Future Insights Using Proton-Beam Therapy. *Int. J. Mol. Sci.* **2020**, *21*, 6492. [CrossRef]
24. Kanakoglou, D.S.; Michalettou, T.-D.; Vasileiou, C.; Gioukakis, E.; Maneta, D.; Kyriakidis, K.V.; Georgakilas, A.G.; Michalopoulos, I. Effects of High-Dose Ionizing Radiation in Human Gene Expression: A Meta-Analysis. *Int. J. Mol. Sci.* **2020**, *21*, 1938. [CrossRef]
25. Carante, M.P.; Aricò, G.; Ferrari, A.; Karger, C.P.; Kozłowska, W.S.; Mairani, A.; Sala, P.R.; Ballarini, F. In Vivo validation of the BIANCA biophysical model: Benchmarking against rat spinal cord RBE data. *Int. J. Mol. Sci.* **2020**, *21*, 3973. [CrossRef] [PubMed]
26. McRobb, L.S.; McKay, M.J.; Gauden, A.J.; Lee, V.S.; Subramanian, S.; Thomas, S.G.; Wiedmann, M.; Moutrie, V.; Grace, M.; Zhao, Z.; et al. Radiation-Stimulated Translocation of CD166 and CRYAB to the Endothelial Surface Provides Potential Vascular Targets on Irradiated Brain Arteriovenous Malformations. *Int. J. Mol. Sci.* **2019**, *20*, 5830. [CrossRef]
27. Nuszkiewicz, J.; Woźniak, A.; Szewczyk-Golec, K. Ionizing radiation as a source of oxidative stress-the protective role of melatonin and vitamin D. *Int. J. Mol. Sci.* **2020**, *21*, 5804. [CrossRef]
28. Fukunaga, H. The Effect of Low Temperatures on Environmental Radiation Damage in Living Systems: Does Hypothermia Show Promise for Space Travel? *Int. J. Mol. Sci.* **2020**, *21*, 6349. [CrossRef] [PubMed]
29. Da Costa, C.; Muniz, G.S.V.; Boduch, P.; Rothard, H.; Da Silveira, E.F. Valine Radiolysis by H^+, He^+, N^+, and S^{15+} MeV Ions. *Int. J. Mol. Sci.* **2020**, *21*, 1893. [CrossRef] [PubMed]
30. Honjo, Y.; Ichinohe, T. Stage-Specific Effects of Ionizing Radiation during Early Development. *Int. J. Mol. Sci.* **2020**, *21*, 3975. [CrossRef]
31. Belli, M.; Tabocchini, M.A. Ionizing Radiation-Induced Epigenetic Modifications and Their Relevance to Radiation Protection. *Int. J. Mol. Sci.* **2020**, *21*, 5993. [CrossRef] [PubMed]

Publisher's Note: MDPI stays neutral with regard to jurisdictional claims in published maps and institutional affiliations.

International Journal of
Molecular Sciences

Article

A Simplified Cluster Analysis of Electron Track Structure for Estimating Complex DNA Damage Yields

Yusuke Matsuya [1,*], Toshiaki Nakano [2], Takeshi Kai [1], Naoya Shikazono [2], Ken Akamatsu [2], Yuji Yoshii [3] and Tatsuhiko Sato [1]

1 Nuclear Science and Engineering Center, Research Group for Radiation Transport Analysis, Japan Atomic Energy Agency, 2-4 Shirakata, Tokai, Ibaraki 319-1195, Japan; kai.takeshi@jaea.go.jp (T.K.); sato.tatsuhiko@jaea.go.jp (T.S.)

2 Department of Quantum life Science, Quantum Beam Science Research Directorate, National Institutes of Quantum and Radiological Science and Technology, 8-1-7 Umemidai, Kizugawa-shi, Kyoto 619-0215, Japan; nakano.toshiaki@qst.go.jp (T.N.); shikazono.naoya@qst.go.jp (N.S.); akamatsu.ken@qst.go.jp (K.A.)

3 Central Institute of Isotope Science, Hokkaido University, Kita-15 Nishi-7, Kita-ku, Sapporo, Hokkaido 060-0815, Japan; yoshii@ric.hokudai.ac.jp

* Correspondence: matsuya.yusuke@jaea.go.jp

Received: 30 January 2020; Accepted: 28 February 2020; Published: 2 March 2020

Abstract: Complex DNA damage, defined as at least two vicinal lesions within 10–20 base pairs (bp), induced after exposure to ionizing radiation, is recognized as fatal damage to human tissue. Due to the difficulty of directly measuring the aggregation of DNA damage at the nano-meter scale, many cluster analyses of inelastic interactions based on Monte Carlo simulation for radiation track structure in liquid water have been conducted to evaluate DNA damage. Meanwhile, the experimental technique to detect complex DNA damage has evolved in recent decades, so both approaches with simulation and experiment get used for investigating complex DNA damage. During this study, we propose a simplified cluster analysis of ionization and electronic excitation events within 10 bp based on track structure for estimating complex DNA damage yields for electron and X-ray irradiations. We then compare the computational results with the experimental complex DNA damage coupled with base damage (BD) measured by enzymatic cleavage and atomic force microscopy (AFM). The computational results agree well with experimental fractions of complex damage yields, i.e., single and double strand breaks (SSBs, DSBs) and complex BD, when the yield ratio of BD/SSB is assumed to be 1.3. Considering the comparison of complex DSB yields, i.e., DSB + BD and DSB + 2BD, between simulation and experimental data, we find that the aggregation degree of the events along electron tracks reflects the complexity of induced DNA damage, showing 43.5% of DSB induced after 70 kVp X-ray irradiation can be classified as a complex form coupled with BD. The present simulation enables us to quantify the type of complex damage which cannot be measured through *in vitro* experiments and helps us to interpret the experimental detection efficiency for complex BD measured by AFM. This simple model for estimating complex DNA damage yields contributes to the precise understanding of the DNA damage complexity induced after X-ray and electron irradiations.

Keywords: Monte Carlo radiation transport; complex DNA damage coupled with base damage; modelling of DNA damage yields

1. Introduction

Ionizing radiation within the human body causes DNA damage [1] both physically (i.e., energy deposition) [2,3] and chemically (i.e., free radical) reacting to the DNA target [4–7]. Among various

DNA damage types [8,9], DNA double-strand breaks (DSBs), defined as two strand breaks within 10 base pairs (bp) [2,10,11] are conventionally recognized as fatal DNA damage, which can lead to cell death with a certain probability [12]. Therefore, the relative biological effectiveness (RBE) at the endpoints of DSBs and cell survival has been investigated *in vitro* and *in silico* in previous reports [2,6,8,9,12–19]. Additionally, complex DNA damage composed of at least three vicinal lesions caused within 10–20 bp, such as DSBs coupled with strand breaks or base damage (BD), is believed to be more lethal to cells than simple DSBs [20,21] due to refractory damage [22,23]. To assess lethality, it is necessary to quantify the clustering degree of DNA damage with both experiments [24,25] and simulations [8,26–29]. It has been pointed out that a considerable amount of complex DNA damage can be induced after irradiation; however, the yield and nature of complex DNA damage are still difficult to be experimentally measured [30]. Due to the difficulty of measuring complex DNA damage, the validity of simulations has not been sufficiently demonstrated yet.

Experimental methods for detecting complex DNA damage have evolved in recent decades [24,30–36]. Among several techniques [24,30–36], microscopic operations coupled with an antibody against γ-H2AX [31–34] enables researchers to obtain spatial distributions of DSB induction in the cell nucleus. The adjacent degree of the DSB site can be evaluated from γ-H2AX foci volume in an assay [35,36]; however, the damage complexity at the nano-meter scale (the scale of DNA) cannot be obtained due to the limited spatial resolution from hundreds of nm to a few µm [35,36]. Meanwhile, complex DNA damage composed of BD can be quantified by means of gel electrophoresis after treatment of base excision repair enzymes [24,30,37,38] and fluorescence resonance energy transfer [39,40]. The structure of complex DNA damage, i.e., the number of lesions per damaged site was recently revealed with atomic force microscopy (AFM), where an individual BD in a complex damage site is specifically labelled with biotin/avidin coupled with an aldehyde reactive probe (ARP) [41]. However, whether all complex BD caused within a few bp can be detected in the AFM experiment [41] or not remains unknown. Thus, it is essential to compare simulations based on track structure with the corresponding experimental data, which contributes to interpreting the detection efficiency of complex BD.

A cluster analysis of inelastic interactions based on a radiation track structure in liquid water has been conducted as a powerful tool for estimating DNA damage yield [2,3,8,16,42]. The aim of this study is to develop a simple model for estimating complex DNA damage (i.e., isolated DSB, DSB + BD, DSB + 2BD, 2BD, 3BD, 4BD) based on previous simulation techniques [17,27,28], and to investigate the simulation accuracy and the nature of X-ray- (and electron-) induced complex damage in comparison with experimental data. This work finally quantifies the complexity of DNA damage under X-ray and electron irradiations.

2. Results and Discussions

2.1. Comparison between the Present Model and Experimental Complex Damage

We first checked the model validity, in detail yield ratio of base damage (BD) and single-strand break (SSB) = 1.3, comparing the simulation results (Equations (2)–(5) defined in Materials and Methods) with experimental data on yield fractions of SSB, double-strand beak (DSB), BD and complex BD (cBD) measured by enzymatic cleavage [37,38]. Figure 1A shows the fractions of SSB, DSB, BD and cBD obtained from our model and experiments [37,38]. Considering the good agreement between the estimation (6.9% total complex damage, 2.6% for DSB and 4.3% for cBD) and experimental data (7.2% total complex damage, 2.5% for DSB and 4.7% for cBD) in Figure 1, the developed model for complex BD considering BD/SSB = 1.3 seems reasonable. Additionally, the assumed induction ratio of BD/SSB = 1.3 could be further validated from the cross sections for impact to the DNA strand and base presented by Bernhardt et al. [43].

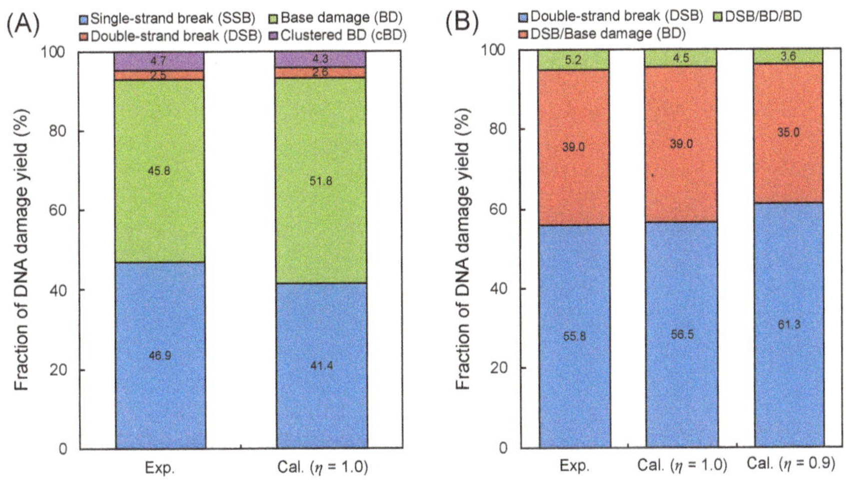

Figure 1. Model validity in comparison with experimental data. (**A**) is the fractions of isolated damage (single-strand break (SSB), base damage (BD)) and complex damage (double-strand break (DSB), complex BD (cBD)); (**B**) is the fractions of simple DSB, DSB/BD and DSB/BD/BD. The experimental data were obtained from the literature with treatment of base excision repair enzymes after 150 kVp X-ray irradiation [37,38] for (**A**) and with the direct observation technique with AFM after 70 kVp X-ray irradiation [41] for (**B**). The calculation represents the present model estimation based on Equations (2)–(5) and cluster analysis based on Table 1, and the determination of isolated or complex damage also follows the summary in Table 1.

Table 1. Classification of complex DNA damage coupled with BD in the present model.

DNA Damage Type	Symbol	Complexity	Event/Cluster N_{cl} *	Model for Yield Calculation
Single-strand breaks	SSB	Isolated	$N_{cl} < 2$	Y_{SSB}, (Equation (2))
Double-strand breaks (+ base damage)	DSB	Complex	$2 \le N_{cl} < 11$	$Y_{DSB} \times f\ (2 \le N_{cl} < 11)$, (Equation (3))
	DSB/BD	Complex	$11 \le N_{cl} < 20$	$Y_{DSB} \times f\ (11 \le N_{cl} < 20)$, (Equation (3))
	DSB/BD/BD	Complex	$20 \le N_{cl} < 29$	$Y_{DSB} \times f\ (20 \le N_{cl} < 29)$, (Equation (3))
Isolated base damage	BD	Isolated	$N_{cl} < 2$	Y_{BD}, (Equations (4))
Complex base damage	BD/BD	Complex	$2 \le N_{cl} < 11$	$Y_{cBD} \times f\ (2 \le N_{cl} < 11)$, (Equation (5))
	BD/BD/BD	Complex	$11 \le N_{cl} < 20$	$Y_{cBD} \times f\ (11 \le N_{cl} < 20)$, (Equation (5))
	BD/BD/BD/BD	Complex	$20 \le N_{cl} < 29$	$Y_{cBD} \times f\ (20 \le N_{cl} < 29)$, (Equation (5))

* The maximum inter-lesion distance, L_c, was set to be 10 bp for sampling the events per cluster.

To directly compare our simulation results with experimental complex damage data [41], we considered the experimental detection efficiency η in our simulation. Focusing on the type of complex damage, as shown in Figure 1B, the present cluster analysis for identifying complex DSB types, in which the efficiency $\eta = 0.9$ was considered, also accurately reproduced the experimental results [41]. It should be noted that the yields of a BD and a cBD, Y_{BD} and Y_{cBD}, are proportional to the efficiency η and the square η^2, respectively (e.g., $Y_{BD} \times \eta$ and $Y_{cBD} \times \eta^2$). Regarding the case of $\eta = 1.0$, the simulation was in better agreement with the experimental data [41] than the estimated value with $\eta = 0.9$, proving that the efficiency for detecting BD under atomic force microscopy (AFM) operation is over 90% which is consistent with the experimental efficiency [41]. Moreover, this agreement shown in Figure 1B suggests (i) that the density of ionizations and electronic excitation events clearly reflects the damage complexity, (ii) that inter-lesion distance in the AFM experiment is within approximately 10 bp and (iii) that 9 ionization and electronic excitation events are needed for inducing one additional BD at the DSB site.

Based on the comparison results shown in Figure 1, the simple methodology in the present DNA damage model is reasonable and sufficient for estimating complex BDs and for identifying the number of BDs at the DSB site. This indicates that the cluster analysis techniques [44,45] also are reasonable for predicting complex DNA damage.

2.2. Testing for Consistency with Other Simulations by Different Codes

We determined the complex damage type (i.e., double-strand break (DSB), DSB/base damage (BD), DSB/BD/BD) using the present model based on a simple cluster analysis using the number of events in a small volume with 10 bp radius, N_{cl} (based on Table 1). The additional comparison between the present model and the previous simulations, such as by Kurbuc [44] and Geant4-DNA [45], thus becomes strong evidence to prove that these assumptions of 9 events for inducing one additional BD at the DSB site and the yield ratio of BD/SSB = 1.3 is reasonable. Added to the comparison between the simulation and the experimental data [37,38,41] (as shown in Figure 1), we also tested if this cluster model can reproduce the previous simulation data on complex DSB [44,45], i.e., DSB+ and DSB++, or not.

The comparison results between the present model coupled with Particle and Heavy Ion Transport System (PHITS) (based on Table 2) and previous simulation data calculated by Kurbuc [44] and Geant4-DNA [45] are summarized in Table 3. The present model with the PHITS calculation indicates that more than 20–30% of DSBs can be classified as complex forms. Among the 0.3 keV, 1 keV, 10 keV and 100 keV electrons, the yield of DSBs in the case of the 0.3 keV electron is highest [17], while the fraction of complex DSBs (cDSBs composed of DSB+ and DSB++) for the case of 1 keV electron is highest. This comparison indicates that the present model can reproduce the simulation results by using a different prediction model from the previous simulations [44,45], suggesting that this identification approach for complex damage type using the number of events in sites with a 10 bp radius (9 and 12 events for inducing a BD and a SB at each DSB site, respectively) is reasonable. Note that the other codes provide the complex damage yield based on the events and energy deposition to the DNA cylinder [2,29,44,45] and ion cluster size [46]. Considering the good agreements with other simulations [44,45] and the experimental data [41], this simplified model with a cluster analysis is, therefore, sufficient for identifying the complexity of DNA damage induced after X-ray and electron irradiations.

Table 2. Classification of complex DSB coupled with strand breaks in the present model.

DNA Damage Type	Symbol	Complexity	Event/Cluster N_{cl} *	Model for Yield Calculation
Single-strand breaks	SSB	Isolated	$N_{cl} < 2$	Y_{SSB}, (Equation (2))
Double-strand breaks (+ strand breaks)	DSB	Simple	$2 \le N_{cl} < 14$	$Y_{DSB} \times f$ $(2 \le N_{cl} < 14)$, (Equation (3))
	DSB+	Complex	$14 \le N_{cl} < 26$	$Y_{DSB} \times f$ $(14 \le N_{cl} < 26)$, (Equation (3))
	DSB++	Complex	$26 \le N_{cl} < 38$	$Y_{DSB} \times f$ $(26 \le N_{cl} < 38)$, (Equation (3))

* The maximum inter-lesion distance, L_c, was set to be 10 bp for sampling the events per cluster.

Table 3. Benchmark test for complex double-strand break (DSB) compared to the other simulations.

Electron Energy (keV)	Type of Simulation Code	DNA Strand Break Yields (%)				Yield Ratio (%)
		SSB (i)	DSB (ii)	DSB+ (iii)	DSB++ (iii)	cDSB/DSB (iv)
0.30	PHITS	87.30	10.13	2.57	0.00	20.24
	Geant4-DNA	93.89	4.89	1.22	0.00	20.00
	KURBUC	87.19	9.18	3.28	0.35	28.31
1.00	PHITS	89.60	7.18	3.19	0.04	31.03
	Geant4-DNA	94.62	4.46	0.87	0.05	17.13
	KURBUC	90.65	6.51	2.55	0.29	30.38
10.0	PHITS	93.94	4.62	1.40	0.04	23.81
	KURBUC	96.59	2.61	0.72	0.08	23.45
100	PHITS	94.69	3.71	1.53	0.07	30.18
	KURBUC	96.64	2.69	0.63	0.05	20.07

(i) SSB = single-strand break; (ii) DSB = double-strand break; (iii) DSB+ and DSB++ = DSB coupled with a strand break and two strand breaks within 10 bp, respectively; (iv) cDSB is the sum of the percentages of DSB yields coupled with strand breaks (DSB+ and DSB++).

2.3. Interpretation of Complex Base Lesions Directly Measured by Atomic Force Microscopy (AFM) Imaging

The most recent atomic force microscopy (AFM) techniques for detecting complex base damage (BD) enables us to quantify the complex damage type, i.e., BD/BD, BD/BD/BD and BD/BD/BD/BD in addition to complex double-strand breaks (DSBs) coupled with BD (e.g., DSB/BD and DSB/BD/BD) [41]. Concerning this detection technique for BD, streptavidin labelling of DNA with aldehyde reactive prob (ARP) and AFM imaging has to be used. To reproduce the experimental data on complex BD directly measured by AFM imaging [41], we must consider several experimental efficiencies, such as the enzymatic reaction efficiency and spatial resolution for ARP. Regarding this, we next estimated the fractions of complex BD and complex DSBs coupled with BD, and tried to reproduce the experimental results measured by AFM imaging [41]. This comparison between simulation and experiments contributes to the interpretation of a direct observation technique for complex BD.

Figure 2 shows the comparison between the simulation results by our model and the experimental data [41], where (A) is the fraction of isolated BD and the complex damage composed of DSB and cBD, and (B) is the fractions of DSB, BD/BD, DSB/BD, BD/BD/BD, DSB/BD/BD and BD/BD/BD/BD. It should be noted that the inter-lesion distance within 10 bp was used for estimating the yield of complex damage. Regarding both cases of $\eta = 1.0$ and 0.9, the calculated fractions of complex damage were in good agreement with the experimental results [41], as shown in Figure 2A. Additionally, as shown in Figure 2B, highly complex forms (i.e., BD/BD/BD) and a high fraction of BD/BD can be seen in the simulation results, while the experiments did not show such tendencies.

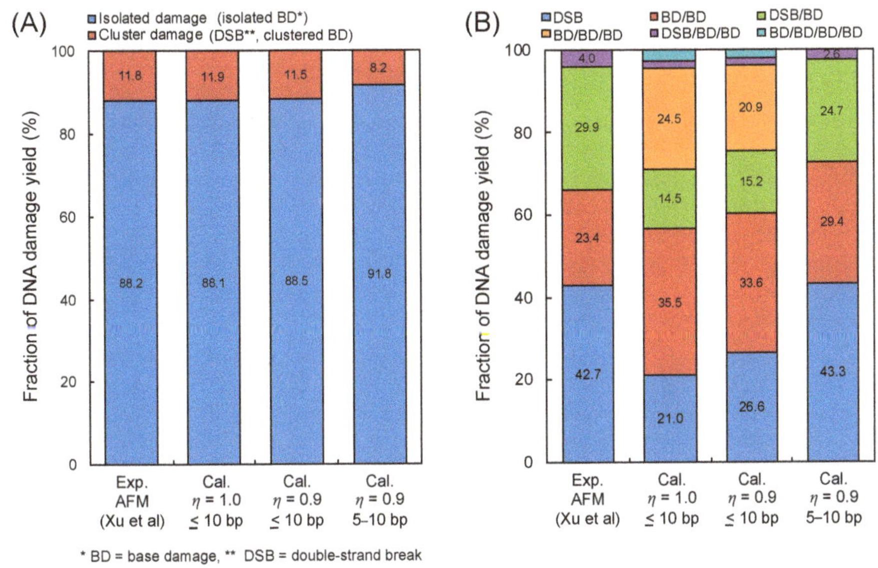

Figure 2. DNA damage complexity obtained by atomic force microscopy (AFM) imaging and simulation. (**A**) is the fraction of isolated base damage (BD) and complex damage (complex BD (cBD), double-strand break (DSB)); (**B**) is the type of complex DNA damage (DSB, BD/BD, DSB/BD, BD/BD/BD, DSB/BD/BD, BD/BD/BD/BD). The experimental results were obtained using AFM imaging for complex damage with BD reported by Xu et al. [41], where the energy of the X-ray is 70 kVp. The calculation represents the estimation based on Equations (2)–(5), where the determination of isolated or complex damage follows the summary in Table 1.

From the comparisons in Figure 2 to reproduce the experimental results, we next considered the loss for detecting complex BD composed of two vicinal lesions caused within 5 bp due to the big ARP size (approximately a 10 bp diameter). Considering these, the yield of complex BD with the detection loss, $Y_{cBD}{}^{*}$, can be expressed as:

$$Y_{cBD}{}^{*} = Y_{cBD}(10\ \text{bp}) - Y_{cBD}(5\ \text{bp}) \quad (1)$$

where $Y_{cBD}(10$ bp) and $Y_{cBD}(5$ bp) are the yields of complex BD caused within 10 bp and 5 bp, respectively. Under this assumption, the detection of complex BD containing more than two BDs within 10 bp was estimated to be completely impossible from the simulation standpoint. The estimations based on Equation (1) also are described as right bars in Figure 2A,B. The estimated fractions of complex damage were in good agreement with the experimental data, compared to the simulation without the detection loss. Based on these results, we found that the detection for complex BD (especially, 3BD and 4BD) was still difficult by means of an *in vitro* experiment. The experimental results for the cBD, i.e., 2BD, 3BD and 4BD, thus should be corrected using this simulation technique.

The experimental process for detecting BD in AFM [41] induced all types of BDs treated with DNA glycosylases (resulting apurinic/apyrimidinic (AP) sites) which can be labelled with an ARP that has both the alkoxyamine for the reaction with the aldehyde group of DNA and the biotin moiety for the subsequent labelling. Since the biotin moiety bound to DNA can be tagged with streptavidin (53 kDa as a large molecule), the resulting ARP-streptavidin complex can be visualized with AFM. During the series of the labelling processes, it was experimentally interpreted that overlapping and uncoupling biotins might result in a reduced efficiency for visualizing complex BDs induced within a few bp, i.e., down to about 70–80% [41]. Regarding this, the evaluation shown in Figure 2 might reflect a 3D structure problem and reduced labelling efficiency.

2.4. Estimation of Complex DNA Damage for Mono-Energetic Electron

The model for estimating the yields of complex DNA damage was tested in these comparisons shown in Figure 1 and Table 3. Using the present model, we further calculated the yields of the complex base damage (BD) including at least two BD, Y_{cBD}, and the ratio of the complex and isolated BD, Y_{cBD}/Y_{BD}, as functions of incident electron energy.

Y_{cBD} and Y_{cBD}/Y_{BD}, estimated by using the maximum inter-lesion distance L_c = 10 bp (used for comparisons) for mono-energetic electron exposure, are shown as the red circle in Figure 3. We also show Y_{DSB} and Y_{DSB}/Y_{SSB} using the blue square in Figure 3 to compare Y_{DSB} with Y_{cBD}. As reported previously [17], the peak of the number of linkages per incident electron energy was found to be around 0.3 keV electrons. The maximum Y_{DSB} and Y_{cBD} for a 10 bp cluster size (L_c = 10 bp) also were found to be 3.24×10^{-11} ($Gy^{-1}Da^{-1}$) (Y_{DSB}/Y_{SSB} = 14.5%) and 4.58×10^{-11} ($Gy^{-1}Da^{-1}$) (Y_{cBD}/Y_{BD} = 16.2%), respectively. The cluster size for complex BD (cBD) was conventionally set to be 3 bp, corresponding to 1.0 nm in other simulations [8]. Regarding this, we also calculated the cBD caused within 3 bp, which is shown as the green triangle in Figure 3. Changing the cluster size down to 3 bp, the maximum Y_{cBD} becomes much lower than that with L_c = 10 bp (Y_{cBD} = 1.37×10^{-11} ($Gy^{-1}Da^{-1}$), Y_{cBD}/Y_{BD} = 3.96% for 0.3 keV electron).

Using the present cluster analysis for predicting damage complexity, we finally estimated yield fractions of all complex DNA damage types (e.g., DSB, BD/BD, DSB/BD, BD/BD/BD, DSB/BD/BD, BD/BD/BD/BD). Figure 4 shows the fractions of complex DNA damage yields, i.e., simple DSB, BD/BD, DSB/BD, BD/BD/BD, DSB/BD/BD, BD/BD/BD/BD. Simple DSB and simple cBD (BD/BD) can be caused in a wide electron energy range of 100 eV–100 keV (shown as the blue and the light blue bar graph), while complex damage including more than two lesions cannot be induced in low energy electrons with 0.1 keV (shown as red, orange, green and purple in the bar graph). The integrated fractions of complex DNA damage including at least three lesions (e.g., DSB/BD, BD/BD/BD, DSB/BD/BD, BD/BD/BD/BD) for 0.3 keV electrons were the highest among the mono-energetic electrons below 100 keV, which was 63.1%. However, complex damage including four lesions (DSB/BD/BD, BD/BD/BD/BD) was induced by higher energy electrons than 0.3 keV, and the yield fraction for 1–100 keV electrons was 5.1% on average.

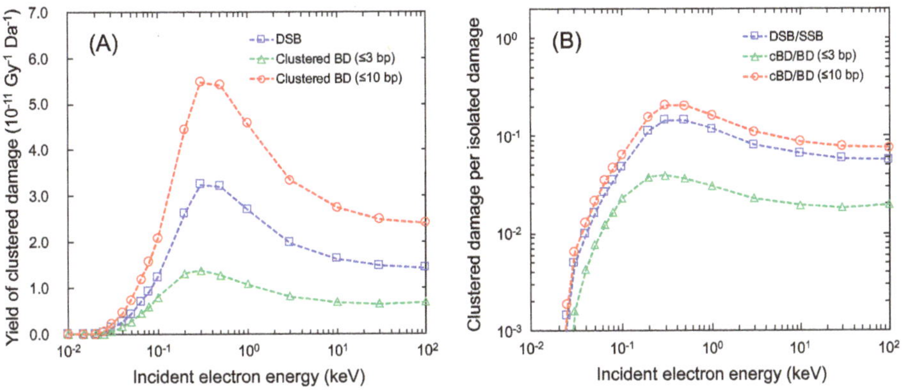

Figure 3. Yields of complex damage for mono-energetic electrons: (**A**) is the yields of complex base damage (BD) for various cluster sizes; (**B**) is the yield ratio of complex damage and isolated damage (complex BD (cBD) per BD and double-strand break (DSB) per single-strand break (SSB)). The blue square is Y_{DSB}/Y_{SSB} for a <10 bp cluster size, the red circle is the Y_{cBD}/Y_{BD} for a <10 bp cluster size, and the green triangle is the Y_{cBD}/Y_{BD} for a <3 bp cluster size.

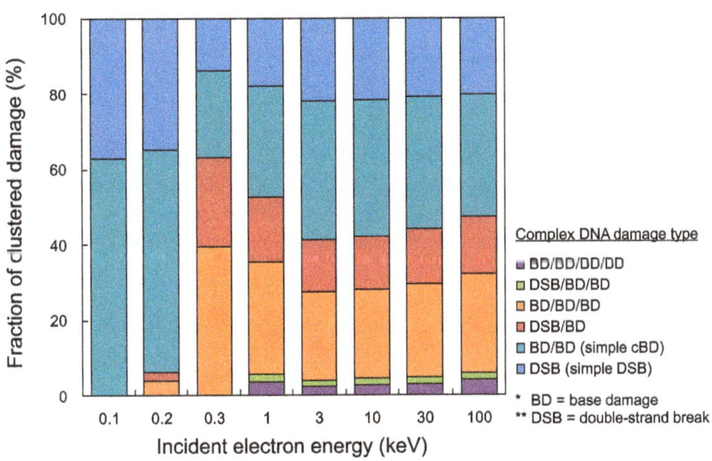

Figure 4. Fractions of cluster DNA damage yields for mono-energetic electron. The blue bar graph is simple double-strand break (DSB), light blue is simple complex base damage (cBD) as BD/BD, red is DSB/BD, orange is BD/BD/BD, green is DSB/BD/BD, and purple is BD/BD/BD/BD. It should be noted that the maximum inter-lesion distance and detection efficiency η were set as 10 bp and 1.0, respectively.

3. Materials and Methods

3.1. Monte Carlo Simulations of X-ray and Electron Processes

To compare our model with the experimental yield of complex DNA damage after X-ray irradiation [37,38,41], we used two types of spectrums of X-rays: 150 kVp [37,38] and 70 kVp [41], both with a 0.2 mm Al filter, and simulated them with the Particle and Heavy Ion Transport System (PHITS) code [47]. The X-ray spectrums were estimated according to the semiempirical model reported by

Tucker et al. [48]. We adapted an electron gamma shower (EGS) [49] mode for photon transport and an electron track structure mode (etsmode) [50–55] for electron transport in the PHITS calculation. It should be noted that the "etsmode" implemented in the PHITS code was verified from various endpoints including range, stopping power, nanodosimetry and double-strand break (DSB) yield [17]. We sampled the secondary electron spectrums generated by 150 kVp and 70 kVp X-rays and transported the electrons in liquid water. The cut-off energy of electrons was set as 1 eV. The coordinates of inelastic interactions were then output using a tally named "t-userdefined", as reported previously [17].

3.2. Model for Estimating Single- and Double-Strand Break Yields

Using the calculated electron spectrum, we estimated strand break yields. According to the DNA damage model previously developed [17], the number of the events, N_{event}, and that of the linkages within 3.4 nm (10 bp), $N_{\mathrm{link}(10)}$, were sampled to calculate the yield of single-strand breaks (SSBs) Y_{SSB} and that of double-strand breaks (DSBs) Y_{DSB} in $\mathrm{Gy}^{-1}\mathrm{Da}^{-1}$. Y_{SSB} and Y_{DSB} can be calculated by:

$$Y_{\mathrm{SSB}} = k_{\mathrm{SSB}} \frac{N_{\mathrm{event}}}{E_{\mathrm{dep}}}, \tag{2}$$

$$Y_{\mathrm{DSB}} = k_{\mathrm{DSB}} \frac{N_{\mathrm{link}(10)}}{E_{\mathrm{dep}}}, \tag{3}$$

where $k_{\mathrm{SSB}} = 5.66 \times 10^{-12}$ (keV $\mathrm{Gy}^{-1}\mathrm{Da}^{-1}$), $k_{\mathrm{DSB}} = 1.61 \times 10^{-13}$ (keV $\mathrm{Gy}^{-1}\mathrm{Da}^{-1}$) [17], and E_{dep} is the energy deposited by electron inelastic events in keV. These coefficients of k_{SSB} and k_{DSB} were found to reproduce the experimental yields of SSB and DSB after exposure to 220 kVp X-rays in our previous report [17]. It should be noted that 10 bp was defined as the classical distance for two SSBs leading to a DSB [11]. Based on Equations (2) and (3), we calculated the DNA strand break yields under 150 kVp, 70 kVp X-rays and mono-energetic electron irradiations.

3.3. Model for Estimating Base Damage Yields

We obtained the coefficient of base damage (BD) induction, k_{BD} (keV $\mathrm{Gy}^{-1}\mathrm{Da}^{-1}$), in the presence of a 10 mM tris (hydroxymethyl) aminomethane-HCl buffer from the experimental literature reporting the yield ratio of BD and single-strand break (SSB), which was given as $k_{\mathrm{BD}}/k_{\mathrm{SSB}} = 1.3$ [37]. It should be noted that the tris-HCl concentration was almost equivalent to liquid water due to the low radical scavenging capacity [37]. Based on this ratio, we deduced the coefficients of isolated BD and complex BD (cBD) to be $k_{\mathrm{BD}} = k_{\mathrm{SSB}} \times 1.3 = 7.36 \times 10^{-12}$ (keV $\mathrm{Gy}^{-1}\mathrm{Da}^{-1}$) and $k_{\mathrm{cBD}} = k_{\mathrm{DSB}} \times 1.3^2 = 2.72 \times 10^{-13}$ (keV $\mathrm{Gy}^{-1}\mathrm{Da}^{-1}$), respectively. Using the same manner as in the strand break case, the yields of BD and cBD can be expressed by:

$$Y_{\mathrm{BD}} = k_{\mathrm{BD}} \frac{N_{\mathrm{event}}}{E_{\mathrm{dep}}}, \tag{4}$$

$$Y_{\mathrm{cBD}}\,(L_c) = k_{\mathrm{cBD}} \frac{N_{\mathrm{link}(L_c)}}{E_{\mathrm{dep}}}, \tag{5}$$

where Y_{BD} and Y_{cBD} are the yields of BD and cBD in $\mathrm{Gy}^{-1}\mathrm{Da}^{-1}$, respectively, and L_c is the maximum distance in bp between two events to sample the linkage. Because the maximum inter-lesion distance, L_c, for cBD depends on the experimental detection conditions, Y_{cBD} should be the yield as a function of L_c. Complex BD can be detected as non-DSB in the enzymatic cleavage technique [37,38], while the diameter of an aldehyde reactive prob (ARP) avidin labelled at a BD site is equal to about 1.5 times the width of a DNA ladder (2.3 nm) in the direct observation technique for BD by atomic force microscopy (AFM) [41]. Based on these, the parameter L_c was set to be 10 bp (3.4 nm) for comparing the simulation experiments with both enzymatic cleavage and AFM techniques. Considering that the distance between two adjacent ARPs can be within 10 bp, we assumed that L_c was equal to 10 bp in the simulation for the AFM experiment. Using this value for parameter L_c, we calculated the yields of isolated and complex BD [37,38,41]. After comparison with the corresponding experimental data, we also estimated Y_{cBD} under the conditions of L_c = 3 bp as an example inducing much toxicity.

3.4. A Cluster Analysis for Determining Complex Damage Type

Because the modelling for the yield estimation of double-strand break (DSB) and complex base damage (cBD) does not enable us to determine the damage complexity, we added a cluster analysis for calculating the event density around a DSB or cBD site. We counted the number of ionization and electronic excitation events within a sampling site with the L_c radius at a DSB or cBD site based on the number of events per cluster, N_{cl}. Then, we determined the type of damage complexity from N_{cl} in reference to the cluster analysis reported by Yoshii et al. [27]. To reproduce the complex DSB coupled with base damage (BD) measured in the literature [41], it was estimated that approximately 9 events per cluster were needed on average to induce a BD within a 10 bp separation from a DSB or 2BD (cBD) site. Additionally, the mean N_{cl} to induce a simple DSB or 2BD site was calculated to be 6 in our previous study [17]. We therefore assumed that the ranges of N_{cl} to induce simple DSB (2BD), DSB+ BD (3BD) and DSB + 2BD (4BD) are $2 \leq N_{cl} < 11$ (6 on average), $11 \leq N_{cl} < 20$ (15 on average), $20 \leq N_{cl} < 29$ (24 on average), respectively. Under these assumptions, the mean deposition energy to cause DSB + BD within 10 bp, i.e., N_{cl} = 15, was estimated to be 121.7 eV, which was within the range of its reference value (102.5–122.6 eV) given by a different calculation [29]. The criteria for determining DNA damage type is summarized in Figure 5 and Table 1. Using the fraction of N_{cl}, $f(N_{cl})$, and the equations listed in Table 1, we calculated the yields of DNA damage, i.e., single-strand break (SSB), DSB (simple DSB), complex DSB (e.g., DSB + BD, DSB + 2BD), isolated BD, and complex BD (e.g., 2BD, 3BD, 4BD).

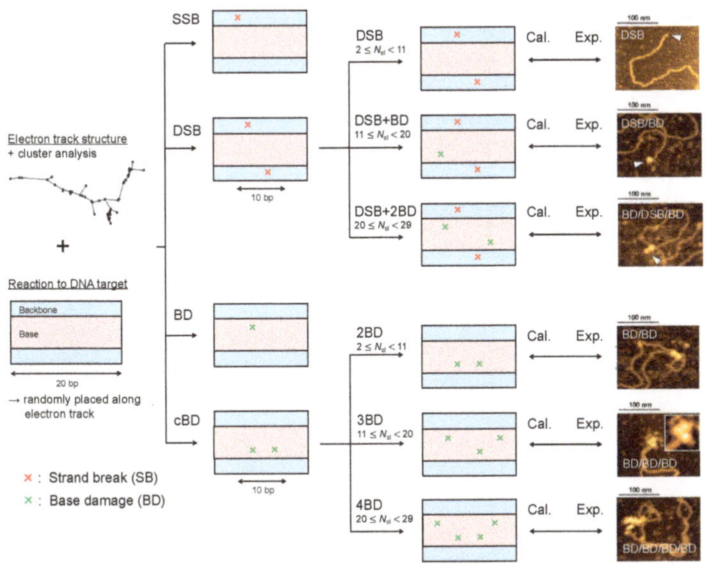

Figure 5. Identification of DNA damage type in the present simulation. The illustration on the left represents an electron track structure and a DNA cylinder. Considering the distance between two ionization and electronic excitation events, we identified single-strand break (SSB), double-strand break (DSB), base damage (BD), complex BD (cBD). The complexities of DSB and cBD then were determined from the number of events per cluster, N_{cl}, at the complex damage site (DSB and cBD). After estimating the yield of each DNA damage type, we compared the estimated results with experimental data [37,38,41] (right images are example pictures obtained with AFM [41]).

3.5. Comparison between Estimation and Experimental Data

We compared the estimated fractions of single-strand break (SSB), double-strand break (DSB), base damage (BD) and complex BD (cBD) with experimental results measured by enzymatic cleavage [37,38]. Using this comparison, we checked the model performance for estimating the yield ratios, BD/SSB and cBD/BD. Then, we compared the fractions of complex DSB (e.g., DSB, DSB/BD, DSB/BD/BD) estimated by this model with experimental data measured by the atomic force microscopy (AFM) imaging [41] to check the performance of this cluster analysis for estimating complex DSB.

After that, to interpret the most recent technique for detecting complex damage type using AFM [41], we compared the estimated fractions of isolated BD and complex damage (cBD and DSB) yields with those measured in the AFM experiment [41]. We also identified all complex damage type (e.g., DSB, BD/BD, DSB/BD, BD/BD/BD, DSB/BD/BD, BD/BD/BD/BD) shown in Figure 5 and calculated the fraction of each complex form. We then compared the estimated complex damage fractions with the experimental data [41]. It should be noted that the inter-lesion distance was set to ≤10 bp throughout this analysis.

3.6. Additional Benchmark Test of Cluster Analysis for Identifying Complex DSBs

Classically, the yields of complex double-strand breaks (DSBs) containing one or more strand breaks (SBs) within a 10 bp separation has been reported. These yields are designated as DSB+ (DSB coupled with one SSB) and DSB++ (DSB coupled with two strand breaks). Added to the comparison of DSB coupled with base damage (BD) between this estimation and the experimental data [37,38,41], we also compared our calculation results of DSB+ and DSB++ with the computational results in the literature [44,45]. Under the assumptions that the mean N_{cl} to induce an additional BD was 9 and the BD/SSB ratio was 1.3 [37], we then deduced that 12 events were needed on average for inducing an additional strand break at a DSB site. Applying the present cluster analysis and the criteria to additionally induce strand breaks (listed in Table 2), we identified the DSB+ from a DSB site on the basis of $14 \leq N_{cl} < 26$ and DSB++ from a DSB site on the basis of $26 \leq N_{cl} < 38$. Regarding the case of mono-energetic electron irradiation of 0.3 keV, 1 keV, 10 keV and 100 keV, we compared the DSB complexity (simple DSB, DSB +, DSB ++) estimated by our simulation to other simulations calculated by the Kurbuc [44] and the Geant4-DNA codes [45].

4. Conclusions

During this work, we compared experimental data of complex DNA damage to computational results based on an electron track structure calculated by the Particle and Heavy Ion Transport System (PHITS) code. Using the comparison between the simulations and the experimental data for complex DNA damage yields, it was confirmed (i) that the yield ratio of base damage (BD) and single-strand break (SSB) is 1.3; (ii) that the spatial pattern (density) of ionization and electronic excitation events reflects the damage complexity; and (iii) that 9 and 12 ionization and electronic excitation events are needed for inducing an additional one BD and one strand break at a double-strand break (DSB) site within a 10 bp separation. The present results indicate that conventional cluster analysis for inelastic interactions is a powerful tool and reasonable for reproducing the experimental complex DNA damage. Additionally, this model estimation can contribute to the interpretation of the experimental efficiency for detecting BD at complex damage sites and presents the fractions of complex DNA damage yields after X-ray and mono-energetic electron irradiations. While further development of this current model for high-LET irradiation is essential, this work can provide a simplified model for estimating the yield of complex DNA lesions which connects experimental and track structure simulations.

Author Contributions: Y.M. designed the study. Y.M. and T.K. performed Monte Carlo simulation for electron track structure. T.S. and Y.Y. supported the modelling of DNA damage yields and this cluster analysis. T.N., K.A. and N.S. provided the raw experimental data of base lesions and supported the comparison between the simulations and experiments for experimental interpretations. Y.M. wrote the manuscript. T.S. and N.S. supervised the study. All authors have read and agreed to the published version of the manuscript.

Funding: This work was supported by the Japan Society for the Promotion of Science KAKENHI (Grant no. 16H02959, 17K07022 and 19K17215).

Acknowledgments: We would like to thank Kenneth L. Sutherland (Graduate School of Medicine, Hokkaido University, Sapporo, Japan) who kindly spared time for the English proofreading of the manuscript.

Conflicts of Interest: The authors declare no conflict of interest.

References

1. Wouters, B.G.; Begg, C.A. Irradiation-induced damage and the DNA damage response. In *Basic Clinical Radiobiology*; Joiner, M., van der Kogel, A.J., Eds.; Hodder Arnold: London, UK, 2009; pp. 11–26.
2. Nikjoo, H.; Goodhead, D.T.; Charlton, D.E.; Paretzke, H.G. Energy deposition in small cylindrical targets by monoenergetic electrons. *Int. J. Radiat. Biol.* **1991**, *60*, 739–756. [CrossRef]
3. Date, H.; Yoshii, Y.; Sutherland, K.L. Nanometer site analysis of electron tracks and dose localization in bio-cells exposed to X-ray irradiation. *Nucl. Instr. Methods Phys. Res.* **2009**, *267*, 1135–1138. [CrossRef]
4. Von Sonntag, C. *Free-Radical-Induced DNA Damage and Its Repair: A Chemical Perspective*; Springer: Berlin/Heidelberg, Germany, 2006; ISBN 9783540305927.
5. Kreipl, M.S.; Friedland, W.; Paretzke, H.G. Time- and space-resolved Monte Carlo study of water radiolysis for photon, electron and ion irradiation. *Radiat. Environ. Biophys.* **2009**, *48*, 11–20. [CrossRef] [PubMed]
6. Yokoya, A.; Cunniffe, S.M.T.; O'Neill, P. Effect of Hydration on the Induction of Strand Breaks and Base Lesions in Plasmid DNA Films by γ-Radiation. *J. Am. Chem. Soc.* **2002**, *124*, 8859–8866. [CrossRef] [PubMed]
7. Ma, J.; Kumar, A.; Muroya, Y.; Yamashita, S.; Sakurai, T.; Denisov, S.A.; Sevilla, M.D.; Adhikary, A.; Seki, S.; Mostafavi, M. Quasi-free electron attachment to nucleoside in solution: Excited anion radical formation and its dissociation. *Nat. Commun.* **2019**, *10*, 102. [CrossRef] [PubMed]
8. Watanabe, R.; Rahmanian, S.; Nikjoo, H. Spectrum of Radiation-Induced Clustered Non-DSB Damage–A Monte Carlo Track Structure Modeling and Calculations. *Radiat. Res.* **2015**, *183*, 525–540. [CrossRef] [PubMed]
9. Nikjoo, H.; Emfietzoglou, D.; Liamsuwan, T.; Taleei, R.; Liljequist, D.; Uehara, S. Radiation track, DNA damage and response—A review. *Rep. Prog. Phys.* **2016**, *79*, 116601. [CrossRef]
10. Hanai, M.; Yazu, K.; Hieda, R. On the experimental distinction between SSBs and DSBs in circular DNA. *Int. J. Radiat. Biol.* **1998**, *73*, 475–479. [CrossRef]
11. Nikitaki, Z.; Nikolov, V.; Mavragani, I.V.; Mladenov, E.; Mangelis, A.; Laskaratou, D.A.; Fragkoulis, G.I.; Hellweg, C.E.; Martin, O.A.; Emfietzoglou, D.; et al. Measurement of complex DNA damage induction and repair in human cellular systems after exposure to ionizing radiations of varying linear energy transfer (LET). *Free Radic. Res.* **2016**, *50*, S64–S78. [CrossRef]
12. Frankenberg-Schwager, M.; Frankenberg, D. DNA double-strand breaks: Their repair and relationship to cell killing in yeast. *Int. J. Radiat. Biol.* **1990**, *58*, 569–575. [CrossRef]
13. Prise, K.M.; Folkard, M.; Davies, S.; Michael, B.D. Measurement of DNA Damage and Cell Killing in Chinese Hamster V79 Cells Irradiated with Aluminum Characteristic Ultrasoft X Rays. *Radiat. Res.* **1989**, *117*, 489–499. [CrossRef] [PubMed]
14. Prise, K.M.; Davies, S.; Michael, B.D. Evidence for induction of DNA double-strand breaks at paired radical sites. *Radiat. Res.* **1993**, *134*, 102–106. [CrossRef] [PubMed]
15. Chen, Z.; Li, J.; Li, C.; Qiu, R.; Wu, Z. A modified microdosimetric kinetic model for relative biological effectiveness calculation. *Phys. Med. Biol.* **2018**, *63*, 015008. [CrossRef] [PubMed]
16. Yachi, Y.; Yoshii, Y.; Matsuya, Y.; Mori, R.; Oikawa, J.; Date, H. Track Structure Study for Energy Dependency of Electrons and X-rays on DNA Double-Strand Break Induction. *Sci. Rep.* **2019**, *9*, 17649. [CrossRef]
17. Matsuya, Y.; Kai, T.; Yoshii, Y.; Yachi, Y.; Naijo, S.; Date, H.; Sato, T. Modelling of yield estimation for DNA strand breaks based on Monte Carlo simulations of electron track structure in liquid water. *J. Appl. Phys.* **2019**, *126*, 124701. [CrossRef]
18. Tang, N.; Bueno, M.; Meylan, S.; Perrot, Y.; Tran, H.N.; Freneau, A.; Santos, M.D.; Vaurijoux, A.; Gruel, G.; Bernal, M.; et al. Assessment of Radio-Induced Damage in Endothelial Cells Irradiated with 40 kVp, 220 kVp, and 4 MV X-rays by Means of Micro and Nanodosimetric Calculations. *Int. J. Mol. Sci.* **2019**, *20*, 6204. [CrossRef]

19. Parisi, A.; Olko, P.; Swakon, J.; Horwacik, T.; Jablonski, H.; Malinowski, L.; Nowak, T.; Struelens, L.; Vanhavere, L. Modeling the radiation-induced cell death in a therapeutic proton beam using thermoluminescent detectors and radiation transport simulations. *Phys. Med. Biol.* **2019**, *65*, 015008. [CrossRef]
20. Ward, J.F. The complexity of DNA damage: Relevance to biological consequences. *Int. J. Radiat. Biol.* **1994**, *66*, 427–432. [CrossRef]
21. Goodhead, D.T. Initial events in the cellular effects of ionizing radiations: Clustered damage in DNA. *Int. J. Radiat. Biol.* **1994**, *65*, 7–17. [CrossRef]
22. Eccles, L.J.; O'Neill, P.; Lomax, M.E. Delayed repair of radiation induced clustered DNA damage: Friend or foe? *Mutat. Res.* **2011**, *711*, 134–141. [CrossRef]
23. Sage, E.; Shikazono, N. Radiation-induced clustered DNA lesions: Repair and mutagenesis. *Free Radic. Biol. Med.* **2017**, *107*, 125–135. [CrossRef] [PubMed]
24. Terato, H.; Tanaka, R.; Nakaarai, Y.; Nohara, T.; Doi, Y.; Iwai, S.; Hirayama, R.; Furusawa, Y.; Ide, H. Quantitative Analysis of Isolated and Clustered DNA Damage Induced by Gamma-rays, Carbon Ion Beams, and Iron Ion Beams. *J. Radiat. Res.* **2008**, *49*, 133–146. [CrossRef] [PubMed]
25. Tokuyama, Y.; Furusawa, Y.; Ide, H.; Yasui, A.; Terato, H. Role of isolated and clustered DNA damage and the post-irradiating repair process in the effects of heavy ion beam irradiation. *J. Radiat. Res.* **2015**, *56*, 446–455. [CrossRef] [PubMed]
26. Mokari, M.; Alamatsaz, M.H.; Moeini, H.; Babaei-Brojeny, A.A.; Taleei, R. Track structure simulation of low energy electron damage to DNA using Geant4-DNA. *Biomed. Phys. Eng. Express.* **2018**, *4*, 065009. [CrossRef]
27. Yoshii, Y.; Sasaki, K.; Matsuya, Y.; Date, H. Cluster analysis for the probability of DSB site induced by electron tracks. *Nucl. Instrum. Methods Phys. Res.* **2015**, *350*, 55–59. [CrossRef]
28. Francis, Z.; Villagrasa, C.; Clairand, I. Simulation of DNA damage clustering after proton irradiation using an adapted DBSCAN algorithm. *Comput. Meth. Prog. Biol.* **2011**, *101*, 265–270. [CrossRef]
29. Liu, W.; Tan, Z.; Zhang, L.; Champion, C. Investigation on the correlation between energy deposition and clustered DNA damage induced by low-energy electrons. *Radiat. Environ. Biophys.* **2018**, *57*, 179–218. [CrossRef]
30. Shikazono, N.; Noguchi, M.; Fujii, K.; Uruushibara, A.; Yokoya, A. The Yield, Processing, and Biological Consequences of Clustered DNA Damage Induced by Ionizing Radiation. *J. Radiat. Res.* **2009**, *50*, 27–36. [CrossRef]
31. Rogakou, E.P.; Pilch, D.R.; Orr, A.H.; Ivanova, V.S.; Bonner, W.M. DNA Double-stranded Breaks Induce Histone H2AX Phosphorylation on Serine 139. *J. Biol. Chem.* **1998**, *273*, 5858–5868. [CrossRef]
32. Matsuya, Y.; Ohtsubo, Y.; Tsutsumi, K.; Sasaki, K.; Yamazaki, R.; Date, H. Quantitative estimation of DNA damage by photon irradiation based on the microdosimetric-kinetic model. *J. Radiat. Res.* **2014**, *55*, 484–493. [CrossRef]
33. Mori, R.; Matsuya, M.; Yoshii, Y.; Date, H. Estimation of the radiation-induced DNA double-strand breaks number by considering cell cycle and absorbed dose per cell nucleus. *J. Radiat. Res.* **2018**, *59*, 253–260. [CrossRef] [PubMed]
34. Matsuya, Y.; Satou, Y.; Hamada, N.; Date, H.; Ishikawa, M.; Sato, T. DNA damage induction during localized chronic exposure to an insoluble radioactive microparticle. *Sci. Rep.* **2019**, *9*, 10365. [CrossRef] [PubMed]
35. Hagiwara, Y.; Niimi, A.; Isono, M.; Yamauchi, M.; Yasuhara, T.; Limsirichaikul, S.; Oike, T.; Sato, H.; Held, K.D.; Nakano, T.; et al. 3D-structured illumination microscopy reveals clustered DNA double-strand break formation in widespread γH2AX foci after high LET heavy-ion particle radiation. *Oncotarget* **2017**, *8*, 109370–109381. [CrossRef]
36. Jezkova, L.; Zadneprianetc, M.; Kulikova, E.; Smirnova, E.; Bulanova, T.; Depes, D.; Falkova, I.; Boreyko, A.; Krasavin, E.; Davidkova, M.; et al. Particles with similar LET values generate DNA breaks of different complexity and reparability: A high-resolution microscopy analysis of γH2AX/53BP1 foci. *Nanoscale* **2018**, *10*, 1162–1179. [CrossRef]
37. Shiina, T.; Watanabe, R.; Shiraishi, I.; Suzuki, M.; Sugaya, Y.; Fujii, K.; Yokoya, A. Induction of DNA damage, including abasic sites, in plasmid DNA by carbon ion and X-ray irradiation. *Radiat. Environ. Biophys.* **2013**, *52*, 99–112. [CrossRef]
38. Shiraishi, I.; Shikazono, N.; Suzuki, M.; Fujii, K.; Yokoya, A. Efficiency of radiation-induced base lesion excision and the order of enzymatic treatment. *Int. J. Radiat. Biol.* **2017**, *93*, 295–302. [CrossRef]

39. Akamatsu, K.; Shikazono, N.; Saito, T. Localization Estimation of Ionizing Radiation-Induced Abasic Sites in DNA in the Solid State Using Fluorescence Resonance Energy Transfer. *Radiat. Res.* **2015**, *185*, 105–113. [CrossRef]
40. Akamatsu, K.; Shikazono, N.; Saito, T. New method for estimating clustering of DNA lesions induced by physical/chemical mutagens using fluorescence anisotropy. *Anal. Biochem.* **2017**, *536*, 78–89. [CrossRef]
41. Xu, X.; Nakano, T.; Tsuda, M.; Kanamoto, R.; Hirayama, R.; Uzawa, A.; Ide, H. Direct observation of damage clustering in irradiated DNA with atomic force microscopy. *Nucleic Acids Res.* **2020**, *48*, e18. [CrossRef]
42. Bäckström, G.; Galassi, M.E.; Tilly, N.; Ahnesjö, A.; Fernández-Varea, J.M. Track structure of protons and other light ions in liquid water: Applications of the L Ion Track code at the nanometer scale. *Med. Phys.* **2013**, *40*, 064101. [CrossRef]
43. Bernhardt, P.; Paretzke, H.G. Calculation of electron impact ionization cross sections of DNA using the Deutsch–Märk and Binary–Encounter–Bethe. *Int. J. Mass Spectrom.* **2003**, *223–224*, 599–611. [CrossRef]
44. Nikjoo, H.; O'Neill, P.; Goodhead, D.T.; Terrissol, M. Computational modelling of low-energy electron-induced DNA damage by early physical and chemical events. *Int. J. Radiat. Biol.* **1997**, *71*, 467–483. [CrossRef] [PubMed]
45. Xie, W.; Li, J.; Li, C.; Qiu, R.; Yan, C.; Zeng, Z. Comparison of direct DNA strand break simulated with different DNA models. *Radiat. Prot. Dos.* **2013**, *156*, 283–288. [CrossRef] [PubMed]
46. Garty, G.; Schulte, R.; Shchemelinin, S.; Leloup, C.; Assaf, G.; Breskin, A.; Chechik, R.; Bashkirov, V.; Milligan, J.; Grosswendt, B. A nanodosimetric model of radiation-induced clustered DNA damage yields. *Phys. Med. Biol.* **2010**, *55*, 761–781. [CrossRef]
47. Sato, T.; Iwamoto, Y.; Hashimoto, S.; Ogawa, T.; Furuta, T.; Abe, S.; Kai, T.; Tsai, P.-E.; Matsuda, N.; Iwase, H.; et al. Features of Particle and Heavy Ion Transport code System (PHITS) version 3.02. *J. Nucl. Sci. Technol.* **2018**, *55*, 684–690. [CrossRef]
48. Tucker, D.M.; Barnes, G.T.; Chakraborty, D.P. Semiempirical model for generating tungsten target X-ray spectra. *Med. Phys.* **1991**, *18*, 211–218. [CrossRef]
49. Hirayama, H.; Namito, Y.; Bielajew, A.F.; Wilderman, S.J.; Nelson, W.R. *The EGS5 Code System, SLAC Report 730*; United States Department of Energy: Washington, DC, USA, 2005.
50. Kai, T.; Yokoya, A.; Ukai, M.; Watanabe, R. Cross sections, stopping powers, and energy loss rates for rotational and phonon excitation processes in liquid water by electron impact. *Radiat. Phys. Chem.* **2015**, *108*, 13–17. [CrossRef]
51. Kai, T.; Yokoya, A.; Ukai, M.; Fujii, K.; Watanabe, R. Dynamics of low energy electrons in liquid water with consideration of Coulomb interaction with positively charged water molecules induced by electron collision. *Radiat. Phys. Chem.* **2014**, *104*, 16–22. [CrossRef]
52. Kai, T.; Yokoya, A.; Ukai, M.; Watanabe, R. Deceleration processes of secondary electrons produced by a high-energy Auger electron in a biological context. *Int. J. Rad. Biol.* **2016**, *92*, 645–659. [CrossRef]
53. Kai, T.; Yokoya, A.; Ukai, M.; Fujii, K.; Watanabe, R. Thermal equilibrium and prehydration processes of electrons injected into liquid water calculated by dynamic Monte Carlo method. *Radiat. Phys. Chem.* **2015**, *115*, 1–5. [CrossRef]
54. Kai, T.; Yokoya, A.; Ukai, M.; Fujii, K.; Watanabe, R. Dynamic Behavior of Secondary Electrons in Liquid Water at the Earliest Stage upon Irradiation: Implications for DNA Damage Localization Mechanism. *J. Phys. Chem. A* **2016**, *120*, 8228–8233. [CrossRef] [PubMed]
55. Kai, T.; Yokoya, A.; Ukai, M.; Fujii, K.; Toigawa, T.; Watanabe, R. A significant role of non-thermal equilibrated electrons in the formation of deleterious complex DNA damage. *Phys. Chem. Chem. Phys.* **2018**, *20*, 2838. [CrossRef] [PubMed]

International Journal of
Molecular Sciences

MDPI

Article

Impact of Target Oxygenation on the Chemical Track Evolution of Ion and Electron Radiation

Daria Boscolo [1,*], Michael Krämer [1], Martina C. Fuss [1] and Marco Durante [1,2,3] and Emanuele Scifoni [3]

[1] Biophysics Department, GSI Helmholtzzentrum für Schwerionenforschung, 64291 Darmstadt, Germany; m.kraemer@gsi.de (M.K.); m.fuss@gsi.de (M.C.F.); m.durante@gsi.de (M.D.)
[2] Institut für Festkörperphysik, TUDarmstadt, 64289 Darmstadt, Germany
[3] Trento Institute for Fundamental Physics and Applications (TIFPA), National Institute for Nuclear Physics, (INFN), 3812 Povo, Italy; emanuele.scifoni@tifpa.infn.it
* Correspondence: d.boscolo@gsi.de

Received: 3 December 2019; Accepted: 7 January 2020; Published: 9 January 2020

Abstract: The radiosensitivity of biological systems is strongly affected by the system oxygenation. On the nanoscopic scale and molecular level, this effect is considered to be strongly related to the indirect damage of radiation. Even though particle track radiolysis has been the object of several studies, still little is known about the nanoscopic impact of target oxygenation on the radical yields. Here we present an extension of the chemical module of the Monte Carlo particle track structure code TRAX, taking into account the presence of dissolved molecular oxygen in the target material. The impact of the target oxygenation level on the chemical track evolution and the yields of all the relevant chemical species are studied in water under different irradiation conditions: different linear energy transfer (LET) values, different oxygenation levels, and different particle types. Especially for low LET radiation, a large production of two highly toxic species ($HO_2^\bullet$ and $O_2^{\bullet-}$), which is not produced in anoxic conditions, is predicted and quantified in oxygenated solutions. The remarkable correlation between the $HO_2^\bullet$ and $O_2^{\bullet-}$ production yield and the oxygen enhancement ratio observed in biological systems suggests a direct or indirect involvement of $HO_2^\bullet$ and $O_2^{\bullet-}$ in the oxygen sensitization effect. The results are in agreement with available experimental data and previous computational approaches. An analysis of the oxygen depletion rate in different radiation conditions is also reported.
The radiosensitivity of biological systems is strongly affected by the system oxygenation. On the nanoscopic scale and molecular level, this effect is considered to be strongly related to the indirect damage of radiation. Even though particle track radiolysis has been the object of several studies, still little is known about the nanoscopic impact of target oxygenation on the radical yields. Here we present an extension of the chemical module of the Monte Carlo particle track structure code TRAX, taking into account the presence of dissolved molecular oxygen in the target material. The impact of the target oxygenation level on the chemical track evolution and the yields of all the relevant chemical species are studied in water under different irradiation conditions: different linear energy transfer (LET) values, different oxygenation levels, and different particle types. Especially for low LET radiation, a large production of two highly toxic species ($HO_2^\bullet$ and $O_2^{\bullet-}$), which is not produced in anoxic conditions, is predicted and quantified in oxygenated solutions. The remarkable correlation between the $HO_2^\bullet$ and $O_2^{\bullet-}$ production yield and the oxygen enhancement ratio observed in biological systems suggests a direct or indirect involvement of $HO_2^\bullet$ and $O_2^{\bullet-}$ in the oxygen sensitization effect. The results are in agreement with available experimental data and previous computational approaches. An analysis of the oxygen depletion rate in different radiation conditions is also reported.

Keywords: radiation track chemistry; chemical track structure; oxygen effect; oxygen depletion; ion beam therapy; ROS; superoxide anion

1. Introduction

The radiosensitivity of biological systems is strongly affected by the system oxygenation level. Based on evidence from *in vitro* experiments [1–3], tissues in hypoxic conditions, or with hypoxic regions, may be up to three times more radioresistant compared to well-oxygenated ones [3]. This effect is one of the main limiting factors for the tumor control in radiotherapy applications, correlating very often with poor prognosis [4] and is generally quantified by the oxygen enhancement ratio (OER). The OER is defined, for a given equal biological effect, as the ratio between the corresponding dose values in fully anoxic and in oxygenated conditions,

$$\mathrm{OER}(pO_2) = \left.\frac{D_{\mathrm{hypoxia}}}{D_{pO_2}}\right|_{\mathrm{same\ effect}} . \quad (1)$$

On the nanoscopic level, the oxygen effect is considered to be strongly related to indirect radiation damage [5] and, in particular, to the $OH^{\bullet}$ radicals [6]. Among the chemical species produced by water radiolysis, $OH^{\bullet}$ radicals are believed to be the most harmful; they have a very short half-life and can react with almost every molecule, including DNA [7]. In oxygenated conditions, the molecular oxygen may react with the damaged molecule stabilizing the damage and making it more difficult to repair [8,9]. Additionally, in oxygenated media, the radiolytic species produced during irradiation can interact with the molecular oxygen dissolved in the target and lead to an enhanced production of highly toxic reactive oxygen species (ROS). Solvated electrons, e^-_{aq}, and hydrogen atoms, $H^{\bullet}$, are generated in large quantities and react to form the superoxide anion, $O_2^{\bullet-}$ and its protonated form $HO_2^{\bullet}$, which have been identified as possibly responsible for the oxygen-driven sensitization effect [10].

$$e^-_{aq} + O_2 \rightarrow O_2^{\bullet-} \quad (2)$$

$$H^{\bullet} + O_2 \rightarrow HO_2^{\bullet}. \quad (3)$$

These species are particularly damaging, since they are involved in the lipid peroxidation chain and play an important role in the production of other toxic species, such as hydroxyl radicals $OH^{\bullet}$ through the Haber-Weiss reaction (catalyzed by the presence of transition metals), peroxynitrite ions $ONOO^-$ through the interaction with nitrogen monoxide $NO^{\bullet}$, and H_2O_2 after scavenging by superoxide dismutase enzymes (SOD). The latter theory is supported by both theoretical approaches [11–15] and chemical and biological experimental observations [16–19] mainly based on studies on the Fenton reaction and on the relation with SOD.

Densely ionizing radiation has been shown to mitigate hypoxia-induced radioresistance [20], motivating a growing interest in ion radiation therapy, especially with high charge, Z, like carbon [21] or better oxygen [22] for the treatment of hypoxic tumors. At the pre-clinical level, new optimization techniques accounting for the oxygenation level and ion linear energy transfer (LET) have recently been developed for particle therapy [3,21].

On the microscopic scale, this effect can be explained as a track density effect. The recombination probability of water-induced free radicals increases with LET, resulting in a lower contribution of the indirect effect of radiation damage and, thus, decreasing the impact of the target oxygenation condition. However, this is not a unique explanation, many other processes might be involved as well and several additional theories have been developed. One of the most accepted theories, the so-called "oxygen in track hypothesis", suggests that the production of O_2 molecule via multiple ionization processes in the track of densely ionizing radiation can cause locally a partially oxygenated response [23–26]. Other possible hypotheses are the interacting radical theory [20], the oxygen depletion in the heavy ion tracks [27], the lesion complexity hypothesis [28], and the radical multiplicity [29].

Though many theories have been developed, the nanoscopic processes involved in the oxygenation effect still have to be clarified and very little experimental data at that scale is available. Monte Carlo track structure codes are particularly suitable for studying the microscopic processes involved in the radiation damage. Among them, several codes are able to describe the transport of particle radiation in a medium including the chemical stage of radiation effect. In most of the cases, however, the chemical evolution of a particle track is described in pure water, without taking into account the impact of any dissolved species, like molecular oxygen, on the chemical reaction chain. Recently the Monte Carlo particle track structure code TRAX has been extended to the pre-chemical and chemical stage of radiation in water. A full description of the code can be found in Boscolo et al. [30]. With the new TRAX-CHEM module, the production, diffusion, and interaction of radiation-induced water-derived radicals can be studied with a step-by-step approach under different irradiation conditions.

A further extension of the code, able to account for different concentrations of dissolved molecular oxygen in the target material, is presented in this work. In order to limit the computational costs of the simulation, the dissolved oxygen molecules are assumed to be uniformly distributed in the target material and are treated as a continuum. Time-dependent and LET-dependent yields of all the considered radiolytic species at different oxygenation levels have been studied for different ion radiations and energies.

2. Results

2.1. Radiolysis of Oxygenated Water

The time-dependent yield of the chemical species has been evaluated for different target oxygenations and radiation qualities in the time interval 10^{-12}–10^{-6} s. Figure 1 shows the chemical evolution of 90 MeV proton radiation in targets under four different oxygenation conditions: 0% (complete anoxia), 3% (compatible with typical tumor oxygenation levels), 7% (in the range of normal tissue oxygenation, a condition called "physioxia") and for a target in fully aerated conditions 21%.

In the early stage of the chemical track evolution (up to ~ 1 ns after the passage of radiation) the radical yields are not affected by the presence of dissolved molecular oxygen in the target and follow the normal water radiolysis behavior. The ion track is very dense and the interaction among the radiation-induced radicals dominates the chemical evolution (independently of the target oxygenation level). At this stage the main products of water radiolysis ($OH^\bullet$, H_3O^+, e^-_{aq}) are the most abundant species; their yield is maximum at the beginning of the chemical stage and decreases with time, as these species are involved in many reaction processes and are consumed during the chemical track evolution. At the same time, the yield of the main reaction products (H_2O_2, H_2, and OH^-) increases during the chemical stage.

After the early stage of the chemical track evolution (1 ns after the passage of radiation), the radical distribution becomes more diffuse and the track dynamics become slower. As a consequence, the interaction of the radiolytic species with the dissolved oxygen becomes more prominent and the chemical track dynamics start to depend strongly on the target oxygenation level.

The main effect of the target oxygenation is the consumption of the e^-_{aq} and $H^\bullet$, which are strongly scavenged by the molecular oxygen (Equations (2) and (3)). For $pO_2 = 7\%$ and $pO_2 = 21\%$ a complete consumption of the e^-_{aq} and $H^\bullet$ can be observed after 0.8 µs and 0.2 µs, respectively. In the case of $pO_2 = 3\%$, only a small decrease with respect to the completely anoxic target in the electron yield can be observed at the end of the chemical stage. However, on a larger timescale, exceeding the range covered by the TRAX-CHEM code, all these e^-_{aq} and $H^\bullet$ will be eventually depleted.

Together with the e^-_{aq} and $H^\bullet$ consumption, the production of $O_2^{\bullet-}$ and $HO_2^\bullet$ is the major effect of dissolved oxygen in the target during the process of water radiolysis. The production yield of HO_2^- is negligible over the time covered by the calculations, for all oxygen concentrations analyzed.

In Figure 2, the calculated time-dependent yield of solvated electrons, produced by a proton track of 5 MeV, in a target with a partial oxygen pressure in air of 21% has been compared with different chemical track structure codes [15,31]. The initial electron yield, simulated by the TRAX-CHEM code, is higher as compared to the other simulation approaches. This can derive from the use of different cross-section sets or different dissociation models adopted by the different codes [32]. However, due to the lack of experimental data, large variability exists in the predicted radical yield at the very early stages of the chemical evolution [33]. At later stages of the track evolution, all codes show good agreement and predict a full electron consumption at about 0.2 μs after the passage of radiation.

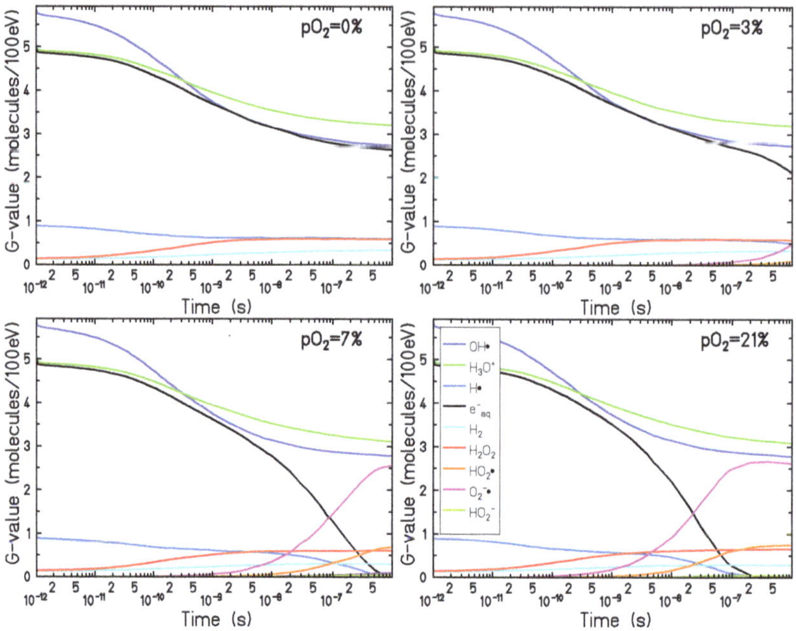

Figure 1. Time dependent yield of the chemical species generated by a 90 MeV proton track (linear energy transfer (LET) = 0.56 keV/μm) in a pure water target (**top left**) and in an oxygenated water target in equilibrium with an atmospheric partial oxygen pressure at the water surface of 3% (**top right**), 7% (**bottom left**), and 21% (**bottom right**).

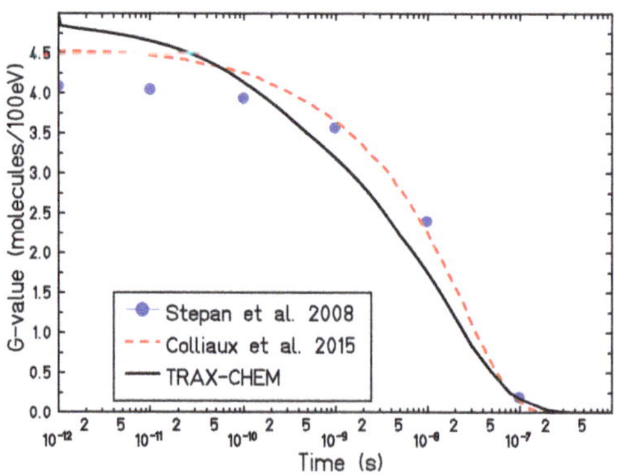

Figure 2. Comparison of different time dependent calculated yield for e^-_{aq} produced by irradiation with 5 MeV protons in a target with a partial oxygen pressure in air of 21%. ——: TRAX-CHEM, - - - -: Colliaux et al. [15], •: Štepán and Davídková [31].

2.2. Time-Dependent Radiolytic Yield for Different Oxygen Concentrations

Since the major effects of target oxygenation have been observed on the time evolution of e^-_{aq}, $H^•$, $HO_2^•$ and $O_2^{•-}$, their time-dependent yields have been studied for a set of oxygen concentrations ranging between $pO_2 = 0\%$ and $pO_2 = 21\%$. Figure 3 shows the simulation results for 500 keV electron tracks.

An increasing production of $O_2^{\bullet -}$ and $HO_2^{\bullet}$ (and a decrease in the yield of e^-_{aq} and $H^{\bullet}$) is observed for all oxygen concentrations. The oxygen reaction dynamics become faster when increasing the oxygen concentration and, for oxygenation levels above $pO_2 = 5\%$, complete consumption of the e^-_{aq} and $H^{\bullet}$ can be observed within a microsecond, leading to a saturation in the production of $O_2^{\bullet -}$ and H_2O . The yield at the saturation level of the $O_2^{\bullet -}$nearly matches the yield of the solvated electrons in anoxic conditions: in fully oxygenated conditions ($pO_2 = 21\%$) the yield of $O_2^{\bullet -}$ $G_{O_2^-}(10^{-6}s, pO_2 = 21\%) = 2.24$ and the electron yield in anoxia is $G_{e^-_{aq}}(10^{-6}s, pO_2 = 0\%) = 2.25$.

The production yield of the $HO_2^{\bullet}$ is slightly larger compared to the maximum yield of the $H^{\bullet}$ in hypoxic conditions: $G_{HO_2^{\bullet}}(10^{-6}s, pO_2 = 21\%) = 0.66$ while $G_{H^{\bullet}}(10^{-6}s, pO_2 = 21\%) = 0.56$. Even though the larger part of the $HO_2^{\bullet}$ is produced through the reaction process described in Equation (3), a smaller but not negligible contribution of the $HO_2^{\bullet}$ yield comes from the recombination of $O_2^{\bullet -}$ with H_3O^+, (reaction (xxiii) in Table 1):

$$H_3O^+ + O_2^{\bullet -} \rightarrow HO_2^{\bullet}. \quad (4)$$

In the present simulations, pH and the acid-base equilibrium of $HO_2^{\bullet}$ and $O_2^{\bullet -}$ are not modeled explicitly, so that all $G_{O_2^-}$ and $G_{HO_2^{\bullet}}$ reflect their production by radiolysis rather than a stable concentration. The pK_a of 4.8 leads to an equilibrium ratio $[O_2^{\bullet -}]/[HO_2^{\bullet}]$ at neutral pH of about 250. Accordingly with what is observed in Figure 1, for targets at oxygenation levels larger than 5% a complete consumption of e^-_{aq} and $H^{\bullet}$ can be observed and the chemical evolution reaches an equilibrium. For less oxygenated targets, however, the chemical dynamic is slower and the equilibrium is not reached within a microsecond and can proceed in a complex way at larger timescales.

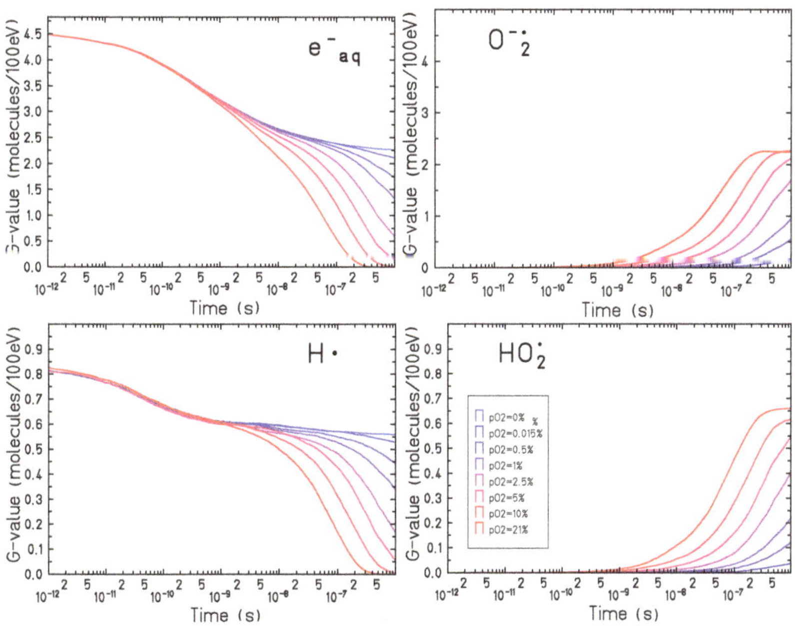

Figure 3. Time dependent yield for e^-_{aq} (**top left**), $H^{\bullet}$ (**bottom left**), $O_2^{\bullet -}$ (**top right**) and $HO_2^{\bullet}$ (**bottom right**) in an oxygenated water target, with oxygen concentration on the water surface between $pO_2 = 0\%$ and $pO_2 = 21\%$, for 500 keV electron irradiation.

2.3. pO_2-Dependent Radiolytic Yield for Different LET

The final radical yield (t = 1 µs, i.e., at the very end of the chemical stage) of all chemical species has been studied as a function of the oxygen concentration. Figure 4 shows the results of the calculations performed for different particles and different LET values: an electron track of 1 MeV, a proton track of 10 MeV and a carbon track of 10 MeV/u. The total yield of the chemical species is larger for low LET radiation and decreases when increasing the LET. For high LET radiation, the reaction kinetics is much faster: the ion track is denser, resulting in a larger recombination probability of the chemical species generated during the water radiolysis. Additionally, the yield of the recombination products (H_2O_2, H_2 and $HO_2^\bullet$) increases when increasing the LET.

The general trend of the radical yield at the different oxygenation conditions is similar for all the radiation qualities investigated. In all cases, the chemical species affected most by the dissolved molecular oxygen are $O_2^{\bullet-}$, $HO_2^\bullet$, e^-_{aq} and $H^\bullet$. A rapid decrease in the yield of the e^-_{aq} and $H^\bullet$ with increasing target oxygenation level can be observed up to $pO_2 \sim 5\%$. For larger oxygenation levels, e^-_{aq} and $H^\bullet$ are completely depleted. Accordingly, a steep increase of the production yield of $O_2^{\bullet-}$ and $HO_2^\bullet$ is observed for oxygen concentrations up to $pO_2 \sim 5\%$, but further increasing this value a saturation level is reached for the $O_2^{\bullet-}$ while the yield of the $HO_2^\bullet$ continues to increase but in a much slower way.

The production yield of all the other radiolytic species is less significantly modified by the water oxygenation level. An increase of the H_2O_2 yield with the oxygen concentration can be observed especially for higher LET radiation. For 10 MeV/u carbon ions, the G-value at 1 µs of the H_2O_2 increases from 0.89 (for the anoxic case, $pO_2 = 0\%$) to 1.19 (for the fully oxygenated target, $pO_2 = 21\%$). For the 1 MeV electron radiation the H_2O_2 G-value goes from 0.6 (in the anoxic target) to 0.67 (in the target with $pO_2 = 21\%$). The time dependent yield of the H_2O_2 is the result of two main processes:

$$OH^\bullet + OH^\bullet \longrightarrow H_2O_2 \tag{5}$$

$$e^-_{aq} + H_2O_2 \longrightarrow OH^\bullet + OH^- . \tag{6}$$

The first process is dominant at the early stages of the chemical track evolution and is the main production channel of the H_2O_2 while the second process becomes significant after 1 ns and removes H_2O_2 from the target. The contribution of the first process is more relevant in the dense primary radical condition after, high LET radiation, and is potentiated by the absence of the second one if sufficient molecular oxygen is present (due to competition with the molecular oxygen scavenging effect). The combined effect results in a larger yield of H_2O_2 at 1 µs after irradiation. A small increase of the $OH^\bullet$ radical yield and a small decrease of the H_3O^+, H_2 and OH^- yield can be observed, but in these cases, changes in the production yield at the microsecond are lower than 10%.

Figure 5 represents the consumption yield of the molecular oxygen at the end of the chemical stage, as a function of the target oxygenation, for different ion radiation qualities: 90 MeV proton (LET = 0.56 keV/µm), 10 MeV proton (LET = 3.9 eV/µm)), 10 MeV/u helium ion (LET = 15.2 keV/µm) and 10 MeV/u carbon ion (LET = 133 keV/µm). The formation of molecular oxygen through second-order recombination processes (see reactions (xvi), (xvii), (xxv) and (xxvi) in Table 1) is also taken into account. In fully oxygenated conditions and for high LET radiation tracks, at 1 µs after the passage of radiation, up to 25% of all the molecular oxygen initially depleted is regenerated in the target through these secondary processes.

The "yield" of oxygen consumption increases with increasing target oxygenation until reaching a plateau at $pO_2 = 5\%$ when all the radiation-induced solvated electrons are scavenged. Though the plateau starts at $pO_2 = 5\%$ for all the radiation quality investigated, the total yield of oxygen consumption is higher for low LET radiation. In contrast, for high LET radiation, the maximum yield of oxygen consumed is lower and the decrease at low oxygenation is more moderate.

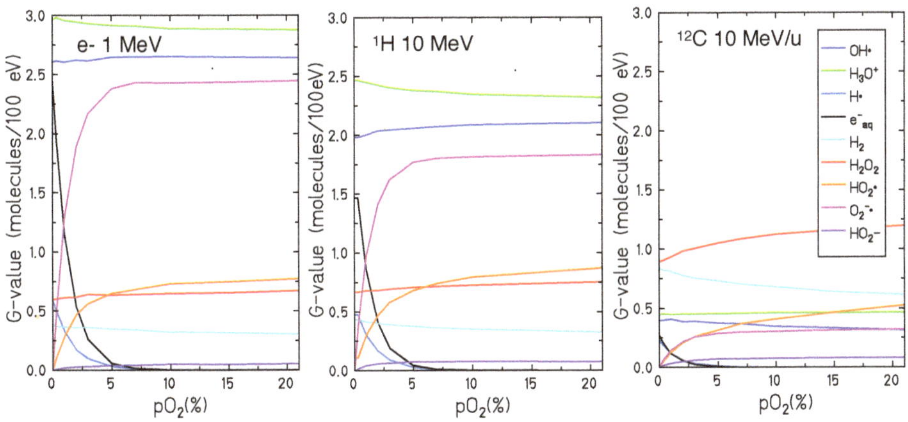

Figure 4. Radiolytic yields, at 1 µs, for all the different chemical species generated by the water radiolysis at different oxygenation conditions by 1 MeV electron irradiation (LET= 0.13 keV/µm) on the left panel and 10 MeV proton (LET= 3.9 keV/µm) in the central panel and 10 MeV/u carbon ions radiation (LET= 133 keV/µm) on the right panel.

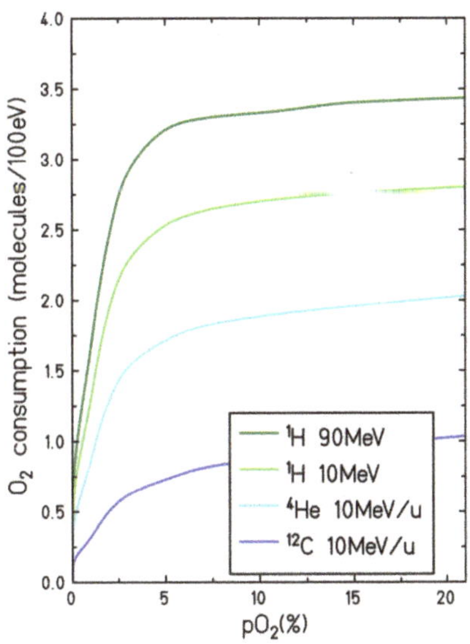

Figure 5. Consumption of the molecular oxygen for different target oxygenation levels and induced by different LET radiations. The consumption yields are calculated at the end of the chemical track evolution, t = 1 µs.

2.4. Radiolytic Yields for Different LET and Particle Type in Oxygenated Water

The impact of dissolved molecular oxygen on the final radical production yield has been studied under different oxygen concentrations for different particle radiation and different energies. Water targets at oxygenation levels of 21%, 3%, 0.2% and 0%, respectively, irradiated by protons, helium, and carbon ions with LET values ranging between 0.14 and 232 keV/μm are investigated here. LET dependent yields at the completion of the chemical stage are shown in Figure 6.

These results are in agreement with what has been already observed in Figures 3 and 4: the solvated electrons and the atomic hydrogen (e_{aq} and $H^{\bullet}$) yields decrease significantly with the increase of the oxygen concentration until full depletion of these species is observed in the case of complete oxygenation for all the radiation qualities investigated. The yields of $O_2^{\bullet-}$ and $HO_2^{\bullet}$, which are the two main indicators of the presence of molecular oxygen in the target, increase when increasing the target oxygenation over the entire range of LETs investigated. Their production yield is maximum for lower LET radiation and decreases for higher LET.

Only minor effects of the target oxygenation are observed for the other chemical species generated by water radiolysis. The scavenging effect of the solvated electrons and atomic hydrogen radicals leads to a general decrease in the production of the H_2 molecule which is mainly generated as a product of the recombination processes described by reactions (vi) and (x) in Table 1:

$$e_{aq}^{-} + e_{aq}^{-} + 2H_2O \longrightarrow H_2 + 2OH^{-} \tag{7}$$

$$H^{\bullet} + H^{\bullet} \longrightarrow H_2. \tag{8}$$

Consistent with what is shown in Figure 4, increased production of H_2O_2 can be observed in oxygenated conditions at high LET, while no effect is apparent at low LET.

The yield of the $OH^{\bullet}$ radical is slightly higher in an oxygenated target for low LET, while at intermediate LET no difference between oxygenated and hypoxic target is observed. A larger yield of $OH^{\bullet}$ is observed in the anoxic case in the high LET region. One of the main processes consuming the $OH^{\bullet}$ radical is its interaction with a solvated electron. For oxygenated targets, this reaction is directly competing with the interaction of the e_{aq}^{-} with the $O_2^{\bullet-}$ and results in a lower amount of scavenged $OH^{\bullet}$. For high LET, however, the track kinetic is faster and the $OH^{\bullet}$ reacts with the e_{aq}^{-} before the interactions with the dissolved oxygen become dominant.

The discontinuities are shown on the LET dependent curves in Figure 6 are due to the different simulated radiation types. This is because the LET is not a unique parameter for describing a particle track structure, and it also depends on the charge and speed of the primary particle. However, the dependence on the particle seems to vary not significantly with the oxygenation level of the target.

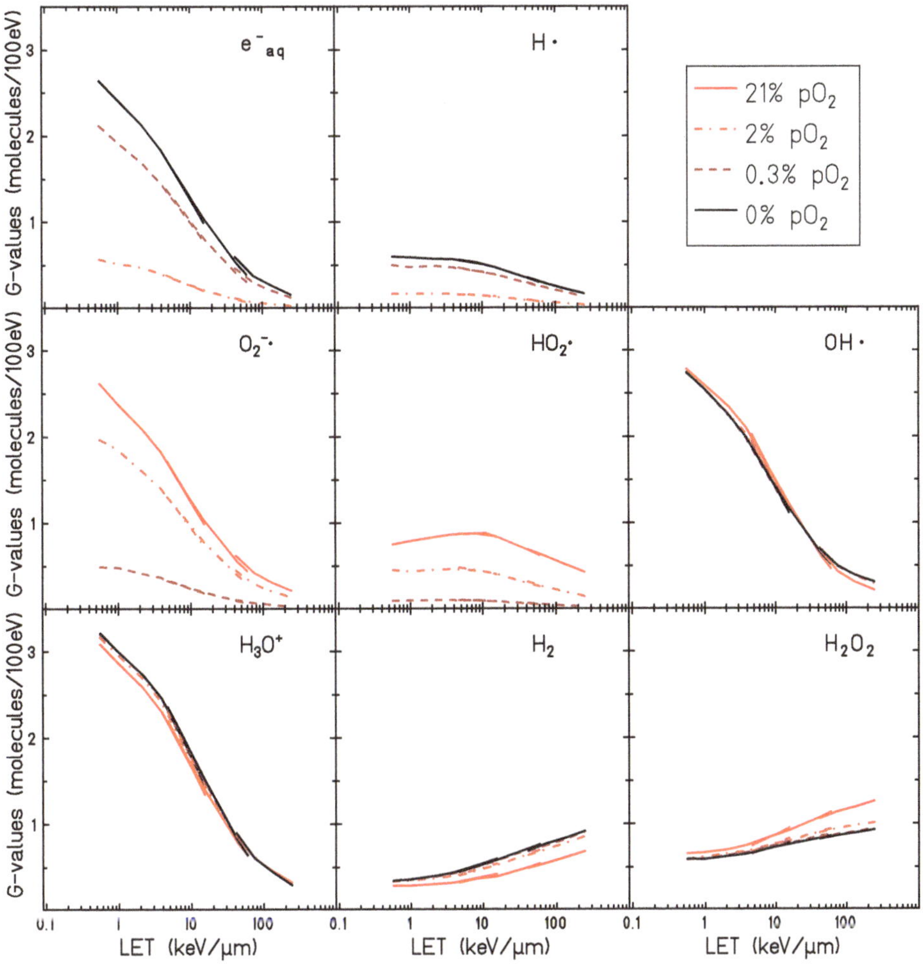

Figure 6. LET dependent radiolytic yields for all chemical species generated by the water radiolysis in water target at a partial oxygen pressure of 0%, 0.3%, 2% and 21%. Calculations were performed with protons, helium ions and carbon ions at one microsecond after the irradiation.

3. Discussion

Motivated by the need for a better understanding of the nanoscopic processes underlying the oxygen-induced radiosensitivity, the chemical track dynamics of the radiolytical species generated by different radiation qualities has been studied for water targets at different oxygenation levels.

Time-dependent radical yields for targets at different oxygenation levels have been calculated for proton and electron radiation, Figures 1 and 3. For all the investigated conditions, the impact of the target oxygenation can be observed only in the later stages of the chemical track evolution, indicating that for the first nanoseconds the radical yields are determined only by the intra-track recombination processes, independent from the target conditions.

The main effect of the dissolved molecular oxygen in the target is the consumption of e^-_{aq} and $H^\bullet$ and a corresponding production of $O_2^{\bullet -}$ and $HO_2^\bullet$ (see reactions (2) and (3)). At 1 μs after the passage of radiation, a complete depletion of e^-_{aq} and $H^\bullet$ is observed for oxygen concentrations larger than

$pO_2 = 5\%$. For oxygen concentrations lower than $pO_2 = 5\%$ the probability of interacting with the dissolved molecular oxygen is lower and the equilibrium on the radical yields is not reached within a microsecond (the time frame covered by the TRAX-CHEM simulations, and normally considered as the end of the chemical stage). The temporal interval of the simulation is chosen in a way that the chemical track evolution can be considered concluded; the radical distribution can be assumed to be uniform and the reaction process is determined only by the reactant concentration and not by their spatial distribution. As shown in Figure 7 the track structure is lost on the microsecond time scale and only a slightly increased radical concentration can be observed on the micrometer scale close to the track center. At the conclusion of the chemical stage, in completely anoxic conditions, when only the intra-track reactions are accounted for, the reaction rates of the different radicals are very low and the yields of the different species become constant. Typical proton and electron track radii corresponding to the end of the chemical stage are in the order of several hundreds of nanometer up to one micrometer.

However, when interactions with target molecules are possible, such as in oxygenated conditions, the radiolytic species will keep interacting with the target even after the track structure is completely lost. The radical yields will not reach equilibrium within the µs time frame and the chemical kinetics can proceed in a complex way for a very long time [15]. In the case of oxygenated water, according to our model (Table 1), the only species able to interact with the dissolved oxygen are e^-_{aq} and $H^\bullet$. Thus, it is to be expected that the whole track reaction kinetics will be limited to the lifetime of these two species in the target material. The chemical evolution of homogeneous systems is beyond the scope of this study; therefore, it has been decided to not extend the simulation time but to limit the study to the accepted time frame of the track evolution.

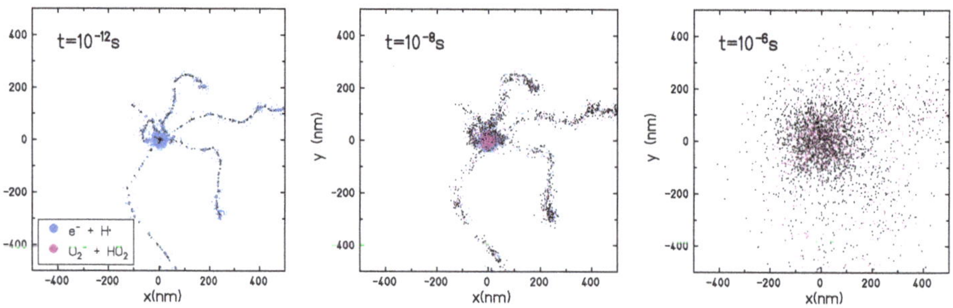

Figure 7. Chemical evolution of a 10 MeV/u carbon ion track in an oxygenated water target with an oxygen partial pressure $pO_2 = 21\%$. Track shown in beam eye view.

The radical production yields at the completion of the chemical track evolution have been studied for different radiation types and oxygenation levels. In Figure 4 the G-values for all the radiolytic species as a function of the oxygen concentration are studied for 1 MeV electron, 10 MeV proton and 10 MeV/u carbon ion radiation, while in Figure 6 the LET dependence of the radical production yield is reported for 21%, 2%, 0.3% and 0% pO_2. The yield of $O_2^{\bullet-}$ and $HO_2^\bullet$ increases with increasing target oxygenation over the whole range of analyzed LET. Their production yield is maximum for lower LET radiation and decreases for higher LET values. This strong dependence on the LET can be explained as a track structure effect: for high LET the ion track is denser and radicals are produced in close proximity. The radiation-induced water radicals will, then, recombine reacting with each other before any significant oxygen scavenging effect. Similar results have been obtained by Colliaux et al. [15] where the LET dependent yield of ($HO_2^\bullet + O_2^{\bullet-}$) has been calculated in an oxygenated water target with $pO_2 = 21\%$ and, as in our calculations, a pronounced decrease in the yield of ($O_2^{\bullet-} + HO_2^\bullet$) with LET has been observed. A significant increase in the production of H_2O_2 has been also observed for oxygenated targets, especially for high LET irradiation.

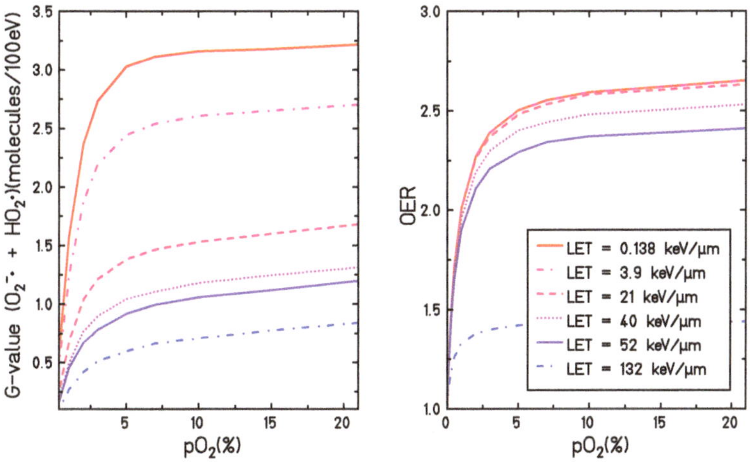

Figure 8. G-values for the production of superoxide and perhydroxyl (**left**) and oxygen enhancement ratio (OER) (**right**) are depicted as a function of target oxygenation. The OER values are calculated according to the parametrization proposed by Tinganelli et al. [3] but using definition (1).

When considering the correlation between the production yield of $O_2^{\bullet -}$ and $HO_2^{\bullet}$ at different LETs, particle types, and dissolved oxygen concentrations, it is reasonable to believe that the interaction of e_{aq}^- and $H^{\bullet}$ with the molecular oxygen leads directly or indirectly to the production of toxic species, able to damage the cell structure or alter cell signaling. This theory is supported by many studies [11–19,34–36] and correlates well with *in vitro* experiments, showing that the oxygen enhancement effect in biological systems has a pronounced dependence on the radiation LET [3,37,38] (it decreases when irradiating with higher LET radiation). This parallelism between the production of $O_2^{\bullet -}$ and $HO_2^{\bullet}$ and the oxygen effects observed in biological systems becomes even more evident when comparing production yields of $O_2^{\bullet -}$ and $HO_2^{\bullet}$ and the OER curve under different irradiation and target conditions. As shown in Figure 8, the general trend of the OER and of $G(O_2^{\bullet -} + H_2O)$ as a function of the target oxygenation level are very similar: a steep increase in both curves can be observed when increasing the target oxygenation until reaching a plateau for partial oxygen pressures larger than 5%. Additionally, a reduction of the entire OER curve and $G(O_2^{\bullet -} + H_2O)$ is observed when increasing the LET for all the oxygenation levels. However, the OER curve shows a maximum sensitivity on the LET for values $\sim$ 100 keV/μm while the radical yield has a maximum sensitivity for LET values $\sim$ 10 keV/μm. Therefore, it is not straightforward to deduce the oxygenation effect in biological systems directly within the present theoretical framework. The present study is focused, indeed, on assessing the role of one possible sensitization mechanism but additional pathways, e.g., the oxygen fixation, must be also taken into account when aiming at a complete explanation of the oxygen-induced radiation sensitivity. Additionally, a water target is a considerably simplified system compared to the cellular environment and all the complex reaction chains taking place with cell medium, including the secondary reactions taking place at further stages, the biological damage and its repair, and the possible cross-talk with signaling pathways caused by altered levels of some ROS, are not accounted for. In this context, further extensions of the model can be considered in order to take into account the presence of additional solutes known to play an important role in the induction of radical damage, such as nitrogen monoxide $NO^{\bullet}$, carbonate or bicarbonate ions, or the presence of metals to catalyze the Fenton chemistry [34,39]. At the same time, radical scavengers such as superoxide dismutase (SOD), catalase (CAT), gluthatione peroxidase (GSH) could be included in the model [40]. However, considering that all these species will only have a role in the system dynamic at later stages of the chemical track evolution, when the primary radiation-induced radicals are already diffused, computationally lighter approaches based on the homogeneous chemistry might be considered appropriate for the implementation of further stages of the system dynamics.

The investigation of the influence of target oxygenation and LET on radiation-induced radical production and oxygen consumption opens the way for applications where these factors are being discussed to enable a differential radioprotective effect. This includes e.g., ultra-high dose rate (FLASH) conditions [39,41,42] where high instantaneous concentrations of ROS are produced and replenishment of oxygen through diffusion is too slow to maintain stable oxygenation.

4. Materials and Methods

4.1. Simulation of Particle Track Evolution in Non-Oxygenated Water

The evolution of charged particle tracks is described in TRAX-CHEM as a three steps process: the physical, the pre-chemical and the chemical stage. These three stages take place subsequently and are characterized by a characteristic time scale. The physical stage of the radiation track evolution consists of the simulation of the ionization and excitation processes along the particle track and is simulated with the standard version of the TRAX code [43]. The electron and ion tracks are simulated with an event-by-event approach, until they reach a cutoff energy of 7.4 eV, which corresponds to the lower electronic excitation level of water. Ion and electron interactions are described through a set of shell specific ionization and excitation cross-section tables, and Auger electron production and electron elastic scattering cross sections are implemented as well. Multiple ionization processes and photon interactions are not implemented in the current version of the TRAX code. At the conclusion of this stage, which is assumed to be at 10^{-15}s after the irradiation, the positions and the shell specific ionization or excitation levels of the target water molecules are provided and can be used as an input for the following stage.

The pre-chemical stage, which lasts up to 10^{-12}s, consists of the dissociation and thermalization of all the products generated during the physical stage: HO_2^+, HO_2^* and e^-_{aq}. The probability of undergoing a dissociation process or relaxing to the ground state depends on the specific ionization or excitation channel. In TRAX-CHEM, all the ionized molecules are thought to dissociate to $OH^•$ and H_3O^+, while four possible dissociation patterns have been considered for excited water molecules: auto-ionization, two dissociative decays ($OH^•$ + $H^•$ and H_2 + H_2O_2) and relaxation to the ground state. During the dissociation process, a fraction of energy is transferred to the dissociation fragments as kinetic energy. These species will then need to release this energy and thermalize with the surrounding medium before starting to behave and interact as chemical species. A complete description of the dissociation and thermalization model has been reported in Boscolo et al. [30]. At the end of the thermalization process, the pre-chemical stage is considered to be concluded and the last and longest stage of the track evolution begins the chemical stage.

During the chemical stage, the radiolytic species diffuse and interact among themselves until reaching the chemical equilibrium. In TRAX-CHEM these two processes are described with a step by step approach, which allows us to determine the position of each chemical species in every step of the simulation. For every time step, the Brownian diffusion process is modeled with a jump in a random direction. The reaction model is described through a proximity parameter; the reaction radius. If two species are closer than the corresponding reaction radius, the reaction is supposed to take place: the two reactants are removed from the chemical list and substituted by the reaction products. Inter-track reactions (i.e., reactions of chemical species originating from different primary particles) are not included in the present TRAX-CHEM extension. Details on the implementation of the chemical model including the stepping algorithm, the reaction models, the calculation of the reaction radii and the diffusion model have been presented in Boscolo et al. [30]. Complete lists of the reactions and the chemical species implemented in TRAX-CHEM are provided in Tables 1 and 2; these tables have been updated and extended with respect to the initial ones [30] in order to take into account the presence of dissolved molecular oxygen in the target, as presented in the next section. In TRAX-CHEM the chemical stage, and thus the track evolution, is supposed to conclude 10^{-6} s after the physical irradiation interactions. After this time, the chemical yields of the different species become constant and the track development can be considered to be finished.

Table 1. List of all the reactions and reaction rate constants, k, used in this work (obtained mainly from [44]). Reactions (14)-(26) arise as a consequence of the presence of dissolved O_2.

	Reaction		Products	$k(10^{10} dm^3 mol^{-1} s^{-1})$
(i)	$OH^\bullet + OH^\bullet$	$\longrightarrow$	H_2O_2	0.6
(ii)	$OH^\bullet + e^-_{aq}$	$\longrightarrow$	OH^-	2.2
(iii)	$OH^\bullet + H^\bullet$	$\longrightarrow$	H_2O	2.0
(iv)	$OH^\bullet + H_2$	$\longrightarrow$	$H^\bullet + H_2O$	0.0045
(v)	$OH^\bullet + H_2O_2$	$\longrightarrow$	$HO_2^\bullet + H_2O$	0.0023
(vi)	$e^-_{aq} + e^-_{aq} + H_2O + H_2O$	$\longrightarrow$	$H_2 + OH^- + OH^-$	0.55
(vii)	$e^-_{aq} + H^\bullet + H_2O$	$\longrightarrow$	$H_2 + OH^-$	2.5
(viii)	$e^-_{aq} + H_3O^+$	$\longrightarrow$	$H^\bullet + H_2O$	1.7
(ix)	$e^-_{aq} + H_2O_2$	$\longrightarrow$	$OH^\bullet + OH^-$	1.0
(x)	$H^\bullet + H^\bullet$	$\longrightarrow$	H_2	1.0
(xi)	$H^\bullet + H_2O_2$	$\longrightarrow$	$OH^\bullet + H_2O$	0.01
(xii)	$H^\bullet + OH^-$	$\longrightarrow$	$e^-_{aq} + H_2O$	0.002
(xiii)	$H_3O^+ + OH^-$	$\longrightarrow$	$H_2O + H_2O$	10.0
(xiv)	$e^-_{aq} + O_2$	$\longrightarrow$	$O_2^{\bullet -}$	1.9
(xv)	$H^\bullet + O_2$	$\longrightarrow$	$HO_2^\bullet$	2.0
(xvi)	$OH^\bullet + HO_2^\bullet$	$\longrightarrow$	O_2	1.0
(xvii)	$OH^\bullet + O_2^{\bullet -}$	$\longrightarrow$	$O_2 + OH^-$	0.9
(xviii)	$OH^\bullet + HO_2^-$	$\longrightarrow$	$HO_2^\bullet + OH^-$	0.5
(xix)	$e^-_{aq} + HO_2^\bullet$	$\longrightarrow$	HO_2^-	2.0
(xx)	$e^-_{aq} + O_2^{\bullet -}$	$\longrightarrow$	$OH^- + HO_2^-$	1.3
(xxi)	$H^\bullet + HO_2^\bullet$	$\longrightarrow$	H_2O_2	2.0
(xxii)	$H^\bullet + O_2^{\bullet -}$	$\longrightarrow$	HO_2^-	2.0
(xxiii)	$H_3O^+ + O_2^{\bullet -}$	$\longrightarrow$	$HO_2^\bullet$	3
(xxiv)	$H_3O^+ + HO_2^-$	$\longrightarrow$	H_2O_2	2.0
(xxv)	$HO_2^\bullet + HO_2^\bullet$	$\longrightarrow$	$H_2O_2 + O_2$	0.000076
(xxvi)	$HO_2^\bullet + O_2^{\bullet -}$	$\longrightarrow$	$O_2 + HO_2^-$	0.0085

Table 2. List of all the chemical species and their diffusion coefficients, D, added to the chemical species list of TRAX to describe the impact of dissolved molecular oxygen in the water target.

Species	D $(10^{-9} m^2 s^{-1})$
$OH^\bullet$	2.8
H_3O^+	9.0
$H^\bullet$	7.0
e^-_{aq}	4.5
H_2	4.8
OH^-	5.0
H_2O_2	2.3
O_2	2.1
$HO_2^\bullet$	2.0
HO_2^-	2.0
$O_2^{\bullet -}$	2.1

4.2. Simulation of Particle Track Evolution in Oxygenated Water

The classical version of the TRAX-CHEM code has been modified and is now able to simulate the chemical evolution of ion tracks in water targets under different oxygen pressure conditions. The particle list and the reaction network of the classical version of the TRAX-CHEM code have been, thus, extended (see Tables 1 and 2) and new species, generated by the interaction of O_2 with the radiation-induced water-free radicals, are now included in the track chemical evolution.

In contrast to all other species, which are explicitly included in the code and treated with the step by step approach mentioned above, the molecular oxygen is assumed to be homogeneously distributed in the target and is treated as a continuum. This approximation, proposed by Pimblott et al. [45], Green et al. [46], is necessary to limit the computational cost of the simulations and has also been adopted by other authors [13,14,47,48]. The explicit introduction of the oxygen in the simulation would dramatically increase the computing time, even for very dilute solutions [45]. As an example, in a chemical simulation of a 10 MeV proton track in a cubic volume of 5 μm side the number of radiolytic species produced per particle track is about 10^3. In a fully oxygenated condition (21% pO_2, corresponding to a concentration of 0.27 mmol/litre) the number of oxygen molecule that have to be explicitly introduced (in the same cubic volume of 5μm side) and followed at every single step of the simulation is about 2×10^7, increasing the simulation time by more than four orders of magnitude. Considering the relatively low radiation-induced oxygen consumption compared to the total amount of molecular oxygen dissolved in the target, a variation of the global oxygen concentration during the track evolution can be excluded for all the oxygenations and radiation conditions examined. When investigating the possibility of noticeable local oxygen depletion in the track cores, it has to be kept in mind that the interaction of the radiolytic species with the target occurs in a later stage of the expanding chemical track evolution since the reaction dynamic between the induced chemical species is slower. For high LET ions (~100 keV/μm), the highest local density of oxygen removal is reached around 10 ns within 10-20 nm of the track core and stays below 250 μM, quickly decreasing as the chemical track diffuses and oxygen conversion to superoxide and perhydroxyl becomes important. For even larger LET values, only a slightly delayed onset of the $HO_2^{\bullet}$ and $O_2^{\bullet -}$ production can, therefore, be expected. Similar conclusions have been reported by Colliaux et al. [15]. Additionally, the molecular oxygen production through multiple ionizations is not accounted for here, but it has been demonstrated that the contribution of this process to the radical yield is very low [15].

Under these conditions, given an oxygen concentration c_s, the probability for a radiolytic species to interact with an oxygen molecule of the target is determined by the rate equation:

$$\frac{\mathrm{d}\Omega(t)}{\mathrm{d}t} = -k(t)c_s\Omega(t), \tag{9}$$

where $\Omega(t)$ is the time-dependent survival probability of the molecule of interest. The time dependent rate coefficient k(t) for the reaction of interest can be calculated according to the Noyes theory [49], as:

$$k(t) = 4\pi D' R_{reac}\left(1 + \frac{R_{reac}}{\sqrt{\pi D' t}}\right), \tag{10}$$

where D' is the relative diffusion coefficient defined, considering the two species A and B, $D' = D_A + D_B$ and R_{reac} is the reaction radius defined according to the Smoluchowski theory, as

$$R_{reac} = \frac{k_{AB}}{4\pi(D_A + D_B)}. \tag{11}$$

The probability for a molecule to react with the dissolved oxygen in a time t will, thus, be:

$$W(t) = 1 - \Omega(t) = 1 - e^{-4\pi D' R_{reac} c_s \left(t + 2R_{reac}\sqrt{\frac{t}{\pi D'}}\right)}. \tag{12}$$

Since TRAX-CHEM uses a variable time step, the probability $W(t)$ that a species will interact is calculated for each time step and is sampled through a uniformly distributed random variable $x \in [0:1]$. When $x \leq W(t)$ the reaction is taking place, the reactants are removed from the simulation and replaced by the corresponding reaction products.

4.3. Calculation

In order to identify a range of oxygen concentrations compatible with the one of a biological system, the concentration of molecular oxygen dissolved in the water target is calculated from the partial pressure of oxygen in the air within standard conditions of pressure by Henry's low. Given a gas partial pressure, p_{gas} (in atmospheres), the solubility of the gas at a fixed temperature in a particular solvent, c_s (in mol per liter):

$$c_s = K_H \cdot P_{gas}, \tag{13}$$

where K_H is Henry's constant and corresponds to the gas solubility in water. For O_2 at a temperature of 20 °C, Henry's constant is $K_H = 1.3 \times 10^{-3}$ mol/(1 atm). In fully oxygenated conditions (partial oxygen pressure $p_{O_2} = 21\%$), the oxygen concentration equals 0.27 mmol/L.

For the results presented in the present simulation results, G values, i.e., numbers of species produced per 100 eV energy deposition, have been calculated for all the chemical species in a simulated volume of $5 \times 5 \times 5$ μm^3 for low LET radiation. In the case of high LET radiation, a geometry reduced in the beam direction of $5 \times 5 \times 0.5$ μm^3 was chosen in order to guarantee the track segment condition. For each simulation, a series of parallel calculations were performed in order to reduce the statistical uncertainty. The number of primary particles was chosen so that a total energy of 2.5 MeV was deposited in the target. The statistical fluctuation in the chemical simulations has been evaluated for every single primary particle and was shown to be within 10% [30].

5. Conclusions

In the presented work, the impact of target oxygenation on radical production yields has been studied for electron and ion radiation in water with the recently implemented extension of the TRAX-CHEM code, able to handle the presence of dissolved molecular oxygen in the target. The molecular oxygen concentrations investigated in this work range from anoxic conditions ($pO_2 = 0\%$) to air-saturated water targets ($pO_2 = 21\%$). Time-dependent and LET dependent yields of all the simulated radiolytic species at different oxygenation levels have been studied for different ion radiation and different energies up to 1 µs after the passage of radiation. In oxygenated conditions, a large production of two highly toxic species ($O_2^{\bullet-}$ and $HO_2^{\bullet}$) has been predicted, especially for low LET radiation. These species are generated as reaction products, from the interaction of respectively e^-_{aq} and $H^{\bullet}$ with the dissolved molecular oxygen. Thus, a decrease in the final (1 µs) yield of the e^-_{aq} and $H^{\bullet}$ is observed in oxygenated targets until reaching a complete consumption for oxygen concentrations at the water surface larger than 5%. Little impact of the dissolved molecular oxygen has been predicted on the production yield of the other radiolytic species, with the exception of the H_2O_2 whose yield is expected to increase with the target oxygenation especially for high LET radiation tracks. Consistent with the LET dependence of the oxygen effect on the macroscopic level, a strong decrease in the production yields of $O_2^{\bullet-}$ and $HO_2^{\bullet}$ is observed with the increase of the LET. The strong correlation between the production yields of ($O_2^{\bullet-}$ and $HO_2^{\bullet}$) and the oxygen radiosensitization effect observed in in vitro cell clonogenic experiments indicate a possible direct or indirect involvement of these species in the indirect radiation damage. Although the code is not yet able to resolve the oxygen effect in biological media and does not account for mechanisms e.g., the oxygen damage fixation, the present implementation provides quantitative insights on the nanoscopic processes involved in the sensitizing effect of oxygen. Further extension of the code to later stages of the chemical dynamics including radiosensitizers, radical scavengers, and specific molecular targets of interest in the cell may be considered.

Author Contributions: D.B. developed and implemented the model, performed the simulation, analyzed the results, and drafted the manuscript based on discussions with E.S., M.K., M.C.F., and M.D.; E.S., M.D., supervised the work, M.C.F. and E.S. and M.K. contributed to the drafting of the text; M.K. contributed to the implementation of the model on TRAX-CHEM. All authors have read and agreed to the published version of the manuscript.

Funding: This research was funded by European Union Seventh Framework Programme (PEOPLE-2013-ITN-ARGENT Marie Curie project), under grant agreement no. 608163.

Conflicts of Interest: The authors declare no conflict of interest.

References

1. Gray, L.; Conger, A.; Ebert, M.; Hornsey, S.; Scott, O. The concentration of oxygen dissolved in tissues at the time of irradiation as a factor in radiotherapy. *Br. J. Radiol.* **1953**, *26*, 638–648. [CrossRef]
2. Evans, S.M.; Koch, C.J. Prognostic significance of tumor oxygenation in humans. *Canc. Lett.* **2003**, *195*, 1–6. [CrossRef]
3. Tinganelli, W.; Durante, M.; Hirayama, R.; Krämer, M.; Maier, A.; Kraft-Weyrather, W.; Furusawa, Y.; Friedrich, T.; Scifoni, E. Kill-painting of hypoxic tumours in charged particle therapy. *Sci. Rep.* **2015**, *5*, 17016. [CrossRef] [PubMed]
4. Fyles, A.; Milosevic, M.; Hedley, D.; Pintilie, M.; Levin, W.; Manchul, L.; Hill, R. Tumor hypoxia has independent predictor impact only in patients with node-negative cervix cancer. *J. Clin. Oncol.* **2002**, *20*, 680–687. [CrossRef] [PubMed]
5. Sonntag, C. *The Chemical Basis of Radiation Biology*; Taylor & Francis: London, UK, 1987.
6. Hirayama, R.; Ito, A.; Noguchi, M.; Matsumoto, Y.; Uzawa, A.; Kobashi, G.; Okayasu, R.; Furusawa, Y. $OH^{\bullet}$ radicals from the indirect actions of X-rays induce cell lethality and mediate the majority of the oxygen enhancement effect. *Radiat. Res.* **2013**, *180*, 514–523. [CrossRef] [PubMed]
7. Reiter, R.J.; Melchiorri, D.; Sewerynek, E.; Poeggeler, B.; Barlow-Walden, L.; Chuang, J.; Ortiz, G.G.; AcuñaCastroviejo, D. A review of the evidence supporting melatonin's role as an antioxidant. *J. Pineal Res.* **1995**, *18*, 1–11. [CrossRef]
8. Johansen, I.; Howard-Flanders, P. Macromolecular repair and free radical scavenging in the protection of bacteria against X-rays. *Radiat. Res.* **1965**, *24*, 184–200. [CrossRef]
9. Ewing, D. The oxygen fixation hypothesis: A reevaluation. *Am. J. Clin. Oncol.* **1998**, *21*, 355–361. [CrossRef]
10. Beuve, M.; Moreau, J.M.; Rodriguez, C.; Testa, E. Biological systems: from water radiolysis to carbon ion radiotherapy. *J. Physics Conf. Ser.* **2015**, *629*, 012004. [CrossRef]
11. Turner, J.E.; Magee, J.L.; Wright, H.A.; Chatterjee, A.; Hamm, R.N.; Ritchie, R.H. Physical and Chemical Development of Electron Tracks in Liquid Water. *Radiat. Res.* **1983**, *96*, 437–449. [CrossRef]
12. Klassen, N.V.; Ross, C.K. Water Calorimetry: The Heat Defect. *J. Res. Natl. Inst. Stand. Technol.* **1997**, *102*, 63–74. [CrossRef] [PubMed]
13. Tomita, H.; Kai, M.; Kusama, T.; Ito, A. Monte Carlo simulation of physicochemical processes of liquid water radiolysis. *Radiat. Environ. Biophys.* **1997**, *36*, 105–116. [CrossRef] [PubMed]
14. Watanabe, R.; Saito, K. Monte Carlo simulation of water radiolysis in oxygenated condition for monoenergetic electrons from 100eV to 1MeV. *Radiat. Phys. Chem.* **2001**, *62*, 217–228. [CrossRef]
15. Colliaux, A.; Gervais, B.; Rodriguez-Lafrasse, C.; Beuve, M. Simulation of ion-induced water radiolysis in different conditions of oxygenation. *Nucl. Instruments Methods Phys. Res. Sect. B Beam Interact. Mater. Atoms* **2015**, *365*, 596–605. [CrossRef]
16. Goldstein, S.; Czapski, G. The reaction of $NO^{\bullet}$ with $O_2^{\bullet-}$ and $HO_2^{\bullet-}$: A pulse radiolysis study. *Free Radic. Biol. Med.* **1995**, *19*, 505–510. [CrossRef]
17. Mikkelsen, R.; Wardman, P. Biological chemistry of reactive oxygen and nitrogen and radiation-induced signal transduction mechanisms. *Oncogene* **2003**, *22*, 5734–5754. [CrossRef]
18. De Grey, A.D. $HO_2^{\bullet}$: The Forgotten Radical. *DNA and Cell Biol.* **2004**, *21*, 251–257. 104454902753759672. [CrossRef]
19. Aikens, J. Perhydroxyl radical ($HO_2^{\bullet}$) initiated lipid peroxidation. The role of fatty acid hydroperoxides. *J. Biol. Chem.* **1991**, *266*, 15091–15098.
20. Alper, T. The Modification of Damage Caused by Primary Ionization of Biological Targets. *Radiat. Res.* **1956**, *5*, 573–586. [CrossRef]
21. Scifoni, E.; Tinganelli, W.; Weyrather, W.; Durante, M.; Maier, A.; Krämer, M. Including oxygen enhancement ratio in ion beam treatment planning: model implementation and experimental verification. *Phys. Med. Biol.* **2013**, *58*, 3871. [CrossRef]

22. Sokol, O.; Scifoni, E.; Tinganelli, W.; Kraft-Weyrather, W.; Wiedemann, J.; Maier, A.; Boscolo, D.; Friedrich, T.; Brons, S.; Durante, M.; et al. Oxygen beams for therapy: advanced biological treatment planning and experimental verification. *Phys. Med. Biol.* **2017**, *62*, 7798–7813. [CrossRef] [PubMed]
23. Meesungnoen, J.; Filali-Mouhim, A.; Ayudhya, N.S.N.; Mankhetkorn, S.; Jay-Gerin, J.P. Multiple ionization effects on the yields of HO_2/O_2^- and H_2O_2 produced in the radiolysis of liquid water with high-LET $^{12}C^{6+}$ ions: A Monte-Carlo simulation study. *Chem. Phys. Lett.* **2003**, *377*, 419–425. [CrossRef]
24. Meesungnoen, J.; Jay-Gerin, J.P. High-LET ion radiolysis of water: oxygen production in tracks. *Radiat. Res.* **2009**, *171*, 379–386. [CrossRef] [PubMed]
25. Gervais, B.; Beuve, M.; Olivera, G.; Galassi, M. Numerical simulation of multiple ionization and high LET effects in liquid water radiolysis. *Radiat. Phys. Chem.* **2006**, *75*, 493–513. [CrossRef]
26. Baverstock, K.; Burns, W. Oxygen as a product of water radiolysis in high-LET tracks. II. Radiobiological implications. *Radiat. Res.* **1981**, *86*, 20–33. [CrossRef]
27. Stuglik, Z. On the "oxygen in heavy-ion tracks" hypothesis. *Radiat. Res.* **1995**, *143*, 343–348. 3579223. [CrossRef]
28. Ward, J.F. The complexity of DNA damage: Relevance to biological consequences. *Int. J. Radiat. Biol.* **1994**, *66*, 427–432. [CrossRef]
29. Michael, B. A multiple-radical model for radiation action on DNA and the dependence of OER on LET. *Int. J. Radiat. Biol.* **1996**, *69*, 351–358. [CrossRef]
30. Boscolo, D.; Krämer, M.; Durante, M.; Fuss, M.; Scifoni, E. TRAX-CHEM: A pre-chemical and chemical stage extension of the particle track structure code TRAX in water targets. *Chem. Phys. Lett.* **2018**, *698*, 11–18. [CrossRef]
31. Štepán, V.; Davídková, M. Impact of oxygen concentration on yields of DNA damages caused by ionizing radiation. *J. Physics Conf. Ser.* **2008**, *101*, 012015. [CrossRef]
32. Shin, W.G.; Ramos-Mendez, J.; Faddegon, B.; Tran, H.; Villagrasa, C.; Perrot, Y.; Okada, S.; Karamitros, M.; Emfietzoglou, D.; Kyriakou, I.; et al. Evaluation of the influence of physical and chemical parameters on water radiolysis simulations under MeV electron irradiation using Geant4-DNA. *J. Appl. Phys.* **2019**, *126*, 114301. [CrossRef]
33. Kreipl, M.S.; Friedland, W.; Paretzke, H.G. Time- and space-resolved Monte Carlo study of water radiolysis for photon, electron and ion irradiation. *Radiat. Environ. Biophys.* **2008**, *48*, 11. [CrossRef]
34. Augusto, O.; Bonini, M.G.; Amanso, A.M.; Linares, E.; Santos, C.C.; De Menezes, S.L. Nitrogen dioxide and carbonate radical anion: Two emerging radicals in biology. *Free Radic. Biol. Med.* **2002**, *32*, 841–859. [CrossRef]
35. Gorman, A.; McGowan, A.; Cotter, T.G. Role of peroxide and superoxide anion during tumour cell apoptosis. *FEBS Lett.* **1997**, *404*, 27–33. [CrossRef]
36. Quijano, C.; Romero, N.; Radi, R. Tyrosine nitration by superoxide and nitric oxide fluxes in biological systems: modeling the impact of superoxide dismutase and nitric oxide diffusion. *Free Radic. Biol. Med.* **2005**, *39*, 728–741. [CrossRef] [PubMed]
37. Hirayama, R.; Ito, A.; Tomita, M.; Tsukada, T.; Yatagai, F.; Noguchi, M.; Matsumoto, Y.; Kase, Y.; Ando, K.; Okayasu, R.; et al. Contributions of direct and indirect actions in cell killing by high-LET radiations. *Radiat. Res.* **2009**, *171*, 212–218. [CrossRef] [PubMed]
38. Scifoni, E. Radiation biophysical aspects of charged particles: from the nanoscale to therapy. *Mod. Phys. Lett. A* **2015**, *30*, 1540019. [CrossRef]
39. Spitz, D.R.; Buettner, G.R.; Petronek, M.S.; St-Aubin, J.J.; Flynn, R.T.; Waldron, T.J.; Limoli, C.L. An integrated physico-chemical approach for explaining the differential impact of FLASH versus conventional dose rate irradiation on cancer and normal tissue responses. *Radiother. Oncol.* **2019**, *139*, 23–27. [CrossRef]
40. Buetler, T.M.; Krauskopf, A.; Ruegg, U.T. Role of superoxide as a signaling molecule. *Physiology* **2004**, *19*, 120–123. [CrossRef]
41. Pratx, G.; Kapp, D.S. A computational model of radiolytic oxygen depletion during FLASH irradiation and its effect on the oxygen enhancement ratio. *Phys. Med. Biol.* **2019**, *64*, 185005. [CrossRef]
42. Montay-Gruel, P.; Acharya, M.M.; Petersson, K.; Alikhani, L.; Yakkala, C.; Allen, B.D.; Ollivier, J.; Petit, B.; Jorge, P.G.; Syage, A.R.; et al. Long-term neurocognitive benefits of FLASH radiotherapy driven by reduced reactive oxygen species. *Proc. Nat. Ac. Sci. USA* **2019**, *116*, 10943–10951. [CrossRef] [PubMed]

43. Krämer, M.; Kraft, G. Calculations of heavy-ion track structure. *Radiat. Environ. Biophys.* **1994**, *33*, 91–109. [CrossRef] [PubMed]
44. Nikjoo, H.; Uehara, S.; Emfietzoglou, D.; Cucinotta, F. Track-structure codes in radiation research. *Radiat. Meas.* **2006**, *41*, 1052–1074. [CrossRef]
45. Pimblott, S.M.; Pilling, M.J.; Green, N.J. Stochastic models of spur kinetics in water. *Int. J. Radiat. Appl. Instrum. Part Radiat. Phys. Chem.* **1991**, *37*, 377–388. [CrossRef]
46. Green, N.J.B.; Pilling, M.J.; Pimblott, S.M.; Clifford, P. Stochastic modeling of fast kinetics in a radiation track. *J. Phys. Chem.* **1990**, *94*, 251–258. [CrossRef]
47. Cobut, V.; Frongillo, Y.; Patau, J.P.; Goulet, T.; Fraiser, M.; Jay-Gerin, J. Monte Carlo Simulation of fast electron and proton tracks in liquid water I. Physical and physicochemical aspects. *Radiat. Phys. Chem.* **1998**, *51*, 229–243. [CrossRef]
48. Colliaux, A.; Gervais, B.; Rodriguez-Lafrasse, C.; Beuve, M. O_2 and glutathione effects on water radiolysis: A simulation study. *J. Physics Conf. Ser.* **2011**, *261*, 012007. [CrossRef]
49. Noyes, R.M. Effects of diffusion rates on chemical kinetics. *Prog. React. Kinet. Mech.* **1961**, *1*, 129–160.

International Journal of
Molecular Sciences

MDPI

Article

Correlative Light and Electron Microscopy (CLEM) Analysis of Nuclear Reorganization Induced by Clustered DNA Damage Upon Charged Particle Irradiation

Susanne Tonnemacher, Mikhail Eltsov *,† and Burkhard Jakob *

Department of Biophysics, GSI Helmholtzzentrum für Schwerionenforschung, 64291 Darmstadt, Germany; s.tonnemacher@gsi.de

* Correspondence: eltsovm@igbmc.fr (M.E.); b.jakob@gsi.de (B.J.)

† Present Address: Institute of Genetics and Molecular and Cellular Biology, University of Strasbourg, 67400 Illkirch-Graffenstaden, France.

Received: 29 January 2020; Accepted: 9 March 2020; Published: 11 March 2020

Abstract: Chromatin architecture plays major roles in gene regulation as well as in the repair of DNA damaged by endogenous or exogenous factors, such as after radiation. Opening up the chromatin might provide the necessary accessibility for the recruitment and binding of repair factors, thus facilitating timely and correct repair. The observed formation of ionizing radiation-induced foci (IRIF) of factors, such as 53BP1, upon induction of DNA double-strand breaks have been recently linked to local chromatin decompaction. Using correlative light and electron microscopy (CLEM) in combination with DNA-specific contrasting for transmission electron microscopy or tomography, we are able to show that at the ultrastructural level, these DNA damage domains reveal a chromatin compaction and organization not distinguishable from regular euchromatin upon irradiation with carbon or iron ions. Low Density Areas (LDAs) at sites of particle-induced DNA damage, as observed after unspecific uranyl acetate (UA)-staining, are thus unlikely to represent pure chromatin decompaction. RNA-specific terbium-citrate (Tb) staining suggests rather a reduced RNA density contributing to the LDA phenotype. Our observations are discussed in the view of liquid-like phase separation as one of the mechanisms of regulating DNA repair.

Keywords: DNA repair; carbon ions; radiation-induced damage; chromatin structure; electron microscopy; CLEM; DNA and RNA cytochemistry; Osmium ammine B; ChromEMT; DNA-specific staining

---

1. Introduction

Genomic DNA is packed in a highly regulated manner within an interphase nucleus to allow gene transcription or duplication of DNA during the S-Phase. To achieve the necessary compaction, the DNA double-helix is wrapped around octamers of core histones forming nucleosomes (reviewed in [1]). The interactions of neighboring nucleosomes, through exposed unfolded regions of core histones enriched in positively-charged residues together with linker histone binding, lead to further compaction (reviewed in [2]). This process is fine-tuned by covalent modifications of histones that encrypt epigenetic codes and recruit specific non-histone regulatory chromatin binders, such as polycomb group proteins (PcG) or heterochromatin protein 1 (HP1), which can result in dense chromatin states known as heterochromatin (reviewed in [3,4]). At the megabase pair scale, longer-distance chromatin interactions segment the interphase chromatids into domains, which can be visualized by imaging techniques and revealed by chromatin conformation capture methods as topologically associated domains (TADs), which eventually cluster, thus forming the interphase chromosome territory (reviewed in [5]). The chromatin domains occupy only a part of the nuclear volume, interspaced by other nuclear compartments, such as the nucleolus, nuclear speckles, and bodies, which are characterized by a low DNA density, whereas they are enriched in RNA and proteins (reviewed in [6]).

Although a detailed understanding of chromatin structure beyond the nucleosome level is still missing, the chromatin compaction plays an important role in the regulation of gene activity. Active genes are located in scattered, low-compacted domains, termed euchromatin (EC), whereas the inactive part of the genome is organized into compact heterochromatin (HC). Besides its role in gene transcription regulation, chromatin structure plays a major role in the repair of damaged DNA as well (reviewed in [7–9]). Efficient and correct DNA repair after, e.g., a radiation insult, is a prerequisite to maintaining genomic stability in mammalian cells [10]. However, especially the repair of double-strand breaks (DSBs) was shown to be heavily impacted by the chromatin environment (reviewed in [8,11]) and that, moreover, dramatic modifications arise in the chromatin surrounding a DSB especially in HC [12–15]. Both the induction of radiation induced DSBs [16,17] as well as the location of repair events and the utilization of repair factors was shown to be influenced by chromatin density, but also—depending on the type of damage induction or DSB density—by the cell cycle and utilized repair pathway [18]. Using various approaches, we have been able to show a localized depletion of heterochromatic DNA staining or changes in the fluorescence lifetime of DNA dyes in mouse chromocenters upon irradiation with charged-particles, indicating a local chromatin decompaction, which was accompanied by a rapid recruitment of DNA repair factors [13,19,20]. Recently, these results have been corroborated using transmission electron microscopy (TEM) showing electron translucent areas in the interphase nuclei of human fibroblasts at putative sites of ion traversals upon irradiation with carbon or calcium ions after embedding into LR White and uranyl acetate (UA) staining [21]. Interestingly, these electron translucent areas (subsequently termed LDA for Low Density Area) show a high density of DSB repair factors like 53BP1 [22] indicated by immunogold-labelling [21], thus confirming the presence of multiple radiation-induced DNA DSBs or clustered DNA lesions. Recent fluorescence microscopy studies on 53BP1 repair foci show properties of liquid-like phase separated domains [23], which could be an underlying mechanism for the observed changes in EM. Nevertheless, addressing the internal organization of these liquid-like phase separated domains and their relation within other nuclear components requires a higher resolution visualization.

The formation of LDAs revealed by the electron microscopy study of Timm et al. [21] matched in time with the formation of focal accumulation of repair factors visualized by immunofluorescence. Although a direct structural overlap between these two events was not provided, these data together with the results of immunogold labelling strongly indicated a distinct structural phenotype of DNA repair foci. It was hypothesized that the LDA pattern was a result of a local chromatin decompaction. Nevertheless, due to unspecific affinity of uranyl acetate to different cellular components, such as RNAs and proteins, this stain could not provide conclusive evidence for DNA reorganization linked to DNA repair.

In the present work, we combine correlative light and electron microscopy (CLEM), the localization of DNA repair foci with two DNA-specific EM staining techniques, ChomEMT, and Feulgen-type reaction with osmium ammine B (OA-B) to explore DNA arrangement therein. We show that the DNA/chromatin compaction state within DNA repair regions generally is lower than in HC, but comparable with euchromatic areas, thus DNA rearrangement cannot be solely responsible for the LDA pattern. We show that, contrary to the rest of nucleoplasm, LDAs have a low density of nuclear RNA-rich components, suggesting that low electron density of DNA repair regions results from the diminution of RNA-rich components rather than DNA/chromatin decompaction. On the basis of our observations, we hypothesize that DNA repair foci (IRIF) encapsulate damaged DNA through phase separation from nucleoplasm to provide favorable conditions for DNA damage repair.

2. Results

2.1. LDA Phenotype is Fixation and Embedding Independent

We applied CLEM to map DNA repair foci in resin embedded samples, using U2OS cells expressing 53BP1-GFP (Figure 1). These cells were irradiated with C or Fe ions and chemically fixed 1 h or 5 h after irradiation. Two fixation protocols were examined and compared: (i) 2% formaldehyde (PFA) and 0.05% glutaraldehyde (GA) in $PBS^{-/-}$ to reproduce conditions of Timm et al. [21], and (ii) 2.5% glutaraldehyde in 0.1 M sodium cacodylate buffer (SCB), which is used in protocols for DNA-specific staining [24,25].

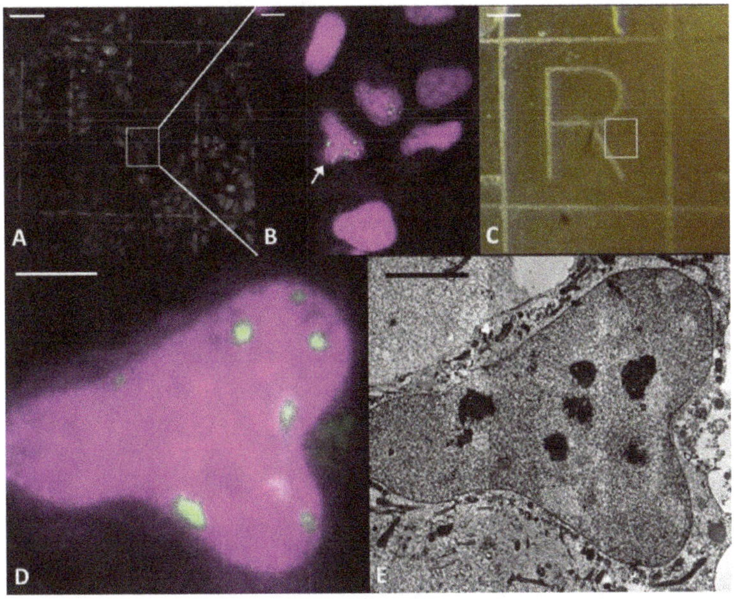

Figure 1. Defining region of interests (ROIs) and relocalization (CLEM): Fluorescence mosaic imaging of the whole grid was done automatically with a 20× dry lens (**A**). ROIs were defined by 53BP1-GFP expressing cells showing radiation induced repair foci (**A**,**B** and **D**) and imaged again at higher resolution using a 63x oil immersion objective (**B**). After the cells were embedded and removed from the glass coverslip, the resin block shows the imprint of the grid (**C**) and the ROI can be relocated (white box). Next, 100 nm sections were stained with UA and by the aid of the imprinted grid cells were relocated in the cut sections allowing imaging of the same nucleus in EM. FM (**B** arrow, **D**) and TEM (**E**). Scale bar: **A&C** 100 µm; **B** 10 µm; **D&E** 5 µm.

Fluorescence imaging of samples performed directly after the different types of fixation revealed the characteristic formation of radiation-induced foci (IRIF) indicated by the accumulation of 53BP1-GFP fluorescence at the DNA damaged domains after irradiation with C or Fe ions (Figure 1B,D, Figure 2B,D,F,H). The characteristics of the 53BP1 IRIF were in full accordance with live cell observations before fixation and with the ones obtained previously after particle irradiation [26,27]. The finding that this distinct fluorescence pattern, as well as sufficient signal intensity, was preserved after fixation indicated that both fixation protocols did not affect the 53BP1 distribution and the nuclear architecture at the detectable level. Fixed samples were imaged using confocal LM at low magnification (20× lens) (Figure 1A) to map cells in relation to the marker grid, for further re-localization at EM level; and at higher magnification, applying a high NA oil immersion lens (63×) (Figure 1B,D) to record three-dimensional maps of 53BP1-GFP foci inside the selected nuclei. After imaging, the samples were dehydrated with a series of acetone dilutions and embedded into Epon resin (see Materials and Methods) (Figure 1C).

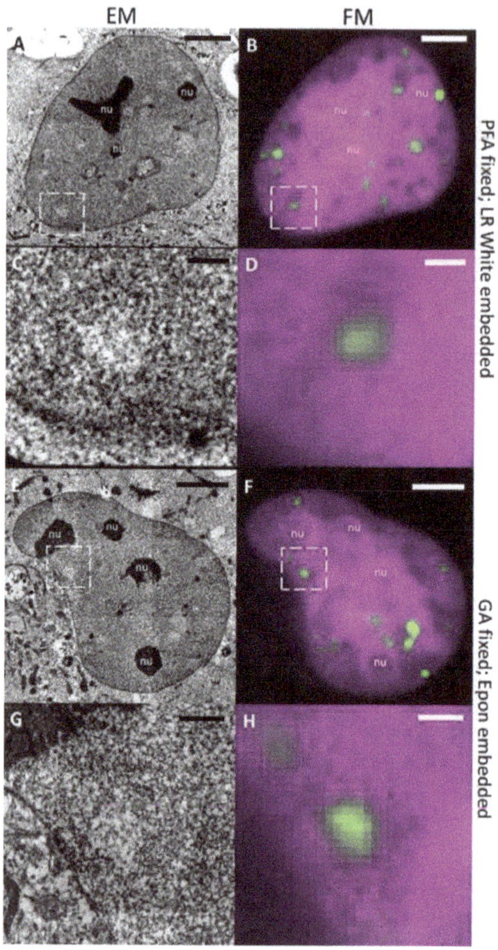

Figure 2. LDA phenotype is tolerant to different fixation and embedding conditions. Samples were PFA fixed and embedded in LR White (**A–D**) or GA fixed and embedded in Epon (**E–H**). Both samples clearly show LDAs in TEM images (left column). The whole nucleus imaged with TEM (**A**,**E**) or with FM (**B**,**F**) is shown. In (**C**,**G**) and (**D**,**H**) a single LDA (dashed white box of (**A**,**E**) and (**B**,**F**)) is magnified. All LDAs in the TEM images correlate with 53BP1-GFP IRIFs (green) of the FM images. Scale bar: (**A**,**B**,**E**&**F**) 5 µm, (**C**,**D**,**G**&**H**) 1 µm.

Using this approach we could show that at the EM level, each 53BP1-GFP focus corresponded to a LDA in sections of resin-embedded samples after carbon ion irradiation stained with UA (Figure 1D/E). Importantly, very similar LDAs were found in samples after both fixation protocols and embedding in LR White as well as in Epon also after longer (5 h) post-irradiative incubation time and irradiation with heavier particles (Fe) (Figure 2). The internal content of LDA in all cases showed a fine fibrillar aspect without visible order. Such fixation and embedding robustness of structural features indicates that LDAs are not an artefact of a given EM protocol, but linked to long-lasting radiation-induced structural rearrangement in the nucleus.

The observed general internal structure of nuclei in EM sections stained with UA appeared to be quite similar after both protocols and is in agreement with previously published observations [21,28]. We could distinguish highly dense nucleoli and granules, whereas the overall density of the remaining nuclear content was relatively uniform, without clearly distinguishable HC domains (Figure 1E, Figure 2A,C,E,G). Interestingly, granules were mostly located inside or at the periphery of LDAs (Supplemental Figure S1). The uniform staining density of the nuclear content visible after UA staining at EM level differs from the signal of the DNA-specific dye DRAQ5 [29] (chromatin) in the FM images. The later shows a complex organization of low-intensity areas (euchromatin) and heterochromatic domains well discernible around nucleoli, which is not visible in EM images stained with UA, thus confirming the unspecific staining of UA (Figure 2). Therefore, we have to conclude that UA staining cannot be used for the unambiguous assessment of DNA compaction.

In order to provide a control of the potential effects of over-expression of 53BP1 on LDAs, WT NIH/3T3 cells expressing only endogenous 53BP1 were irradiated with carbon ions using targeted irradiation with a cross pattern at the GSI microbeam setup [30]. LDA patterns similar to the ones described before for the U2OS cells were obtained in NIH/3T3 cells fixed with protocol (ii) at the sites of ion traversals (Supplemental Figure S2B). This observation, together with the previously published data [21], supports the general occurrence of LDAs in different cell types after irradiation with high linear energy transfer (LET) radiation.

2.2. DNA-Specific Stains Reveal EC-Like Areas within DNA-Repair Foci, but No LDA Pattern

In order to overcome the limitations of UA staining for judging on chromatin density, irradiated U2OS cells fixed with glutaraldehyde were further processed for ChromEMT, a recently developed method for chromatin-specific contrasting [24]. For this purpose, DNA was firstly stained with the DNA intercalating dye DRAQ5, then diaminobenzidine (DAB) polymerization was specifically induced around DNA molecules via photooxidation and DAB precipitations were stained by reduced osmium [24]. In contrast to UA staining, this DNA-specific stain revealed non-homogeneous distribution of DNA within the cell nuclei (Figure 3). It consists of dense heterochromatic patches located at the nuclear and nucleolar periphery, and in part within the nucleoplasm as well as lower density areas indicating euchromatic regions. The chromatin density correlates well with the DRAQ5 signal. ChromEMT staining revealed some background in the nucleoli and cell membranes, which is in agreement with published data [24] and results from some unspecific affinity of reduced osmium.

Surprisingly, irradiated U2OS cells show no LDA pattern after ChromEMT staining. Using the CLEM approach described before, we could precisely locate areas corresponding to the 53BP1 IRIF in serial thin sections, thus allowing a dedicated analysis of the irradiation sites (Figure 3, Supplemental Figure S3, Supplemental Movie SM1). Sites of 53BP1 IRIF showed a relatively low density of DNA in comparison to HC patches, however their density was very similar to other euchromatic areas (Figure 3E,F).

To independently verify our results after ChromEMT staining, we applied DNA-specific osmium ammine B (OA-B). This stain is based on the Feulgen reaction, which is specific to DNA [31], thus avoiding unspecific contrasting of cellular components with the reduced osmium used in ChromEMT staining. The general contrast of chromatin after OA-B is lower compared to ChromEMT staining, because the reaction occurs only at the section surface. Nevertheless, the results of OA-B stain on irradiated U2OS cells were quite similar to ones obtained in ChromEMT stained samples: IRIFs showed a low DNA density, which was again comparable to non-irradiated euchromatin areas (EC-like (EC-l) structures, Figure 4).

Additionally, these observations could be reproduced again using WT NIH/3T3 cells (Supplemental Figure S2C), thus corroborating our findings. Importantly, these confirming results were obtained from the same samples which were used for UA staining showing a clear LDA pattern (Supplemental Figure S2). This strongly supports our hypothesis that the observed LDAs are not purely DNA decondensation patterns.

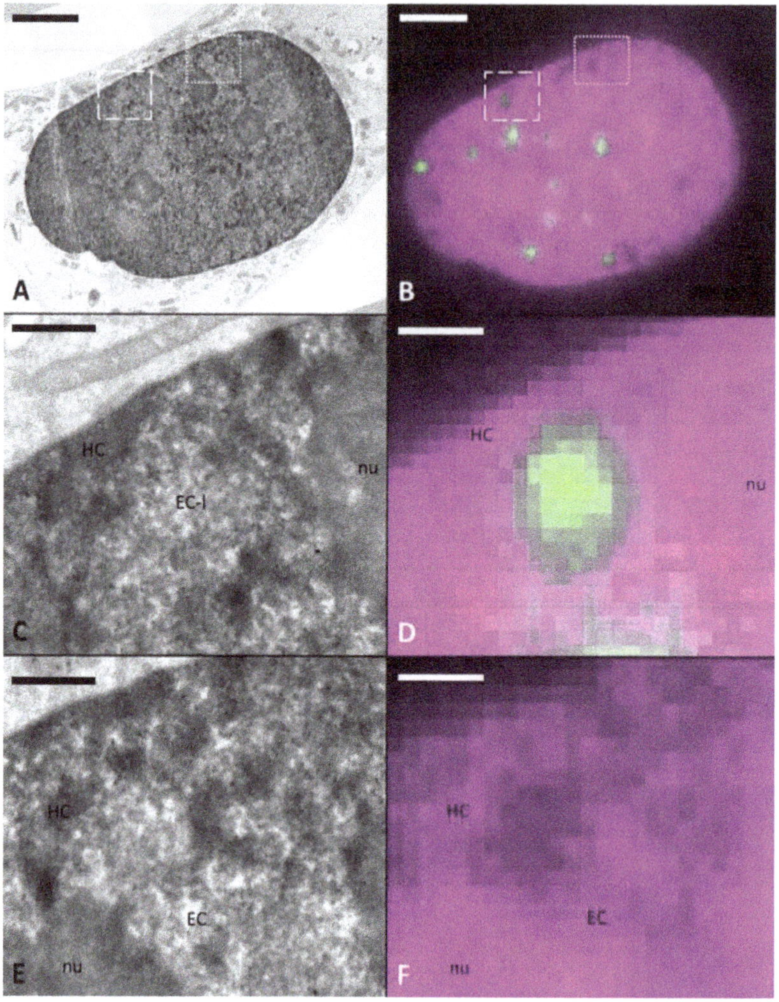

Figure 3. Absence of LDAs at damage site after chromatin-specific contrasting (ChromEMT). TEM image of a DNA-specific (ChromEMT) stained sample (**A**) correlated with FM image (**B**) with DNA (magenta) and 53BP1 (green). With ChromEMT staining, no LDA phenotype is visible. However, the irradiated area (dashed box) shows a structure (**C** with correlated FM image (**D**)) similar to euchromatin (EC) (doted box) of a non-irradiated area (**E**, correlated FM image (**F**)). Heterochromatin (HC) is clearly visible as more dark and dense structures mainly along the border of the nucleus and the nucleoli (nu) (**A**,**C**,**E**). Scale bar: (**A**,**B**) 5 μm, (**C**–**F**) 1 μm.

2.3. Tomographic Analysis of Chromatin Organisation within DNA Repair Foci

To better understand the chromatin organization linked to DNA repair we performed dual-axis tomography on semi thin (2500 nm) sections of irradiated cells stained for ChromEMT. The OA-B method is not suitable for this task because of the location of the stain, preferentially at the section surface.

The areas for tomography data collection was targeted by confocal imaging using DRAQ5 DNA densities as fiducial markers (Supplemental Figure S4). ChromEMT visualized a 3D-network of heterogeneous fibers and roundish dense bodies. The fibers appear very similar in irradiated areas and non-irradiated EC areas, whereas roundish domains prevailed in HC (Figure 5, Supplemental Figure S4). The position and shape of 53BP1 IRIF were extracted from confocal fluorescence images, converted into two dimensional masks and overlapped with the slices of the EM-tomograms after adjustment of the magnification (Figure 5 white outline, Supplemental Movie SM2). Then, the tomogram was cut along the z-axis around the mask contour, the corresponding volume extracted, and the fiber thickness analyzed. To compare irradiated and non-irradiated EC, the IRIF contour masks (shape and area) were randomly placed over non-irradiated EC and volumes extracted, accordingly (Supplemental Figure S5). For the HC analysis a similarly sized mask was used confining only HC-volumes (Supplemental Figure S6).

To compare chromatin features quantitatively, we segmented the chromatin elements in the extracted volumes and measured their diameters using custom software developed in-house (see Materials and Methods, Supplemental Figure S7). Diameter distributions obtained from three irradiated and non-irradiated EC areas from three individual cells were compared to each other and to HC areas from the same tomograms. The obtained fiber diameter distributions were nearly identical for irradiated areas and non-irradiated EC (Figure 6 green/red). Both showed a prevalence of 30–60 nm fibrillar elements and a median of ~46.5 nm for irradiated and of ~45.5 nm for non-irradiated EC. HC areas were enriched with elements of large diameter, thus increasing the median value to ~66.2 nm (Figure 6 purple).

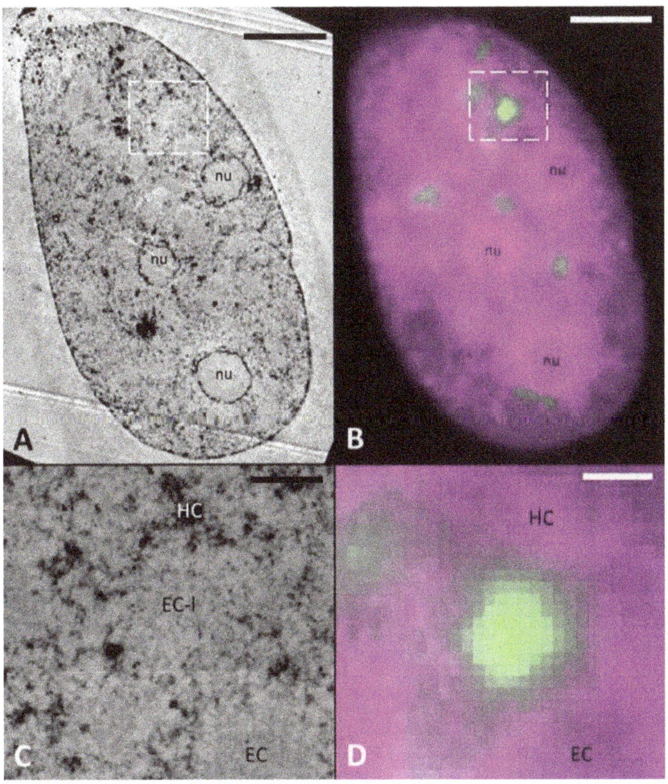

Figure 4. DNA-specific contrasting with osmium ammine B (OA-B). TEM image of OA-B stained sample (**A**). (**B**) Shows the same cell in fluorescence microscopy (53BP1 (green) and DNA (magenta)). Similar to ChromEMT stained sample (Figure 3) no LDA phenotype is visible. 53BP1 focus (**D**, green) correlates to a EC-like (EC-l) structure (**C**) and shows a similar DNA density to regular EC. Scale bar: (**A**,**B**) 5 μm, (**C**,**D**) 1 μm.

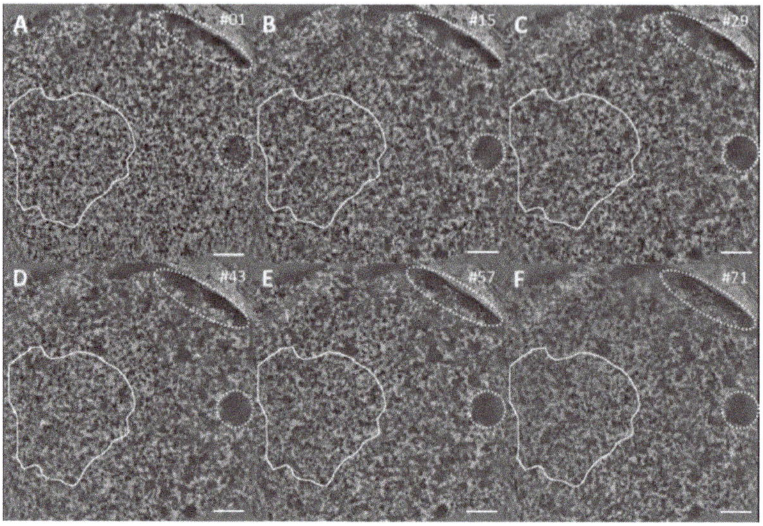

Figure 5. Tomography of an irradiated area stained by ChromEMT. A gallery of 2 nm thick tomographic slices extracted from a dual axis tomographic reconstruction of 250 thick section of an irradiated U2OS cell. The chromatin structure shows no obvious difference between the inside and outside of the damage site (white outline) through the whole volume. Roundish domains (inside doted areas) indicating HC. For further comparison see Supplemental Figure S5 which shows the tomography of a non-irradiated area. Slice numbers shown in the upper right corner. All scale bars: 500 nm. Slices thickness **2** nm.

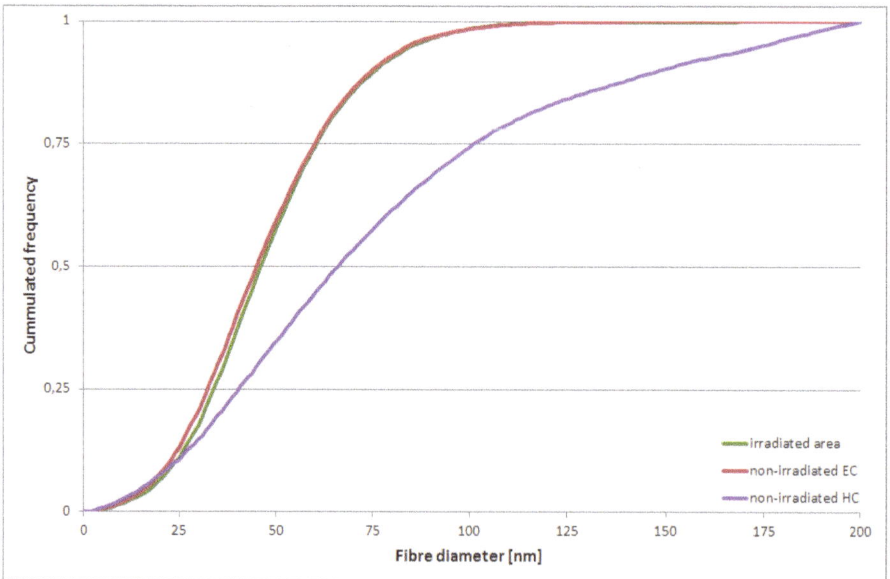

Figure 6. Cumulative histogram of fiber diameter of irradiated areas (green), non-irradiated EC (red), and HC (purple): Irradiated area shows a distribution of fiber diameters very similar to non-irradiated EC. The HC curve clearly deviates; it shows a higher proportion of fibers with a larger diameter. Each curve contains pooled data and was normalized to the volume. $N = 3$ images.

2.4. Reduced RNA Density within DNA Repair Regions

Since chromatin within DNA repair foci was not different in density and organization from regular euchromatic regions, we decided to explore the contribution of another, non-chromatin nucleic acid constituent (RNA) to LDA formation. To map the distribution of RNA-rich nuclear components, we used terbium-citrate (Tb) staining introduced by the group of Fakan [32], who demonstrated selective affinity to RNA. Accordingly, the stain showed a high density accumulation within nucleoli (Figure 7A), whereas the areas corresponding to high DNA density at the border of nucleoli and close to the nuclear envelope showed lower density in comparison to UA staining (Supplemental Figure S8). Importantly, the LDAs were present in terbium-stained sections and corresponded again to 53BP1 IRIF 1h after irradiation, as we could visualize using our CLEM approach (Figure 7A,B; see also Supplemental Figures S8 and S9). RNA diminution at LDAs was also clearly visible at 5 h post-irradiation after LR White embedding (Supplemental Figure S10). These results give a clear indication for the exclusion of RNA-rich components from the DNA repair regions. In agreement with that, we also noticed areas of a higher Tb contrast, which corresponded to low DNA density visualized by DRAQ5 staining, indicating again a selective binding to RNA (Supplemental Figure S8). These results are corroborated by the staining of irradiated cells using the fluorescent RNA-specific dye Syto RNASelect Green, which supported the observation of a reduced density of RNA in 53BP1 IRIFs (Supplementary Figure S11).

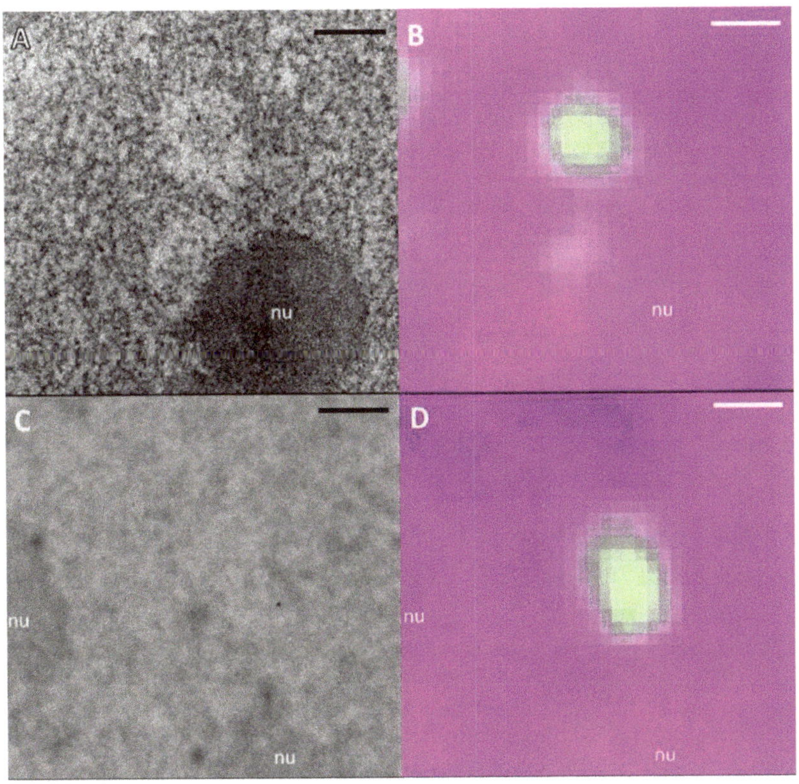

Figure 7. LDA phenotype at damage sites of RNA-specific stained (terbium-citrate; Tb) sections vanish after RNase treatment. Sections of U2OS nuclei irradiated with carbon ions, fixed 1 h after irradiation and stained with Tb (**A**) show LDAs correlating with 53BP1 signal (**B**, green). After 24 h treatment with RNase no LDAs are visible (**C**) at the damage site (**D**). All scale bars: 1 μm.

To assess this staining specificity in our preparations, we pre-treated sections with RNase prior to incubation with Tb. Even though these samples still show some staining density in nucleoli indicating an incomplete digestion of RNA in chemically fixed and resin embedded samples, LDAs completely vanished or appear strongly diminished after RNase treatment (Figure 7C&D). Together, these results provide clear evidence for the participation of RNA-rich components in UA stained samples and the contribution of RNA diminution in the formation of radiation induced LDAs.

3. Discussion

In this study, we explored the radiation-induced reorganization of cell nuclei after charged particle irradiation by applying CLEM, a combination of fluorescent light microscopy detection of IRIF at irradiation sites and electron microscopy techniques for selective staining of DNA and RNA. Our CLEM results clearly demonstrated that radiation-induced DNA repair foci, visualized by the fluorescence of 53BP1-GFP in U2OS cells, directly correspond to the LDAs visible in EM sections after non-selective UA staining (Figures 1 and 2). We could confirm LDA formation in ion-irradiated WT NIH/3T3 cells, and thus demonstrate that these LDAs are not an artefact of overexpression of 53BP1-GFP. This is in agreement with the recent observation of LDAs in human dermal fibroblasts after irradiation with carbon or calcium ions which were interpreted as decondensed chromatin regions (DCRs) [21]. Although the LDA phenotype was initially reported after formaldehyde fixation and embedding into acrylic resin (LR White) [21], we show that the LDA phenotype also occurs after glutaraldehyde fixation and Epon embedding (Figure 2), the sample preparation protocol used for DNA selective staining ChromEMT and OA-B. Interestingly, the observed LDA phenotype after irradiation (1 to 5 h) is very similar to the one of persistent DNA damage foci (PDDF), which was described in mouse neurons after DSB generation [33]. PDDFs also accumulate 53BP1 and their structural appearance in TEM images is visually indistinguishable from our LDAs, although they were visualized after low temperature embedding into Lowicryl K4M, which is a much milder treatment than we used [33]. This observation further outlines the structural consistency and robustness of DSB DNA damage repair structures in relation to cell type and the sample preparation protocol.

To check the previous models of LDAs, linking their structural appearance to chromatin decondensation (termed as DCRs in [21]), we applied DNA-specific stains to assess the density and structural organization of chromatin within DNA repair foci. EM sample preparation protocols generally are based on chemical fixation and dehydration prone to aggregation of chromatin fibers, which hampers their applicability in interpretation of native chromatin organization, especially at the nucleosome fiber scale [34–36]. Nevertheless, bearing this limitation in mind, these methods still can be used for comparative analysis of chromatin organization in response of a treatment, in our case for radiation-induced DNA damage, in cells fixed with the same protocol. Importantly, a comparison of distributions and shapes of HC patches visible at the EM level in the 3D reconstruction of nuclear volume from serial thin sections (Supplemental Figure S3), and confocal fluorescence images before DAB polymerization and embedding, did not reveal detectable chromatin rearrangements during ChromEMT procedure, thus underpinning the validity of our approach.

Our results from ChromEMT, while generally supporting open chromatin after high-LET irradiation, did not reveal any difference in density and chromatin organization between irradiated area and non-irradiated euchromatic areas of the cell (Figure 3). However, we cannot rule out that in addition to the absence of significant chromatin changes in damaged EC, heterochromatic areas are "euchromatized" and thus decompacted at damage sites. This would be in agreement with the observation in light microscopy where, on the basis of fluorescence intensity or fluorescence lifetime imaging (FLIM), only radiation-induced chromatin decompaction at heterochromatic sites after charged-particle traversals had been observed [13,20].

We acknowledge that the fiber thickness distribution assessed by ChromEMT in irradiated U2OS cells (median of around 45 nm) differs from previously reported values (mainly 12–24 nm fibers) for non-irradiated primary human small airway epithelial cells [24]. This disparity might be partially linked to cell type specific differences; however we attribute it to a stronger photooxidation and deposition of DAB in our experimental setup. The resulting stronger DAB deposition and accordingly larger diameters of chromatin fibers in our experiments did not hamper comparative analysis, and was also beneficial for an unambiguous discrimination of DNA structures from the general background of reduced osmium unavoidable in ChromEMT. Importantly, our ChromEMT results are corroborated by independent OA-B staining based on a completely different reaction chemistry [25,31,37] and pointing again to a euchromatin-like (EC-l) structure at the sites of particle traversals (Figure 4). To summarize, DNA-specific EM staining clearly showed that DNA decompaction cannot be solely responsible for the LDA phenotype.

Our results from terbium staining strongly suggest that LDAs mainly originate from a substantially lower density of RNA at IRIF. This observation and interpretation agree with a low density of RNA in PDDF, demonstrated by Mata-Garrido et al. [33], using a different cytochemical technique, EDTA regressive stain [38]. In this respect, our observations provide a new interpretation for the results of previous electron spectroscopy imaging (ESI) of 53BP1 IRIF [39,40] that revealed a lower phosphorus density and correspondingly, a lower concentration of nucleic acids therein in comparison to surrounding nuclear structures. Since, according to our results, the DNA density of DNA damage domains is very similar to the surrounding EC, the low phosphorus density observed in ESI must have resulted rather from RNA diminution. Supporting that view, immunofluorescence confocal microscopy showed no colocalization between IRIF and RNA-rich speckle domains or Cajal bodies, although they can be adjacent [41,42]. Even so, it has been shown that nascent transcript RNA might be utilized in error-free DSB repair at actively transcribed genes [43,44], the low RNA concentration within IRIF suggests a general sequestration of splicing and transcription components from these domains, thus leading to a local environment fully dedicated to DNA repair. In line with this view, DSB induction distal to the promoter of a reporter gene was shown to repress its transcription [45], and DSB induced an inhibition of RNA polymerase I [46], both in an ATM dependent manner. This would be beneficial for the protection of damaged DNA from further deterioration during transcription. Manfrini et al. [47] concluded that the inhibition of local transcription is a general response to DSB formation and is mediated by the resection of the break ends, which is a hallmark after high-LET radiation not only in G2 but also in G1-phase cells [48]. In this case it might prevent hybridization of putative exposed single stranded DNA with RNA, thus facilitating alternative end joining pathways. Interestingly, the initial formation of 53BP1 foci has been described to require the recruitment of RNA polymerase II and transcription of long non coding DNA in the vicinity of a DSB [49–51] and processing them into small non-coding DNA-damage response RNA (DDRNA, reviewed in [52]). Although the evidence of these processes after high-LET irradiation have not yet been presented, such transcriptional activity is fully compatible with the open EC-l DNA organization within DNA damage domains. However, further research of the temporal aspects will be required to understand it in the context of the low RNA content, as observed in this study. We see several possible explanations: RNA synthesis occurs (a) at the beginning of IRIF formation and RNA is degraded or diffused away later or (b) at a much lower intensity in comparison to a transcriptional activity of the regular EC; (c) it is localized at the border of DNA damage domain or (d) besides the RNA specifically produced or utilized in the damage response, transcription is generally inhibited (at later times) in the larger domain of the IRIF [44], leading to a net loss of RNA. Although the detection of the transcriptional activity in these domains was out of the scope of the current study, results from Mata-Garrido et al. [33] indicated the absence of RNA polymerase II (phosphorylated on Ser 2, as a marker of the elongation) as well as transcription inside neuronal PDDF. Nevertheless, the transcription was shown to be intense at the border of PDDFs, thus supporting the third hypothesis (c).

Several recent studies revealed that IRIF show liquid-like properties of biomolecular condensates [13,23,50]. Formation of those condensates can be driven by liquid-liquid phase separation (LLPS) of the 53BP1 protein [23] and might enhance DSB end-joining probability by isolation of the break ends in confined space. The phase separation can further contribute to DNA damage repair by the modulation of macromolecular diffusion, thus favoring the accumulation of DNA repair factors inside IRIF and the exclusion of undesired components, such as transcription and splicing machineries. The observed roundish shape of LDAs, and the low RNA content of IRIF, are fully compatible with this hypothesis. Indeed, the strikingly "empty" appearance distinguishing IRIFs in EM images suggests a diffusion barrier, preventing the uncontrolled entrance of mobile macromolecules, which would otherwise equilibrate the IRIF density with the one of the crowded nucleoplasm. The nature of this barrier remains to be unraveled; however, our EM results suggest that it is unlikely to rely solely on the mere accumulation of 53PB1 or other proteins. The "empty" phenotype in TEM indicates the abundance of molecular components, which cannot be crosslinked by aldehydes during fixation and, thus, are extracted during dehydration and embedding steps of sample preparation for EM. Phosphoinositide lipids (PPIs) can be one candidate for such "extractable" components. Due to the absence of amino groups, they cannot be crosslinked by aldehydes, are enriched in the nucleus and, according to recent studies [53], accumulate quickly at DNA damage sites. Although RNA can in principle react with aldehydes, its crosslinking is very low in comparison to proteins [54], thus chemical fixation may have a low efficiency for the immobilization of small RNA species. Thus, the extraction of DDRNAs provides us with another hypothesis explaining LDA phenotype. In addition to extraction, conventional sample preparation for TEM might lead to the precipitation of poorly fixed components within IRIFs, we speculated that it might explain the abundance of the dense granules within them. Testing of this hypothesis and the further analysis of IRIF will require the application of cryo-electron microscopy methods to explore the IRIF structure in the absence of the putative extraction/precipitation artefacts.

Conclusion: Using CLEM and chromatin-specific contrasting, we demonstrated that the chromatin density and organization within ionizing radiation-induced DNA damage domains is very similar to EC, thus arguing against the previous models of the particular chromatin decondensation during DNA DSB repair, at least in euchromatic regions. The here described LDA phenotype of IRIFs is characterized by low RNA content and supporting models of LLPS in DNA repair domains after charged particle irradiation.

4. Materials and Methods

4.1. Cell Culture, Irradiation, and Fixation

Human osteosarcoma cells (U2OS) stably expressing 53BP1-GFP (kindly provided by C. Lukas, Copenhagen, Denmark) were grown on round photo-etched coverslips (28 mm in diameter, ibidi, Gräfelfing, Germany) in 35 mm petri dishes at 37 °C with 95% humidity and 5% CO_2 in DMEM (Gibco, Carlsbad, CA, USA) + 10% FCS (Biochrom, Berlin, Germany) + 1 μg/mL Puromycin. The irradiation was done at the linear accelerator UNILAC of the GSI Helmholzzentrum für Schwerionenforschung GmbH with carbon (9.8 MeV/u, 170 keV/μm) or iron ions (7.0 MeV/u, 2875 keV/μm) at a fluence of 5×10^6 particles/cm^2. For the microbeam experiment, NIH/3T3 cells (ATCC, Manassas, VA, USA) were prepared and irradiated according to [30], using 4.8 MeV/u carbon ions (LET: 285 keV/μm). Fixation was done 1 h (carbon ions) or 5 h (iron ions) after irradiation, respectively. Two fixation methods were used: (i) 2% formaldehyde (PFA; Sigma-Aldrich, Saint-Louis, MO, USA) + 0.05% EM grade glutaraldehyde (GA; Science Services, München, Germany) in PBS$^{-/-}$ (w/o calcium and magnesium; Biochrom, Berlin, Germany), 5min RT + 1h on ice; (ii) 2.5% GA + 5 mM $CaCl_2$ in 0.1 M sodium cacodylate buffer (SCB; Serva, Heidelberg, Germany), 5 min RT + 1 h on ice, 3 × 5 min blocking with 10 mM glycine (Roth, Karlsruhe, Germany) in 0.1 M SCB.

4.2. Fluorescence Microscopy

Fluorescence microscopy was done after fixation, but before DAB polymerization, dehydration, and embedding. Mosaic screening of the GFP-signal of the whole coverslip was done automatically on a Nikon TiE spinning disc confocal microscope, using a 20× dry lens. Two channels (DRAQ5 and GFP) of the selected nuclei were recorded again at a higher resolution, using a 63x oil immersion lens, NA 1.3 at a Leica DMI 4000 B.

4.3. Electron Microscopy

4.3.1. Embedding and Cutting for EM

After FM imaging with Leica DMI 4000 B, the samples were dehydrated and embedded into a resin. For LR White (Science Services, München, Germany) embedding, dehydration was done with ethanol (Roth, Karlsruhe, Germany) (5 min 30%, 10 min 50%, 15 min 70%, 20 min 90%, 2× 30 min 100%), followed by incubation with LR White/100% ethanol (1:1 mixture) overnight, 2x changes 100% LR White for 2 h. Then, the coverslips with cells were placed into a 35 mm Petri dish filled with fresh LR White, the dish cover was sealed by the application of a thin layer of LR White resin mixed with accelerator to reduce oxygen penetration inside the dish. After 30 min of the sealing resin layer polymerization at RT, the dishes were placed into an oven and cured at 55° for 48 h. After polymerization the coverslip was removed from the resin blocks through a careful cooling by contact with a metal rod that was pre chilled with liquid nitrogen.

For Epon embedding, dehydration was done with acetone (5 min 30%, 10 min 50%, 15 min 70%, 20 min 90%, 2× 30 min 100%; Electron Microscopy Sciences, Hatfield, PA, USA) and 2x 10 min with 100 % propylene oxide (Science Services, München, Germany). After dehydration, samples were incubated once in a 1:1 mixture of propylene oxide and Epon812 substitute (Sigma-Aldrich, Sent-Lois, MO, USA; 30 min) and twice with pure Epon812 substitute (2h and overnight). Polymerization of Epon was done at 65 °C for 48 h. Resin blocks were trimmed using diamond trimming knives (Diatome, Nidau, Switzerland). For TEM 100 nm sections and for TEM tomography, 250 nm sections were directly cut from the top side of the trimmed block with an ultra 35° diamond knife (Diatome, Nidau, Switzerland). Sections were collected on Formvar-coated (Electron Microscopy Sciences, Hatfield, PA, USA), palladium-copper slot grids (Science Services, München, Germany) and stained in 2% Uranyl acetate (Science Services, München, Germany) solution in methanol.

4.3.2. ChromEMT Staining

The DNA-specific staining was done according to the original protocol published by Ou et al. [24]. Living cells were stained with 1 μM DRAQ5 (Thermo Fisher Scientific, Waltham, MA, USA) in OptiMEM (Gibco, Carlsbad, CA, USA) at 37 °C for 10 min, washed 2× 5 min with $PBS^{-/-}$ and fixed. The coverslip with cells was transferred into a new plastic petri dish and bathed in 2.5 mM DAB (Sigma-Aldrich, Saint-Louis, MO, USA) in 0.1 M SCB. For the photooxidation of a larger area, DRAQ5 was excited at 615 nm provided by a CoolLED pE light source (BFI Optilas, Dietzenbach, Germany) set at 100% and applied via a light guide placed directly below the petri dish for up to 1 h at RT. A round area with a diameter of 5 mm was illuminated simultaneously. Cells were washed for 1 h (3× 20 min) in ice-cold 0.15 M SCB + 2 mM $CaCl_2$ on ice and stained 1 h in 0.15 M SCB + 2 mM $CaCl_2$ + 1.5% potassium ferrocyanide + 2% aqueous OsO_4 (Science Services, München, Germany) on ice and subsequently washed 5× 3 min with ddH_2O at RT, followed by dehydration and embedding in Epon, as described above (see 4.3.1).

4.3.3. Staining with Osmium Ammine B (OA-B)

100 nm thick sections of Epon-embedded samples were collected on 200 mesh gold grids (Plano, Wetzlar, Germany). The grids were dried and incubated for 3 h at 70 °C to improve the section attachment, and cooled to RT. The grids were incubated with 5N HCl for 30 min at RT within 1.5 mL Eppendorf tubes, washed with ddH_2O, air dried, and immersed into the osmium ammine B (Polysciences, Hirschberg, Germany) reagent activated exactly as described by [25]. After staining for 1 h at 37 °C, the grids were washed with ddH_2O and air dried.

4.3.4. RNA-Specific Staining with Terbium-Citrate (Tb)

The staining reagent was prepared as described by [32]. 100 nm thick sections of LR White embedded samples prepared according to protocol (i) were collected on Formvar-coated, palladium-copper slot grids. The grids were incubated on the reagent for 1 h. For control of the staining specificity, selected grids with sections were incubated on drops of 10 mg/mL RNase A (Sigma-Aldrich, Saint-Louis, MO, USA) in 50 mM Tris-HCl pH 7.0 at 37 °C for 24 h and washed with distilled water before Tb staining. The solution of RNase was boiled for 5 min to eliminate a risk of possible proteinase contamination and cooled down to 37 °C before use.

4.3.5. Transmission EM

100 nm thick sections were collected on Formvar-coated slot grids and imaged in a transmission electron microscope (Jeol 2100Plus, 120 kV). For serial sections, Serial EM software (University of Colorado [55]) was used.

4.3.6. EM Tomography

250 nm thick sections were collected on Formvar-coated slot grids for TEM tomography. Next, 15 nm colloidal gold particles (Sigma-Aldrich, Saint-Louis, MO, USA) were applied on the supporting film to serve as fiducial markers for tomogram alignment. The tomographic data collection was performed using Tecnai F30 transmission electron microscope (Thermo Fisher Scientific, Waltham, MA, USA) equipped with a US4000 CCD camera (Gatan, Pleasanton, CA, USA). Regions of interests (ROIs) were pre-exposed at 3000 e/$Å^2$ to minimize the section drift and warping [56,57]. The dual-axis tomographic acquisition was done at 12,000× magnification (1.96 nm/pixel, camera binning 2) using Serial EM software (University of Colorado [55]) running a continuous tilt-scheme from −60° to +60° and a constant angular increment of 1.5°. Tilt series alignment and tomogram reconstruction were performed using ETOMO program of IMOD software package [58].

4.4. Data Analysis

The open source programs ImageJ (version 1.52a, NIH and LOCI, University of Wisconsin, USA) and IMOD (version 4.5.0, University of Colorado, USA) were used for general image analysis. Fiber thickness in the tomograms was determined using in-house written software, ImageD (David Eilenstein) (Supplemental Figure S7). Briefly, image grey values were inverted and a 3D median filter applied to remove shot noise. A ROI was defined and selected. For chromatin fiber detection, a threshold was applied automatically inside the ROI and the resulting fibers binarized. Binarized fibers were skeletonized, center-points determined, and the diameter of the fibers analyzed on the basis of the shortest distance between center-points and boundary.

Supplementary Materials: Supplementary materials can be found at http://www.mdpi.com/1422-0067/21/6/1911/s1.

Author Contributions: S.T. performed the experiments and data evaluation and wrote the manuscript. M.E. conceived and performed the experiments and wrote the manuscript. B.J. designed the project, conceived, and supervised the experiments and wrote the manuscript. All authors have read and agreed to the published version of the manuscript.

Funding: This work was supported by the Bundesministerium für Bildung und Forschung (Germany) Grant 02NUK037A. The research presented is in part a result of a R&D project experiment UBio_Jakob in the frame of FAIR Phase-0 supported by the GSI Helmholtzzentrum für Schwerionenforschung in Darmstadt (Germany).

Acknowledgments: The authors wish to thank Gudrun Becker and Henrieke Förster for cell culture and excellent technical assistance. We are indebted to David Eilenstein for the development of the fiber analysis software. We thank Yannick Schwab and Mandy Börmel for assistance at the Electron Microscopy Core Facility at EMBL. The support by Achilleas Frangakis, Diana Grewe, and Fabrizio Porrati at Goethe University of Frankfurt (IFB) / Buchmann Institute for Molecular Life Sciences (BMLS) and by the Frankfurt Center for Electron Microscopy (FCEM) is gratefully acknowledged. We are grateful to our colleagues at GSI for operating the accelerator, Kay-Obbe Voss for the irradiation at the GSI microprobe, and to C. Lukas (Danish Cancer society, Copenhagen, Denmark) for providing the U2OS-53BP1-GFP cell line.

Conflicts of Interest: The authors declare no conflict of interest. The funders had no role in the design of the study; in the collection, analyses, or interpretation of data; in the writing of the manuscript, or in the decision to publish the results.

Abbreviations

CLEM	correlative light and electron microscopy
LDA	low density area
PcG	polycomb group protein
HP1	heterochromatin protein 1
TAD	topologically associated domain
DSB	double-strand break
UA	uranyl acetate
TEM	transmission electron microscope
EM	electron microscope
OA-B	osmium ammine B
IRIF	ionizing radiation induced focus
PFA	(para) formaldehyde
GA	glutaraldehyde
53BP1	p53-binding protein 1
GFP	green fluorescent protein
FM	fluorescence microscope
LET	linear energy transfer
EC	euchromatin
EC-l	euchromatin-like
HC	heterochromatin
DCR	decondensed chromatin region
PDDF	persistent DNA damage focus
FLIM	fluorescence lifetime imaging microscopy
DAB	3,3′-Diaminobenzidine
LLPS	liquate-liquate phase separation
SCB	sodium cacodylate buffer
$PBS^{-/-}$	phosphate buffer saline without calcium and magnesium
Tb	terbium-citrate
FCS	fetal calf serum
DMEM	Dulbecco's Modified Eagle Medium
ROI	region of interest

References

1. Zhou, K.; Gaullier, G.; Luger, K. Nucleosome structure and dynamics are coming of age. *Nat. Struct. Mol. Biol.* **2019**, *26*, 3–13. [CrossRef] [PubMed]
2. Hansen, J.C.; Connolly, M.; McDonald, C.J.; Pan, A.; Pryamkova, A.; Ray, K.; Seidel, E.; Tamura, S.; Rogge, R.; Maeshima, K. The 10-nm chromatin fiber and its relationship to interphase chromosome organization. *Biochem. Soc. Trans.* **2018**, *46*, 67–76. [CrossRef] [PubMed]
3. Janssen, A.; Colmenares, S.U.; Karpen, G.H. Heterochromatin: Guardian of the genome. *Annu. Rev. Cell Dev. Biol.* **2018**, *34*, 265–288. [CrossRef] [PubMed]
4. Cheutin, T.; Cavalli, G. The multiscale effects of polycomb mechanisms on 3D chromatin folding. *Crit. Rev. Biochem. Mol. Biol.* **2019**, *54*, 399–417. [CrossRef] [PubMed]
5. Szabo, Q.; Bantignies, F.; Cavalli, G. Principles of genome folding into topologically associating domains. *Sci. Adv.* **2019**, *5*, eaaw1668. [CrossRef]
6. Cremer, T.; Cremer, M.; Hübner, B.; Silahtaroglu, A.; Hendzel, M.; Lanctôt, C.; Strickfaden, H.; Cremer, C. The Interchromatin Compartment Participates in the Structural and Functional Organization of the Cell Nucleus. *Bioessays* **2020**, *42*, 1900132. [CrossRef]

7. Goodarzi, A.A.; Jeggo, P.; Löbrich, M. The influence of heterochromatin on DNA double strand break repair: Getting the strong, silent type to relax. *DNA Repair (Amst.)* **2010**, *9*, 1273–1282. [CrossRef]
8. Goodarzi, A.A.; Jeggo, P.A. The heterochromatic barrier to DNA double strand break repair: How to get the entry visa. *Int. J. Mol. Sci.* **2012**, *13*, 11844–11860. [CrossRef]
9. Fortuny, A.; Polo, S.E. The response to DNA damage in heterochromatin domains. *Chromosoma* **2018**, *127*, 291–300. [CrossRef]
10. Jeggo, P.A.; Pearl, L.H.; Carr, A.M. DNA repair, genome stability and cancer: A historical perspective. *Nat. Rev. Cancer* **2016**, *16*, 35–42. [CrossRef]
11. Lemaître, C.; Soutoglou, E. Double strand break (DSB) repair in heterochromatin and heterochromatin proteins in DSB repair. *DNA Repair (Amst.)* **2014**, *19*, 163–168. [CrossRef] [PubMed]
12. Lans, H.; Marteijn, J.A.; Vermeulen, W. ATP-dependent chromatin remodeling in the DNA-damage response. *Epigenetics Chromatin* **2012**, *5*, 4. [CrossRef] [PubMed]
13. Jakob, B.; Splinter, J.; Conrad, S.; Voss, K.-O.; Zink, D.; Durante, M.; Löbrich, M.; Taucher-Scholz, G. DNA double-strand breaks in heterochromatin elicit fast repair protein recruitment, histone H2AX phosphorylation and relocation to euchromatin. *Nucleic Acids Res.* **2011**, *39*, 6489–6499. [CrossRef] [PubMed]
14. Ryu, T.; Spatola, B.; Delabaere, L.; Bowlin, K.; Hopp, H.; Kunitake, R.; Karpen, G.H.; Chiolo, I. Heterochromatic breaks move to the nuclear periphery to continue recombinational repair. *Nat. Cell Biol.* **2015**, *17*, 1401–1411. [CrossRef] [PubMed]
15. Chagin, V.O.; Reinhart, B.; Becker, A.; Mortusewicz, O.; Jost, K.L.; Rapp, A.; Leonhardt, H.; Cardoso, M.C. Processive DNA synthesis is associated with localized decompaction of constitutive heterochromatin at the sites of DNA replication and repair. *Nucleus* **2019**, *10*, 231–253. [CrossRef] [PubMed]
16. Falk, M.; Lukásová, E.; Kozubek, S. Chromatin structure influences the sensitivity of DNA to gamma-radiation. *Biochim. Biophys. Acta* **2008**, *1783*, 2398–2414. [CrossRef]
17. Takata, H.; Hanafusa, T.; Mori, T.; Shimura, M.; Iida, Y.; Ishikawa, K.; Yoshikawa, K.; Yoshikawa, Y.; Maeshima, K. Chromatin compaction protects genomic DNA from radiation damage. *PLoS ONE* **2013**, *8*, e75622. [CrossRef]
18. Tsouroula, K.; Furst, A.; Rogier, M.; Heyer, V.; Maglott-Roth, A.; Ferrand, A.; Reina-San-Martin, B.; Soutoglou, E. Temporal and Spatial Uncoupling of DNA Double Strand Break Repair Pathways within Mammalian Heterochromatin. *Mol. Cell* **2016**, *63*, 293–305. [CrossRef]
19. Müller, I.; Merk, B.; Voss, K.O.; Averbeck, N.; Jakob, B.; Durante, M.; Taucher-Scholz, G. Species conserved DNA damage response at the inactive human X chromosome. *Mutat. Res. Genet. Toxicol. Environ. Mutagen.* **2013**, *756*, 30–36. [CrossRef]
20. Abdollahi, E.; Taucher-Scholz, G.; Durante, M.; Jakob, B. Upgrading the GSI beamline microscope with a confocal fluorescence lifetime scanner to monitor charged particle induced chromatin decondensation in living cells. *Nucl. Instrum. Methods Phys. Res. B* **2015**, *365*, 626–630. [CrossRef]
21. Timm, S.; Lorat, Y.; Jakob, B.; Taucher-Scholz, G.; Rübe, C.E. Clustered DNA damage concentrated in particle trajectories causes persistent large-scale rearrangements in chromatin architecture. *Radiother. Oncol.* **2018**, *129*, 600–610. [CrossRef] [PubMed]
22. Panier, S.; Boulton, S.J. Double-strand break repair: 53BP1 comes into focus. *Nat. Rev. Mol. Cell Biol.* **2014**, *15*, 7–18. [CrossRef] [PubMed]
23. Kilic, S.; Lezaja, A.; Gatti, M.; Bianco, E.; Michelena, J.; Imhof, R.; Altmeyer, M. Phase separation of 53BP1 determines liquid-like behavior of DNA repair compartments. *EMBO J.* **2019**, *38*, e101379. [CrossRef] [PubMed]
24. Ou, H.D.; Phan, S.; Deerinck, T.J.; Thor, A.; Ellisman, M.H.; O'Shea, C.C. ChromEMT: Visualizing 3D chromatin structure and compaction in interphase and mitotic cells. *Science* **2017**, *357*, eaag0025. [CrossRef] [PubMed]
25. Vázquez-Nin, G.H.; Biggiogera, M.; Echeverría, O.M. Activation of osmium ammine by SO2-generating chemicals for EM Feulgen-type staining of DNA. *Eur. J. Histochem.* **1995**, *39*, 101–106. [PubMed]
26. Jakob, B.; Splinter, J.; Durante, M.; Taucher-Scholz, G. Live cell microscopy analysis of radiation-induced DNA double-strand break motion. *Proc. Natl. Acad. Sci. USA* **2009**, *106*, 3172–3177. [CrossRef]
27. Jakob, B.; Dubiak-Szepietowska, M.; Janiel, E.; Schmidt, A.; Durante, M.; Taucher-Scholz, G. Differential repair protein recruitment at sites of clustered and isolated DNA double-strand breaks produced by high-energy heavy ions. *Sci. Rep.* **2020**, *10*, 1443. [CrossRef]

28. Lorat, Y.; Timm, S.; Jakob, B.; Taucher-Scholz, G.; Rübe, C.E. Clustered double-strand breaks in heterochromatin perturb DNA repair after high linear energy transfer irradiation. *Radiother. Oncol.* **2016**, *121*, 154–161. [CrossRef]
29. Smith, P.J.; Blunt, N.; Wiltshire, M.; Hoy, T.; Teesdale-Spittle, P.; Craven, M.R.; Watson, J.V.; Amos, W.B.; Errington, R.J.; Patterson, L.H. Characteristics of a novel deep red/infrared fluorescent cell-permeant DNA probe, DRAQ5, in intact human cells analyzed by flow cytometry, confocal and multiphoton microscopy. *Cytometry* **2000**, *40*, 280–291. [CrossRef]
30. Heiß, M.; Fischer, B.E.; Jakob, B.; Fournier, C.; Becker, G.; Taucher-Scholz, G. Targeted irradiation of mammalian cells using a heavy-ion microprobe. *Radiat. Res.* **2006**, *165*, 231–239. [CrossRef]
31. Gautier, A. Ultrastructural localization of DNA in ultrathin tissue sections. *Int. Rev. Cytol.* **1976**, *44*, 113–191. [CrossRef] [PubMed]
32. Biggiogera, M.; Fakan, S. Fine structural specific visualization of RNA on ultrathin sections. *J. Histochem. Cytochem.* **1998**, *46*, 389–395. [CrossRef] [PubMed]
33. Mata-Garrido, J.; Casafont, I.; Tapia, O.; Berciano, M.T.; Lafarga, M. Neuronal accumulation of unrepaired DNA in a novel specific chromatin domain: Structural, molecular and transcriptional characterization. *Acta Neuropathol. Commun.* **2016**, *4*, 41. [CrossRef] [PubMed]
34. Maeshima, K.; Eltsov, M.; Laemmli, U.K. Chromosome structure: Improved immunolabeling for electron microscopy. *Chromosoma* **2005**, *114*, 365–375. [CrossRef] [PubMed]
35. Fussner, E.; Ching, R.W.; Bazett-Jones, D.P. Living without 30 nm chromatin fibers. *Trends Biochem. Sci.* **2011**, *36*, 1–6. [CrossRef] [PubMed]
36. Maeshima, K.; Hihara, S.; Eltsov, M. Chromatin structure: Does the 30-nm fibre exist In Vivo? *Curr. Opin. Cell Biol.* **2010**, *22*, 291–297. [CrossRef]
37. Olins, A.L.; Moyer, B.A.; Kim, S.H.; Allison, D.P. Synthesis of a more stable osmium ammine electron-dense DNA stain. *J. Histochem. Cytochem.* **1989**, *37*, 395–398. [CrossRef]
38. Bernhard, W. A new staining procedure for electron microscopical cytology. *J. Ultrastruct. Res.* **1969**, *27*, 250–265. [CrossRef]
39. Kruhlak, M.J.; Celeste, A.; Dellaire, G.; Fernandez-Capetillo, O.; Müller, W.G.; McNally, J.G.; Bazett-Jones, D.P.; Nussenzweig, A. Changes in chromatin structure and mobility in living cells at sites of DNA double-strand breaks. *J. Cell Biol.* **2006**, *172*, 823–834. [CrossRef]
40. Strickfaden, H.; Xu, Z.Z.; Hendzel, M.J. Visualization of miniSOG Tagged DNA Repair Proteins in Combination with Electron Spectroscopic Imaging (ESI). *J. Vis. Exp.* **2015**, *103*, e52893. [CrossRef]
41. Harrigan, J.A.; Belotserkovskaya, R.; Coates, J.; Dimitrova, D.S.; Polo, S.E.; Bradshaw, C.R.; Fraser, P.; Jackson, S.P. Replication stress induces 53BP1-containing OPT domains in G1 cells. *J. Cell Biol.* **2011**, *193*, 97–108. [CrossRef] [PubMed]
42. Legartová, S.; Sehnalová, P.; Malyšková, B.; Küntziger, T.; Collas, P.; Cmarko, D.; Raška, I.; Sorokin, D.V.; Kozubek, S.; Bártová, E. Localized Movement and Levels of 53BP1 Protein Are Changed by γ-irradiation in PML Deficient Cells. *J. Cell. Biochem.* **2016**, *117*, 2583–2596. [CrossRef] [PubMed]
43. Meers, C.; Keskin, H.; Storici, F. DNA repair by RNA: Templated, or not templated, that is the question. *DNA Repair (Amst.)* **2016**, *44*, 17–21. [CrossRef] [PubMed]
44. Chakraborty, A.; Tapryal, N.; Venkova, T.; Horikoshi, N.; Pandita, R.K.; Sarker, A.H.; Sarkar, P.S.; Hazra, T.K. Classical non-homologous end-joining pathway utilizes nascent RNA for error-free double-strand break repair of transcribed genes. *Nat. Commun.* **2016**, *7*, 13049. [CrossRef]
45. Shanbhag, N.M.; Rafalska-Metcalf, I.U.; Balane-Bolivar, C.; Janicki, S.M.; Greenberg, R.A. ATM-dependent chromatin changes silence transcription in cis to DNA double-strand breaks. *Cell* **2010**, *141*, 970–981. [CrossRef]
46. Kruhlak, M.; Crouch, E.E.; Orlov, M.; Montaño, C.; Gorski, S.A.; Nussenzweig, A.; Misteli, T.; Phair, R.D.; Casellas, R. The ATM repair pathway inhibits RNA polymerase I transcription in response to chromosome breaks. *Nature* **2007**, *447*, 730–734. [CrossRef]
47. Manfrini, N.; Clerici, M.; Wery, M.; Colombo, C.V.; Descrimes, M.; Morillon, A.; Di d'Adda Fagagna, F.; Longhese, M.P. Resection is responsible for loss of transcription around a double-strand break in Saccharomyces cerevisiae. *Elife* **2015**, *4*, e08942. [CrossRef]
48. Averbeck, N.B.; Ringel, O.; Herrlitz, M.; Jakob, B.; Durante, M.; Taucher-Scholz, G. DNA end resection is needed for the repair of complex lesions in G1-phase human cells. *Cell Cycle* **2014**, *13*, 2509–2516. [CrossRef]

49. Pryde, F.; Khalili, S.; Robertson, K.; Selfridge, J.; Ritchie, A.-M.; Melton, D.W.; Jullien, D.; Adachi, Y. 53BP1 exchanges slowly at the sites of DNA damage and appears to require RNA for its association with chromatin. *J. Cell Sci.* **2005**, *118*, 2043–2055. [CrossRef]
50. Pessina, F.; Giavazzi, F.; Yin, Y.; Gioia, U.; Vitelli, V.; Galbiati, A.; Barozzi, S.; Garre, M.; Oldani, A.; Flaus, A.; et al. Functional transcription promoters at DNA double-strand breaks mediate RNA-driven phase separation of damage-response factors. *Nat. Cell Biol.* **2019**, *21*, 1286–1299. [CrossRef]
51. Michelini, F.; Pitchiaya, S.; Vitelli, V.; Sharma, S.; Gioia, U.; Pessina, F.; Cabrini, M.; Wang, Y.; Capozzo, I.; Iannelli, F.; et al. Damage-induced lncRNAs control the DNA damage response through interaction with DDRNAs at individual double-strand breaks. *Nat. Cell Biol.* **2017**, *19*, 1400–1411. [CrossRef] [PubMed]
52. Bader, A.S.; Hawley, B.R.; Wilczynska, A.; Bushell, M. The roles of RNA in DNA double-strand break repair. *Br. J. Cancer* **2020**, *122*, 613–623. [CrossRef] [PubMed]
53. Wang, Y.-H.; Hariharan, A.; Bastianello, G.; Toyama, Y.; Shivashankar, G.V.; Foiani, M.; Sheetz, M.P. DNA damage causes rapid accumulation of phosphoinositides for ATR signaling. *Nat. Commun.* **2017**, *8*, 2118. [CrossRef]
54. Griffiths, G. *Fine Structure Immunocytochemistry*; Springer: Berlin/Heidelberg, Germany, 1993; pp. 26–80. ISBN 978-3-642-77095-1.
55. Mastronarde, D.N. Automated electron microscope tomography using robust prediction of specimen movements. *J. Struct. Biol.* **2005**, *152*, 36–51. [CrossRef] [PubMed]
56. Luther, P.K.; Lawrence, M.C.; Crowther, R.A. A method for monitoring the collapse of plastic sections as a function of electron dose. *Ultramicroscopy* **1988**, *24*, 7–18. [CrossRef]
57. Braunfeld, M.B.; Koster, A.J.; Sedat, J.W.; Agard, D.A. Cryo automated electron tomography: Towards high-resolution reconstructions of plastic-embedded structures. *J. Microsc.* **1994**, *174*, 75–84. [CrossRef]
58. Kremer, J.R.; Mastronarde, D.N.; McIntosh, J.R. Computer visualization of three-dimensional image data using IMOD. *J. Struct. Biol.* **1996**, *116*, 71–76. [CrossRef]

International Journal of
Molecular Sciences

MDPI

Article

Modeling the Effect of Hypoxia and DNA Repair Inhibition on Cell Survival after Photon Irradiation

Hans Liew [1,2,3,4,5,6], Carmen Klein [2,3,4,5], Frank T. Zenke [7], Amir Abdollahi [2,3,4,5], Jürgen Debus [1,2,3,4,5,6], Ivana Dokic [2,3,4,5,*,†] and Andrea Mairani [2,3,4,5,*,†]

1 Clinical Cooperation Unit Radiation Oncology, German Cancer Research Center (DKFZ), 69120 Heidelberg, Germany; h.liew@dkfz-heidelberg.de (H.L.); Juergen.Debus@med.uni-heidelberg.de (J.D.)

2 Division of Molecular and Translational Radiation Oncology, National Center for Tumor Diseases (NCT), Heidelberg University Hospital, 69120 Heidelberg, Germany; carmen.klein@dkfz-heidelberg.de (C.K.); a.amir@dkfz-heidelberg.de (A.A.)

3 Heidelberg Institute of Radiation Oncology (HIRO), German Cancer Research Center (DKFZ), 69120 Heidelberg, Germany

4 German Cancer Consortium (DKTK), 69120 Heidelberg, Germany

5 Heidelberg Ion-Beam Therapy Center (HIT), 69120 Heidelberg, Germany

6 Faculty of Physics and Astronomy, Heidelberg University, 69120 Heidelberg, Germany

7 Merck KGaA, 64293 Darmstadt, Germany; Frank.Zenke@merckgroup.com

* Correspondence: i.dokic@dkfz-heidelberg.de (I.D.); andrea.mairani@med.uni-heidelberg.de (A.M.)

† These authors contributed equally to this work.

Received: 24 October 2019; Accepted: 28 November 2019; Published: 30 November 2019

Abstract: Mechanistic approaches to modeling the effects of ionizing radiation on cells are on the rise, promising a better understanding of predictions and higher flexibility concerning conditions to be accounted for. In this work we modified and extended a previously published mechanistic model of cell survival after photon irradiation under hypoxia to account for radiosensitization caused by deficiency or inhibition of DNA damage repair enzymes. The model is shown to be capable of describing the survival data of cells with DNA damage repair deficiency, both under norm- and hypoxia. We find that our parameterization of radiosensitization is invariant under change of oxygen status, indicating that the relevant parameters for both mechanisms can be obtained independently and introduced freely to the model to predict their combined effect.

Keywords: ionizing radiation; DNA repair; hypoxia; modeling; radiosensitizer

1. Introduction

Radiation therapy is one of the cornerstones of cancer care, where ~50% of patents receive radiation during the course of disease [1]. Radiobiological modeling is an integral part of radiation oncology and radiation therapy, used to predict normal and tumor tissue response, which is of particular significance when moving towards personalized radiation treatment, including treatment gap corrections, normal tissue tolerance predictions, optimization of therapy determined by predictive assays, multi-modality schedule design, and the simulation of clinical trials [2]. In this work, we present the first step towards development of our modeling platform called "UNIfied and VERSatile Engine" (UNIVERSE), within which we aim to integrate multiple biological responses and mechanisms relevant for describing radiation response of different cell types. In this manuscript, we focus on describing the cellular response of a particularly radioresistant tumor sub-population—hypoxic cells. Cells under hypoxic conditions exhibit increased radioresistance, and tumors containing hypoxic regions have significantly worse chances of successful treatment with radiotherapy [3–5]. The increased radiosensitivity of cells in the presence of free oxygen is usually explained by the oxygen fixation hypothesis: Molecular oxygen has the ability to react with radicals produced in the DNA, thereby fixating the damage, preventing the direct chemical restoration of the DNA radical by reacting with H^+ [3]. The concept of the oxygen enhancement ratio (OER) is classically used to quantify the dependence of cell survival on the oxygenation status. The OER is usually defined as the ratio between the doses needed to induce the same survival fraction in a hypoxic and a normoxic environment [6]. One of the key strategies to overcome the radioresistance in hypoxic cells is a dual treatment, i.e., a combined treatment with photon irradiation and administration of radiosensitizing drugs, such as DNA damage response (DDR) inhibitors [7–10].

To model the response of hypoxic tumor cells to dual treatment, we based our approach on a previously published model of our group [11]. Similarly to models described by other groups [12–14], the here presented model divides the cells' nucleus into equally sized subvolumes containing about 2 Mbp, dubbed giant loops [15–17]. Those giant loops containing either exactly one or two and more double strand breaks (DSB) are classified as isolated DSB (iDSB) or complex DSB (cDSB), respectively. Giant loops have been identified as possible critical targets, inside which multiple lesions resist swift repair [18–21]. It was shown that computed numbers of iDSB (N_{iDSB}) and cDSB (N_{cDSB}) matched well with observed frequencies of quickly (iDSB) and slowly (cDSB) repaired DSB in rejoining studies [22,23]. The core model of our choice [11,13] associates both classes of lesions with so-called lethality parameters, K_{iDSB} and K_{cDSB}, respectively. These parameters are the corresponding probabilities for lesions of a given class to become lethal, meaning the cell loses its potential to further proliferate. Complex lesions are considered to be significantly more lethal than isolated lesions, as each of them poses a high risk for chromatin loss [14] and they remain unrepaired for a prolonged time [24–26]. It is indeed found that K_{cDSB} is several magnitudes larger than K_{iDSB} in cases derived from experimental data [11,26].

Regarding the hypoxic cell population, Carlson et al. [6] made compelling arguments for an interpretation of the OER as the ratio of doses needed to induce the same total amount of DSBs in hypoxic and normoxic cells. They further suggested the replacement of the OER term with the name hypoxia reduction factor ($HRF_{DSB}^{O_2}$) at a given oxygen concentration $[O_2]$ within this context [27]. In line with the above ideas, we introduced the $HRF_{DSB}^{O_2}$ as a parameter into our model, which solely modifies the initial total yield of DSB (N_{tDSB}). Following evidence that the oxygen status has no effect on the DSB rejoining rates [6], the lethality parameters were assumed to be invariant under the change of oxygenation. We could show in our previous work, that the sole introduction of the $HRF_{DSB}^{O_2}$ parameter was sufficient to describe survival data from literature and that the derived $HRF_{DSB}^{O_2}$ values were well described by a parameterization suggested by Carlson et al. [6] (a function of oxygen concentration $[O_2]$, inspired by the initial studies of Alper and Howard-Flanders [28]) [11]. In the work presented here, we investigate whether our model is capable of describing an experimental set of cell survival data containing five cell lines at three different oxygen levels each, and whether the derived $HRF_{DSB}^{O_2}$ values are in accordance with the formerly introduced parameterization.

To further extend the model and describe the particular case of hypoxic tumor cells response to dual treatment, radiotherapy, and DDR inhibition, we introduced a so-called radiosensitization factor (*RSF*) that modifies K_{iDSB}. This is based on observations by Hufnagl et al. [26], that the increased radiosensitivity of repair-deficient cell lines could be accounted for by increasing the lethality parameter of isolated DSB, while keeping the lethality parameter for complex DSB constant. Complex DSB are argued to pose such severe challenge to the DDR that any change in the repair capabilities of the cell has no effect on their lethality parameter. To validate this extension, we benchmarked our model in two scenarios, using experimental data obtained from cells in which one of the two key radiation-induced DDR molecules, DNA-dependent protein kinase (DNA-PK) or ataxia-telangiectasia mutated (ATM), was impaired [29]. First, we studied the robustness of the model to predict survival data of DNA-PK-deficient mutants of CHO cells. Secondly, we tested the model to predict the survival of two human lung cancer cell lines with pharmacologically-inhibited ATM. Both scenarios were investigated under normoxia and hypoxia.

2. Results

2.1. Modeling Hypoxia-Induced Radioresistence

Our survival data for five different cell lines (A549, H460, H1437, B16, Renca) exposed to three distinct oxygenation levels (normoxia 20% $[O_2]$, 1% $[O_2]$, 0.5% $[O_2]$) were fitted using our model (Figure 1). K_{iDSB} and K_{cDSB} were fitted for each cell line as free parameters to the normoxic data. Keeping these parameters fixed, for each cell line, the specific $HRF^{O_2}_{DSB}$ for the two hypoxic conditions were fitted. The determined numerical values of each parameter can be found in Table 1. Our model shows excellent capability in describing the acquired data, indicating consistency with our earlier work [11]. Furthermore, the $HRF^{O_2}_{DSB}$ values derived from our data were in accordance with the parametrization of $HRF^{O_2}_{DSB}$ as a function of oxygen concentration published in [11], as shown in the bottom right panel of Figure 1.

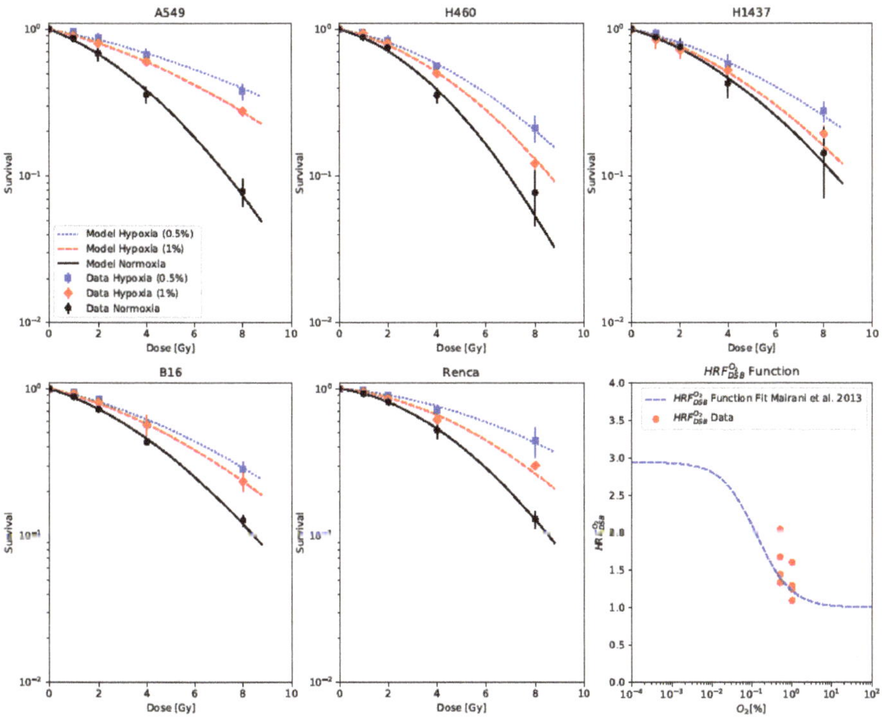

Figure 1. Cell survival data of five cell lines irradiated under normoxia (black) and under two hypoxia levels (0.5% and 1% $[O_2]$, blue and red) fitted by the model. Lower-right panel: The derived $HRF^{O_2}_{DSB}$ values compared to a parametrization introduced in [11].

Table 1. Model parameters derived from cell survival data of five cell lines irradiated under normoxia and under two hypoxia levels (0.5% and 1% $[O_2]$).

Cell Line	K_{iDSB}	K_{cDSB}	$HRF^{O_2}_{DSB}$ 1% $[O_2]$	$HRF^{O_2}_{DSB}$ 0.5% $[O_2]$
A549	$4.83 \times 10^{-3} \pm 0.88 \times 10^{-3}$	$1.69 \times 10^{-1} \pm 0.31 \times 10^{-1}$	1.60	2.04
H460	$3.28 \times 10^{-3} \pm 1.20 \times 10^{-3}$	$2.41 \times 10^{-1} \pm 0.86 \times 10^{-1}$	1.24	1.44
H1437	$3.83 \times 10^{-3} \pm 0.83 \times 10^{-3}$	$1.37 \times 10^{-1} \pm 0.38 \times 10^{-1}$	1.09	1.33
B16F10	$4.05 \times 10^{-3} \pm 0.44 \times 10^{-3}$	$1.34 \times 10^{-1} \pm 0.18 \times 10^{-1}$	1.29	1.44
Renca	$1.67 \times 10^{-3} \pm 0.18 \times 10^{-3}$	$2.04 \times 10^{-1} \pm 0.06 \times 10^{-1}$	1.28	1.67

2.2. Modeling Cell Survival of DNA-PK-Impaired Cell Lines

Survival data of the CHO cell line and two of its DNA-PK response-deficient mutants (V3 cell line is DNA-PKcs-deficient and xrs-5 cell line is Ku80-deficient) under normoxia and hypoxia were gathered from Cartwright et al. (Figure 2) [30]. The cell line-specific parameters K_{iDSB} and K_{cDSB} were found by fitting our model to the normoxic survival data of the wild-type cells. The $HRF^{O_2}_{DSB}$ of the cell line was then derived by fitting the model to the hypoxic survival data, keeping the aforementioned lethality parameters (K_{iDSB} and K_{cDSB}) fixed. By increasing only the lethality parameter of isolated DSB (K_{iDSB}) by a radiosensitization factor (RSF) for each of the mutant cell lines, their survival under normoxia was able to be fitted accurately. The numerical values of the parameters can be found in Table 2. By applying the $HRF^{O_2}_{DSB}$ derived from the wild-type data and the RSF derived for each mutant under normoxia, the survival of the two mutant cell lines under hypoxia could be predicted satisfactorily. However, deviations can be observed at the highest reported doses.

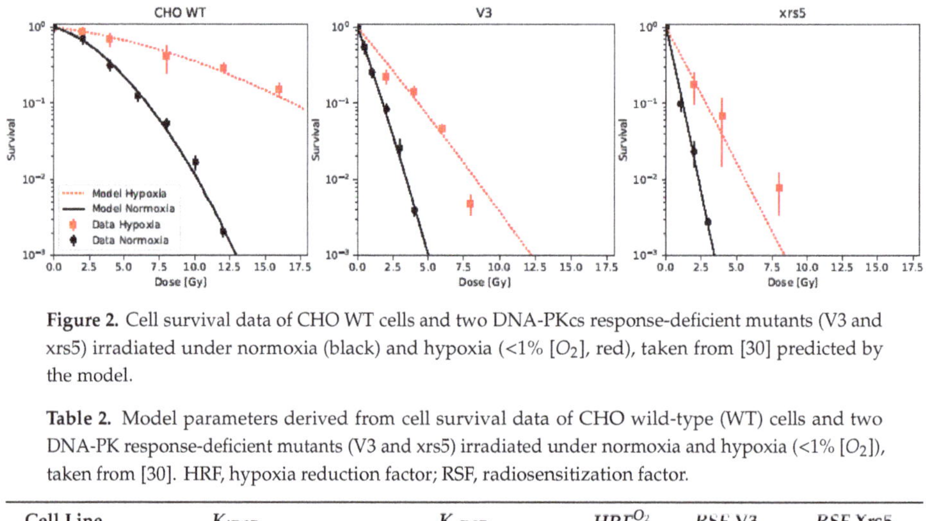

Figure 2. Cell survival data of CHO WT cells and two DNA-PKcs response-deficient mutants (V3 and xrs5) irradiated under normoxia (black) and hypoxia (<1% [O_2], red), taken from [30] predicted by the model.

Table 2. Model parameters derived from cell survival data of CHO wild-type (WT) cells and two DNA-PK response-deficient mutants (V3 and xrs5) irradiated under normoxia and hypoxia (<1% [O_2]), taken from [30]. HRF, hypoxia reduction factor; RSF, radiosensitization factor.

Cell Line	K_{iDSB}	K_{cDSB}	$HRF^{O_2}_{DSB}$	RSF V3	RSF Xrs5
CHO WT	$4.38 \times 10^{-3} \pm 1.37 \times 10^{-3}$	$2.33 \times 10^{-1} \pm 0.27 \times 10^{-1}$	2.44	9.60 ± 0.19	14.85 ± 0.50

2.3. Modeling Cell Survival of Cell Lines with Pharmacologically-Inhibited ATM

The same approach was applied to our data containing two of the initially presented cell lines (H460 and H1437) exposed to different concentrations of an ATM inhibitor (ATMi) and irradiated under normoxia and hypoxia (Figure 3). First, the cell line-specific lethality parameters (K_{iDSB} and K_{cDSB}) were derived by fitting our model to the data of cells irradiated under normoxia and without drug treatment. Second, the $HRF^{O_2}_{DSB}$ of each cell line was derived by fitting the model to the data of cells receiving no drug but irradiated under hypoxia, keeping the lethality parameters K_{iDSB} and K_{cDSB} fixed. Third, RSFs for the lethality parameters of the isolated lesions (K_{iDSB}) were derived for each cell line and drug concentration by fitting the model to the data of cells at each drug concentration under normoxia. The numerical values of the derived parameters can be found in Table 3. Again, by applying the $HRF^{O_2}_{DSB}$ derived from the non-treated cells and the RSF values found for each drug concentration irradiated under normoxia, the survival of the two cell lines exposed to the combination of different drug concentrations and hypoxia could be predicted very well.

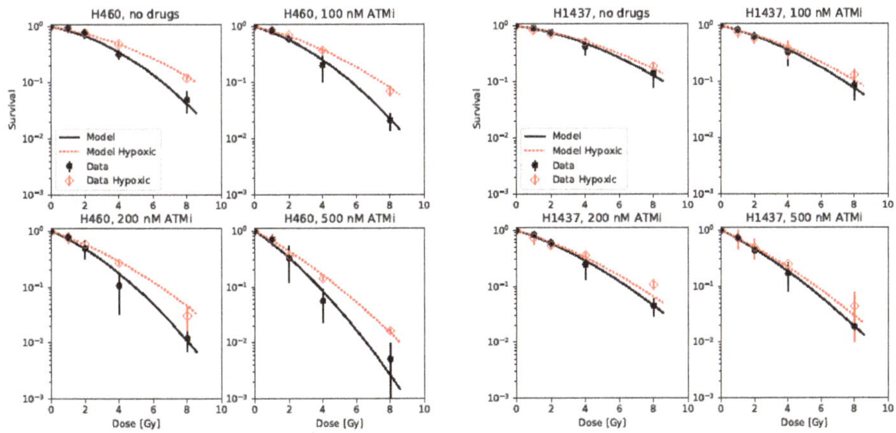

(a) (b)

Figure 3. Cell survival data of two cell lines, (**a**) H460 and (**b**) H1437, irradiated under normoxia (black) and hypoxia (1% $[O_2]$, red) after exposure to different concentrations of an ataxia-telangiectasia mutated inhibitor (ATMi) fitted by the model.

Table 3. Model parameters derived from cell survival data of H1437 and H460 cells irradiated under normoxia and hypoxia (1% $[O_2]$) after exposure to different concentrations of an ATM inhibitor.

Cell Line	K_{iDSB}	K_{cDSB}	$HRF^{O_2}_{DSB}$	RSF 100 nM	RSF 200 nM	RSF 500 nM
H460	$3.88 \times 10^{-3} \pm 2.19 \times 10^{-3}$	$2.55 \times 10^{-1} \pm 0.85 \times 10^{-1}$	1.31	1.73 ± 0.15	2.56 ± 0.27	4.21 ± 0.59
H1437	$3.11 \times 10^{-3} \pm 0.86 \times 10^{-3}$	$1.50 \times 10^{-1} \pm 0.35 \times 10^{-1}$	1.10	1.77 ± 0.12	2.52 ± 0.13	3.77 ± 0.15

3. Discussion

Our model provided an excellent description of the survival data of five cell lines and three oxygen levels presented in Figure 1, confirming the applicability of the previously published framework [11]. It underlines the hypothesis introduced in our former work, that cell survival under hypoxic conditions can be described in a first approximation by keeping the defined lethality of isolated and complex lesions invariant and only modifying the overall induction of DSBs by a given factor ($HRF^{O_2}_{DSB}$). In contrast to our former publication, we determined the lethality parameters K_{iDSB} and K_{cDSB} by fitting them to the experimental data. This practice might lead to better predictions, as up to now, these values were recalculated from provided or fitted LQ model parameters (α and β values), based on a Taylor expansion at low doses of our model equations. However such approximate recalculations remain to be crucial in cases in which only the α and β values are available but not the full set of cell survival data. Furthermore, we were able to show that the derived $HRF^{O_2}_{DSB}$ values from our data coincided well with the $HRF^{O_2}_{DSB}$ parametrization as a function of oxygen concentration introduced in our former publication. The highest $HRF^{O_2}_{DSB}$ value for both oxygen levels was obtained from A549 cells. Including these data points, the mean of the derived $HRF^{O_2}_{DSB}$ values deviated by 0.08 and 0.19 for 1% $[O_2]$ and 0.5% $[O_2]$, respectively. If one excludes the A549 data, the mean of the derived $HRF^{O_2}_{DSB}$ values only deviated from the prediction by 0.01 and 0.07 for 1% $[O_2]$ and 0.5% $[O_2]$, respectively. This is further evidence for this parameterization to be a widely applicable estimate for the $HRF^{O_2}_{DSB}$ in cases where the data to derive the exact value are not available.

The idea of increasing the lethality of isolated lesions K_{iDSB} under the presented model in order to describe the increased cell killing observed for repair deficient cell lines has already been expressed and successfully demonstrated by Hufnagl et al. within the GLOBLE model [26]. They consider the lethality of complex lesions K_{cDSB} as being fixed, as each complex lesion poses a "significant burden for the cell", irrespective of the DNA damage repair capabilities of the cell. Further, they argue, that in NHEJ deficient cell lines, the lethality of isolated lesions increases depending on the cell cycle status of the cells. While we adapt and fully agree with the notion of a fixed lethality of complex damages independent of the repair capabilities of a cell, we had no information on the cell cycle distributions underlying the data taken from Cartwright et al. [30] and found different radiosensitizing factors (*RSF*) for the two DNA-PK response-deficient mutants analyzed. Therefore we decided to introduce the *RSF* as a free parameter in UNIVERSE, fitted to each mutant cell line. The *RSF* values found for the two CHO mutants V3 and xrs5 indicate that the probability of an isolated lesion to become a lethal lesion increases through DNA-PK response deficiency in these cell lines by a factor of about 10 and 15, respectively. Even though both *RSF* factors are related to the functional DNA-PK repair activity, the difference in the *RSF* of both cell lines might be retraced to the fact, that both are deficient of different enzymes taking part in the DNA-PK response: While xrs-5 cells are deficient of the Ku80 DNA-PK subunit [30,31], V3 cells lack the catalytic subunit for DNA-PK [30,32]. It is, however, unclear if and how the *RSF* is coupled to the activity of both proteins. Furthermore, different extents of remaining expression or compensation by other repair proteins might also lead to the observed difference in the *RSF* values of both mutants. One would have to compare several groups of cell lines with the exact same deficiencies to gain a better understanding of the underlying dependencies. More importantly, we could show that the survival of the mutant cell lines under hypoxia were well predicted by our model by combining the $HRF_{DSB}^{O_2}$ determined based on the wild-type data and keeping the *RSF* values derived from the normoxic data of each mutant invariant.

Since both DNA-PK and ATM are essential molecules for irradiation-induced DNA damage repair, recruited to the DNA damage sites [33–35], we assumed that pharmacological inhibition of ATM should also lead to an increased lethality of isolated damage sites, similarly to the genetic models in which DNA-PK is deficient, with a concentration-dependent effect. Indeed, the observed survival in cells exposed to different concentrations of the ATM inhibitor under normoxia was described with high fidelity by introducing an *RSF* for isolated lesions. Further, the values of the derived *RSF* illustrate the increasing lethality of isolated lesions with increasing drug concentrations, leading to stronger inhibition of the repair, while staying below the value derived for a pathway deficiency. It is also to be noticed that the *RSF* values for both H460 and H1437 cells are fairly similar. However, more cell lines have to be analyzed to investigate whether this can be extended to a general trend and if the presented method is generally applicable to other repair-inhibiting drugs. Nevertheless, we could show that the *RSF* values derived from normoxic data can accurately describe the survival in hypoxic conditions by introducing the $HRF_{DSB}^{O_2}$ derived from cells without drug treatment.

Taken together, we could show that impairment of DNA damage repair, both in repair-deficient cell lines and cells treated with a DDR inhibitor, could be accounted for by a manipulation of the lethality of isolated lesions with an *RSF* in our model. Moreover, this modified lethality could stay invariant under change of oxygen supply, while sustaining good predictive capabilities. Thus, the uniqueness of our approach lies in its capability to describe two separate cell response mechanisms in any combination using minimal input parameters, which can be separately derived. Based on this, we believe that our approach has a high potential to implement further cellular mechanisms in order to produce predictions tailored to diverse clinical scenarios.

4. Materials and Methods

4.1. Experimental Data from Literature

Experimental data used to benchmark the model on survival of genetically DDR-deficient cell lines were taken from [30]. The data used to benchmark the model on survival of NCI-H460 (H460) and NCI-H1437 (H1437) cells in which the DDR was pharmacologically inhibited using an ATM inhibitor under normoxia and hypoxia, as well as the survival data of A549 cells under hypoxia and normoxia, were taken from [10].

4.2. Cell Culture, Clonogenic Survival Assay, and Irradiation

For the validation of our model under hypoxic conditions, additional experiments were performed using Renca (murine renal carcinoma; American Type Culture Collection, Manassas, VA, USA) and B16-Blue ISG (murine melanoma; Invitrogen, Thermo Fischer, Waltham, MA, USA) (B16) cells. Both cell lines were grown in RPMI 1640 Medium (Gibco, Thermo Fischer, Waltham, MA, USA) supplemented with 10% fetal bovine serum (FBS)(Merck Millipore, Darmstadt, Germany) at 37 °C and 5% CO_2 atmosphere. Experiments in hypoxic conditions were performed at 0.5 or 1% O_2 and 5% CO_2 using a custom hypoxic chamber (C-chamber; Biospherix, Parish, NY, USA), including an online monitoring controller for O_2 and CO_2 concentrations (ProOx and ProCO2 model; Biospherix, Parish, NY, USA). Fifty cells per well in a 96-well format were seeded not more than 16 h before irradiation. Hypoxic irradiation was performed after incubation for 4 h under respective oxygen conditions. Cells were irradiated in the sealed hypoxia chamber with a dose series of photons of 1, 2, 4, or 8 Gy and thereafter incubated under normoxic conditions. The ATM inhibitor was kindly provided by Merck KGaA, and dissolved in DMSO (PAN-Biotech, Aidenbach, Germany) and diluted in RPMI 1640 medium. The inhibitor was added to H460 and H1437 cells at 100, 200, or 500 nM just before incubation under hypoxia or normoxia started. Controls also contained < 0.1% DMSO. Inhibitors were left in the media for 24 h and then replaced with fresh RPMI 1640 medium and the plates were returned to the incubator for colony formation. After 4 days (A549), 5 days (H460, Renca, and B16), or 7 days (H1437), plates were imaged by an online microscopy system at 4x magnification (IncuCyte, Essen Bioscience, Sartorius, Göttingen, Germany). The images were analyzed by the IncuCyte Zoom Software (ver. 2016a) (Essen Bioscience, Sartorius, Göttingen,) and colony counts were confirmed by manual curation.

4.3. Dose Planning and Simulations

The irradiation plan was carried out as a step and shoot intensity-modulated radiotherapy (IMRT) plan, describing the different dose levels as separate target regions. Planning was done with Raystation treatment planning system (RaySearch Laboratories, Stockholm, Sweden) based on a CT scan of the hypoxia chamber containing 96-well plates filled with water. Irradiation was performed on a Siemens Artiste (6 MV) (Siemens, München, Germany).

4.4. Modeling Approach

Large parts of the general model and its derivation were presented and discussed in detail in [11] Computationally, the code was fully rewritten in Python and elements of GPU computation were introduced. In short, for low LET radiation a homogeneous deposition of dose throughout the cell nucleus with a cell line independent DSB induction rate $\alpha_{DSB} = 5 \cdot 10^{-3} DSB/(Mbp \cdot Gy)$, constant over the clinical dose range, wass assumed [36–39]. Thus, the expected number of DSB in the nucleus ($\langle N_{tDSB} \rangle$) can be expressed as:

$$\langle N_{tDSB} \rangle = \alpha_{DSB} \cdot D \cdot DNA_c \tag{1}$$

where *DNAc* is the DNA content of a cell in Mbp and *D* the applied dose in Gy. The total number of giant loops (N_{gl}) with a DNA content of DNA_{gl} inside the nucleus is then given by:

$$N_{gl} = \frac{DNA_c}{DNA_{gl}} \tag{2}$$

In this work we assumed *DNAc* and DNA_{gl} to be 6 Gbp and 2 Mbp, respectively.

We implemented a Monte Carlo routine in which, at each iteration, the number of total DSB in the nucleus (N_{tDSB}) was sampled following a Poisson distribution with the expectation value given by Equation (1). After randomly distributing the sampled amount of DSBs over the giant loops, the number of giant loops without any DSB (N_0), with an isolated DSB (N_{iDSB}), or a complex DSB (N_{cDSB}) were scored. With the lethality parameters K_{iDSB} and K_{cDSB}, which represent the probabilities of an isolated lesion and a complex lesion leading to cell death, respectively, the probability of the cell to survive (S) is given by [13]:

$$S = (1 - K_{iDSB})^{N_{iDSB}} \cdot (1 - K_{cDSB})^{N_{cDSB}} \tag{3}$$

We obtained the expected fraction of a cell population surviving an irradiation by meaning S values obtained from the Monte Carlo algorithm. The lethality parameters can be determined by fitting the result of this routine to survival data.

Experimental evidence suggests that one can assume a homogeneous distribution of oxygen inside the nucleus, no change in DNA content or nucleus volume under reduction of oxygen supply [40], and that the oxygen concentration in the cell does effect the initially induced total number of DSB but not their repair rate [6]. Thus, in our model, a change in oxygenation solely leads to an introduction of a modified DSB induction rate $\alpha_{DSB}^{O_2}$, which is given by:

$$\alpha_{DSB}^{O_2} = \frac{\alpha_{DSB}}{HRF_{DSB}^{O_2}} \tag{4}$$

where α_{DSB} is the rate under normoxia.

The modification of α_{DSB} leads through Equation (1) to a change of N_{tDSB} to $N_{tDSB}^{O_2}$, which subsequently leads to alterations of N_{iDSB} and N_{cDSB} to $N_{iDSB}^{O_2}$ and $N_{cDSB}^{O_2}$, respectively. Keeping the lethality parameters invariant, as implied above, the survival probability of a cell under hypoxic conditions can then finally be written as:

$$S_{O_2} = (1 - K_{iDSB})^{N_{iDSB}^{O_2}} \cdot (1 - K_{cDSB})^{N_{cDSB}^{O_2}} \tag{5}$$

The $HRF_{DSB}^{O_2}$ value can be determined by fitting the model to hypoxic data, while keeping K_{iDSB} and K_{cDSB} at the values derived from normoxic data. If either one, hypoxic or normoxic data, is not available, the $HRF_{DSB}^{O_2}$ can be estimated from the parameterization:

$$HRF_{DSB}^{O_2} = \frac{m \cdot K + [O_2]}{K + [O_2]} \tag{6}$$

introduced in our previous work [11], proposed by Carlson et al. [6] and inspired by the initial works of Alper and Howard-Flanders [28]. Fitting this parameterization to data available in literature, we found the values m = 2.94 and K = 0.129% [11].

In order to model the increased cell killing of repair-deficient mutant cell lines or cells exposed to different concentrations of a repair inhibiting drug, a radiosensitization factor (RSF) was introduced into the model. The RSF modifies the lethality parameter of isolated damages K_{iDSB} only, so that the survival probability of a repair impaired cell reads:

$$S_{-Repair} = (1 - RSF \cdot K_{iDSB})^{N_{iDSB}} \cdot (1 - K_{cDSB})^{N_{cDSB}} \tag{7}$$

The RSF is introduced as a free parameter, which is determined by fitting the result of the modified survival probability given in Equation (7) to repair impaired data, while the lethality parameters K_{iDSB} and K_{cDSB} remain set to the values derived from wild type/non-treated cells. As argued above, no interaction between the oxygenation status and the repair capacity is assumed. Therefore, also in the case of a modification of the isolated lesions lethality, it is assumed that the RSF value is set to be invariant under change of the oxygen concentration.

Author Contributions: Conceptualization, H.L., I.D., and A.M.; methodology, H.L., A.M., and I.D.; software, H.L. and A.M.; formal analysis, H.L., I.D., and A.M.; investigation, I.D., C.K., and H.L.; resources, F.T.Z., A.A., and J.D.; data curation, H.L., A.M., I.D., and C.K.; writing—original draft preparation, H.L. and I.D.; writing—review and editing, H.L., C.K., F.T.Z., A.A., J.D., I.D., and A.M.; visualization, H.L.; supervision, I.D. and A.M.; project administration, A.A. and J.D.; funding acquisition, A.A. and J.D.

Funding: This work was supported by the National Center for Tumor Diseases (NCT 3.0-2015.22 BioDose), as well as a PhD stipend from the Helmholtz International Graduate School for Cancer Research in Heidelberg to H.L.

Conflicts of Interest: A.A. and J.D. received research funds from Merck KGaA, Darmstadt, Germany, and are involved as academic mentors in the DNA-Repair Cluster initiative with Merck KGaA-BioMedX. Frank T. Zenke is an employee of Merck KGaA, Darmstadt, Germany. The other authors declare no conflicts of interest. The funders had no role in the design of the study; in the collection, analyses, or interpretation of data; in the writing of the manuscript; or in the decision to publish the results.

Abbreviations

LET	Linear energy transfer
UNIVERSE	Unique and versatile engine
DSB	(DNA) Double strand break
iDSB	Isolated DSB
cDSB	Complex DSB
OER	Oxygen enhancement ratio
HRF	Hypoxia reduction factor
ATM	Ataxia-telangiectasia mutated
NHEJ	Non-homologous end joining
DDR	DNA damage repair
CHO	Chinese hamster ovary
GLOBLE	Giant loop binary lesion
CT	Computed tomography
LQ	Linear quadratic (model)

References

1. Delaney, G.; Jacob, S.; Featherstone, C.; Barton, M. The role of radiotherapy in cancer treatment: Estimating optimal utilization from a review of evidence-based clinical guidelines. *Cancer* **2005**, *104*, 1129–1137. [CrossRef] [PubMed]
2. Dale, R.G.; Jones, B. *Radiobiological Modelling in Radiation Oncology*; British Institute of Radiology: London, UK, 2007; ISBN 978-0-905749-60-0.
3. Joiner, M.; van der Kogel, A. *Basic Clinical Radiobiology*, 5th ed.; CRC Press/Taylor & Francis Group: Boca Raton, FL, USA, 2018; ISBN 978-1-4441-7963-7.
4. Tawk, B.; Schwager, C.; Deffaa, O.; Dyckhoff, G.; Warta, R.; Linge, A.; Krause, M.; Weichert, W.; Baumann, M.; Herold-Mende, C.; et al. Comparative analysis of transcriptomics based hypoxia signatures in head-and neck squamous cell carcinoma. *Radiother. Oncol.* **2016**, *118*, 350–358. [CrossRef] [PubMed]
5. Rofstad, E.K.; Sundfør, K.; Lyng, H.; Tropé, C.G. Hypoxia-induced treatment failure in advanced squamous cell carcinoma of the uterine cervix is primarily due to hypoxia-induced radiation resistance rather than hypoxia-induced metastasis. *Br. J. Cancer* **2000**, *83*, 354–359. [CrossRef] [PubMed]
6. Carlson, D.J.; Stewart, R.D.; Semenenko, V.A. Effects of oxygen on intrinsic radiation sensitivity: A test of the relationship between aerobic and hypoxic linear-quadratic (LQ) model parameters. *Med. Phys.* **2006**, *33*, 3105–3115. [CrossRef]
7. Batey, M.A.; Zhao, Y.; Kyle, S.; Richardson, C.; Slade, A.; Martin, N.M.B.; Lau, A.; Newell, D.R.; Curtin, N.J. Preclinical evaluation of a novel ATM inhibitor, KU59403, in vitro and in vivo in p53 functional and dysfunctional models of human cancer. *Mol. Cancer Ther.* **2013**, *12*, 959–967. [CrossRef]
8. Dohmen, A.J.C.; Qiao, X.; Duursma, A.; Wijdeven, R.H.; Lieftink, C.; Hageman, F.; Morris, B.; Halonen, P.; Vens, C.; van den Brekel, M.W.M.; et al. Identification of a novel ATM inhibitor with cancer cell specific radiosensitization activity. *Oncotarget* **2017**, *8*, 73925–73937. [CrossRef]
9. Durant, S.T.; Zheng, L.; Wang, Y.; Chen, K.; Zhang, L.; Zhang, T.; Yang, Z.; Riches, L.; Trinidad, A.G.; Fok, J.H.L.; et al. The brain-penetrant clinical ATM inhibitor AZD1390 radiosensitizes and improves survival of preclinical brain tumor models. *Sci. Adv.* **2018**, *4*, eaat1719. [CrossRef] [PubMed]

10. Klein, C.; Dokic, I.; Mairani, A.; Mein, S.; Brons, S.; Häring, P.; Haberer, T.; Jäkel, O.; Zimmermann, A.; Zenke, F.; et al. Overcoming hypoxia-induced tumor radioresistance in non-small cell lung cancer by targeting DNA-dependent protein kinase in combination with carbon ion irradiation. *Radiat. Oncol.* **2017**, *12*, 208. [CrossRef]
11. Mairani, A.; Böhlen, T.T.; Dokic, I.; Cabal, G.; Brons, S.; Haberer, T. Modelling of cell killing due to sparsely ionizing radiation in normoxic and hypoxic conditions and an extension to high LET radiation. *Int. J. Radiat. Biol.* **2013**, *89*, 782–793. [CrossRef]
12. Elsässer, T.; Weyrather, W.K.; Friedrich, T.; Durante, M.; Iancu, G.; Krämer, M.; Kragl, G.; Brons, S.; Winter, M.; Weber, K.-J.; et al. Quantification of the Relative Biological Effectiveness for Ion Beam Radiotherapy: Direct Experimental Comparison of Proton and Carbon Ion Beams and a Novel Approach for Treatment Planning. *Int. J. Radiat. Oncol. Biol. Phys.* **2010**, *78*, 1177–1183. [CrossRef]
13. Friedrich, T.; Durante, M.; Scholz, M. Modeling cell survival after photon irradiation based on double-strand break clustering in megabase pair chromatin loops. *Radiat. Res.* **2012**, *178*, 385–394. [CrossRef] [PubMed]
14. Friedrich, T.; Scholz, U.; Elsässer, T.; Durante, M.; Scholz, M. Calculation of the biological effects of ion beams based on the microscopic spatial damage distribution pattern. *Int. J. Radiat. Oncol. Biol. Phys.* **2012**, *88*, 103–107. [CrossRef] [PubMed]
15. Yokota, H.; van den Engh, G.; Hearst, J.E.; Sachs, R.K.; Trask, B.J. Evidence for the organization of chromatin in megabase pair-sized loops arranged along a random walk path in the human G0/G1 interphase nucleus. *J. Cell Biol.* **1995**, *130*, 1239–1249. [CrossRef] [PubMed]
16. Sachs, R.K.; van den Engh, G.; Trask, B.; Yokota, H.; Hearst, J.E. A random-walk/giant-loop model for interphase chromosomes. *Proc. Natl. Acad. Sci. USA* **1995**, *92*, 2710–2714. [CrossRef] [PubMed]
17. Solovjeva, L.; Svetlova, M.; Stein, G.; Chagin, V.; Rozanov, Y.; Zannis-Hadjopoulos, M.; Price, G.; Tomilin, N. Conformation of Replicated Segments of Chromosome Fibres in Human S-phase Nucleus. *Chromosome Res.* **1998**, *6*, 595–602. [CrossRef] [PubMed]
18. Johnston, P.J.; Bryant, P.E. A component of DNA double-strand break repair is dependent on the spatial orientation of the lesions within the higher-order structures of chromatin. *Int. J. Radiat. Biol.* **1994**, *66*, 531–536. [CrossRef]
19. Johnston, P.J.; Olive, P.L.; Bryant, P.E. Higher-order chromatin structure-dependent repair of DNA double-strand breaks: Modeling the elution of DNA from nucleoids. *Radiat. Res.* **1997**, *148*, 561–567. [CrossRef]
20. Johnston, P.J.; MacPhail, S.H.; Banáth, J.P.; Olive, P.L. Higher-order chromatin structure-dependent repair of DNA double-strand breaks: Factors affecting elution of DNA from nucleoids. *Radiat. Res.* **1998**, *149*, 533–542. [CrossRef]
21. Gauter, B.; Zlobinskaya, O.; Weber, K.-J. Rejoining of radiation-induced DNA double-strand breaks: Pulsed-field electrophoresis analysis of fragment size distributions after incubation for repair. *Radiat. Res.* **2002**, *157*, 721–733. [CrossRef]
22. Tommasino, F.; Friedrich, T.; Scholz, U.; Taucher-Scholz, G.; Durante, M.; Scholz, M. A DNA double-strand break kinetic rejoining model based on the local effect model. *Radiat. Res.* **2013**, *180*, 524–538. [CrossRef]
23. Tommasino, F.; Friedrich, T.; Jakob, B.; Meyer, B.; Durante, M.; Scholz, M. Induction and Processing of the Radiation-Induced Gamma-H2AX Signal and Its Link to the Underlying Pattern of DSB: A Combined Experimental and Modelling Study. *PLoS ONE* **2015**, *10*, e0129416. [CrossRef]
24. Banáth, J.P.; Klokov, D.; MacPhail, S.H.; Banuelos, C.A.; Olive, P.L. Residual gammaH2AX foci as an indication of lethal DNA lesions. *BMC Cancer* **2010**, *10*, 4. [CrossRef] [PubMed]
25. Koch, U.; Höhne, K.; von Neubeck, C.; Thames, H.D.; Yaromina, A.; Dahm-Daphi, J.; Baumann, M.; Krause, M. Residual γH2AX foci predict local tumour control after radiotherapy. *Radiother. Oncol.* **2013**, *108*, 434–439. [CrossRef] [PubMed]
26. Hufnagl, A.; Herr, L.; Friedrich, T.; Durante, M.; Taucher-Scholz, G.; Scholz, M. The link between cell-cycle dependent radiosensitivity and repair pathways: A model based on the local, sister-chromatid conformation dependent switch between NHEJ and HR. *DNA Repair (Amst.)* **2015**, *27*, 28–39. [CrossRef] [PubMed]
27. Carlson, D.J.; Keall, P.J.; Loo, B.W.; Chen, Z.J.; Brown, J.M. Hypofractionation results in reduced tumor cell kill compared to conventional fractionation for tumors with regions of hypoxia. *Int. J. Radiat. Oncol. Biol. Phys.* **2011**, *79*, 1188–1195. [CrossRef] [PubMed]

28. Alper, T.; Howard-Flanders, P. Role of Oxygen in Modifying the Radiosensitivity of E. Coli B. *Nature* **1956**, *178*, 978–979. [CrossRef] [PubMed]
29. Blackford, A.N.; Jackson, S.P. ATM, ATR, and DNA-PK: The Trinity at the Heart of the DNA Damage Response. *Mol. Cell* **2017**, *66*, 801–817. [CrossRef]
30. Cartwright, I.M.; Su, C.; Haskins, J.S.; Salinas, V.A.; Sunada, S.; Yu, H.; Uesaka, M.; Hirakawa, H.; Chen, D.J.; Fujimori, A.; et al. DNA Repair Deficient Chinese Hamster Ovary Cells Exhibiting Differential Sensitivity to Charged Particle Radiation under Aerobic and Hypoxic Conditions. *Int. J. Mol. Sci.* **2018**, *19*, 2228. [CrossRef]
31. Jeggo, P.A.; Kemp, L.M. X-ray-sensitive mutants of Chinese hamster ovary cell line isolation and cross-sensitivity to other DNA-damaging agents. *Mutat. Res. DNA Repair Rep.* **1983**, *112*, 313–327. [CrossRef]
32. Whitmore, G.F.; Varghese, A.J.; Gulyas, S. Cell Cycle Responses of Two X-ray Sensitive Mutants Defective in DNA Repair. *Int. J. Radiat. Biol.* **1989**, *56*, 657–665. [CrossRef]
33. Kastan, M.B.; Lim, D.S. The many substrates and functions of ATM. *Nat. Rev. Mol. Cell Biol.* **2000**, *1*, 179–186. [CrossRef] [PubMed]
34. Shiloh, Y. ATM and ATR: Networking cellular responses to DNA damage. *Curr. Opin. Genet. Dev.* **2001**, *11*, 71–77. [CrossRef]
35. Shiloh, Y.; Kastan, M.B. ATM: Genome stability, neuronal development, and cancer cross paths. *Adv. Cancer Res.* **2001**, *83*, 209–254. [PubMed]
36. Frankenberg-Schwager, M. Review of repair kinetics for DNA damage induced in eukaryotic cells in vitro by ionizing radiation. *Radiother. Oncol.* **1989**, *14*, 307–320. [CrossRef]
37. Prise, K.M.; Ahnström, G.; Belli, M.; Carlsson, J.; Frankenberg, D.; Kiefer, J.; Löbrich, M.; Michael, B.D.; Nygren, J.; Simone, G.; et al. A review of dsb induction data for varying quality radiations. *Int. J. Radiat. Biol.* **1998**, *74*, 173–184. [CrossRef]
38. Prise, K.M.; Pinto, M.; Newman, H.C.; Michael, B.D. A review of studies of ionizing radiation-induced double-strand break clustering. *Radiat. Res.* **2001**, *156*, 572–576. [CrossRef]
39. Asaithamby, A.; Chen, D.J. Cellular responses to DNA double-strand breaks after low-dose γ-irradiation. *Nucleic Acids Res.* **2009**, *37*, 3912–3923. [CrossRef]
40. Freyer, J.P.; Jarrett, K.; Carpenter, S.; Raju, M.R. Oxygen enhancement ratio as a function of dose and cell cycle phase for radiation-resistant and sensitive CHO cells. *Radiat. Res.* **1991**, *127*, 297–307. [CrossRef]

International Journal of
Molecular Sciences

MDPI

Article

Modeling Direct and Indirect Action on Cell Survival After Photon Irradiation under Normoxia and Hypoxia

Hans Liew [1,2,3,4,5,6], Stewart Mein [2,3,4,5], Jürgen Debus [1,2,3,4,5,6], Ivana Dokic [2,3,4,5] and Andrea Mairani [2,3,4,5,*]

1 Clinical Cooperation Unit Radiation Oncology, German Cancer Research Center (DKFZ), 69120 Heidelberg, Germany; h.liew@dkfz-heidelberg.de (H.L.); Juergen.Debus@med.uni-heidelberg.de (J.D.)
2 Division of Molecular and Translational Radiation Oncology, National Center for Tumor Diseases (NCT), Heidelberg University Hospital, 69120 Heidelberg, Germany; s.mein@dkfz-heidelberg.de (S.M.); i.dokic@dkfz-heidelberg.de (I.D.)
3 Heidelberg Institute of Radiation Oncology (HIRO), German Cancer Research Center (DKFZ), 69120 Heidelberg, Germany
4 German Cancer Consortium (DKTK), 69120 Heidelberg, Germany
5 Heidelberg Ion-Beam Therapy Center (HIT), 69120 Heidelberg, Germany
6 Faculty of Physics and Astronomy, Heidelberg University, 69120 Heidelberg, Germany
* Correspondence: andrea.mairani@med.uni-heidelberg.de; Tel.: +49-0-6221-56-37535

Received: 24 April 2020; Accepted: 11 May 2020; Published: 14 May 2020

Abstract: The demand for personalized medicine in radiotherapy has been met by a surge of mechanistic models offering predictions of the biological effect of ionizing radiation under consideration of a growing number of parameters. We present an extension of our existing model of cell survival after photon irradiation to explicitly differentiate between the damage inflicted by the direct and indirect (radicals-mediated) action of ionizing radiation. Within our approach, we assume that the oxygenation status affects the indirect action. The effect of different concentrations of dimethyl sulfoxide (DMSO), an effective radical scavenger, has been simulated at different dose levels in normoxic and hypoxic conditions for various cell lines. Our model is found to accurately predict experimental data available in literature, validating the assumptions made in our approach. The presented extension adds further flexibility to our model and could act as basis for further developments of our model.

Keywords: ionizing radiation; direct and indirect damage; hypoxia; modeling; DMSO; radicals; radical scavengers

1. Introduction

Nearly 50% of cancer patients are treated with some form of radiation therapy during the course of the disease [1] with recent trends shifting towards more personalized planning and delivery. Innovative treatments, however, require development, validation and clinical translation of highly detailed and accurate physical and biological models for normal tissue and tumor response, considering various bio-factors based on both measurable quantities, discovered mechanisms and theory [2].

In this manuscript, we present an extension of the "UNIfied and VERSatile bio response Engine" (UNIVERSE) biological modeling environment [3], which is progressively extended by mechanistic implementations of biological processes relevant for the ultimate radiation response of cells. In a recent publication [3], the cellular response of hypoxic cells in combination with the administration of radio-sensitizing drugs, such as DNA damage response (DDR) inhibitors was implemented into

UNIVERSE [4–7]. In this work, consequences of direct and indirect (radicals-mediated) damage in both conditions of normoxia and hypoxia are explicitly considered and incorporated into the UNIVERSE framework. In doing so, we obtain an improved understanding and predictability of the effect of hypoxia, which is known to be highly relevant to treatment outcomes in radiation clinics [8–10].

It is generally accepted, that the induction of DNA damage by photons can be separated into two pathways: direct actions due to the deposition of radiation energy directly on the DNA molecule and indirect actions, which are mediated by formation of free radicals and their interaction with cellular structures [11,12]. Due to this distinct role of free radicals in radiation induced damage, chemical compounds that act as free radical scavengers are under consideration as potential agents to minimize side effects in radio therapy [13,14] or reduce the probability of carcinogenesis after exposure to ionizing radiation from diagnostic imaging [15]. In particular, dimethyl sulfoxide (DMSO) is an effective scavenger of the OH radical, one of the main mediators of low linear energy transfer (LET) radiation induced cellular damage, and has been widely used to investigate the indirect action of ionizing radiation [12,16–18].

The increased radioresistence of cells under absence of free oxygen (oxygen effect) is most commonly explained by the oxygen fixation hypothesis: molecular oxygen can react with radicals produced in the DNA, which directly competes with its chemical restoration by reaction with H+, and thus fixing part of the damage [8,19,20]. However, the results of several studies investigating both cell survival [12,21] and extent of DNA damage [21,22], have suggested that the oxygen effect predominantly affects the indirect action of damage induction, while the direct action is only weakly modified.

In the previous version of UNIVERSE, predictions were based on a general double strand break (DSB) yield (α_{DSB}), independent of the underlying mechanism. Now, we implement a cell line independent separation of this yield into a direct fraction and an indirect fraction. Furthermore, we consider a free radical scavenger such as DMSO to reduce solely the yield of the indirect fraction. In the first part of this work, we determined an empirical parametrization of the fraction of the indirect damage quenched by a given DMSO concentration (f_{DMSO}), based on data from three different cell lines in the literature. In order to validate the approach, we compared the predicted relative DSB damage yield over a range of DMSO concentrations to measurements from the literature.

In an earlier publication [3], it was demonstrated, that one can successfully account for the oxygen effect by simply modifying the total DSB yield by a hypoxia reduction factor ($HRF_{DSB}^{O_2}$) resembling the classical oxygen enhancement factor (OER) for a given oxygen concentration. Based on the evidence mentioned above we assume that, while the total DSB yield is still being reduced as before, the yield corresponding to the direct pathway of damage induction is unaffected by the oxygen status. This assumption is benchmarked in the second part of this work by applying the previously derived f_{DMSO} to the reduced yield corresponding to the indirect damage induction for two cell lines under hypoxia at three different dose levels over a range of DMSO concentrations.

2. Results

2.1. Parametrization of the Effect of DMSO on the Indirect Damage in Our Model

Fitting the normoxic survival measurements without application of DMSO (data of Hirayama et al. [12] and Chapman et al. [16]), the cell line specific lethality parameters of our model K_{iDSB} and K_{cDSB} were derived for the V79, CHO and AA8 cell-line. These parameters indicate the probability of an isolated (iDSB) and complex DSB (cDSB) leading to the cell's inactivation (see Materials and Methods). The numerical results are summarized in Table 1. Based on the contribution of indirect action found by Hirayama et al. [12] we assume the fraction of DSB yield based on the direct action (fr_{DIR}) to be 20%. Based on this assumption, the fraction of indirect damage quenched by DMSO (f_{DMSO}) was tuned for each DMSO concentration measured by Chapman et al. [16] and Hirayama et al. [12] so that our model would reproduce the observed survival trends under normoxia (data in Figure 2 and data in left panel of Figure 3 for CHO cells at 4 Gy under normoxia). Values determined for f_{DMSO} are shown in the left panel of Figure 1 together with a cell-line independent empirical fit of the data. The model predictions shown in Figure 2 were calculated based on this parametrization. As further validation of our approach, we calculated the relative yield of DSBs as the ratio between the DSB without application of DMSO to values of measurements over a range of DMSO concentrations taken from literature (right panel, Figure 1), which have been measured using a variety of techniques [21,23,24].

Table 1. Model parameters derived from cell survival data of three cell lines irradiated with different DMSO concentrations and in hypoxic condition when available.

Cell Line	K_{iDSB}	K_{cDSB}	$HRF^{O_2}_{DSB}$	Reference
CHO	$5.56 \times 10^{-3} \pm 1.12 \times 10^{-3}$	$7.65 \times 10^{-1} \pm 0.44 \times 10^{-1}$	2.90	Hirayama et al. 2013 [12]
AA8	$14.00 \times 10^{-3} \pm 1.48 \times 10^{-3}$	$9.16 \times 10^{-1} \pm 0.89 \times 10^{-1}$	2.85	Hirayama et al. 2013 [12]
V79	$4.79 \times 10^{-3} \pm 0.52 \times 10^{-3}$	$3.17 \times 10^{-1} \pm 0.13 \times 10^{-1}$	–	Chapman et al. 1979 [16]

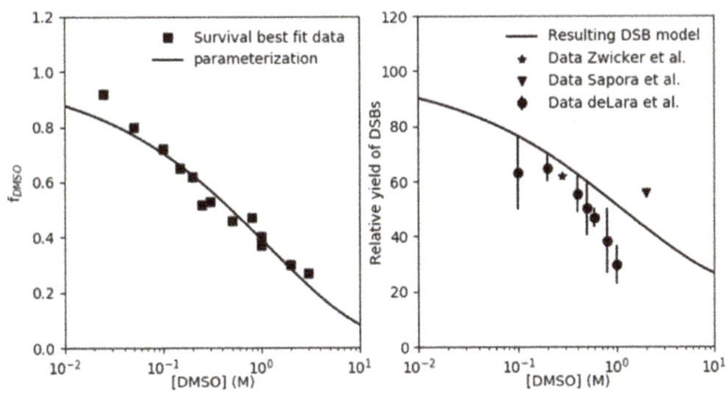

Figure 1. Left panel: f_{DSMO} parameterization as function of DMSO in units of M is depicted with line together with best-fit f_{DSMO} data (squares) obtained by reproducing the cell survival data in normoxia. Right panel: predicted initial normoxic yield of DSBs relative to the yield in absence of DMSO (R_{DSB}) calculated using Equation (9) is compared against the experimental data from the literature.

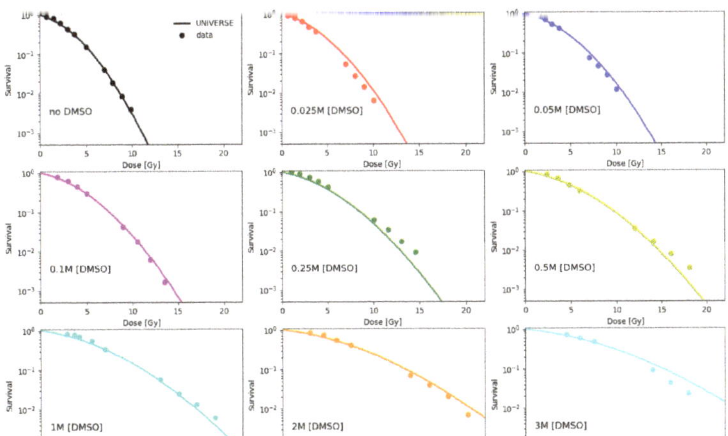

Figure 2. Model predictions (lines) based on the parametrization found in Figure 1 (lines) are plotted together with cell survival data (dots) of V79 cell line irradiated under normoxia without and with different DMSO concentrations ranging from 0.025 M to 3 M, taken from Chapman et al. [16].

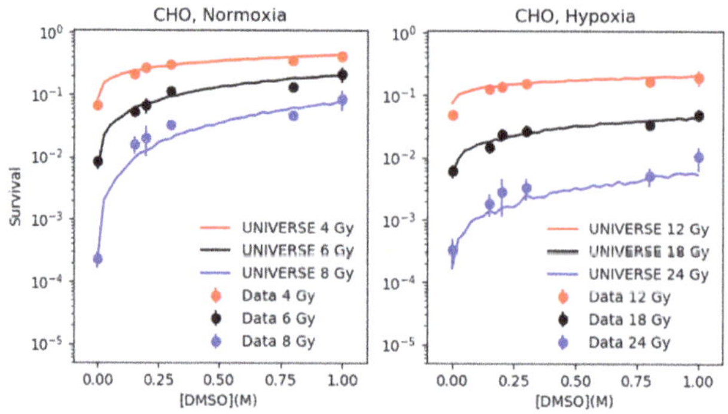

Figure 3. CHO cell survival data (dots with error bars) in normoxia (left panel) and in hypoxia (right panel) as a function of DMSO concentration in units of M for different dose levels as shown in the legends, taken from Hirayama et al. [12], are compared against model predictions (lines).

2.2. Modeling the Effect of DMSO under Normoxia and Hypoxia

Based on the parametrization of f_{DMSO} determined in the previous section, the survival of CHO and AA8 cells were predicted by our model over the range of 0.0 M to 1.0 M of DMSO at three different dose levels and compared to the data taken from Hirayama et al. [12] (left panels of Figures 3 and 4). In order to extend our predictions to the hypoxic case, the $HRF_{DSB}^{O_2}$ was tuned to resemble the survival observed without application of DMSO under hypoxia (Supplementary Figure S1). The numerical values found for $HRF_{DSB}^{O_2}$ for both cell-lines can be found in Table 1. The derived $HRF_{DSB}^{O_2}$ was applied to the total yield of DSB, from which subsequently the fixed yield based on the direct action ($\alpha_{DSB} \cdot fr_{DIR}$) is subtracted to determine the yield based on the indirect action. The reduction determined by the parametrization of f_{DMSO} is applied only to the latter. Finally, the survival is predicted based on the sum of the unchanged yield based on the direct action and the reduced yield based on the indirect action and compared to the data taken from Hirayama et al. [12].

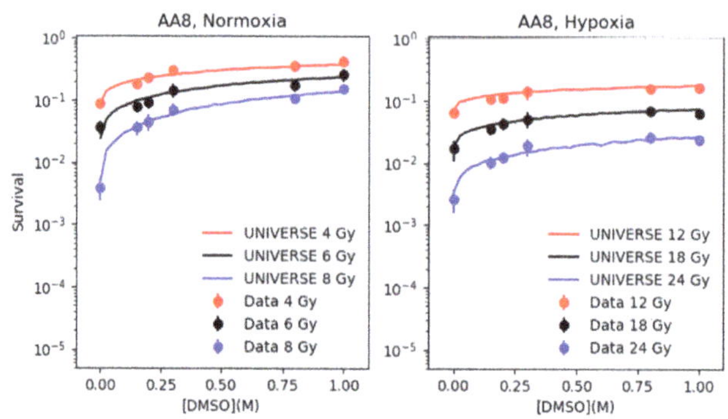

Figure 4. AA8 cell survival data (dots with error bars) in normoxia (left panel) and in hypoxia (right panel) as a function of DMSO concentration in units of M for different dose levels as shown in the legends, taken from Hirayama et al. [12], are compared against model predictions (lines).

3. Discussion

The parametrization for the fraction of indirect damage quenched by DMSO (f_{DMSO}) carried out in the first part of the analysis (left panel, Figure 1) and its implementation into our model described, in good agreement, the cell survival trends over a wide range of DMSO concentrations at different dose levels for three different cell lines (Figures 2–4). In addition, the findings in this work suggest that the rough estimate made for the DSB yield by direct action of ~20% sufficiently described cell survival under the given settings. The selected f_{DMSO} parametrization underestimates the DMSO best-fit scaling value for about 10% at the lowest DMSO concentration. Of course, an improved fitting could be achieved increasing the number of free parameters of f_{DMSO} parametrization; however, in this study we opted for model simplicity (2 free parameters), which itself was capable of replicating the general trends of the f_{DMSO}. Furthermore, the f_{DMSO} data point discussed above corresponds to cell survival measurements at 0.025 M, which are still satisfactorily estimated by our model (upper middle panel of Figure 2). The prediction for the ratio of DSB measured with and without DMSO (right panel, Figure 1) generally lies above the values observed in literature [21,23,24]. At low DMSO concentrations (below 0.5 M), the prediction overestimates the measured R_{DSB} by about 10%. At concentrations above 0.6 M, the data of deLara et al. indicate a steeper decline in DSB yields, which they account to unspecified "additional modes of protection" [23,25]. We could not observe this effect in the investigated cell survival data set. The absence of such steep decline could indicate an underlying cell type specific response to irradiation. The work by Kashino et al. [26] implies possible interaction of DMSO with the DSB repair mechanisms, which we possibly could account for by a modification of the lethality parameter of the isolated DSB (K_{iDSB}) as we had demonstrated in an earlier publication [3]. On the other hand, Sapora et al. [21] observed significantly higher DSB yields at higher concentrations of DMSO. Based on this, we believe that our predictions are within the range represented by the experimental data.

The extension of our approach introduced here, which includes the effect of hypoxia, results in excellent predictions of the experimental survival data over a range of DMSO concentration at three different dose levels each for two different cell lines (Figures 3 and 4). It shall be emphasized here, that the parametrization of f_{DMSO} is based only on the cell survival data from Figure 2 and the data points of the left panel of Figure 3 (CHO) at 4 Gy without any further adjustments to predict measured cell survival data in hypoxia. The ability of our extension to predict the cell survival under hypoxia with such precision for both cell lines at all three dose levels and over the observed range of DMSO concentrations, suggests that our simplistic assumption, which completely excludes the direct action from the oxygen effect, is valid.

In conclusion, we could demonstrate, that we can introduce a simple mechanistic differentiation of direct and indirect action into our existing model and accurately predict cell survival for different levels of free radicals reduction via DMSO. Not only could we show this for different cell-lines and irradiation dose levels but also under different cell oxygenation states (i.e., normoxia and hypoxia). Ultimately, this adds to the existing versatility of UNIVERSE to consider multiple radio biologically relevant parameters in its predictions of cell survival, as it can now account for manipulations of the indirect action for a given oxygenation status, e.g., by administered radical scavengers. Furthermore, the assumption that the direct action was unaffected by the oxygenation status, which was here successfully shown for photons, will be considered as a basis for a hypoxia model for the planned heavy-ion extensions of UNIVERSE.

4. Materials and Methods

4.1. Experimental Data from Literature

The experimental survival data used to benchmark the extensions of our model were taken from Hirayama et al. [12] and Chapman et al. [16]. The experimental data of the DSB yield as function of [DMSO] were collected from Sapora et al. [21], deLara et al. [23] and Zwicker et al. [24].

4.2. Modeling Approach

The modeling approach of the basic model has been extensively described and discussed in previous publications [3,27]. However, we must reiterate the key points in an abridged version as folows: for an irradiation with photons we assume the dose deposition to be homogeneous throughout the cell nucleus and a cell line independent DSB yield of $\alpha_{DSB} = 5 \times 10^{-3} DSB/(Mbp \times Gy)$ (bp = base pairs), which is constant over the dose range typically applied in the clinic [28–31]. The expected total number of DSB in the cell nucleus ($< N_{tDSB} >$) is therefore given by:

$$< N_{tDSB} >= \alpha_{DSB} \times D \times DNA_c \tag{1}$$

where DNA_c is the DNA content of a cell in Mbp and D the applied dose in units of Gy. In our base model, as in similar models of other groups [32–34], so called giant loops, a type of chromatin sub-structure [35–37], are believed to be the critical targets in the cell. The total number of giant loops (N_{gl}) with a DNA content of DNA_{gl} inside the nucleus is given by:

$$N_{gl} = \frac{DNA_c}{DNA_{gl}} \tag{2}$$

As in previous works $DNAc$ and DNA_{gl} were assumed to be 6 Gbp and 2 Mbp, respectively.

The number of total DSB in the nucleus (N_{tDSB}) is sampled following a Poisson distribution by a Monte Carlo routine, with the expectation value given by Equation (1). Thereafter, the sampled amount of DSBs is randomly distributed over the giant loops in the nucleus and the number of giant loops with one DSB (isolated DSB; N_{iDSB}), or two or more DSBs (complex DSB; N_{cDSB}) are scored. The lethality parameters, K_{iDSB} and K_{cDSB}, quantify the probabilities of an isolated lesion and a complex lesion leading to the inactivation of the cell, respectively. The overall probability of a cell to survive (S) can then be calculated using [33]:

$$S = (1 - K_{iDSB})^{N_{iDSB}} \times (1 - K_{cDSB})^{N_{cDSB}} \tag{3}$$

The survival fraction of a cell population after irradiation was determined by the average S value determined by the Monte Carlo algorithm. The cell line dependent lethality parameters, K_{iDSB} and K_{cDSB}, can be determined by fitting the result of this algorithm to experimental survival data.

In previous works it was shown that in our model a change in oxygenation solely leads to a reduction of the DSB yield, by a hypoxia reduction factor ($HRF^{O_2}_{DSB}$) resembling the classical oxygen enhancement factor (OER), while the lethality parameters could be assumed constant [3,27]. The reduced DSB yield, $\alpha^{O_2}_{DSB}$, is given by:

$$\alpha^{O_2}_{DSB} = \frac{\alpha_{DSB}}{HRF^{O_2}_{DSB}} \tag{4}$$

where α_{DSB} is the rate under normoxia.

Based on Equation (1), the alteration of α_{DSB} leads to a change of N_{tDSB} to $N^{O_2}_{tDSB}$. This again leads to alterations of N_{iDSB} and N_{cDSB} to $N^{O_2}_{iDSB}$ and $N^{O_2}_{cDSB}$, respectively. With lethality parameters invariant under the oxygenation status, the probability of a cell under hypoxic conditions to survive is expressed in our model as:

$$S_{O_2} = (1 - K_{iDSB})^{N^{O_2}_{iDSB}} \times (1 - K_{cDSB})^{N^{O_2}_{cDSB}} \tag{5}$$

The $HRF^{O_2}_{DSB}$ value for a given oxygenation level can be determined by fitting the model to the hypoxic survival data, while keeping K_{iDSB} and K_{cDSB} constant. However, if either hypoxic or normoxic data are not available, the $HRF^{O_2}_{DSB}$ for a given oxygen concentration $[O_2]$ can be estimated using the following formula:

$$HRF^{O_2}_{DSB} = \frac{m \times K + [O_2]}{K + [O_2]} \tag{6}$$

where m and K are the maximum value and the turning point of the function, respectively. The parametrization was introduced in a previous work [27], first proposed by Carlson et al. [38] and originally inspired by works of Alper and Howard-Flanders [39]. Our current best fit of available data with this parametrization yields m = 2.94 and K = 0.129% [27].

In our extension of this model for this work, we introduced a distinction of direct and indirect (radical-mediated) DSB induction rate. Based on the work of Hirayama et al. [12], we assumed that the fraction of DSB yield based on direct action (fr_{DIR}) is 20% of the total DSB yield. Further, we consider the reduction of the DSB yield through DMSO as an OH radical scavenger, by a given factor f_{DSMO}, only to apply to the fraction of DSB yield based on indirect action. Equation (1) can be then rewritten as:

$$< N_{tDSB} >= fr_{DIR} \times \alpha_{DSB} \times D \times DNA_c + (1 - fr_{DIR}) \times \alpha_{DSB} \times D \times DNA_c \times f_{DMSO} \quad (7)$$

The f_{DSMO} values in the left panel of Figure 1 were obtained by tuning the predicted cell survival resulting from Equation (7) to reproduce the experimental survival data shown in Figure 2 and the 4 Gy normoxic data of the CHO cells (left panel, Figure 3). Limiting the number of free parameters (<3) for model simplicity, the parametrization used to fit these values was:

$$f_{DMSO} = \exp\left(-a_{DMSO} \times [DMSO]^{b_{DMSO}}\right) \quad (8)$$

where a_{DMSO} = 0.9065 and b_{DMSO} = 0.4172 and [DMSO] is the DMSO concentration in units of M. The function approaches unity and zero for vanishing and for increasing DMSO concentrations, respectively. The predicted initial yield of DSBs in normoxia relative to the yield in absence of DMSO (R_{DSB}) is calculated as:

$$R_{DSB} = (fr_{DIR} + (1 - fr_{DIR}) \times f_{DMSO}) \times 100 \quad (9)$$

Various experimental investigations have implied, that both for cell survival [12,21] and induction of DNA damage itself [21,22], the oxygen effect primarily interferes with the indirect pathway of damage induction, while the direct pathway is only weakly affected. In our model, we assume that the fraction of DSB yield based on the direct action is completely unaffected by the oxygen affect. In the hypoxic case, Equation (7) therefore transforms to:

$$< N^{O_2}_{tDSB} >= fr_{DIR} \times \alpha_{DSB} \times D \times DNA_c + \left(\alpha^{O_2}_{DSB} - \alpha_{DSB} \times fr_{DIR}\right) \times D \times DNA_c \times f_{DMSO} \quad (10)$$

Throughout the calculations, the lethality parameters K_{iDSB} and K_{cDSB} are kept constant and the simulation of the survival fraction is carried out analogously to the procedure described above, only replacing Equation (1) with Equation (7) or Equation (10), for normoxia or hypoxia, respectively.

Supplementary Materials: Supplementary materials can be found at http://www.mdpi.com/1422-0067/21/10/3471/s1.

Author Contributions: Conceptualization, A.M.; methodology, H.L., A.M. and I.D.; software, H.L. and A.M.; formal analysis, H.L., I.D. and A.M.; investigation, A.M., I.D. and H.L.; resources, J.D.; data curation, S.M., A.M.; writing—original draft preparation, A.M, H.L. and I.D.; writing—review and editing, H.L., S.M., J.D., I.D. and A.M.; visualization, A.M.; supervision, A.M.; project administration, A.M and J.D.; funding acquisition, J.D. All authors have read and agreed to the published version of the manuscript.

Funding: This work was supported by Deutsche Forschungsgemeinschaft (DFG) collaborative research center (CRC/SFB-1389, Project number 404521405) to ID, as well as a PhD stipend from the Helmholtz International Graduate School for Cancer Research in Heidelberg to H.L.

Conflicts of Interest: The authors declare no conflicts of interest.

Abbreviations

LET	Linear energy transfer
DMSO	Dimethyl sulfoxide
UNIVERSE	UNIfied and VERSatile bio response Engine
DSB	(DNA) Double strand break
iDSB	Isolated DSB

cDSB	Complex DSB
OER	Oxygen enhancement ratio
HRF	Hypoxia reduction factor
CHO	Chinese hamster ovary

References

1. Delaney, G.; Jacob, S.; Featherstone, C.; Barton, M. The role of radiotherapy in cancer treatment: Estimating optimal utilization from a review of evidence-based clinical guidelines. *Cancer* **2005**, *104*, 1129–1137. [CrossRef] [PubMed]
2. *Radiobiological Modelling in Radiation Oncology*; Dale, R.G.; Jones, B. (Eds.) British Institute of Radiology: London, UK, 2007; ISBN 978-0-905749-60-0.
3. Liew, H.; Klein, C.; Zenke, F.T.; Abdollahi, A.; Debus, J.; Dokic, I.; Mairani, A. Modeling the Effect of Hypoxia and DNA Repair Inhibition on Cell Survival After Photon Irradiation. *Int. J. Mol. Sci.* **2019**, *20*, 6054. [CrossRef]
4. Batey, M.A.; Zhao, Y.; Kyle, S.; Richardson, C.; Slade, A.; Martin, N.M.B.; Lau, A.; Newell, D.R.; Curtin, N.J. Preclinical evaluation of a novel ATM inhibitor, KU59403, in vitro and in vivo in p53 functional and dysfunctional models of human cancer. *Mol. Cancer* **2013**, *12*, 959–967. [CrossRef] [PubMed]
5. Dohmen, A.J.C.; Qiao, X.; Duursma, A.; Wijdeven, R.H.; Lieftink, C.; Hageman, F.; Morris, B.; Halonen, P.; Vens, C.; van den Brekel, M.W.M.; et al. Identification of a novel ATM inhibitor with cancer cell specific radiosensitization activity. *Oncotarget* **2017**, *8*, 73925–73937. [CrossRef] [PubMed]
6. Durant, S.T.; Zheng, L.; Wang, Y.; Chen, K.; Zhang, L.; Zhang, T.; Yang, Z.; Riches, L.; Trinidad, A.G.; Fok, J.H.L.; et al. The brain-penetrant clinical ATM inhibitor AZD1390 radiosensitizes and improves survival of preclinical brain tumor models. *Sci. Adv.* **2018**, *4*, eaat1719. [CrossRef] [PubMed]
7. Klein, C.; Dokic, I.; Mairani, A.; Mein, S.; Brons, S.; Häring, P.; Haberer, T.; Jäkel, O.; Zimmermann, A.; Zenke, F.; et al. Overcoming hypoxia-induced tumor radioresistance in non-small cell lung cancer by targeting DNA-dependent protein kinase in combination with carbon ion irradiation. *Radiat. Oncol.* **2017**, *12*, 208. [CrossRef]
8. *Basic Clinical Radiobiology*, 5th ed.; Joiner, M.; van der Kogel, A. (Eds.) CRC Press/Taylor & Francis Group: Boca Raton, FL, USA, 2018; ISBN 978-1-4441-7963-7.
9. Tawk, B.; Schwager, C.; Deffaa, O.; Dyckhoff, G.; Warta, R.; Linge, A.; Krause, M.; Weichert, W.; Baumann, M.; Herold-Mende, C.; et al. Comparative analysis of transcriptomics based hypoxia signatures in head- and neck squamous cell carcinoma. *Radiother. Oncol.* **2016**, *118*, 350–358. [CrossRef]
10. Rofstad, E.K.; Sundfør, K.; Lyng, H.; Tropé, C.G. Hypoxia-induced treatment failure in advanced squamous cell carcinoma of the uterine cervix is primarily due to hypoxia-induced radiation resistance rather than hypoxia-induced metastasis. *Br. J. Cancer* **2000**, *83*, 354–359. [CrossRef]
11. Azzam, E.I.; Jay-Gerin, J.-P.; Pain, D. Ionizing radiation-induced metabolic oxidative stress and prolonged cell injury. *Cancer Lett.* **2012**, *327*, 48–60. [CrossRef]
12. Hirayama, R.; Ito, A.; Noguchi, M.; Matsumoto, Y.; Uzawa, A.; Kobashi, G.; Okayasu, R.; Furusawa, Y. OH radicals from the indirect actions of X-rays induce cell lethality and mediate the majority of the oxygen enhancement effect. *Radiat. Res.* **2013**, *180*, 514–523. [CrossRef]
13. Johnke, R.M.; Sattler, J.A.; Allison, R.R. Radioprotective agents for radiation therapy: Future trends. *Future Oncol.* **2014**, *10*, 2345–2357. [CrossRef] [PubMed]
14. Jones, J.B.; Cramer, H.M.; Inch, W.R.; Lampe, H.B. Radioprotective effect of free radical scavenging enzymes. *J. Otolaryngol.* **1990**, *19*, 299–306. [PubMed]
15. Smith, T.A.; Kirkpatrick, D.R.; Smith, S.; Smith, T.K.; Pearson, T.; Kailasam, A.; Herrmann, K.Z.; Schubert, J.; Agrawal, D.K. Radioprotective agents to prevent cellular damage due to ionizing radiation. *J. Transl. Med.* **2017**, *15*, 232. [CrossRef] [PubMed]
16. Chapman, J.D.; Doern, S.D.; Reuvers, A.P.; Gillespie, C.J.; Chatterjee, A.; Blakely, E.A.; Smith, K.C.; Tobias, C.A. Radioprotection by DMSO of mammalian cells exposed to X-rays and to heavy charged-particle beams. *Radiat. Env. Biophys.* **1979**, *16*, 29–41. [CrossRef] [PubMed]

17. Bishayee, A.; Rao, D.V.; Bouchet, L.G.; Bolch, W.E.; Howell, R.W. Protection by DMSO against Cell Death Caused by Intracellularly Localized Iodine-125, Iodine-131 and Polonium-210. *Radiat. Res.* **2000**, *153*, 416–427. [CrossRef]
18. Kim, S.E.; Moos, W.S. Radiation protection by topical DMSO application. *Health Phys.* **1967**, *13*, 601–606. [CrossRef]
19. Willson, R.L. The reaction of oxygen with radiation-induced free radicals in DNA and related compounds. *Int. J. Radiat. Biol. Relat. Stud. Phys. Chem. Med.* **1970**, *17*, 349–358. [CrossRef]
20. Alexander, P. Division of Biophysics: On the Mode of Action of Some Treatments That Influence the Radiation Sensitivity of Cells. *Trans. New York Acad. Sci.* **1962**, *24*, 966–978. [CrossRef]
21. Sapora, O.; Barone, F.; Belli, M.; Maggi, A.; Quintiliani, M.; Tabocchini, M.A. Relationships Between Cell Killing, Mutation Induction and DNA Damage in X-irradiated V79 Cells: The Influence of Oxygen and DMSO. *Int. J. Radiat. Biol.* **1991**, *60*, 467–482. [CrossRef]
22. Frankenberg, D.; Frankenberg-Schwager, M.; Harbich, R. Mechanisms of Oxygen Radiosensitization in Irradiated Yeast. *Int. J. Radiat. Biol.* **1993**, *64*, 511–521. [CrossRef]
23. deLara, C.M.; Jenner, T.J.; Townsend, K.M.; Marsden, S.J.; O'Neill, P. The effect of dimethyl sulfoxide on the induction of DNA double-strand breaks in V79-4 mammalian cells by alpha particles. *Radiat. Res.* **1995**, *144*, 43–49. [CrossRef] [PubMed]
24. Zwicker, F.; Hauswald, H.; Debus, J.; Huber, P.E.; Weber, K.-J. Impact of dimethyl sulfoxide on irradiation-related DNA double-strand-break induction, -repair and cell survival. *Radiat. Env. Biophys.* **2019**, *58*, 417–424. [CrossRef]
25. Ewing, D.; Walton, H.L. Radiation protection of in vitro mammalian cells: Effects of hydroxyl radical scavengers on the slopes and shoulders of survival curves. *Radiat. Res.* **1991**, *126*, 187–197. [CrossRef] [PubMed]
26. Kashino, G.; Liu, Y.; Suzuki, M.; Masunaga, S.; Kinashi, Y.; Ono, K.; Tano, K.; Watanabe, M. An alternative mechanism for radioprotection by dimethyl sulfoxide; possible facilitation of DNA double-strand break repair. *J. Radiat. Res.* **2010**, *51*, 733–740. [CrossRef] [PubMed]
27. Mairani, A.; Böhlen, T.T.; Dokic, I.; Cabal, G.; Brons, S.; Haberer, T. Modelling of cell killing due to sparsely ionizing radiation in normoxic and hypoxic conditions and an extension to high LET radiation. *Int. J. Radiat. Biol.* **2013**, *89*, 782–793. [CrossRef] [PubMed]
28. Frankenberg-Schwager, M. Review of repair kinetics for DNA damage induced in eukaryotic cells in vitro by ionizing radiation. *Radiother. Oncol.* **1989**, *14*, 307–320. [CrossRef]
29. Prise, K.M.; Ahnström, G.; Belli, M.; Carlsson, J.; Frankenberg, D.; Kiefer, J.; Löbrich, M.; Michael, B.D.; Nygren, J.; Simone, G.; et al. A review of dsb induction data for varying quality radiations. *Int. J. Radiat. Biol.* **1998**, *74*, 173–184. [CrossRef] [PubMed]
30. Prise, K.M.; Pinto, M.; Newman, H.C.; Michael, B.D. A review of studies of ionizing radiation-induced double-strand break clustering. *Radiat. Res.* **2001**, *156*, 572–576. [CrossRef]
31. Asaithamby, A.; Chen, D.J. Cellular responses to DNA double-strand breaks after low-dose gamma-irradiation. *Nucleic Acids Res.* **2009**, *37*, 3912–3923. [CrossRef]
32. Elsässer, T.; Weyrather, W.K.; Friedrich, T.; Durante, M.; Iancu, G.; Krämer, M.; Kragl, G.; Brons, S.; Winter, M.; Weber, K.-J.; et al. Quantification of the relative biological effectiveness for ion beam radiotherapy: Direct experimental comparison of proton and carbon ion beams and a novel approach for treatment planning. *Int. J. Radiat. Oncol. Biol. Phys.* **2010**, *78*, 1177–1183. [CrossRef]
33. Friedrich, T.; Durante, M.; Scholz, M. Modeling cell survival after photon irradiation based on double-strand break clustering in megabase pair chromatin loops. *Radiat. Res.* **2012**, *178*, 385–394. [CrossRef]
34. Friedrich, T.; Scholz, U.; Elsässer, T.; Durante, M.; Scholz, M. Calculation of the biological effects of ion beams based on the microscopic spatial damage distribution pattern. *Int. J. Radiat. Biol.* **2012**, *88*, 103–107. [CrossRef] [PubMed]
35. Yokota, H.; van den Engh, G.; Hearst, J.E.; Sachs, R.K.; Trask, B.J. Evidence for the organization of chromatin in megabase pair-sized loops arranged along a random walk path in the human G0/G1 interphase nucleus. *J. Cell Biol.* **1995**, *130*, 1239–1249. [CrossRef] [PubMed]
36. Sachs, R.K.; van den Engh, G.; Trask, B.; Yokota, H.; Hearst, J.E. A random-walk/giant-loop model for interphase chromosomes. *Proc. Natl. Acad. Sci. USA* **1995**, *92*, 2710–2714. [CrossRef] [PubMed]

37. Solovjeva, L.; Svetlova, M.; Stein, G.; Chagin, V.; Rozanov, Y.; Zannis-Hadjopoulos, M.; Price, G.; Tomilin, N. Conformation of replicated segments of chromosome fibres in human S-phase nucleus. *Chromosome Res.* **1998**, *6*, 595–602. [CrossRef] [PubMed]
38. Carlson, D.J.; Stewart, R.D.; Semenenko, V.A. Effects of oxygen on intrinsic radiation sensitivity: A test of the relationship between aerobic and hypoxic linear-quadratic (LQ) model parameters. *Med. Phys.* **2006**, *33*, 3105–3115. [CrossRef]
39. Alper, T.; Howard-Flanders, P. Role of Oxygen in Modifying the Radiosensitivity of E. Coli B. *Nature* **1956**, *178*, 978–979. [CrossRef]

International Journal of *Molecular Sciences*

Int. J. Mol. Sci. 2019, 20, 6204

MDPI

Article

Assessment of Radio-Induced Damage in Endothelial Cells Irradiated with 40 kVp, 220 kVp, and 4 MV X-rays by Means of Micro and Nanodosimetric Calculations

Nicolas Tang [1], Marta Bueno [1], Sylvain Meylan [2], Yann Perrot [1], Hoang N. Tran [1], Amélie Freneau [1], Morgane Dos Santos [1], Aurélie Vaurijoux [1], Gaëtan Gruel [1], Mario A. Bernal [3], Marie-Claude Bordage [4,5], Dimitris Emfietzoglou [6], Ziad Francis [7], Susanna Guatelli [8], Vladimir Ivanchenko [9,10], Mathieu Karamitros [11], Ioanna Kyriakou [6], Wook-Geun Shin [11], Sébastien Incerti [11,12] and Carmen Villagrasa [1,*]

1 IRSN, Institut de Radioprotection et de Sûreté Nucléaire, BP17, 92262 Fontenay aux Roses, France; nicolas.tang78@gmail.com (N.T.); mbuenovizcarra@gmail.com (M.B.); yann.perrot@irsn.fr (Y.P.); ngoc-hoang.tran@irsn.fr (H.N.T.); amelie.freneau@gmail.com (A.F.); morgane.dossantos@irsn.fr (M.D.S.); aurelie.vaurijoux@irsn.fr (A.V.); gaetan.gruel@irsn.fr (G.G.)
2 SymAlgo Technologies, 75 rue Léon Frot, 75011 Paris, France; sylvain.meylan@symalgo-tech.com
3 Instituto de Fisica Gleb Wataghin, Universidade Estadual de Campinas, Campinas 13083-859, SP, Brazil; mbernalrod@gmail.com
4 INSERM, Université Paul Sabatier, UMR 1037, CRCT, 31330 Toulouse, France; marie-claude.bordage@inserm.fr
5 Université Toulouse III-Paul Sabatier, UMR 1037, CRCT, 31330 Toulouse, France
6 Medical Physics Laboratory, University of Ioannina, Medical School, GR-45110 Ioannina, Greece; demfietz@gmail.com (D.E.); ikyriak@cc.uoi.gr (I.K.)
7 Department of Physics, Faculty of Sciences, Université Saint Joseph, 1104 2020 Beirut, Lebanon; ziad.francis@gmail.com
8 Centre for Medical Radiation Physics, University of Wollongong, 2522 Wollongong, Australia; susanna@uow.edu.au
9 Geant4 Associates International Ltd., Hebden Bridge HX7 7BT, UK; vnivanch@mail.cern.ch
10 LEPHE, Tomsk State University, 634050 Tomsk, Russia
11 Université de Bordeaux, CNRS/IN2P3, Centre Nucléaires de Bordeaux Gradignan, CENBG, chemin du solarium, BP120, 33175 Gradignan, France; matkara@googlemail.com (M.K.); ukguen@gmail.com (W.-G.S.); incerti@cenbg.in2p3.fr (S.I.)
12 CNRS/IN2P3, Centre d'Etudes Nucléaires de Bordeaux Gradignan, CENBG, chemin du solarium, BP120, 33175 Gradignan, France
* Correspondence: carmen.villagrasa@irsn.fr; Tel.: +33-1583-586-88

Received: 28 October 2019; Accepted: 4 December 2019; Published: 9 December 2019

Abstract: The objective of this work was to study the differences in terms of early biological effects that might exist between different X-rays energies by using a mechanistic approach. To this end, radiobiological experiments exposing cell monolayers to three X-ray energies were performed in order to assess the yields of early DNA damage, in particular of double-strand breaks (DSBs). The simulation of these irradiations was set in order to understand the differences in the obtained experimental results. Hence, simulated results in terms of microdosimetric spectra and early DSB induction were analyzed and compared to the experimental data. Human umbilical vein endothelial cells (HUVECs) were irradiated with 40, 220 kVp, and 4 MV X-rays. The Geant4 Monte Carlo simulation toolkit and its extension Geant4-DNA were used for the simulations. Microdosimetric calculations aiming to determine possible differences in the variability of the energy absorbed by the irradiated cell population for those photon spectra were performed on 10,000 endothelial cell nuclei representing a cell monolayer. Nanodosimetric simulations were also carried out using a computation chain that allowed the simulation of physical, physico-chemical, and chemical stages on a single realistic endothelial cell nucleus model including both heterochromatin and euchromatin. DNA damage was scored in terms of yields of prompt DSBs per Gray (Gy) and per giga (10^9) base pair (Gbp) and DSB complexity was derived in order to be compared to experimental data expressed as numbers of histone variant H2AX (γ-H2AX) foci per cell. The calculated microdosimetric spread in the irradiated cell population was similar when comparing between 40 and 220 kVp X-rays and higher when comparing with 4 MV X-rays. Simulated yields of induced DSB/Gy/Gbp were found to be equivalent to those for 40 and 220 kVp but larger than those for 4 MV, resulting in a relative biological effectiveness (RBE) of 1.3. Additionally, DSB complexity was similar between the considered photon spectra. Simulated results were in good agreement with experimental data obtained by IRSN (Institut de radioprotection et de sûreté nucléaire) radiobiologists. Despite differences in photon energy, few differences were observed when comparing between 40 and 220 kVp X-rays in microdosimetric and nanodosimetric calculations. Nevertheless, variations were observed when comparing between 40/220 kVp and 4 MV X-rays. Thanks to the simulation results, these variations were able to be explained by the differences in the production of secondary electrons with energies below 10 keV.

Keywords: Monte Carlo simulation; Geant4/Geant4-DNA; X-rays; microdosimetry; nanodosimetry; DSB yield

---

1. Introduction

For the evaluation of DNA damage induced by photon irradiation, most radiobiological experiments rely on the mean absorbed dose [1]. Hence, these experiments assume that a homogeneous dose equal to the macroscopic absorbed dose D is delivered to the cell population. However, due to the stochastic nature of energy depositions and the small volume of a given cell nucleus, the energy received from one cell to another may be significantly different in the irradiated population.

Moreover, although X-rays are considered low-linear energy transfer (LET) radiation [2], a higher relative biological effectiveness (RBE) at very low-energy (≤30 keV) than at high energy has been reported in the literature. Indeed, very low energy X-rays may lead to higher risk estimates at low doses for many biological endpoints such as double-strand breaks (DSBs), chromosome aberrations, micronucleus formation, and cell survival [3–22]. This could be explained by the fact that energy depositions vary between these different photon energies since the type of interaction (photoelectric effect, Compton effect, or pair production), which causes the energy deposition depends on the photon energy [23]. In addition, the lower the photon energy, the lower the energy of the secondary electrons, meaning the stopping power of the latter is higher. The stochastic nature of the radiation–tissue interaction leads to differences in terms of energy received by each cell nucleus in a cell population [1], which can be analyzed through microdosimetry.

Differences in track structure and patterns of energy depositions can also impact DNA DSB induction and location [24,25], which itself can lead to differences in final cell fate. Indeed, it is well known that DSBs are among the most deleterious forms of DNA lesions and can lead to potential chromosome aberrations and cell death if unrepaired or misrepaired [26]. Yields of DSBs, as well as their spatial distribution and especially their proximity, play an important role in the formation of chromosome aberrations and other cellular endpoints such as cell death [24,25,27,28]. However, it should be noted that for the same biological endpoint of interest and radiation quality, RBE values from the literature can present great variability [29].

In this work, simulations were performed to reproduce experimental irradiations using different photon sources carried out at the IRSN (Institut de Radioprotection et de Sûreté Nucléaire), Fontenay aux Roses, France [30]. In Freneau et al. [30], human umbilical vein endothelial cells (HUVECs) were irradiated with 40 kVp, 220 kVp, and 4 MV X-rays. The experiments showed that the yield of histone variant H2AX (γ-H2AX) foci was similar for 40 kVp and 220 kVp but higher when compared to 4 MV for a dose of 2 Gray (Gy). It has been pointed out that the secondary electron energy spectrum is more likely to explain differences in the yields of γ-H2AX foci rather than the initial photon energy spectrum. Measurements of γ-H2AX foci can be related to the yield of DSB, although certain precautions must be taken. Indeed, the usual ratio of 1 DSB:1 γ-H2AX focus [31] is not entirely accurate when high-LET radiation is considered and given the size of the γ-H2AX foci, which is about 0.2 μm^2 as reported by Rothkam and Horn [32]. In this case, several DSBs can be contained within a single γ-H2AX focus [33]. However, this ratio remains more or less accurate for low-LET radiation such as photon irradiation, as DSBs are less close [34].

The calculation of energy depositions by Monte Carlo simulation can then be useful to understanding the observations made in radiobiological experiments since DNA damage and energy depositions are closely linked.

The simulations in this work were performed by means of a computation chain that has been extensively described in [35] and which is based on the Geant4 [36–38] Monte Carlo simulation toolkit (version 10.1) and its Geant4-DNA extension [39–42].

As a first approach, microdosimetric quantities were calculated, such as the frequency distribution of the specific energy and the microdosimetric spread, which reflects the spread in specific energy per cell in an irradiated cell population, since these quantities could be of use to study the possible difference between radiation qualities in terms of RBE [43].

In order to evaluate the influence of radiation quality on DSB induction, nanodosimetric calculations were performed using the computation chain in order to assess the yields of DSB per unit absorbed dose and per Gbp, taking into account direct and indirect damage as well as DSB complexity [35]. Then, the number of simulated DSBs was correlated with measurements of γ-H2AX foci. For these calculations, a model representing an endothelial cell nucleus filled with around 6 Gbp [44] of DNA composed of heterochromatin and euchromatin in the Gap0/Gap1 (G0/G1) phase of the cell cycle was generated with DnaFabric [45,46] software and used in the simulation.

2. Results

2.1. Secondary Electron Spectra

Proportions in photoelectric effect (PE), Compton effect (CE), and pair production (CONV) were scored and compared between the three photon energy spectra. For 40 kVp X rays, the photoelectric effect was found to be dominant (68.9%), the whereas Compton effect represented 31.1% of the photon interactions, as presented in Table 1. As the photon energy increases, the photoelectric effect decreases and the Compton effect increases, becoming dominant (the two latter effects represented 19.6% and 80.4%, respectively, for 220 kVp X-rays). At higher energies, i.e., 4 MV here, the conversion process can be observed, although at a very low rate (PE = 3.3%, CE = 96.3%, and CONV = 0.4%). In principle, it could be assumed that these variations could lead to differences in terms of energy depositions because this could give rise to different secondary electron spectra. Nevertheless, it can be seen in Figure 1 that the secondary electron spectra calculated at the cell positions did not appear as expected from our first hypothesis. Indeed, the photon spectra are different, but we can observe that the distributions of secondary electrons resulting from both 40 and 220 kVp X-rays are quite similar, as well as their mean energies (9.8 and 18.4 keV, respectively), if we account for the fact that the acceleration potential increases almost six times. Conversely, the secondary electron spectrum for the 4 MV X-rays shows relatively large differences, especially in its mean energy, which is 858.4 keV. As a consequence, from these three secondary electron spectra, we were able to expect variations in energy depositions, in particular between 40 or 220 kVp and 4 MV. Furthermore, a sudden decrease can be seen in Figure 1 in the frequency of secondary electrons at 100 eV for 4 MV. This is not a physical phenomenon but a bias resulting from the simulation and is due to the default production threshold of 100 eV of Livermore. Indeed, when an electron reached 100 eV in the simulation, its energy was deposited locally and its transport was stopped. As a result, electrons with an energy of less than 100 eV that had not reached yet the cell layer or had not been created directly inside it were not recorded. However, this decrease was not observed for 40 and 220 kVp, considering that the majority of electrons (85% and 68%, respectively) were created directly inside the cell layer volume. By contrast, in the case of the 4 MV X-ray configuration, 98% of the electrons were created at the bottom surface of the cell layer. As a result, the vast majority of electrons with an energy of less than 100 eV were not recorded in the phase space file, except for those created directly inside the cell layer. Nevertheless, this cut-off energy of 100 eV did not have any impact on the calculation of the energy deposition in the cell nuclei. Indeed, cell nuclei of 2 μm in thickness were placed in the middle of the layer which was 5 μm in height (in z). The distance separating them from each surface was then 1.5 μm. As this distance was larger than the range of 100 eV electrons, which was found to be 4 nm ± 2 nm in our calculation, these electrons could not reach the cell nuclei.

Table 1. Proportion of photoelectric effect (PE), Compton effect (CE), and pair production (CONV) as well as the mean energy of secondary electrons (in keV), for each photon energy spectrum.

Radiation Quality	40 kVp X-rays	220 kVp X-rays	4 MV X-rays
% PE	68.9	19.6	3.3
% CE	31.1	80.4	96.3
% CONV	0	0	0.4
Mean energy of secondary electrons (keV)	9.8	18.4	858.4

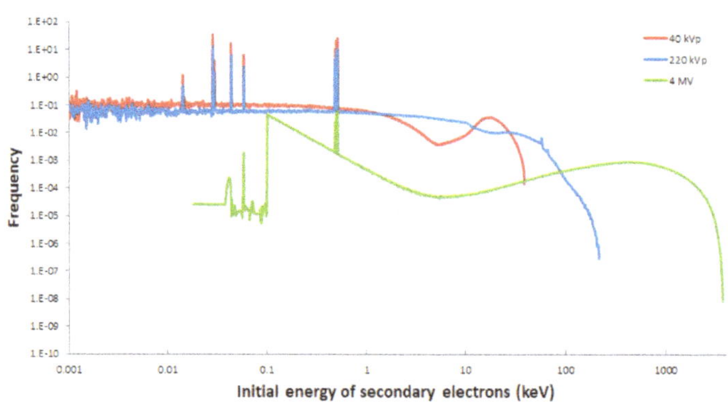

Figure 1. Frequency of secondary electrons per unit energy in log scale resulting from 40 kVp (red line), 220 kVp (blue line), and 4 MV (green line) X-rays.

2.2. Microdosimetric Calculations

Distributions of the specific energy f (z; D) in the irradiated cell population for a dose of 0.25 Gy for 40, 220 kVp, and 4 MV are shown in Figure 2. In this figure it is possible to observe that the f (z; D) distributions correspond to normal distributions centered at the simulated dose of 0.25 Gy. The distributions are very much alike between 40 and 220 kVp, while the distribution is narrower in the case of the 4 MV X-rays.

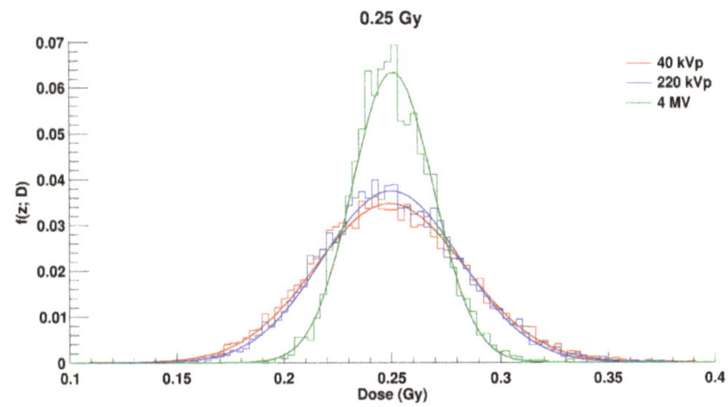

Figure 2. f (z; D) distributions in the irradiated cell population for a dose of 0.25 Gray (Gy) for 40 kVp (in red), 220 kVp (in blue), and 4 MV (in green) X-rays. Bars correspond to the histograms obtained in the simulations and the solid curves correspond to the distributions fitted with a Gaussian function.

Figure 3 shows the microdosimetric spread (σ_z^{rel}) calculated with Equation (4) (see Section 4.3.1) in the irradiated cell population for the three photon energy spectra as a function of the simulated doses 0.25 Gy, 0.5 Gy, 1 Gy, and 2 Gy, as used in the experiments [30]. As can be observed, the σ_z^{rel} decreases when the dose increases. This was expected, since the total number of individual energy depositions increases in each of the cell nucleus volume.

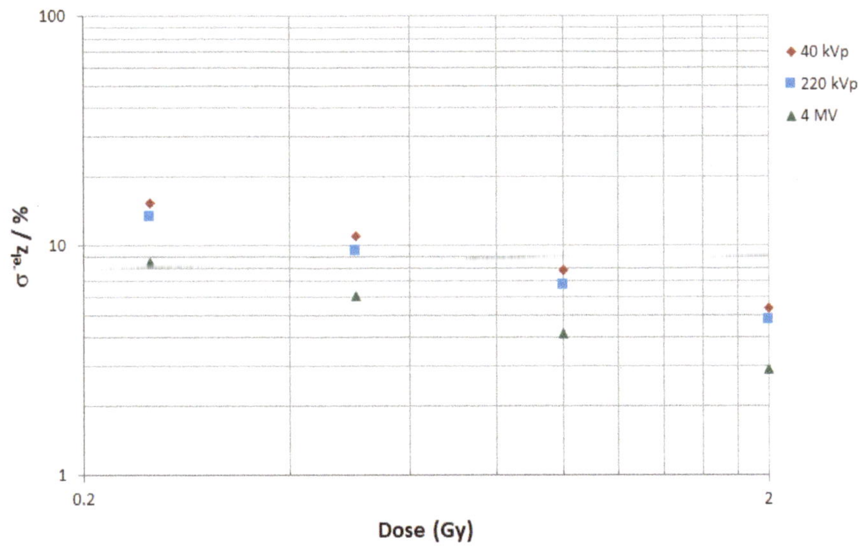

Figure 3. Microdosimetric spread σ_z^{rel} (in %) for 40 kVp (red diamonds), 220 kVp (blue squares), and 4 MV (green triangles) at 0.25, 0.5, 1, and 2 Gy.

2.3. Yields of DSB/Gy/Gbp and DSB Complexity

The yields of DSB/Gy/Gbp computed by the number of DSB divided by the dose received and by the 6 Gbp contained in the cell nucleus are presented in Table 2 for chemistry stage simulation end-times of 2.5 ns and 10 ns. Errors correspond to the standard deviation (SD) of the mean over 10 batches.

Table 2. Simulated results in this work: mean number of double-strand breaks (DSB)/Gy/giga base pairs (Gbp) and proportion of DSB complexity for 40, 220 kVp, and 4 MV X-rays. Simple DSBs only contain two strand breaks while complex DSBs contain three or more strand breaks with at least one of them in an opposite strand to the others. Errors from simulations correspond to the standard deviation of the mean over 10 batches.

Radiation Quality	40 kVp X-rays		220 kVp X-rays		4 MV X-rays	
Duration of the chemical stage	2.5 ns	10 ns	2.5 ns	10 ns	2.5 ns	10 ns
DSB/Gy/Gbp	3.5 ± 0.3	4.7 ± 0.3	3.5 ± 0.3	4.7 ± 0.2	2.8 ± 0.3	3.6 ± 0.3
Simple DSB (%)	86.0 ± 3.3	86.3 ± 1.8	86.4 ± 2.1	86.2 ± 2.1	87.7 ± 2.9	86.5 ± 2.3
Complex DSB (%)	14.0 ± 3.3	13.7 ± 1.8	13.6 ± 2.1	13.8 ± 2.1	12.3 ± 2.9	13.5 ± 2.3

2.4. Comparison between Simulated Results and Experimental Data

The experimental number of γ-H2AX foci per nucleus, determined by image analysis as shown in Figure 4, is compared to the simulated number of DSBs for chemistry stage simulation end-times of 2.5 and 10 ns in Table 3 for a dose of 1 Gy. The mean number of γ-H2AX foci per endothelial cell nucleus was observed experimentally 30 min post-irradiation for 0.25, 0.5, 1, and 2 Gy doses and obtained from at least three replicate experiments [30]. Errors correspond to the standard error. Endothelial cells were all synchronized in the G0/G1 phase of the cell cycle in order to minimize the dispersion in the cell DNA content during irradiation and to keep it close to the 6 Gbp used in the simulated cell nucleus phantom.

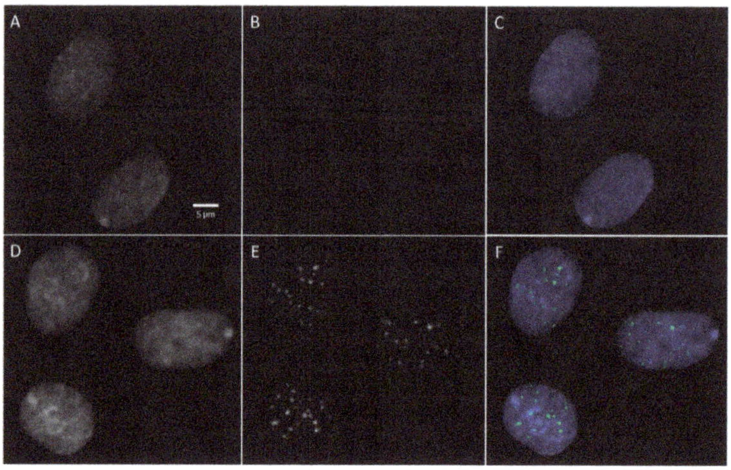

Figure 4. Labeling by immunofluorescence of the phosphorylation of the serine 139 of histone variant H2AX (γ-H2AX) for non-irradiated cells (**A**,**B**,**C**), and 30 min after irradiation of human umbilical vein endothelial cells (HUVECs) with 1 Gy of 4 MV X-rays delivered by medical linear accelerator (Elekta Synergy®) (**D**,**E**,**F**). (**A**) and (**D**) correspond to nuclear DNA labelled with 4′,6-diamidino-2-phenylindole (DAPI). (**B**) and (**E**) correspond to the immunostaining of the phosphorylation of serine 139 of γ-H2AX. (**C**) and (**F**) correspond to the merging of the images obtained for DNA and γ-H2AX labelling.

Table 3. Simulated results and experimental data obtained at IRSN [30]: mean number of γ-H2AX foci per endothelial cell nucleus in Gap0/Gap1 (30 min post-irradiation) and mean number of simulated DSB per nucleus for chemical simulation end-times of 2.5 and 10 ns for a dose of 1 Gy related to the experimental data at 1 Gy for 40 kVp, 220 kVp, and 4 MV X-rays. Errors from simulations correspond to the standard deviation over 10 batches while experimental ones correspond to the standard error.

Experimental Data	40 kVp X-rays	220 kVp X-rays	4 MV X-rays
Mean number of γ-H2AX foci per nucleus (30 min post-irradiation) [30]	0.25 Gy: 5.35 ± 1.13	0.25 Gy: 7.35 ± 2.17	0.25 Gy: 4.35 ± 0.21
	0.5 Gy: 9.88 ± 0.87	0.5 Gy: 10.24 ± 1.73	0.5 Gy: 8.54 ± 1.42
	1 Gy: 18.59 ± 0.43	1 Gy: 18.64 ± 2.33	1 Gy: 16.46 ± 1.63
	2 Gy: 30.30 ± 2.21	2 Gy: 30.59 ± 2.96	2 Gy: 26.42 ± 0.87
Simulated DSBs and Experimental Foci at 1 Gy	**40 kVp X-rays**	**220 kVp X-rays**	**4 MV X-rays**
Mean number of DSBs per nucleus (sim.) for a chemical simulation end-time of 2.5 ns	21.0 ± 0.3	21.0 ± 0.3	16.8 ± 0.3
Mean number of DSBs per nucleus (sim.) for a chemical simulation end-time of 10 ns	28.2 ± 0.3	28.2 ± 0.2	21.6 ± 0.3
Mean number of γ-H2AX foci per nucleus (exp.)	18.59 ± 0.43	18.64 ± 2.33	16.46 ± 1.63

3. Discussion

The variations that can be seen in the secondary electron spectra between 40 kVp, 220 kVp, and 4 MV X-rays seem to explain the origin of the difference in RBE for DSB induction. The fact that electron spectra are very much alike between 40 and 220 kVp but different when compared to 4 MV could indicate a similar RBE value between 40 and 220 kVp but a higher RBE for the latter with respect to 4 MV. RBE for DSB induction increases drastically for electron energies around 1 keV and below for a maximum value of around 5, while it remains constant and close to 1 for electron energies above 10 keV [20]. Considering this, it is more convenient to examine the proportion of electrons below 10 keV than the ones above 10 keV. In Figure 1, we can clearly observe that the proportion of low-energy electrons ($\lesssim$ 10 keV) is almost the same between 40 and 220 kVp and much lower at 4 MV, indicating a possible difference in RBE.

From the microdosimetric point of view, σ_z^{rel}, which reflects the spread of the specific energy received per nucleus in a cell population, was relatively similar when comparing between 40 and 220 kVp. Indeed, for a macroscopic dose of 0.25, 0.5, 1, or 2 Gy, the difference when comparing between 40 and 220 kVp was always below 2%. However, this difference was higher when comparing between 40/220 kVp and 4 MV, e.g., 7% at 0.25 Gy. This can be explained by the fact that the mean energy deposited by an electron track in a cell nucleus is lower for 4 MV X-rays than for 40 and 220 kVp X-rays (1.8 mGy, 1.6 mGy, and 0.4 mGy for 40 kVp, 220 kVp, and 4 MV, respectively) since electrons from 4 MV X-rays are more energetic, and, thus, their mean free path is larger. As a result, more tracks are required for 4 MV X-rays to obtain the same macroscopic dose, which leads to a reduction in statistical dispersion.

DSB induction was equivalent between 40 and 220 kVp. Indeed, as can be seen in Table 2, the simulated DSB yield was 3.5 ± 0.3 DSB/Gy/Gbp for both photon spectra. However, the DSB yield decreased to 2.8 ± 0.3 DSB/Gy/Gbp for 4 MV X-rays. In a review of experimental DSB yield carried out by Prise et al. [18], data obtained for mammalian cells using pulsed-field gel electrophoresis (PFGE) techniques ranged from 5.8 to 6.0 DSB/Gy/Gbp for 225 and 250 kV X-rays, and from 4.2 to 6.9 DSB/Gy/Gbp for ^{60}Co, which is consistent with our results, especially since for the γ-H2AX technique, the repair processes had already been initiated. However, the simulation of DSB yield has led to higher values in other studies. Indeed, in Friedland et al. [4], the authors' calculation was found to be 8.8 ± 1.4 DSB/Gy/Gbp using 220 kVp X-rays.

In our simulation, an RBE of 1.3 for DSB induction was observed for 40 or 220 kVp with respect to 4 MV, which is similar to simulated RBEs reported earlier [29].

It should be noted that the scoring method for DNA damage can strongly influence the final yield of simulated DSB induction. As mentioned in our previous works [35,46,47], a change in the parameters to score direct and/or indirect strand breaks can greatly increase the yield of DSB. Moreover, a chemical simulation end-time of 10 ns has been suggested in our computation chain instead of the 2.5 ns currently used when considering the cell nucleus model filled with heterochromatin and euchromatin [46]. Hence, simulations were also performed with a parameter of 10 ns and the number of DSBs showed an increase by a factor of ~1.3 with respect to those obtained with 2.5 ns (Table 2). Although the chemical simulation end-time was changed, these DSB yields (3.6 to 4.7 DSB/Gy/Gbp for 10 ns) are still in agreement with published data. Furthermore, the RBE value did not change, which indicates that the number of DSBs increased in the same manner for the three radiation qualities.

Another important point to underline here is that, as mentioned in the Materials and Methods section, in these simulations only the strand breaks of the DNA were explicitly scored. Nevertheless, as reported in [44,48–51], complex DNA damage that is at the origin of high mutagenic or carcinogenic potential also includes non-DSB oxidative clustered DNA lesions (OCDL). Moreover, as reported in a previous work [48], post-irradiation repair of some sugar and base residues can produce additional strand breaks that can convert some of these non-DSB clustered lesions into DSBs, changing the total number of detected DSBs after irradiation compared to those predicted by the simulation.

From the experiments led at IRSN [30], we observed in Table 3 that the mean number of γ-H2AX foci per nucleus 30 min post irradiation increased with the dose for each photon spectrum. No significant difference was observed in that study between 40 and 220 kVp. However, significant differences were observed between 40 or 220 kVp relative to 4 MV for a dose of 2 Gy, which is in line with the results showed in the simulation. It is also important to note that the number of DSBs per nucleus at 1 Gy obtained using the simulation was always higher than the number of γ-H2AX foci obtained experimentally (Table 3). In addition, as mentioned above, γ-H2AX foci were observed 30 min post irradiation. Between the time of irradiation and observation of the γ-H2AX foci, DSB repair has already been initiated and some foci have already disappeared [31]. The kinetics of foci formation and disappearance may also differ depending on the level of compaction of the chromatin. Falk et al. [31] have even suggested that DSBs occurring in euchromatin could be repaired without causing the formation of γ-H2AX foci, and, even if this assumption is quite questionable, rapid repair of DSBs in euchromatin regions tends to decrease the number of foci detected with respect to the number of DSBs that can be obtained by simulation. Besides, from our simulation results, it can be seen that about 55% of the DSBs occurred in the euchromatin regions and 45% in the heterochromatin ones for the three X-rays energies considered in the present work. It should also be noted that overlapping phenomena can occur between different foci during microscopy detection [52]. In the end, it appears that the number of foci is experimentally underestimated [53]. We can see in Table 3 that the number of DSBs simulated per nucleus at 2.5 ns seemed to be close to the number of γ-H2AX foci. The increase in the number of simulated DSBs for a chemical simulation end-time of 10 ns showed that this time would be better adapted in the simulation for the calculation of DSBs. Despite difficulties in the direct comparison of the absolute number of simulated DSBs and experimental γ-H2AX foci, we can observe that the relative comparison of the number of DSBs between the three X-rays energies is in good agreement with that of γ-H2AX foci. Our RBE value of 1.3 between 40/220 kVp and 4 MV is similar to that in other works [12,54] where the RBE values for DSB induction were 1.15 and 1.1 for 29 kVp X-rays and 125 kVp X-rays with respect to ^{60}Co. Nevertheless, an RBE value for DSB induction of 1.1 with respect to ^{60}Co was reported in this study, which is in line with our 1.3 RBE value when comparing 220 kVp with 4 MV X-rays. In general, RBE values between 40–250 kVp X-rays and higher energies, e.g., ^{60}Co, have been found to be in the range ~1.12–1.53 [55–58].

Concerning DSB complexity as defined in the Materials and Methods section, i.e., only taking into account the number of strand breaks included in the cluster, DSB complexity was found to be similar between the considered X-ray energies. Indeed, as can be seen in Table 2, the proportions of simple and complex DSB were about 86% and 14% for all the radiation qualities. In Liang et al. [21] it was reported that DSB complexity increases with decreasing photon energy with a maximum of around 1 keV. Nevertheless, it was also mentioned that DSB complexity was similar for soft and conventional X-rays as well as for ^{60}Co γ-rays when base damage was not considered. However, DSB complexity presented a larger difference when base damage simulated in the chemical stage was taken into account. In that sense, as has been pointed out above, an important improvement in the computation chain will consist of being able to better simulate and consider base damage as well as different oxygen concentrations in the cell nucleus medium. These improvements are being currently developed by the Geant4-DNA collaboration and will be available in future public releases. These improvements are needed in order to simulate not only DSBs but also non-DSB clustered damage in more realistic cellular media and will allow for extending simulations to repair processes in a correct manner. With the same objective, the use of a new standard to record DNA damage (standard DNA damage) [59] which has been recently proposed can also simplify inter-code comparisons of DNA damage induction and facilitate simulation extension to chromosome aberration computations or other late effects.

4. Materials and Methods

4.1. Simulation of Experimental Cell Irradiations

4.1.1. Experimental Data on γ-H2AX Foci

Primary HUVECs from the Lonza Group (ref. C2519A, lot0000394986) and isolated by Lonza from human tissue (from two females and two males) were donated after permission and obtained for their use in research applications by informed consent or legal authorization. HUVECs were handled as described in [30]. Briefly, G0/G1-phase synchronized HUVECs were obtained by contact inhibition induced in confluent culture. Synchronized cells were seeded 5 h prior to irradiation at a density of 30,000 cells/cm^2 on plastic dishes (1-well Permanox® in Nunc® Lab-Tek® chamber slide systems; Thermo Fisher Scientific, Waltham, Massachusetts, USA) and incubated at 37 °C. Then, cells were irradiated using two irradiation facilities at IRSN, Fontenay aux Roses, France; these were a small animal radiation research platform (SARRP, XSTRAHL Ltd., Camberley, England) and a medical linear accelerator (Elekta Synergy®, Stockholm, Sweden). With the SARRP, the irradiations were performed with 40 kVp X-rays and 220 kVp X-rays while the medical linear accelerator was used to deliver 4 MV X-rays, as detailed elsewhere [30]. Both simulated setups are described hereafter. After exposure to X-rays, HUVECs were immunostained for the in situ detection of phosphorylation of the serine 139 of γ-H2AX, and DNA was stained with 4′,6-diamidino-2-phenylindole (DAPI). A detailed process for this has been previously described in [30]. In addition to the synchronization of cells in the G0/G1 phase prior to radiation exposure, the average number of γ-H2AX foci per endothelial cell nucleus was determined specifically in the nucleus in G0/G1 using the methodology described in [1]. Briefly, thousands of images were acquired on an inverted Olympus IX81 fluorescence microscope with a UPLSAPO 100XO oil immersion objective (Olympus, Tokyo, Japan). For each channel, images were acquired as 3 z-stack layers around and including the focus plane with a step size of 0.5 µm between planes. The images of the 3D stack were projected to 2D xy images using maximum intensity projection. Image analysis was then performed with ScanR analysis software (Olympus, version 2.8.1). An edge segmentation algorithm was used to detect nuclei and γ-H2AX foci. A first selection of relevant nuclei was based on their area and circularity. This step allowed us to consider only isolated nuclei by removing from the analysis objects corresponding to nucleus clusters. To isolate nuclei in the G0/G1 phase, a second level of selection was based on the DNA content of each nucleus (related to the integrated intensity levels of DAPI fluorescence) and the whole nucleus level of γ-H2AX (associated with the S-phase of the cell cycle). The average number of γ-H2AX foci per endothelial cell nucleus was measured for doses of 0.25, 0.5, 1, and 2 Gy 30 min after irradiation. These doses were selected to remain in the range of values where the fluorescence technique showed linearity with the dose and where the intrinsic noise of foci in G0/G1 cells remained low compared to the radio-induced signal. On the other hand, only experimental data at 30 min post irradiation were compared here to simulated data. Indeed, it is assumed that the number of detectable γ-H2AX foci reach a maximum at 30 min after irradiation [1]. Therefore, and as the simulation tool did not take into consideration any repair process, we focused on the post-irradiation time which maximizes the number of detectable DNA damage on the basis of a γ-H2AX foci methodology. The number of γ-H2AX foci was obtained from at least three identical experiments (replicates) for each configuration with approximately 4000 cells analyzed each time [30].

4.1.2. SARRP Configuration

As mentioned in our previous work [30], the SARRP configuration makes use of an inherent filtration of 0.8 mm of beryllium and an additional filtration of 1 mm of aluminum, which results in the suppression of low-energy photons. The initial photon energy spectra after inherent and additional filtrations corresponding to 40 and 220 kVp X-rays used in this work are shown in Figure 5. The spectra were calculated using SpekCalc software [60–62]. The half-value layer (HVL) values of aluminum were 0.852 mm and 5.420 mm [63] and the resulting mean photon energies were 25.6 keV and 70.2 keV for 40 kVp and 220 kVp, respectively. The dose rate used was about 1 Gy.min^{-1} expressed in terms of Kerma in air at 30.5 cm from the source.

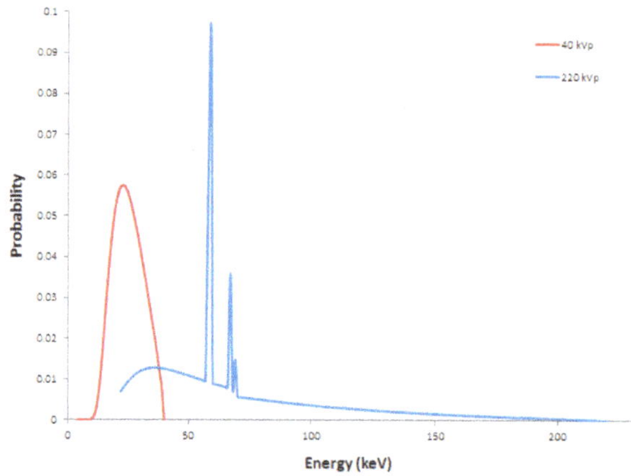

Figure 5. Photon energy spectra after inherent and additional filtrations for 40 kVp (in red) and 220 kVp (in blue) X-rays corresponding to small animal radiation research platform (SARRP) irradiations. Spectra were obtained with SpekCalc software [60–62].

The cell culture chamber (1-well Permanox® in Nunc® Lab-Tek® chamber slide systems; Thermo Fisher Scientific, Waltham, Massachusetts, USA) was modeled by a polystyrene volume (1.06 g.cm^{-3}) with a 3 mm height and a liquid water volume of 5 µm in height representing the cell layer, followed by a liquid water volume of 3 mm in height representing the cell culture medium, as depicted in Figure 6a. The dimensions in the plane of the cell culture (x–y plane) of each layer were 125 mm × 85 mm. In the SARRP configuration, photons were generated in a parallel beam emitted from above and directed perpendicular to the cell monolayer.

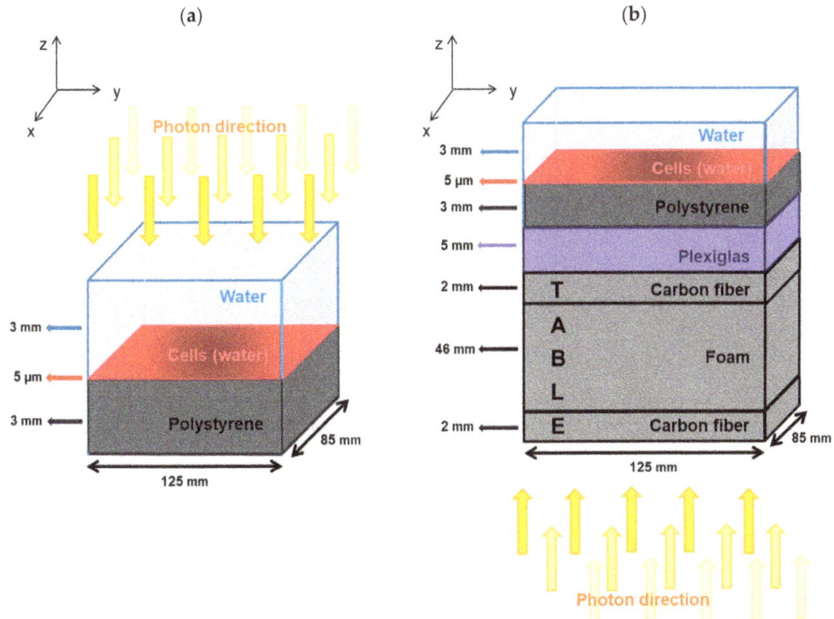

Figure 6. (**a**) Setup used for simulation for SARRP irradiation. The cell culture chamber was modeled by three layers of material (from top to bottom): 3 mm of liquid water representing the cell culture medium, 5 µm of liquid water defining the cell layer, and 3 mm of polystyrene modeling the chamber material. The dimensions in the x–y plane of each layer were 125 mm × 85 mm. Photons were generated in a parallel beam emitted from above and directed towards the cell monolayer. (**b**) Setup for simulation of medical linear accelerator irradiation. The cell chamber (same as the one used in the SARRP simulation) was placed on a table made of carbon fiber and foam (the carbon fiber was 4 mm high and the foam was 46 mm high). Between the cell chamber and the table, a layer of 5 mm in height of Plexiglas® was placed to ensure an electronic equilibrium. Photons were generated in a parallel beam from below the table towards the cell chamber.

4.1.3. Medical Linear Accelerator Configuration

For the medical linear accelerator simulation, a chamber containing cells (same as the one used in the SARRP configuration) was placed on a table made of a layer of carbon fiber (0.55 g.cm^{-3}, 2 mm in height), then a layer of plastic foam (0.03 g.cm^{-3}, 46 mm in height), and another layer of carbon fiber for the irradiation (2 mm in height). An additional layer of Plexiglas® (1.19 g.cm^{-3}, 5 mm in height) was placed between the cell chamber and the table to ensure electronic equilibrium in the cell layer. The dimensions in the x–y plane of each layer were set to 125 mm × 85 mm. In this configuration, photons were generated in a parallel beam from below the table towards the cell chamber, as represented in Figure 6b. The dose rate measured with an ionizing chamber calibrated in free-air Kerma was about 1 Gy.min^{-1} at a distance of 120 cm from the source and had a 30 cm × 30 cm irradiation field [30]. Concerning the source description, the 4 MV X-ray energy spectrum after the collimator used in the simulation is represented in Figure 7 and was taken from Sheikh-Bagheri et al. [64], leading to a mean photon energy of 1.3 MeV and a beam quality of 0.626 in terms of percentage depth dose at 10 cm in liquid water.

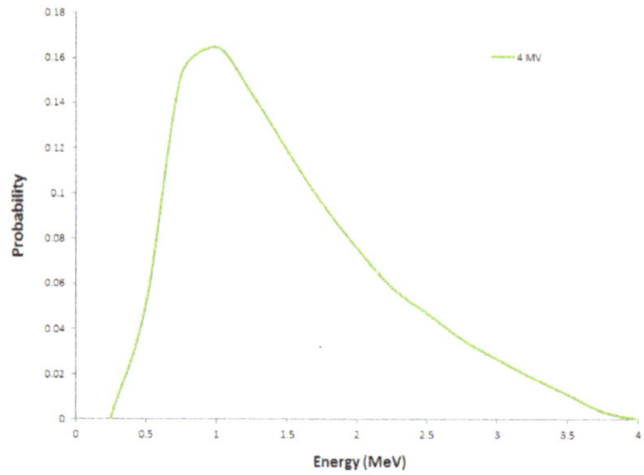

Figure 7. Photon energy spectrum for 4 MV X-rays [63].

4.2. *Phase Space of Secondary Electrons*

The simulations in this work were constructed in two steps. The first one consisted of simulating the photon source at the macroscopic level as represented in Figure 6. The aim of this stage was to score a phase space in the 5 mm high cell layer, recording the position, momentum, and energy of every secondary electron reaching the cell layer. For this purpose, the Livermore low energy physics library of Geant4 was used to simulate photons and electrons. Electron transport was simulated down to a minimum energy of 100 eV by default. In addition, the Auger process was activated as its effect could have an impact on DSB damage, although this was not major [19]. This first stage is a crucial step in simulating radiation-induced DNA damage by photons, given the fact that the secondary electron spectrum will determine the energy depositions in the cell layer or cell nuclei [54].

The second part of the simulation corresponded to micro- or nanodosimetric simulation in the irradiated cells. The electrons that were scored during the first part were used as a particle source.

4.3. *Microdosimetric and Nanodosimetric Approaches*

4.3.1. Microdosimetric Calculations in the Cell Population

The stochastic nature of energy depositions at the cell level leads to a spread of the dose per cell around the macroscopic irradiation dose D. The beam quality can have an effect on the proportion of cells that do not receive exactly the macroscopic dose, which can lead to possible differences in cell fate such as senescence or cell death.

This is why biological effects should be linked to the energy actually deposited in each cell nucleus rather than the macroscopic dose. Analysis of the energy deposition at the corresponding spatial scale can be achieved using microdosimetry, which gives the formalism [65,66] to describe the distribution of energy depositions that occur stochastically. Indeed, energy depositions in cell nuclei result from the combination of the number of tracks traversing the volume of interest as well as the energy deposited per track, with both having a statistical dispersion. From this stochastic property, different values of deposited energy are obtained in each cell nucleus, leading to different specific energies, z. The absorbed dose D corresponds to the ratio between the mean value of the deposited energy in the matter and the mass of the volume where the depositions occur.

Unlike D, z is a stochastic quantity with a probability density f(z). As a matter of fact, D represents the expected value of the specific energy, i.e.,

$$D = \bar{z} = \int z f(z; D) dz \tag{1}$$

The probability density function used to obtain the specific energy z for a given macroscopic dose D in the volume of interest is represented by f(z; D), i.e.,

$$f(z;D) = \sum_{\nu=0}^{\infty} p(\nu).f_{\nu}(z) \tag{2}$$

where

$$p(\nu) = e^{-n}\frac{n^{\nu}}{\nu!} \tag{3}$$

The number of tracks ν leading to energy depositions in the target volume follows a Poisson distribution with mean value n, with n being the number of tracks going through the target volume.

It is therefore appropriate to consider the width of the distribution f(z) because it represents a variability of the energy absorbed by the irradiated cell population.

The relative standard deviation is equal to

$$\sigma_z^{rel} = \frac{\sigma_z}{D} \tag{4}$$

where σ_z is the standard deviation of the distribution f(z).

In this work, microdosimetric simulations were computed using the Geant4 Monte Carlo toolkit version 10.2 as well as its extension, Geant4-DNA. In particular, the G4EmDNAPhysics_option2 physics constructor from the Geant4-DNA extension was used which enables electron transport from 1 MeV down to 7.4 eV.

The cell layer dimensions were reduced and modeled by a 4 mm × 4 mm × 5 μm volume (x, y, and z respectively). The x–y coordinates of the electrons from the first part of the simulation that were outside 4 mm × 4 mm were uniformly sampled in order to be always contained in the cell layer volume. In this volume, 10,000 phantoms of endothelial cell nuclei were placed. Each phantom was modeled as an elliptical cylinder (semi-major axis = 9.5 μm, semi-minor axis = 5.1 μm, and height = 2.0 μm) with a total volume of 304.4 μm^3 made of liquid water (without DNA constituents). These cell nucleus dimensions corresponded to the mean values of the irradiated cells as determined from the biological experiments. Individual energy depositions in each cell nucleus were scored and cumulated to derive the distribution f (z; D). The mean and the standard deviation σ_z of the specific energy were then calculated. To compute the σ_z^{rel} value (in %), the standard deviation σ_z was divided by the mean specific energy, which corresponded to the macroscopic dose D (Equation (4)).

To do this, during the simulation of the electron transport in the cell layer modeled by 10,000 endothelial cell nuclei, several quantities were recorded: the "event identifier", which corresponded to a track generated by an electron, the nucleus identifier in the cell population, and the energy deposition for each electron inelastic interaction. From this data, it was then possible to calculate f (z; D) and the statistical spread of the specific energy z. To compute f (z; D), all energy depositions in each of the cell nuclei were cumulated. The initial macroscopic dose was 0.25 Gy. Then, the specific energy was cumulated by pairs of different cell nuclei relatively far enough to ensure superimposing independent tracks in order to get the f (z; D) distribution and the σ_z^{rel} value for higher doses (0.5 Gy, 1 Gy, and 2 Gy) without having to run additional and time-consuming simulations. However, this method resulted in decreasing the number of cell nuclei available in the cell population at each cumulating step and therefore could not be applied for higher doses.

4.3.2. Nanodosimetric Simulations for the Calculation of Double-Strand Breaks

In this work, nanodosimetric calculations were also performed to compute DNA damage yields in order to study the possible difference in RBE in terms of DSB induction and characteristics between the three photon energy spectra. Then, the simulated results were compared to the experimental data obtained at IRSN, Fontenay aux Roses, France.

Simulation Configuration

In this part of the study we focused on a single endothelial cell nucleus model in the G0/G1 phase of the cell cycle containing the geometry representing ~6 Gbp that had been generated with DnaFabric software [45]. This cell nucleus model was filled with DNA with the proportions (derived from experimental measurements on endothelial cells) 52% euchromatin and 48% heterochromatin distributed with a uniform spatial distribution as described in our previous work [46].

For the irradiations at 40 and 220 kVp, 10,000 electrons taken from the phase space were generated in order to be used as the source, whereas 50,000 electrons were generated for 4 MV in order to always maintain statistical uncertainties (root mean square error (RMSE)) below 6% of the mean number of DSB per track. The electron properties (position in z, momentum, and kinetic energy) were derived from the first stage of the simulation as explained in Section 4.2. Coordinates in the x–y plane were restricted to the cell nucleus dimensions.

Computation Chain to Score Strand Breaks

A computation chain [35] using the Geant4-DNA extension of the Geant4 toolkit (version 10.1) was used to perform simulations of physical, physico-chemical, and chemical stages within the DNA structure.

In brief, direct damage to the backbone of the DNA structure (direct simple strand breaks (SSB)) is scored from the addition of the energy deposited by inelastic collisions within the 2-deoxyribose, phosphate, and hydration shell volumes of the same nucleotide. If the cumulated energy is higher than a threshold value of 17.5 eV [67–69], a direct SSB is registered at that location of the DNA molecule.

After simulation of the physical interactions (inelastic and elastic collisions) of the electrons in the endothelial cells, simulation of the physico-chemical stage transforms the ionized and excited water molecules surrounding the DNA structure into water radicals, as described in Karamitros et al. [70,71].

In the simulation of the chemical stage, the seven types of water radicals ($H^{\bullet}$, $OH^{\bullet}$, H_2, H_2O_2, OH^-, H_3O^+, and solvated e^-) are taken into account and interact chemically with each other or with DNA molecules (deoxyribose, phosphate, base, and histone volumes) according to the list of possible reactions given by default in the code as shown in Table 4. In our simulation all these interactions were taken into account and indirect damage to the backbone was scored when an $OH^{\bullet}$ radical interacted with a deoxyribose molecule of the target in 40% of the cases [72,73]. Although reactions between some radicals and the DNA bases were also included in these reactions, these were not detailed enough for a correct description of the base damage, the goal being here to achieve a better evaluation of the available radical concentrations for strand breaks production.

As can be seen from Table 4, the simulation considers anoxic conditions since no reaction involving oxygen is considered. This is an unrealistic assumption that should be corrected in further developments of the computation chain that are currently ongoing. Another important feature of the chemical stage simulation is the way scavengers in the cell medium are taken into account. Indeed, the scavenging capacity of histone is simulated by introducing a specific reaction by which all types of radicals arriving to the sphere representing the histone proteins are removed from the simulation. The second important factor is the duration of the chemical stage that prevents radicals far from the DNA structure from being at the origin of indirect effects. This parameter was set by default to 2.5 ns. However, in our recent work [46] introducing the nucleus model filled with heterochromatin and euchromatin, it was suggested that an end-time of 10 ns would be better adapted in order to reproduce the mean distance travelled by $OH^{\bullet}$ radicals in a cell medium after irradiation. Hence, simulations with a chemistry stage simulation end-time of 10 ns were also performed in this work.

Table 4. Reactions and reaction rates used in the simulation [46]. (**a**) default reactions present in the Geant4-DNA chemistry module and (**b**) reactions added in the simulation.

(a)

Reaction	Reaction Rate ($10^{10}\cdot M^{-1}s^{-1}$)
$H^{\bullet} + e^{-}_{aq} + H_2O \rightarrow OH^{-} + H_2$	2.65
$H^{\bullet} + OH^{\bullet} \rightarrow H_2O$	1.44
$H^{\bullet} + H^{\bullet} \rightarrow H_2$	1.20
$H_2 + OH^{\bullet} \rightarrow H^{\bullet} + H_2O$	4.17×10^{-3}
$H_2O_2 + e^{-}_{aq} \rightarrow OH^{-} + OH^{\bullet}$	1.41
$H_3O^{+} + e^{-}_{aq} \rightarrow H^{\bullet} + H_2O$	2.11
$H_3O^{+} + OH^{-} \rightarrow 2H_2O$	14.3
$OH^{\bullet} + e^{-}_{aq} \rightarrow OH^{\bullet}$	2.95
$OH^{\bullet} + OH^{\bullet} \rightarrow H_2O_2$	0.44
$e^{-}_{aq} + e^{-}_{aq} + 2H_2O \rightarrow 2OH^{-} + H_2$	0.50

(b)

Reaction	Reaction Rate ($10^{9}\cdot M^{-1}s^{-1}$)
2-deoxyribose + $OH^{\bullet}$	1.8
Adenine + $OH^{\bullet}$	6.1
Guanine + $OH^{\bullet}$	9.2
Thymine + $OH^{\bullet}$	6.4
Cytosine + $OH^{\bullet}$	6.1
2-deoxyribose + e^{-}_{aq}	0.01
Adenine + e^{-}_{aq}	9.0
Guanine + e^{-}_{aq}	14.0
Thymine + e^{-}_{aq}	18.0
Cytosine + e^{-}_{aq}	13.0
2-deoxyribose + $H^{\bullet}$	0.029
Adenine + $H^{\bullet}$	0.10
Thymine + $H^{\bullet}$	0.57
Cytosine + $H^{\bullet}$	0.092

Definition of a DSB and DSB Complexity

In this work, a focus was placed on the calculation of simple and double-strand breaks of the DNA backbone and their complexity, which can be more easily correlated to the experimental measurements of foci used for comparison. A double-strand break was defined as at least two strand breaks separated by less than 10 base pairs (bp) and with one of them in an opposite strand to the other(s). Concerning definitions of DSB complexity, different authors have used different definitions [48,74]. As in previous work by our group, the DBScan [75] clustering algorithm was used in order to look at the relative position of all the strand breaks produced in the simulation either in the physical or chemical stage and to calculate the DSBs. Using this algorithm, any other SB separated by less than 10 bp from the initial DSB was merged to create a complex DSB. The DSB complexity was thus given by the number of SBs inside a complex DSB. Our algorithm is therefore not intended to provide non-DSB clusters including other DNA lesions such as base damage.

5. Conclusions

In this work, microdosimetric calculations illustrated the distributions of specific energy in an irradiated cell population, and, thus, that the microdosimetric spread σ_z^{rel} was similar when considering 40 and 220 kVp but different when looking at 4 MV. With regard to the nanodosimetric aspect, DSB induction in the form of DSB/Gy/Gbp was very similar among 40 kVp and 220 kVp X-rays and higher in comparison with 4 MV by a factor of 1.3, as could be expected from their calculated secondary electron spectra. Nevertheless, DSB complexity was relatively similar. The simulation results presented in this study confirmed results observed in previous experiments led by IRSN radiobiologists, who found no significant differences between 40 kVp and 220 kVp but a higher number of γ-H2AX foci from these radiation qualities compared to 4 MV X-rays for a dose of 2 Gy. From the simulation results, these observations could be explained by the differences in the proportion of low-energy electrons (≤10 keV) between the three X-rays energies.

Author Contributions: N.T. developed the simulation, analyzed the results, and drafted the manuscript based on discussions with C.V., S.I., Y.P., and H.N.T.; C.V., Y.P., and S.I. supervised the work and contributed to the drafting of the text; C.V., M.B., and S.M. contributed to the simulation design and analysis of the results; S.I., C.V., Y.P., H.N.T., M.A.B., M. C.B., D.E., Z.F., S.C., V.I., M.K., I.K., and W. G.S. contributed to the developments of Geant4-DNA needed in this work; and A.F., M.D.S., A.V., and G.G. performed the radiobiological experiments and contributed to the analysis.

Funding: M.A. Bernal thanks the Conselho Nacional para o Desenvolvimento Científico e Tecnológico (CNPq), Brazil, for financing his research activities through the project 306298/2018-0. He also acknowledges the support received from the FAPESP foundation in Brazil through the 2011/51594-2, 2015/21873-8, and 2018/15316-7 projects.

Acknowledgments: M.A. Bernal thanks the Conselho Nacional para o Desenvolvimento Científico e Tecnológico (CNPq), Brazil, for financing his research activities through the project 306298/2018-0. He also acknowledges the support received from the FAPESP foundation in Brazil through the 2011/51594-2, 2015/21873-8, and 2018/15316-7 projects.

Conflicts of Interest: The authors declare no conflict of interest.

References

1. Gruel, G.; Villagrasa, C.; Voisin, P.; Clairand, I.; Benderitter, M.; Bottollier-Depois, J.F.; Barquinero, J.F. Cell to Cell Variability of Radiation-Induced Foci: Relation between Observed Damage and Energy Deposition. *PLoS ONE* **2016**, *11*, e0145786. [CrossRef] [PubMed]
2. Hunter, N.; Muirhead, C.R. Review of relative biological effectiveness dependence on linear energy transfer for low-LET radiations. *J. Radiol. Prot.* **2009**, *29*, 5–21. [CrossRef] [PubMed]
3. Friedland, W.; Jacob, P.; Paretzke, H.G.; Stork, T. Monte Carlo Simulation of the Production of Short DNA Fragments by Low-Linear Energy Transfer Radiation Using Higher-Order DNA Models. *Radiat. Res.* **1998**, *150*, 170–182. [CrossRef] [PubMed]
4. Friedland, W.; Jacob, P.; Paretzke, H.G.; Merzagora, M.; Ottolenghi, A. Simulation of DNA fragment distributions after irradiation with photons. *Radiat Environ. Biophys.* **1999**, *38*, 39–47. [CrossRef] [PubMed]
5. Friedland, W.; Bernhardt, P.; Jacob, P.G.; Paretzke, H. Simulation of DNA Damage after Proton and Low LET Irradiation. *Radiat. Prot. Dosim.* **2002**, *99*, 99–102. [CrossRef]
6. Göggelmann, W.; Jacobsen, C.; Panzer, W.; Walsh, L.; Roos, H.; Schmid, E. Re-evaluation of the RBE of 29 kV X-rays (mammography X-rays) relative to 220 kV X-rays using neoplastic transformation of human CGL1-hybrid cells. *Radiat Environ. Biophys.* **2003**, *42*, 175–182. [CrossRef] [PubMed]
7. Brenner, D.J.; Amols, H.I. Enhanced risk from low-energy screen—Film mammography X rays. *BJR* **1989**, *62*, 910–914. [CrossRef]
8. Frankenberg, D.; Brede, H.J.; Schrewe, U.J.; Steinmetz, C.; Frankenberg-Schwager, M.; Kasten, G.; Pralle, E. Induction of DNA double-strand breaks in mammalian cells and yeast. *Adv. Space Res.* **2000**, *25*, 2085–2094. [CrossRef]
9. Frankenberg, D.; Kelnhofer, K.; Bär, K.; Frankenberg-Schwager, M. Enhanced Neoplastic Transformation by Mammography X Rays Relative to 200 kVp X Rays: Indication for a Strong Dependence on Photon Energy of the RBEM for Various End Points. *Rare* **2002**, *157*, 99–106. [CrossRef]
10. Lehnert, A.; Lessmann, E.; Pawelke, J.; Dörr, W. RBE of 25 kV X-rays for the survival and induction of micronuclei in the human mammary epithelial cell line MCF-12A. *Radiat Environ. Biophys.* **2006**, *45*, 253–260. [CrossRef]
11. Lehnert, A.; Lessmann, E.; Pawelke, J.; Dörr, W. RBE of 10 kV X rays determined for the human mammary epithelial cell line MCF-12A. *Radiat Res.* **2008**, *169*, 330–336. [CrossRef] [PubMed]
12. Kühne, M.; Urban, G.; Frankenberg, D.; Löbrich, M. DNA Double-Strand Break Misrejoining after Exposure of Primary Human Fibroblasts to CK Characteristic X Rays, 29 kVp X Rays and 60Co γ Rays. *Rare* **2005**, *164*, 669–677. [CrossRef]
13. Mestres, M.; Caballín, M.R.; Barrios, L.; Ribas, M.; Barquinero, J.F. RBE of X Rays of Different Energies: A Cytogenetic Evaluation by FISH. *Radiat. Res.* **2008**, *170*, 93–100. [CrossRef] [PubMed]

14. Mestres, M.; Benkhaled, L.; Caballín, M.R.; Barrios, L.; Ribas, M.; Barquinero, J.F. Induction of Incomplete and Complex Chromosome Aberrations by 30 kVp X Rays. *Rare* **2010**, *175*, 201–208. [CrossRef]
15. Schmid, E.; Regulla, D.; Kramer, H.M.; Harder, D. The Effect of 29 kV X Rays on the Dose Response of Chromosome Aberrations in Human Lymphocytes. *Radiat. Res.* **2002**, *158*, 771–777. [CrossRef]
16. Schmid, E.; Krumrey, M.; Ulm, G.; Roos, H.; Regulla, D. The Maximum Low-Dose RBE of 17.4 and 40 keV Monochromatic X Rays for the Induction of Dicentric Chromosomes in Human Peripheral Lymphocytes. *Radiat. Res.* **2003**, *160*, 499–504. [CrossRef]
17. Goodhead, D.T.; Nikjoo, H. Current Status of Ultrasoft X Rays and Track Structure Analysis as Tools for Testing and Developing Biophysical Models of Radiation Action. *Radiat. Prot. Dosim.* **1990**, *31*, 343–350. [CrossRef]
18. Prise, K.M. A review of dsb induction data for varying quality radiations. *Int. J. Radiat. Biol.* **1998**, *74*, 173–184. [CrossRef]
19. Kellerer, A.M. Electron Spectra and the RBE of X Rays. *Rare* **2002**, *158*, 13–23. [CrossRef]
20. Nikjoo, H.; Lindborg, L. RBE of low energy electrons and photons. *Phys. Med. Biol.* **2010**, *55*, R65–R109. [CrossRef]
21. Liang, Y.; Fu, Q.; Wang, X.; Liu, F.; Yang, G.; Luo, C.; Ouyang, Q.; Wang, Y. Relative biological effectiveness for photons: Implication of complex DNA double-strand breaks as critical lesions. *Phys. Med. Biol.* **2017**, *62*, 2153–2175. [CrossRef] [PubMed]
22. Buch, T.; Scifoni, E.; Krämer, M.; Durante, M.; Scholz, M.; Friedrich, T. Modeling Radiation Effects of Ultrasoft X Rays on the Basis of Amorphous Track Structure. *Radiat Res.* **2018**, *189*, 32–43. [CrossRef] [PubMed]
23. Hill, M.A. The variation in biological effectiveness of X-rays and gamma rays with energy. *Radiat. Prot. Dosim.* **2004**, *112*, 471–481. [CrossRef] [PubMed]
24. Friedrich, T.; Durante, M.; Scholz, M. Simulation of DSB yield for high LET radiation. *Radiat. Prot. Dosim.* **2015**, *166*, 61–65. [CrossRef]
25. Prise, K.M.; Pinto, M.; Newman, H.C.; Michael, B.D. A Review of Studies of Ionizing Radiation-Induced Double-Strand Break Clustering. *Rare* **2001**, *156*, 572–577. [CrossRef]
26. Davis, A.J.; Chen, D.J. DNA double strand break repair via non-homologous end-joining. *Transl. Cancer Res.* **2013**, *2*, 130–143. [CrossRef]
27. Ballarini, F.; Carante, M.P. Chromosome aberrations and cell death by ionizing radiation: Evolution of a biophysical model. *Radiat. Phys. Chem.* **2016**, *128*, 18–25. [CrossRef]
28. Elsässer, T.; Scholz, M. Cluster effects within the local effect model. *Radiat. Res.* **2007**, *167*, 319–329. [CrossRef]
29. Bernal, M.A.; de Almeida, C.E.; David, M.; Pires, E. Estimation of the RBE of mammography-quality beams using a combination of a Monte Carlo code with a B-DNA geometrical model. *Phys. Med. Biol.* **2011**, *56*, 7393–7403. [CrossRef]
30. Freneau, A.; Dos Santos, M.; Voisin, P.; Tang, N.; Bueno Vizcarra, M.; Villagrasa, C.; Roy, L.; Vaurijoux, A.; Gruel, G. Relation between DNA double-strand breaks and energy spectra of secondary electrons produced by different X-ray energies. *Int. J. Radiat. Biol.* **2018**, *9*, 1–10. [CrossRef]
31. Falk, M.; Lukasova, E.; Kozubek, S. Higher-order chromatin structure in DSB induction, repair and misrepair. *Mutat. Res.* **2010**, *704*, 88–100. [CrossRef] [PubMed]
32. Rothkamm, K.; Horn, S. gamma-H2AX as protein biomarker for radiation exposure. *Ann. Ist. Super. Sanita* **2009**, *45*, 265–271. [PubMed]
33. Tommasino, F.; Friedrich, T.; Jakob, B.; Meyer, B.; Durante, M.; Scholz, M. Induction and Processing of the Radiation-Induced Gamma-H2AX Signal and Its Link to the Underlying Pattern of DSB: A Combined Experimental and Modelling Study. *PLoS ONE* **2015**, *10*, e0129416. [CrossRef] [PubMed]
34. Friedland, W.; Kundrát, P. Track structure based modelling of chromosome aberrations after photon and alpha-particle irradiation. *Mutat. Res. Genet. Toxicol. Environ. Mutagen.* **2013**, *756*, 213–223. [CrossRef] [PubMed]
35. Meylan, S.; Incerti, S.; Karamitros, M.; Tang, N.; Bueno, M.; Clairand, I.; Villagrasa, C. Villagrasa. Simulation of early DNA damage after the irradiation of a fibroblast cell nucleus using Geant4-DNA. *Sci. Rep.* **2017**, *7*. [CrossRef]
36. Agostinelli, S.; Allison, J.; Amako, K.A.; Apostolakis, J.; Araujo, H.; Arce, P.; Asai, M.D.; Axen, M.; Banerjee, S.; Barrand, G.; et al. Geant4—A simulation toolkit. *Nucl. Instrum. Methods Phys. Res. Sect. A Accel. Spectrom. Detect. Assoc. Equip.* **2003**, *506*, 250–303. [CrossRef]

37. Allison, J.; Amako, K.; Apostolakis, J.E.A.; Araujo, H.A.A.H.; Dubois, P.A.; Asai Barrand, G.; Capra, R.; Chauvie, S.; Chytracek, R.; Cirrone, G.A. Geant4 developments and applications. *IEEE Trans. Nucl. Sci.* **2006**, *53*, 270–278. [CrossRef]
38. Allison, J.; Amako, K.; Apostolakis, J.; Arce, P.; Asai, M.; Aso, T.; Bagli, E.; Bagulya, A.; Banerjee, S.; Barrand, G.; et al. Recent developments in Geant4. *Nucl. Instrum. Methods Phys. Res. Sect. A Accel. Spectrom. Detect. Assoc. Equip.* **2016**, *835*, 186–225. [CrossRef]
39. Incerti, S.; Kyriakou, I.; Bernal, M.A.; Bordage, M.C.; Francis, Z.; Guatelli, S.; Ivanchenko, V.; Karamitros, M.; Lampe, N.; Lee, S.B.; et al. Geant4-DNA example applications for track structure simulations in liquid water: A report from the Geant4-DNA Project. *Med. Phys.* **2018**, *6*. [CrossRef]
40. Bernal, M.A.; Bordage, M.C.; Brown, J.M.C.; Davídková, M.; Delage, E.; El Bitar, Z.; Enger, S.A.; Francis, Z.; Guatelli, S.; Ivanchenko, V.N.; et al. Track structure modeling in liquid water: A review of the Geant4-DNA very low energy extension of the Geant4 Monte Carlo simulation toolkit. *Phys. Med.* **2015**, *31*, 861–874. [CrossRef]
41. Incerti, S.; Ivanchenko, A.; Karamitros, M.; Mantero, A.; Moretto, P.; Tran, H.N.; Mascialino, B.; Champion, C.; Ivanchenko, V.N.; Bernal, M.A.; et al. Comparison of GEANT4 very low energy cross section models with experimental data in water. *Med. Phys.* **2010**, *37*, 4692–4708. [CrossRef] [PubMed]
42. Incerti, S.; Baldacchino, G.; Bernal, M.; Capra, R.; Champion, C.; Francis, Z.; Guèye, A.; Mantero, B.; Mascialino, P.; Moretto, P.; et al. THE GEANT4-DNA PROJECT. *Int. J. Model. Simul. Sci. Comput.* **2010**, *1*, 157–178. [CrossRef]
43. Villegas, F.; Tilly, N.; Ahnesjö, A. Microdosimetric spread for cell-sized targets exposed to 60Co, 192Ir and 125I sources. *Radiat. Prot. Dosim.* **2015**, *166*, 365–368. [CrossRef] [PubMed]
44. Stewart, R.D.; Yu, V.K.; Georgakilas, A.G.; Koumenis, C.; Park, J.H.; Carlson, D.J. Effects of radiation quality and oxygen on clustered DNA lesions and cell death. *Radiat. Res.* **2011**, *176*, 587–602; [CrossRef] [PubMed]
45. Meylan, S.; Vimont, U.; Incerti, S.; Clairand, I.; Villagrasa, C. Geant4-DNA simulations using complex DNA geometries generated by the DnaFabric tool. *Comput. Phys. Commun.* **2016**, *204*, 159–169. [CrossRef]
46. Tang, N.; Bueno, M.; Meylan, S.; Incerti, S.; Tran, H.N.; Vaurijoux, A.; Gruel, G.; Villagrasa, C. Influence of chromatin compaction on simulated early radiation-induced DNA damage using Geant4-DNA. *Med. Phys.* **2019**, *46*, 1501–1511. [CrossRef]
47. Villagrasa, C.; Meylan, S.; Gonon, G.; Gruel, G.; Giesen, U.; Bueno, M.; Rabus, H. Geant4-DNA simulation of DNA damage caused by direct and indirect radiation effects and comparison with biological data. *EPJ. Web. Conf.* **2017**, *153*, 04019. [CrossRef]
48. Georgakilas, A.G.; O'Neill, P.; Stewart, R.D. Induction and repair of clustered DNA lesions: What do we know so far? *Radiat. Res.* **2013**, *180*, 100–109. [CrossRef]
49. Sutherland, B.M.; Bennett, P.V.; Sutherland, J.C.; Laval, J. Clustered DNA damages induced by x rays in human cells. *Radiat. Res.* **2002**, *157*, 611–616. [CrossRef]
50. Hada, M.; Sutherland, B.M. Spectrum of complex DNA damages depends on the incident radiation. *Radiat. Res.* **2006**, *165*, 223–230. [CrossRef]
51. Hada, M.; Georgakilas, A.G. Formation of clustered DNA damage after high-LET irradiation: A review. *J. Radiat. Res.* **2008**, *49*, 203–210. [CrossRef] [PubMed]
52. Barbieri, S.; Baiocco, G.; Babini, G.; Morini, J.; Friedland, W.; Buonanno, M.; Grilj, V.; Brenner, D.J.; Ottolenghi, A. Modelling γ-H2AX foci induction to mimic limitations in the scoring technique. *Radiat. Prot. Dosim.* **2019**, *183*, 121–125. [CrossRef] [PubMed]
53. Gonon, G.; Villagrasa, C.; Voisin, P.; Meylan, S.; Bueno, M.; Benadjaoud, M.A.; Tang, N.; Langner, F.; Rabus, H.; Barquinero, J.-F.; et al. From Energy Deposition of Ionizing Radiation to Cell Damage Signaling: Benchmarking Simulations by Measured Yields of Initial DNA Damage after Ion Microbeam Irradiation. *Radiat. Res.* **2019**, *191*, 566–584. [CrossRef] [PubMed]
54. Kirkby, C.; Ghasroddashti, E.; Poirier, Y.; Tambasco, M.; Stewart, R.D. RBE of kV CBCT radiation determined by Monte Carlo DNA damage simulations. *Phys. Med. Biol.* **2013**, *58*, 5693–5704. [CrossRef] [PubMed]
55. Bistrović, M.; Bisćan, M.; Viculin, T. RBE of 20 kV and 70 kV X-rays determined for survival of V 79 cells. *Radiother. Oncol.* **1986**, *7*, 175–180. [CrossRef]
56. Hoshi, M.; Antoku, S.; Nakamura, N.; Russell, W.J.; Miller, R.C.; Sawada, S.; Mizuno, M.; Nishio, S. Soft X-ray dosimetry and RBE for survival of Chinese hamster V79 cells. *Int. J. Radiat. Biol.* **1988**, *54*, 577–591. [CrossRef]

57. Spadinger, I.; Palcic, B. The relative biological effectiveness of 60Co gamma-rays, 55 kVp X-rays, 250 kVp X-rays, and 11 MeV electrons at low doses. *Int. J. Radiat. Biol.* **1992**, *61*, 345–353. [CrossRef]
58. Sinclair, W.K. The relative biological effectiveness of 22-Mevp X-rays, cobalt-60 gamma rays, and 200-Kvcp X-rays. VII. Summary of studies for five criteria of effect. *Radiat. Res.* **1962**, *16*, 394–398. [CrossRef]
59. Schuemann, J.; McNamara, A.L.; Warmenhoven, J.W.; Henthorn, N.T.; Kirkby, K.J.; Merchant, M.J.; Ingram, S.; Paganetti, H.; Held, K.D.; Ramos-Mendez, J.; et al. A New Standard DNA Damage (SDD) Data Format. *Radiat. Res.* **2019**, *191*, 76–92; [CrossRef]
60. Poludniowski, G.G.; Evans, P.M. Calculation of X-ray spectra emerging from an X-ray tube. Part, I. Electron penetration characteristics in X-ray targets. *Med Phys.* **2007**, *34*, 2164–2174. [CrossRef]
61. Poludniowski, G.G. Calculation of X-ray spectra emerging from an X-ray tube. Part II. X-ray production and filtration in X-ray targets. *Med Phys.* **2007**, *34*, 2175–2186. [CrossRef] [PubMed]
62. Poludniowski, G.; Landry, G.; DeBlois, F.; Evans, P.M.; Verhaegen, F. SpekCalc: A program to calculate photon spectra from tungsten anode X-ray tubes. *Phys. Med. Biol.* **2009**, *54*, N433–N438. [CrossRef] [PubMed]
63. Dos Santos, M.; Paget, V.; Ben Kacem, M.; Trompier, F.; Benadjaoud, M.A.; François, A.; Guipaud, O.; Benderitter, M.; Milliat, F. Importance of dosimetry protocol for cell irradiation on a low X-rays facility and consequences for the biological response. *Int. J. Radiat. Biol.* **2018**, *94*, 597–606. [CrossRef] [PubMed]
64. Sheikh-Bagheri, D.; Rogers, D.W.O. Monte Carlo calculation of nine megavoltage photon beam spectra using the BEAM code. *Med Phys.* **2002**, *29*, 391–402. [CrossRef] [PubMed]
65. Kellerer, A.M.; Chmelevsky, D. Concepts of microdosimetry II. Probability distributions of the microdosimetric variables. *Radiat. Environ. Biophys.* **1975**, *12*, 205–216. [CrossRef]
66. Santa Cruz, G.A. Microdosimetry: Principles and applications. *Rep. Pract. Oncol. Radiother.* **2016**, *21*, 135–139. [CrossRef]
67. Lobachevsky, P.N.; Martin, R.F. DNA strand breakage by 125I-decay in a synthetic oligodeoxynucleotide–2. Quantitative analysis of fragment distribution. *Acta. Oncol.* **1996**, *35*, 809–815. [CrossRef]
68. Prise, K.M.; Folkard, M.; Michael, B.D.; Vojnovic, B.; Brocklehurst, B.; Hopkirk, A.; Munro, I.H. Critical energies for SSB and DSB induction in plasmid DNA by low-energy photons: Action spectra for strand-break induction in plasmid DNA irradiated in vacuum. *Int. J. Radiat. Biol.* **2000**, *76*, 881–890.
69. Nikjoo, H.; O'Neill, P.; Wilson, W.E.; Goodhead, D.T. Computational approach for determining the spectrum of DNA damage induced by ionizing radiation. *Radiat. Res.* **2001**, *156*, 577–583. [CrossRef]
70. Karamitros, M.; Mantero, A.; Incerti, S.; Friedland, W.; Baldacchino, G.; Barberet, P.; Bernal, M.; Capra, R.; Champion, C.; Bitar, Z.E.; et al. Modeling Radiation Chemistry in the Geant4 Toolkit. *Prog. Nucl. Sci. Technol.* **2011**, *2*, 503–508. [CrossRef]
71. Karamitros, M.; Luan, S.; Bernal, M.A.; Allison, J.; Baldacchino, G.; Davidkova, M.; Francis, Z.; Friedland, W.; Ivantchenko, V.; Ivantchenko, A.; et al. Diffusion-controlled reactions modeling in Geant4-DNA. *J. Comput. Phys.* **2014**, *274*, 841–882. [CrossRef]
72. Balasubramanian, B.; Pogozelski, W.K.; Tullius, T.D. DNA strand breaking by the hydroxyl radical is governed by the accessible surface areas of the hydrogen atoms of the DNA backbone. *PNAS* **1998**, *95*, 9738–9743. [CrossRef] [PubMed]
73. Dizdaroglu, M.; Jaruga, P. Mechanisms of free radical-induced damage to DNA. *Free Radic. Res.* **2012**, *46*, 382–419. [CrossRef] [PubMed]
74. Nikjoo, H.; O'Neill, P.; Goodhead, D.T.; Terrissol, M. Computational modelling of low-energy electron-induced DNA damage by early physical and chemical events. *Int. J. Radiat. Biol.* **1997**, *71*, 467–483. [CrossRef]
75. Ester, M.; Kriegel, H.-P.; Sander, J.; Xu, X. A density-based algorithm for discovering clusters in large spatial databases with noise. *KDD* **1996**, *96*, 226−231.

International Journal of
Molecular Sciences

MDPI

Article

Effects of High-Dose Ionizing Radiation in Human Gene Expression: A Meta-Analysis

Dimitrios S. Kanakoglou [1,2,†], Theodora-Dafni Michalettou [1,2,3,†], Christina Vasileiou [1,3], Evangelos Gioukakis [1,3], Dorothea Maneta [1,3], Konstantinos V. Kyriakidis [1,4], Alexandros G. Georgakilas [3] and Ioannis Michalopoulos [1,*]

1 Centre of Systems Biology, Biomedical Research Foundation, Academy of Athens, 115 27 Athens, Greece; kanakogloud@biol.uoa.gr (D.S.K.); daphnelettos@biol.uoa.gr (T.-D.M.); christinevasil@mail.ntua.gr (C.V.); ge13010@central.ntua.gr (E.G.); ge13097@central.ntua.gr (D.M.); kyriakidk@pharm.auth.gr (K.V.K.)
2 Section of Cell Biology and Biophysics, Department of Biology, School of Sciences, National and Kapodistrian University of Athens, 157 01 Athens, Greece
3 DNA Damage Laboratory, Physics Department, School of Applied Mathematical and Physical Sciences, National Technical University of Athens, 157 80 Athens, Greece; alexg@mail.ntua.gr
4 Laboratory of Pharmacology, School of Pharmacy, Aristotle University of Thessaloniki, 541 24 Thessaloniki, Greece
* Correspondence: imichalop@bioacademy.gr; Tel.: +30-210-659-7127
† These authors contributed equally to this work.

Received: 13 February 2020; Accepted: 9 March 2020; Published: 12 March 2020

Abstract: The use of high-dose Ionizing Radiation (IR) is currently one of the most common modalities in treatment of many types of cancer. The objective of this work was to investigate the effects of high-dose ionizing radiation on healthy human tissue, utilizing quantitative analysis of gene expression. To this end, publicly available transcriptomics datasets from human samples irradiated with a high dose of radiation and non-irradiated (control) ones were selected, and gene expression was determined using RNA-Seq data analysis. Raw data from these studies were subjected to quality control and trimming. Mapping of RNA-Seq reads was performed by the partial selective alignment method, and differential gene expression analysis was conducted. Subsequently, a meta-analysis was performed to select differentially expressed genes across datasets. Based on the differentially expressed genes discovered by meta-analysis, we constructed a protein-to-protein interaction network, and we identified biological pathways and processes related to high-dose IR effects. Our findings suggest that cell cycle arrest is activated, supported by our top down-regulated genes associated with cell cycle activation. DNA repair genes are down-regulated in their majority. However, several genes implicated in the nucleotide excision repair pathway are upregulated. Nevertheless, apoptotic mechanisms seem to be activated probably due to severe high-dose-induced complex DNA damage. The significant upregulation of CDKN1A, as a downstream gene of TP53, further validates programmed cell death. Finally, down-regulation of TIMELESS, signifies a correlation between IR response and circadian rhythm. Nonetheless, high-dose IR exposure effects regarding normal tissue (radiation toxicity) and its possible long-term outcomes should be studied to a greater extend.

Keywords: high-dose ionizing radiation; RNA-Seq; differential gene expression; DNA damage response

1. Introduction

Regarding human exposure to Ionizing Radiation (IR), doses below 0.1Gy are classified as "low" [1], while doses normally used in medical procedures, such as Radiation Therapy (RT) (2-3Gy) are classified as high [2,3]. As such, apart from understanding the biological consequences of rare very high-dose exposures like in the case of nuclear accidents, a major field of radiobiological and clinical interest is the optimization of RT which usually involves moderate to high fraction doses.

The main therapeutic modality employed in RT are photon beams in the form of low Linear Energy Transfer (LET) radiation (X-rays, Gamma-rays), although high LET radiation (protons, alpha particles, and other heavy ions) are sometimes incorporated due to their precise dose localization. Radiation particles deposit more energy on the targeted tumor areas (a phenomenon known as the "Bragg peak" [4]) and have a higher Relative Biological Effectiveness (RBE) [5–7], while photon beams deposit a relatively small quantity of energy that disperses further to the surrounding healthy tissue due to scattering phenomena. In general, the idea behind RT is that the rapidly proliferating cancer cells are usually more sensitive to radiation than normal cells, while normal cells can usually repair themselves at a faster rate and retain their normal function. Therefore, the goal is to inhibit cancer cell multiplication potential, eventually leading to cell death, while minimizing dosage absorption in normal tissue, to prevent toxicity [5,8]. Nonetheless, IR exposure effects regarding healthy tissue (radiation toxicity) and its possible long-term outcomes should be studied to a greater extend.

Cancer and healthy cells are targeted alike, either directly through damage on their cellular molecules and especially on DNA strands or indirectly via the formation of free radicals, a phenomenon referred to as oxidative stress [9]. In essence, oxidative stress is a procedure of water radiolysis and involves the formation of intermediate, partially reduced oxygen species, collectively termed as Reactive Oxygen Species (ROS), that give rise to the formation of hydroxyl radicals that produce a number of adverse biological reactions by attacking structural and functional molecules [10], thus resulting in generalized cellular stress. Hydroxyl radicals can sometimes indirectly produce Single-Strand Breaks (SSBs) and a plethora of base and sugar lesions in DNA molecules, which can be cytotoxic or mutagenic [11], as well as crosslinks between two complementary DNA strands [12]. On the other hand, direct DNA damage primarily involves the induction of Double-Strand Breaks (DSBs) that represent the most lethal types of DNA damage, leading to cell death or genomic instability if left unrepaired. Finally, closely spaced DNA lesions (referred to as complex or clustered DNA damage) that may occur after IR exposure have been suggested to be highly repair-resistant or non-repairable. Therefore, they are considered highly significant biological lesions [13]. This continuously challenging process may lead to genomic instability and cancer [14], concurrently fueling DNA Damage Response (DDR) activation [15] which constitutes the main component of IR effects on a cellular level.

In general, DDR can be defined as the synthesis of functions (sensors, transducers, effectors) that orchestrate DNA damage sensing and signal transduction, triggering either DNA repair, cell survival, or cell death (apoptosis). Furthermore, pathways of cell cycle checkpoint control are also essential components of DDR [16,17]. Main pathways of DNA repair include Base Excision Repair (BER) and Nucleotide Excision Repair (NER), which repair DNA base damages, and Mismatch Repair (MMR), which corrects base mispairs and small loops that are often found in repetitive sequence DNA. In addition, Homology-dependent Recombination (HR) and Non-Homologous End Joining (NHEJ) act alone or together to repair DSBs and complex events such as inter-strand crosslinks [18,19]. Dysregulation of DDR mechanisms can cause several human disorders that are associated with cancer susceptibility, accelerated aging, and developmental abnormalities [20]. Moreover, like other types of stress, radiation exposure affects the development of the immune system through radiation-induced apoptosis, differentiation, and induction of inflammatory environment via different components of DDR [21,22].

The complete pattern of biological responses to different doses and radiation types is unclear and currently one of the most important questions in radiation biology. The general consensus is that results of IR exposure in any living organism involve a topical and/or systemic stress. A variety of responses is induced, including—but not limited to—oxidative stress in the irradiated area or in the whole body (through systemic non-targeted effects), DDR, DNA repair, and pro-inflammatory pathway initiation [23]. From a systems biology perspective, the aforementioned cellular mechanisms, as well as other related ones, can be examined through altered gene expression. Thus, in this work, we performed a Differential Gene Expression Analysis (DGEA), in human tissues exposed to high-dose IR, taking advantage of the wealth of publicly available RNA-Sequencing (RNA-Seq) data, as previously suggested [24–26]. To this end, we carefully selected five datasets of healthy human cell samples, and in each of them, we identified Differentially Expressed Genes (DEGs) between irradiated and non-irradiated cells. Finally, we performed a meta-analysis, highlighting the common ground of high-dose effects.

2. Results

2.1. Data Collection, Filtering, Pre-Processing and Mapping

ENA queries identified 71 projects that fulfilled our search criteria. After manual curation, the Bioproject [27] accession numbers of the selected datasets (Table 1) were PRJNA494581 [28], PRJNA450083 [29], PRJNA421022 [30], PRJNA436999 [31], and PRJNA396832. We downloaded RNA-Seq-related FASTQ files of those Bioprojects. After the initial quality control, we performed soft trimming on raw RNA-Seq data, choosing PHRED score Q = 20 [32,33], removing on average ~2.4% of the nucleotide reads with more than 1% probability of an incorrect base call. PRJNA396832 trimming rate was 7.7% with soft trimming, due to adapter contamination and poor quality of reads.

Table 1. Information of experiment accessions and sample description.

Bioproject	Geo	IR Type	Tissue	Condition	Sample Count
PRJNA494581	GSE120805	X-rays	Human Lens Epithelial Cells	Control 2 Gy | 20 h 5 Gy | 20 h	5 5 5
PRJNA421022	GSE107685	X-rays	iPSC-Derived Cardiomyocytes	Control 5 Gy| 48 h	3 3
PRJNA436999	GSE111437	X-rays	Primary Human Lung Fibroblasts (IMR90)	Control | 6 h Control | 24 h 2 Gy | 6 h 2 Gy | 24 h	3 3 3 3
PRJNA396832	GSE102145	X-rays	Skin Fibroblasts (WS1)	Control 5 Gy | 24 h	1 1
PRJNA450083	GSE113125	Gamma-rays	Skin Fibroblasts	Control 5 Gy | 1 h	1 1
			iPSC-Fibroblasts	Control 5 Gy | 1 h	1 1
			iPSC-Neural Progenitor Cells	Control 5 Gy | 1 h	1 1

Mapping of the trimmed reads to the reference genome and transcriptome (gentrome) was estimated at a rate of ~83%. There were distinct outliers during the quality control, trimming, and alignment processes, and this served as the first indicator that PRJNA396832 and PRJNA450083 were not of the desired quality for conducting our analysis. Without the outliers, mapping rate was at ~87% across the remaining 3 datasets (PRJNA421022, PRJNA436999, and PRJNA494581).

2.2. Differential Gene Expression

It was suggested that the minimal number of biological replicates required for RNA-Seq based DGEA is 6 [34]. PRJNA396832 and PRJNA450083 had a single biological replicate for each condition, and were therefore excluded for subsequent analysis, as statistical significance could not be estimated. Studies that contained combinations of experimental conditions (PRJNA436999 2 Gy for 6 and 24 h and PRJNA494581 2 Gy and 5 Gy for 20 h), were split into distinct studies (Table 2) for DGEA. DESeq2 was used to identify DEGs in PRJNA421022 (Supplementary Materials, Table S1) in one of the distinct studies of PRJNA436999 (Supplementary Materials, Table S2) and in one of the distinct studies of

PRJNA494581 (Supplementary Materials, Table S3). From 2006 genes that were considered DEGs in at least one dataset, 542 overlapped (Figure 1).

Table 2. Statistically significant gene counts derived from the differential gene expression analysis for each comparison within each dataset. The table includes information about experimental sample parameters, dataset accession numbers, and total mapped gene counts for each dataset.

Bioproject Accession	PRJNA421022	PRJNA436999		PRJNA494581	
GEO Accession	GSE107685	GSE111437		GSE120805	
IR Type	X-rays	X-rays		X-rays	
Cell Type	iPSC-Derived Cardiomyocytes	Primary Human Lung Fibroblasts (IMR90)		Human Lens Epithelial Cells (HLE)	
Dose	5 Gy	2 Gy	2 Gy	2 Gy	5 Gy
Time Point	48 h	6 h	24 h	20 h	20 h
DEG Counts	721	353	908	59	1003

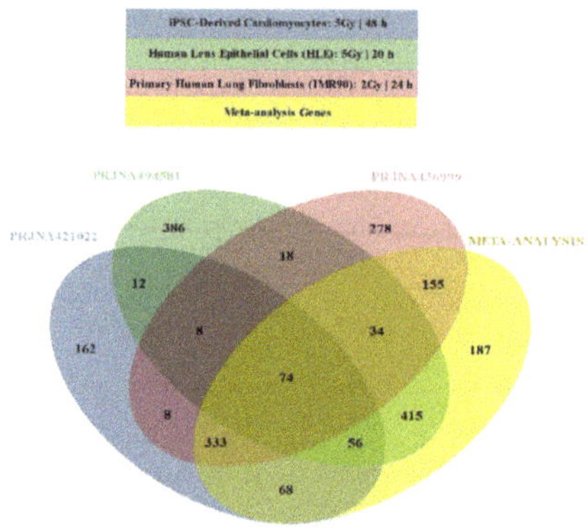

Figure 1. Venn diagram illustrating overlapping Differentially Expressed Genes (DEGs) across selected DEG lists and meta-analysis results.

Meta-analysis was conducted using the three lists that derived from the DESeq2 analysis, yielding 1322 DEGs (Supplementary Materials, Table S4): 872 DEGs were found in at least one DESeq2 analysis but not in the meta-analysis; 187 genes that were not characterized as DEGs in any DESeq2 analyses were considered DEGs by meta-analysis (Figure 1).

Some of the genes that were identified as DEGs in the original RNA-Seq analyses were PCR-validated by their own groups. For PRJNA494581 dataset [28], ddPCR validated CDKN1A, AREG, H2BC13 (HIST1H2BL), GDF15, H3C11 (HIST1H31), H1-4 (HIST1H1E), H2BC10 (HIST1H2BF), and TP53INP1 as DEGs and B2M and RPL13A as non-differentially expressed genes. For PRJNA421022 dataset [30], qPCR validated PLK1, BIRC5, AURKB, KIF20A, TOP2A, and CCNA2 as down-regulated and CDKN1A and FDXR as up-regulated genes. Moreover, qPCR did not validate ANGPTL4 and SOGA3 as up-regulated genes, even if those genes were estimated as DEGs in their RNA-Seq data analysis. The DEG list of our meta-analysis agrees with all PCR validations.

2.3. Functional Enrichment Results

Functional enrichment analysis of up- and down-regulated genes resulting from meta-analysis produced lists of statistically significant Gene Ontology (GO) [35] biological processes, KEGG [36] biological pathways (Table 3), and gene-targeting transcription factors (Table 4). A Protein–Protein Interaction (PPI) network of DEGs was also constructed (Figure 2).

Table 3. Enriched biological processes (Gene Ontology—GO) and enriched biological pathways (KEGG) for up- and down-regulated genes after meta-analysis.

Gene Set	Description	Source	FDR
	Up-Regulated Genes ↑		
GO:0072331	Signal transduction by p53 class mediator	GO	0.0022959
hsa04115	p53 signaling pathway	KEGG	2.24×10^{-9}
GO:0042770	Signal transduction in response to DNA damage	GO	0.009620632
GO:0097193	Intrinsic apoptotic signaling pathway	GO	0.010992521
GO:0071496	Cellular response to external stimulus	GO	0.01930771
GO:0104004	Cellular response to environmental stimulus	GO	0.032524732
GO:0008643	Carbohydrate transport	GO	0.037630683
hsa01524	Platinum drug resistance	KEGG	0.032524732
	Down-Regulated Genes ↓		
GO:0000075	Cell cycle checkpoint	GO	0
hsa04110	Cell cycle	KEGG	0
GO:0051321	Meiotic cell cycle	GO	0
GO:0044772	Mitotic cell cycle phase transition	GO	0
GO:0045930	Negative regulation of mitotic cell cycle	GO	0
GO:1902850	Microtubule cytoskeleton organization involved in mitosis	GO	0
GO:0044839	Cell cycle G2/M phase transition	GO	0
GO:0044843	Cell cycle G1/S phase transition	GO	0
GO:1901987	Regulation of cell cycle phase transition	GO	0
GO:0010948	Negative regulation of cell cycle process	GO	0
GO:0045787	Positive regulation of cell cycle	GO	0
GO:0007050	Cell cycle arrest	GO	7.77×10^{-4}
hsa04115	p53 signaling pathway	KEGG	0.002014026
GO:0006260	DNA replication	GO	0
hsa03030	DNA replication	KEGG	0
hsa00240	Pyrimidine metabolism	KEGG	5.67×10^{-10}
hsa00230	Purine metabolism	KEGG	4.39×10^{-5}
GO:0042769	DNA damage response, detection of DNA damage	GO	1.88×10^{-6}
GO:0006310	DNA recombination	GO	0
GO:0006302	Double-strand break repair	GO	0
GO:0036297	Inter-strand cross-link repair	GO	1.35×10^{-12}
hsa03430	Mismatch repair	KEGG	3.11×10^{-11}
hsa03440	Homologous recombination	KEGG	3.11×10^{-11}
GO:0006284	Base excision repair	GO	6.54×10^{-6}
hsa03410	Base excision repair	KEGG	1.91×10^{-7}
GO:0006298	Mismatch repair	GO	4.02×10^{-4}
GO:0006289	Nucleotide excision repair	GO	5.10×10^{-4}
hsa03420	Nucleotide excision repair	KEGG	2.91×10^{-6}
GO:0006333	Chromatin assembly or disassembly	GO	0
GO:0007051	Spindle organization	GO	0
GO:0071103	DNA conformation change	GO	0
GO:0007059	Chromosome segregation	GO	0
GO:0061641	CENP-A containing chromatin organization	GO	0
GO:0048285	Organelle fission	GO	0
GO:0051052	Regulation of DNA metabolic process	GO	0
GO:0071824	Protein-DNA complex subunit organization	GO	0
hsa03013	RNA transport	KEGG	5.66×10^{-5}
hsa03008	Ribosome biogenesis in eukaryotes	KEGG	4.13×10^{-5}
hsa04114	Oocyte meiosis	KEGG	0.001082137
hsa04914	Progesterone-mediated oocyte maturation	KEGG	0.008441906
hsa05322	Systemic lupus erythematosus	KEGG	0
hsa05203	Viral carcinogenesis	KEGG	1.46×10^{-11}
hsa05206	MicroRNAs in cancer	KEGG	0.008110817
hsa03460	Fanconi anemia pathway	KEGG	7.71×10^{-11}
hsa05166	Human T-cell leukemia virus 1 infection	KEGG	5.47×10^{-4}
hsa04217	Necroptosis	KEGG	1.19×10^{-4}
hsa04218	Cellular senescence	KEGG	0.0014615

Table 4. Enriched gene-targeting transcription factors for down-regulated genes after meta-analysis.

Gene Set	Gene Symbol	Gene Name	FDR
		Down-Regulated Genes ↓	
E2F_Q3_01;E2F_Q4_01; E2F_Q6_01;E2F1_Q4_01	TFDP1	Transcription factor Dp-1	0
E2F1_Q3;E2F1_Q6; E2F1_Q6_01	E2F1	E2F transcription factor 1	0
E2F1DP1_01	E2F1;TFDP1	E2F transcription factor 1; transcription factor Dp-1	0
E2F1DP1RB_01	E2F1;TFDP1; RB1	E2F transcription factor 1; transcription factor Dp-1; RB transcriptional corepressor 1	0
E2F1DP2_01	TFDP2	Transcription factor Dp-2	0
E2F4DP1_01	E2F4;TFDP1	E2F transcription factor 4; transcription factor Dp-1	0
E2F4DP2_01	E2F4;TFDP2	E2F transcription factor 4; transcription factor Dp-2	0
E2F1_Q4;	E2F1	E2F transcription factor 1	9.80×10^{-12}
E2F1_Q3_01	E2F1	E2F transcription factor 1	4.39×10^{-5}

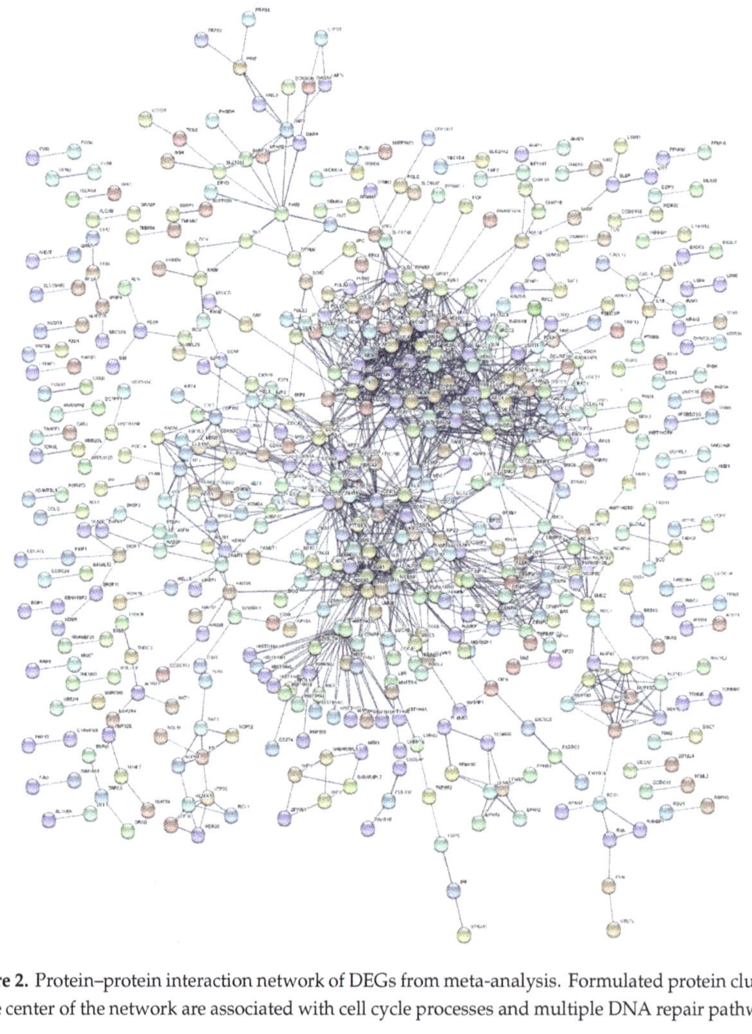

Figure 2. Protein–protein interaction network of DEGs from meta-analysis. Formulated protein clusters in the center of the network are associated with cell cycle processes and multiple DNA repair pathways.

3. Discussion

RNA-Seq is gradually becoming the predominant technique for transcriptome analysis, superseding microarrays. RNA-Seq technology is more sensitive in detecting genes with low expression levels, it lacks the associated background noise of hybridization-based techniques, and it is more reproducible [24]. Furthermore, a crucial limitation of microarrays is their inability to study the expression of genes for which no probe is available on the chip. So far, there is no gold-standard methodology for analyzing the transcriptome using RNA-Seq technology. Poorly designed pipelines for differential gene expression can have detrimental effects, compromising the experimental results. Suboptimal pre-processing of raw data, directly affects the mapping process, resulting in poor mapping rates (<60%) [37].

An extensive debate [38–40] on the effects of the pre-processing step of trimming, suggests that reads should be carefully trimmed. Illumina Next-Generation Sequencing (NGS) platforms produce sequences of between 25–250 nucleotides, the colorimetric signals of which are translated by an internal Illumina software (CASAVA) to base calls and are represented in FASTQ [41] file format. Minimal trimming ($Q < 10$) keeps low quality base calls in NGS analyses, adding unreliable and random sequences to the final dataset [38]. However, hard trimming ($Q > 30$) of reads can have a particularly strong negative impact on RNA-Seq-based gene expression estimates, as it introduces unpredictable and unwanted biases [40]. Hence, we used soft trimming ($Q = 20$) as a balance, preserving biological information that was not ideally recorded, while discarding non-sense information.

The expected mapping rate of RNA-Seq reads is between 70% and 90% when mapped against the human genome and slightly less when mapped against the transcriptome [37]. In conventional RNA-Seq mapping, multiple reads are mapped across splice junctions [42]. The inability of conventional algorithms, like BWA [43], to handle spliced transcripts renders them obsolete. Splice-aware aligners [44], such as Hisat2 [45], handle mapping in a more efficient manner. However, these traditional approaches require significant computational resources amidst an explosively growing storage-hungry environment [46]. Alignment-independent methods, such as Salmon [47], bypass the mapping step and proceed to quantify directly transcript abundance, boasting a more lightweight and significantly faster novel approach. Our mapping rate, minus the outliers, was calculated at a satisfactory level of ~87% using Salmon's selective alignment method.

The resulting datasets combined with DESeq2 which utilizes estimates of dispersions and logarithmic fold changes by incorporating data-driven prior distributions [48], yielded substantial statistically significant results. Instead of using transcript levels, as in a classical meta-analysis, we chose Mosteller–Bush, a method which is based on the combination of weighted z-values for the independent studies, which are calculated from the p-values produced by DESeq2. PCR-based validations showed that our meta-analysis outperformed original RNA-Seq data analyses.

Although a fraction of differentially expressed genes overlap among the 3 DESeq2 studies, a meta-analysis was conducted to enhance the validity of our DGEA (Figure 1). In all three DESeq2 DEG lists (Supplementary Materials, Table S1–S3) and in the meta-analysis DEG list (Supplementary Materials, Table S4), the number of statistically significant under-expressed genes considerably surpasses the number of over-expressed genes. Such a result may indicate that cell cycle arrest has been activated, which is supported by the top down-regulated genes (i.e., MKI67, CCNA2, CDK1, PLK1, and CDCA3) identified by the meta-analysis and associated with cell cycle activation [49]. This difference between up- and down-regulated genes, as well as cell cycle arrest activation, coincides with the primary results for PRJNA421022 dataset [30]. At the same time, observed up-regulation of CDKN1A, responsible for cell cycle G1 phase arrest, in response to a variety of stress stimuli further upholds this suggestion. p21 protein coded by this gene (known to be interdependent with tumor suppressor protein TP53) is also responsible for inhibiting cellular proliferation in response to DNA damage [50]. In addition, it is highly correlated with DNA repair, while also being instrumental in the execution of apoptosis. Although p21 is assumed to play a key role as "genome guardian", it can alternatively act as the mediator of genomic instability, cellular senescence, and carcinogenesis under certain circumstances, like IR exposure and TP53 deficiency [51]. Another substantial result is the significant up-regulation of GDF15, associated with the response to oxidative stress and induction of inflammatory environment, thus coinciding with DDR and IR response in general [49,52].

TP53 is essential in DDR mechanisms through its downstream responses, which include cell cycle arrest, DNA repair, and apoptosis. The accurate transition from G1 phase of the cell cycle to S phase is crucial for a controlled cell proliferation, and its misregulation promotes oncogenesis [53]. G1 arrest provides the cell adequate time to repair the DNA damage. Should repair be unsuccessful, TP53 levels drop and CDK-cyclin protein kinase activity resumes, leading to entry into S phase and possible apoptosis triggering. [54]. In our results, for samples selected over 20 h post irradiation, TP53 indeed shows no altered expression, while MDM2, as a TP53 downstream gene transcriptionally activated by it [55], is over-expressed.

Regarding enrichment results for down-regulated genes (Table 3), cell cycle checkpoint as well as various DNA repair mechanisms are over-represented. These results may indicate that DNA repair genes (i.e., MSH2, MSH6, XRCC3, and POLA2) are suppressed, due to programmed cell death. However, NER-associated genes DDB1, DDB2, and XPC where found up-regulated. DNA repair in general, due to its complexity, requires balanced expression of its genes in order to avoid erroneous repairs [56]. Over-represented gene-targeting factors E2F1, E2F2, RB1, TFDP1, and TFDP2 (Table 4) are connected through involvement in cell cycle G1/S phase transition, TP53 regulation, and cellular senescence [49]. Moreover, under-expression of DNA repair genes, mediated by the RB/E2F pathway, may play a causal role in senescence induction [57]. In the case of up-regulated genes, p53 signaling pathway, DDR, and apoptotic mechanisms seem to be activated. In addition, platinum drug resistance (Table 3) could arise from increased DNA repair, decreased mismatch repair, defective apoptosis, and altered oncogene expression [58].

Regarding the PPI network (Figure 2), edges corresponding to protein interactions, were constructed with sizeable restrictions regarding the validity of their sources and their assigned score. Distinct clusters were formulated, coinciding with the main components of high-dose IR response. Simultaneously, through these clusters, the network validates the biological processes and pathways derived from the enrichment results. Two major, densely packed clusters are formed, representing multiple cell cycle processes. These clusters are indicative of a collection of DNA repair pathways (GO:0006281), such as DSBs repair (RAD51, BLM, DNA2), Mismatch Repair (MSH6, MSH2), Homologous Recombination (XRCC3), and NER (DDB1, DDB2, XPC, POLE). Furthermore, TIMELESS, a gene found in the center of the network due to its contribution to DSB repair, also acts as a circadian rhythm pathway regulator [49]. Additional circadian rhythm-related results are the under-expression of TYMS and the over-expression of CRY2.

Circadian genes are known to regulate a variety of cellular processes, including cell cycle, apoptosis, and DNA damage repair [59]. Both oxidative defense mechanisms and repair of X-ray induced DSBs in DNA are synchronized by circadian rhythms; thus, RT timing needs to be coordinated as we enter personalized medicine [60]. Furthermore, disruption in circadian gene expression is associated with increased incidence of cancers and gliomas [61]. Finally, the circadian clock system also controls various parameters of the immune system and its biological defense functions [62]. This correlation between circadian clock dysregulation and IR response may reveal possible underlying mechanisms of chronic inflammatory disease development. In addition, understanding the interplay between circadian rhythm, cell cycle, cell proliferation, and DNA repair will deliver benefits in RT by reducing its side effects on healthy tissues.

Biological response to IR (high doses), especially at the organism level, is complicated and partially unknown. RNA-Seq as an -omics methodology provides information on gene expression for several thousand proteins. This opens a unique opportunity to approach this difficult task of delineating the mechanisms triggered after radiation-induced stress, at a systems biology level. Therefore, in this study, after critical screening of several RNA-Seq datasets and applying state-of-the-art bioinformatics and meta-analysis, we were able to identify: (A) 1322 DEGs (371 up-regulated, 951 down-regulated,

Supplementary Materials, Table S4); (B) cell cycle checkpoint activation, apoptosis, and various down-regulated repair genes. The last probably relates to late post-irradiation time points (20–48 h), where repair is expected to be completed and the cell is sent to apoptosis. Another suggestion is that transcriptional up-regulation of DNA repair genes by genotoxic stress (p53 activation, Tables 3 and 4) is counteracted by possible DNA damage that blocks transcription [56]; (C) indication of cellular senescence; (D) association of IR response with the circadian clock.

4. Materials and Methods

Our RNA-Seq analysis involves a pipeline (Figure 3) of in silico processes, each with its own specific parameters (Appendix A, Table A1), where files undergo a series of transformations. In each step, data are manipulated in a way that information is retained and expanded by additional meta-data.

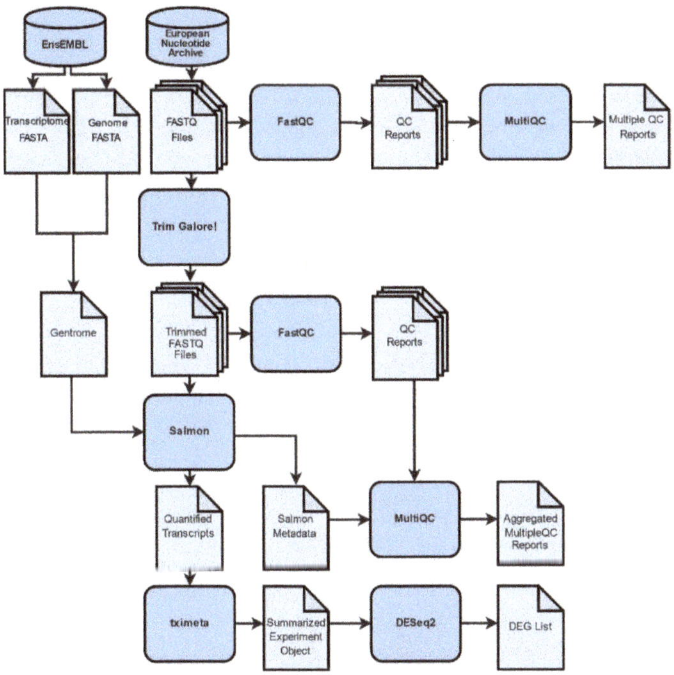

Figure 3. Differential gene expression analysis workflow is comprised of 4 distinct steps: (A) Data collection from the online repositories ENA and EnsEMBL. (B) Quality control and trimming. A conventional analysis pipeline of RNA-Seq data starts with the pre-processing of the raw reads with FastQC, MultiQC, and Trim Galore! (C) Gene abundance quantification with Salmon and tximeta by mapping of the reads to a reference genome and/or transcriptome. (D) Differential gene expression analysis with DESeq2, where gene expression levels of all mapped transcripts are quantified and normalized in order to define differentially expressed genes.

4.1. Datasets

We searched for datasets available in public repositories to identify the studies that performed RNA-Seq in normal and ionized tissues. The appropriate datasets were identified using the European Nucleotide Archive (ENA) advanced search engine [63]. We narrowed our search down to human Illumina RNA-Seq studies which involved ionizing radiation. More specifically, in the read domain of ENA advanced search, our query was: *"instrument_platform = "ILLUMINA" AND library_strategy = "RNA-Seq" AND tax_eq (9606)"*, and in the study domain, our query was: *"(study_name = "*ionizing*"*

*OR study_title = "*ionizing*" OR study_description = "*ionizing*" OR study_name = "*alpha particle*" OR study_title = "*alpha particle*" OR study_description = "*alpha particle*" OR study_name = "*irradiation*" OR study_title = "*irradiation*" OR study_description = "*irradiation*" OR study_name = "*X-ray*" OR study_title = "*X-ray*" OR study_description = "*X-ray*" OR study_name = "*X ray*" OR study_title = "*X ray*" OR study_description = "*X ray*" OR study_name = "*gamma ray*" OR study_title = "*gamma ray*" OR study_description = "*gamma ray*" OR study_name = "*positron*" OR study_title = "*positron*" OR study_description = "*positron*" OR study_name = "*radiotherapy*" OR study_title = "*radiotherapy*" OR study_description = "*radiotherapy*" OR study_name = "*ionising*" OR study_title = "*ionising*" OR study_description = "*ionising*" OR study_name = "*IR-induced*" OR study_title = "*IR-induced*" OR study_description = "*IR-induced*") AND tax_tree(9606)"*. The first query yielded 10,541 studies and the second query yielded 504 studies. Afterwards, we selected the overlapping study names. The studies were then manually curated in order to exclude non-irradiated, UV-irradiated, and tumor samples.

4.2. Raw Read Evaluation

Extensive quality control was performed on RNA-Seq data of each sample using FastQC (*version 0.11.8*) [64], and summaries were produced by MultiQC (*version 1.8*) [65] to evaluate the integrity of RNA-Seq experiments. Quality control reports mainly involved the analysis of sequence accuracy, presence of PCR artifacts, and adaptor sequences that were not automatically cleaned by Illumina platforms, GC content, k-mer levels, etc. Surgical elimination of low-quality regions, known as "trimming", was performed—when necessary—by Trim Galore! (*version 0.6.4*) [66], a wrapper package around Cutadapt (*version 2.8*) [67], and FastQC. Consequently, results where re-evaluated with FastQC and MultiQC to verify that the quality of raw data had improved after the trimming process.

4.3. Sequence Alignment

Mapping aligns trimmed sequence reads against a known genome and transcriptome. Its efficiency mainly depends on the bioinformatics tools used and the quality of the sequences. Reads were directly mapped into *Homo sapiens* reference genome and transcriptome FASTA-formatted sequences. To this end, we used the latest release of Salmon (*version 1.1.0)* [47] which adopts a selective-alignment algorithm in order to overcome the shortcomings of lightweight approaches, without the additional computational burden of traditional alignment [68]. We produced the transcriptome index for Salmon via the partial selective alignment method, mapping the transcriptome to the genome, extracting the relevant portion out of the genome, and, finally, indexing it along with the transcriptome.

4.4. Transcript Quantification

The entirety of the statistical analysis was performed using packages provided by Bioconductor (*BiocManager version 3.10*) [69,70], a suite for analyzing high-throughput genomic data in R (*version 3.6.2*) [71] statistical programming language. All R code was executed through RStudio server (*version 1.2.5033*) [72]. Transcript-level quantification was estimated using tximeta (*version 1.4.3*) [73], an expansion of tximport (*version 1.14.0*) [74].

4.5. Differential Gene Expression Analysis

Transcript-level quantification data were processed for DGEA, using DESeq2 (*version 1.26.0*) [48]. Studies with more than two distinct conditions were split in order to analyze genes that were differentially expressed between cells exposed to a specific dose for a specific time point and their corresponding control samples. Exported lists containing statistically significant differentially expressed genes include metrics such as Log_2 Fold Change (Log_2FC), *p*-values, and False Discovery Rate (FDR)-adjusted [75] *p*-values for each gene. The lists were further annotated using org.Hs.eg.db (*version 3.8.2*) [76] to include HGNC [77] gene symbols and gene names. The threshold for statistical significance was set at the adjusted p-value < 0.05, as non-adjusted *p*-values are not to be considered [75].

4.6. Meta-Analysis

DEG lists were further combined in a meta-analysis (Figure 4) to identify genes of differential expression across studies. In order to achieve optimal results, only DEG lists derived from similar condition groups regarding dose and time of post-irradiation collection were considered. To this end, we excluded the DEG list derived from the PRJNA436999 samples that were collected 6 h post-irradiation, leaving only the DEG list from the samples that were collected after more than 20 h. Finally, the DEG list acquired from the PRJNA494581 samples after 2 Gy irradiation was also excluded, as it was considered an outlier, due to its low DEG number.

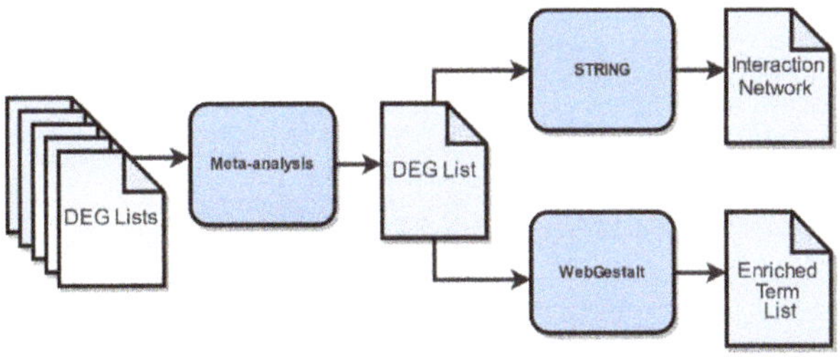

Figure 4. Functional enrichment analysis workflow. Differential Gene Expression Analysis (DGEA)-derived genes, subdued to meta-analysis, yielded the final DEGs of utmost statistical significance. Inputting these DEGs into WebGestalt and STRING generated the resulting enriched terms and PPI network, respectively.

Our meta-analysis combined unadjusted p-values of each study for every gene, using a weighted version of Stouffer meta-analysis [78], proposed by Mosteller and Bush [79]. For each gene and study, its two-tail unadjusted p-value was converted into an one-tail p-value, based on the sign of the corresponding Log_2FC. For each one-tailed p-value, the corresponding z-score was calculated using the inverse normal distribution function (Φ^{-1}). Meta-analysis p-value for each gene was calculated from the weighted z-score sum, using the normal distribution function (Φ):

$$p = \Phi\left(\frac{\sum_{i=1}^{k} n_i \Phi^{-1}(p_i)}{\sqrt{\sum_{i=1}^{k} n_i^2}}\right)$$

where p_i is the DESeq2-derived p-value, and n_i is the number of samples of study i and k the number of studies. Finally, p-values underwent FDR adjustment, and 0.05 was selected as threshold for statistical significance.

To evaluate the efficiency of our method, we compared the experimentally validated DEGs derived from the studies the RNA-Seq data of which we chose for analysis with our own estimations.

4.7. Functional Enrichment Analysis and Gene Network Construction

To highlight the biological background of DEGs, a functional enrichment analysis (Figure 4) was performed using the WEB-based Gene Set Analysis Toolkit (WebGestalt) [80]. We selected Over-Representation Analysis (ORA) [81] method which performs a statistical evaluation of the fraction of genes in a particular pathway found among the set of genes showing changes in expression. Our terms of interest include biological process GO terms [35], KEGG biological pathways [58], and gene-targeting transcription factors. The statistical significance of each over-representation of biological terms was estimated using hyper-geometric distribution. p-values were FDR-adjusted, and terms with adjusted p-values < 0.05 were considered statistically significant.

In order to investigate the interactome of the DEGs and identify possible underlying cell mechanisms, we constructed their Protein–Protein Interaction (PPI) network, using STRING (*version 11.0*) [82]. Edges of the network, corresponding to protein interactions, were determined solely based on text mining sources with high confidence (Table A1).

Supplementary Materials: Supplementary materials can be found at http://www.mdpi.com/1422-0067/21/6/1938/s1. Table S1: Statistically significant DEGs (Adj.*p*-value<0.05) derived from DGEA of RNA-Seq data from human iPSC-Derived Cardiomyocytes after X-ray irradiation, using DESeq2. The experiment consists of 3 control samples and 3 irradiated with 5 Gy X-ray radiation for 48 hours [Bioproject: PRJNA421022]; Table S2: Statistically significant DEGs (Adj.*p*-value<0.05) derived from DGEA of RNA-Seq data from Primary Human Lung Fibroblasts (IMR90) after X-ray irradiation, using DESeq2. The experiment consists of 3 control samples and 3 irradiated with 2 Gy X-ray radiation for 24 hours [Bioproject: PRJNA436999]; Table S3: Statistically significant DEGs (Adj.*p*-value<0.05) derived from DGEA of RNA-Seq data from Human Lens Epithelial Cells after X-ray irradiation, using DESeq2. The experiment consists of 5 control samples and 5 irradiated with 5 Gy X-ray radiation for 20 hours [Bioproject: PRJNA494581]; Table S4: Statistically significant DEGs derived from a meta-analysis comparing DESeq2 outputs from irradiated and control samples of three datasets [PRJNA421022, PRJNA436999, PRJNA494581].

Author Contributions: Conceptualization, I.M. and A.G.G.; methodology, I.M., D.S.K., T.-D.M., C.V., E.G., D.M. and K.V.K.; software, I.M., D.S.K., T.-D.M.; validation, D.S.K. and T.-D.M.; formal analysis, I.M., K.V.K. and T.-D.M.; investigation, D.S.K. and E.G.; data curation, D.S.K.; writing—original draft preparation, I.M., A.G.G., D.S.K. and T.-D.M.; writing—review and editing, I.M., A.G.G., D.S.K. and T.-D.M.; visualization, D.S.K., T.-D.M.; C.V., D.M., E.G. and K.V.K.; supervision, I.M.; project administration, I.M. All authors have read and agreed to the published version of the manuscript.

Funding: This research received no external funding.

Conflicts of Interest: The authors declare no conflict of interest.

Abbreviations

BER	Base Excision Repair
DDR	DNA Damage Response
DEGs	Differentially Expressed Genes
DGE	Differential Gene Expression
DGEA	Differential Gene Expression Analysis
DSBs	Double Strand Breaks
ENA	European Nucleotide Archive
FDR	False Discovery Rate
GSEA	Gene Set Enrichment Analysis
GO	Gene Ontology
HR	Homology-dependent Recombination
IR	Ionizing Radiation
LET	Linear Energy Transfer
Log_2FC	Log_2 Fold Change
MMR	Mismatch Repair
NER	Nucleotide Excision Repair
NGS	Next-Generation Sequencing
NHEJ	Non-Homologous End Joining
ORA	Over-Representation Analysis
PPI	Protein to Protein Interaction
Q	PHRED Score
RBE	Relative Biological Effectiveness
RNA-Seq	RNA-Sequencing
RNS	Reactive Nitrogen Species
ROS	Reactive Oxygen Species
RT	Radiation Therapy
SSBs	Single-Strand Breaks
WebGestalt	WEB-based Gene Set Analysis Toolkit

Appendix A

Table A1. Specific arguments used during the analysis. Unless explicitly specified otherwise, default parameters were used.

Trim Galore!	trim_galore –cores 4 –illumina -q 20 –phred33 –paired –fastqc <fastq files>
Salmon	salmon quant -i salmon_index –libType A -1 <forward_1.fq.gz> -2 <reverse_1.fq.gz> –gcBias –validateMappings -o <transcripts_directory>
WebGestalt	**Basic Parameters:**
	Organism of Interest: Homo sapiens Method of Interest: ORA Functional Database: geneontology + (Biological Process: no redundant), pathway + (KEGG), Network + (Transcription Factor target) Gene List: Select Gene ID Type: EnsEMBL Gene IDReference Gene List: Upload: Mappings per study: EnsEMBL Gene ID
	Advanced Parameters
	minimum number of genes for category: 2 Multiple Test Adjustment: Benjamini-Hochberg Significance level: FDR (0.05) Number of categories visualized in the report: 100
STRING	**Basic Settings:**
	meaning of network edges: confidence active interaction sources: textmining minimum required interaction score: high confidence (0.7)
	Advanced Settings:
	hide disconnected nodes in the network disable structure previous inside network bubbles

References

1. UNSCEAR (United Nations Scientific Committee on the Effects of Atomic Radiation). *Biological Mechanisms of Radiation Actions at Low Doses*; United Nations; New York, NY, USA, 2012.
2. Leuraud, K.; Richardson, D.B.; Cardis, E.; Daniels, R.D.; Gillies, M.; O'Hagan, J.A.; Hamra, G.B.; Haylock, R.; Laurier, D.; Moissonnier, M.; et al. Ionising radiation and risk of death from leukaemia and lymphoma in radiation-monitored workers (INWORKS): An international cohort study. *Lancet. Haematol.* **2015**, *2*, e276–e281. [CrossRef]
3. Ray, M.; Yunis, R.; Chen, X.; Rocke, D.M. Comparison of low and high dose ionising radiation using topological analysis of gene coexpression networks. *Bmc Genom.* **2012**, *13*, 190. [CrossRef]
4. Bragg, W.H.; Kleeman, R. LXXIV. On the ionization curves of radium. *Lond. Edinb. Dublin Philos. Mag. J. Sci.* **1904**, *8*, 726–738. [CrossRef]
5. Baskar, R.; Dai, J.; Wenlong, N.; Yeo, R.; Yeoh, K.W. Biological response of cancer cells to radiation treatment. *Front. Mol. Biosci.* **2014**, *1*, 24. [CrossRef]
6. Mehta, S.R.; Suhag, V.; Semwal, M.; Sharma, N. Radiotherapy: Basic Concepts and Recent Advances. *Med. J. Armed. Forces India* **2010**, *66*, 158–162. [CrossRef]
7. Kjellberg, R.N.; Hanamura, T.; Davis, K.R.; Lyons, S.L.; Adams, R.D. Bragg-Peak Proton-Beam Therapy for Arteriovenous Malformations of the Brain. *N. Engl. J. Med.* **1983**, *309*, 269–274. [CrossRef]
8. Bernier, J.; Hall, E.J.; Giaccia, A. Radiation oncology: A century of achievements. *Nat. Rev. Cancer* **2004**, *4*, 737–747. [CrossRef]
9. Georgakilas, A.G. Bystander and non-targeted effects: A unifying model from ionizing radiation to cancer. *Cancer Lett.* **2015**, *356*, 3–4. [CrossRef]
10. Riley, P.A. Free radicals in biology: Oxidative stress and the effects of ionizing radiation. *Int. J. Radiat. Biol.* **1994**, *65*, 27–33. [CrossRef]

11. Wallace, S.S. Enzymatic processing of radiation-induced free radical damage in DNA. *Radiat. Res.* **1998**, *150*, S60–S79. [CrossRef]
12. Mavragani, I.V.; Nikitaki, Z.; Souli, M.P.; Aziz, A.; Nowsheen, S.; Aziz, K.; Rogakou, E.; Georgakilas, A.G. Complex DNA Damage: A Route to Radiation-Induced Genomic Instability and Carcinogenesis. *Cancers* **2017**, *9*, 91. [CrossRef]
13. Georgakilas, A.G. Processing of DNA damage clusters in human cells: Current status of knowledge. *Mol. Biosyst.* **2008**, *4*, 30–35. [CrossRef]
14. Nikitaki, Z.; Hellweg, C.E.; Georgakilas, A.G.; Ravanat, J.L. Stress-induced DNA damage biomarkers: Applications and limitations. *Front. Chem.* **2015**, *3*, 35. [CrossRef]
15. Ogrunc, M.; Di Micco, R.; Liontos, M.; Bombardelli, L.; Mione, M.; Fumagalli, M.; Gorgoulis, V.G.; d'Adda di Fagagna, F. Oncogene-induced reactive oxygen species fuel hyperproliferation and DNA damage response activation. *Cell Death Differ.* **2014**, *21*, 998–1012. [CrossRef]
16. Saini, D.; Shelke, S.; Mani Vannan, A.; Toprani, S.; Jain, V.; Das, B.; Seshadri, M. Transcription profile of DNA damage response genes at G(0) lymphocytes exposed to gamma radiation. *Mol. Cell. Biochem.* **2012**, *364*, 271–281. [CrossRef]
17. Nikitaki, Z.; Pavlopoulou, A.; Hola, M.; Dona, M.; Michalopoulos, I.; Balestrazzi, A.; Angelis, K.J.; Georgakilas, A.G. Bridging Plant and Human Radiation Response and DNA Repair through an In Silico Approach. *Cancers* **2017**, *9*, 65. [CrossRef]
18. Knijnenburg, T.A.; Wang, L.; Zimmermann, M.T.; Chambwe, N.; Gao, G.F.; Cherniack, A.D.; Fan, H.; Shen, H.; Way, G.P.; Greene, C.S.; et al. Genomic and Molecular Landscape of DNA Damage Repair Deficiency across The Cancer Genome Atlas. *Cell Rep.* **2018**, *23*, 239–254 e236. [CrossRef]
19. Friedberg, E.C. A history of the DNA repair and mutagenesis field: The discovery of base excision repair. *Dna Repair* **2016**, *37*, A35–A39. [CrossRef]
20. Pan, M.R.; Li, K.; Lin, S.Y.; Hung, W.C. Connecting the Dots: From DNA Damage and Repair to Aging. *Int. J. Mol. Sci.* **2016**, *17*, 685. [CrossRef]
21. Georgakilas, A.G.; Pavlopoulou, A.; Louka, M.; Nikitaki, Z.; Vorgias, C.E.; Bagos, P.G.; Michalopoulos, I. Emerging molecular networks common in ionizing radiation, immune and inflammatory responses by employing bioinformatics approaches. *Cancer Lett.* **2015**. [CrossRef]
22. Nakad, R.; Schumacher, B. DNA Damage Response and Immune Defense: Links and Mechanisms. *Front. Genet.* **2016**, *7*, 147. [CrossRef]
23. Hatzi, V.I.; Laskaratou, D.A.; Mavragani, I.V.; Nikitaki, Z.; Mangelis, A.; Panayiotidis, M.I.; Pantelias, G.E.; Terzoudi, G.I.; Georgakilas, A.G. Non-targeted radiation effects in vivo: A critical glance of the future in radiobiology. *Cancer Lett.* **2015**, *356*, 34–42. [CrossRef]
24. Zhao, S.; Fung-Leung, W.P.; Bittner, A.; Ngo, K.; Liu, X. Comparison of RNA-Seq and microarray in transcriptome profiling of activated T cells. *Plos ONE* **2014**, *9*, e78644. [CrossRef]
25. Hrdlickova, R.; Toloue, M.; Tian, B. RNA-Seq methods for transcriptome analysis. *Wiley Interdiscip Rev. Rna.* **2017**, *8*. [CrossRef]
26. Romero, J.P.; Ortiz-Estevez, M.; Muniategui, A.; Carrancio, S.; de Miguel, F.J.; Carazo, F.; Montuenga, L.M.; Loos, R.; Pio, R.; Trotter, M.W.B.; et al. Comparison of RNA-seq and microarray platforms for splice event detection using a cross-platform algorithm. *Bmc Genom.* **2018**, *19*, 703. [CrossRef]
27. Barrett, T.; Wilhite, S.E.; Ledoux, P.; Evangelista, C.; Kim, I.F.; Tomashevsky, M.; Marshall, K.A.; Phillippy, K.H.; Sherman, P.M.; Holko, M.; et al. NCBI GEO: Archive for functional genomics data sets–update. *Nucleic Acids Res.* **2013**, *41*, D991–D995. [CrossRef]
28. Chauhan, V.; Rowan-Carroll, A.; Gagne, R.; Kuo, B.; Williams, A.; Yauk, C.L. The use of in vitro transcriptional data to identify thresholds of effects in a human lens epithelial cell-line exposed to ionizing radiation. *Int. J. Radiat. Biol.* **2019**, *95*, 156–169. [CrossRef]
29. Shimada, M.; Tsukada, K.; Kagawa, N.; Matsumoto, Y. Reprogramming and differentiation-dependent transcriptional alteration of DNA damage response and apoptosis genes in human induced pluripotent stem cells. *J. Radiat. Res.* **2019**, *60*, 719–728. [CrossRef]
30. Becker, B.V.; Majewski, M.; Abend, M.; Palnek, A.; Nestler, K.; Port, M.; Ullmann, R. Gene expression changes in human iPSC-derived cardiomyocytes after X-ray irradiation. *Int. J. Radiat. Biol.* **2018**, *94*, 1095–1103. [CrossRef]

31. Becker, B.V.; Kaatsch, L.; Obermair, R.; Schrock, G.; Port, M.; Ullmann, R. X-ray irradiation induces subtle changes in the genome-wide distribution of DNA hydroxymethylation with opposing trends in genic and intergenic regions. *Epigenetics* **2019**, *14*, 81–93. [CrossRef]
32. Ewing, B.; Hillier, L.; Wendl, M.C.; Green, P. Base-calling of automated sequencer traces using phred. I. Accuracy assessment. *Genome Res.* **1998**, *8*, 175–185. [CrossRef]
33. Ewing, B.; Green, P. Base-calling of automated sequencer traces using phred. II. Error probabilities. *Genome Res.* **1998**, *8*, 186–194. [CrossRef] [PubMed]
34. Schurch, N.J.; Schofield, P.; Gierlinski, M.; Cole, C.; Sherstnev, A.; Singh, V.; Wrobel, N.; Gharbi, K.; Simpson, G.G.; Owen-Hughes, T.; et al. How many biological replicates are needed in an RNA-seq experiment and which differential expression tool should you use? *RNA* **2016**, *22*, 839–851. [CrossRef] [PubMed]
35. Gene Ontology Consortium. Gene Ontology Consortium: Going forward. *Nucleic Acids Res.* **2015**, *43*, D1049–D1056. [CrossRef]
36. Kanehisa, M.; Goto, S. KEGG: Kyoto encyclopedia of genes and genomes. *Nucleic Acids Res.* **2000**, *28*, 27–30. [CrossRef]
37. Conesa, A.; Madrigal, P.; Tarazona, S.; Gomez-Cabrero, D.; Cervera, A.; McPherson, A.; Szczesniak, M.W.; Gaffney, D.J.; Elo, L.L.; Zhang, X.; et al. A survey of best practices for RNA-seq data analysis. *Genome Biol.* **2016**, *17*, 13. [CrossRef]
38. Del Fabbro, C.; Scalabrin, S.; Morgante, M.; Giorgi, F.M. An extensive evaluation of read trimming effects on Illumina NGS data analysis. *PLoS ONE* **2013**, *8*, e85024. [CrossRef]
39. Liao, Y.; Shi, W. Read trimming is not required for mapping and quantification of RNA-seq reads. *bioRxiv* **2019**. [CrossRef]
40. Williams, C.R.; Baccarella, A.; Parrish, J.Z.; Kim, C.C. Trimming of sequence reads alters RNA-Seq gene expression estimates. *Bmc Bioinform.* **2016**, *17*, 103. [CrossRef]
41. Cock, P.J.; Fields, C.J.; Goto, N.; Heuer, M.L.; Rice, P.M. The Sanger FASTQ file format for sequences with quality scores, and the Solexa/Illumina FASTQ variants. *Nucleic Acids Res.* **2010**, *38*, 1767–1771. [CrossRef]
42. Kukurba, K.R.; Montgomery, S.B. RNA Sequencing and Analysis. *Cold Spring Harb Protoc* **2015**, *2015*, 951–969. [CrossRef]
43. Li, H.; Durbin, R. Fast and accurate long-read alignment with Burrows-Wheeler transform. *Bioinformatics* **2010**, *26*, 589–595. [CrossRef]
44. Krizanovic, K.; Echchiki, A.; Roux, J.; Sikic, M. Evaluation of tools for long read RNA-seq splice-aware alignment. *Bioinformatics* **2018**, *34*, 748–754. [CrossRef]
45. Kim, D.; Paggi, J.M.; Park, C.; Bennett, C.; Salzberg, S.L. Graph-based genome alignment and genotyping with HISAT2 and HISAT-genotype. *Nat. Biotechnol.* **2019**, *37*, 907–915. [CrossRef]
46. Kodama, Y.; Shumway, M.; Leinonen, R.; International Nucleotide Sequence Database, C. The Sequence Read Archive: Explosive growth of sequencing data. *Nucleic Acids Res.* **2012**, *40*, D54–D56. [CrossRef]
47. Patro, R.; Duggal, G.; Love, M.I.; Irizarry, R.A.; Kingsford, C. Salmon provides fast and bias-aware quantification of transcript expression. *Nat Methods* **2017**, *14*, 417–419. [CrossRef]
48. Love, M.I.; Huber, W.; Anders, S. Moderated estimation of fold change and dispersion for RNA-seq data with DESeq2. *Genome Biol.* **2014**, *15*, 550. [CrossRef]
49. Belinky, F.; Nativ, N.; Stelzer, G.; Zimmerman, S.; Iny Stein, T.; Safran, M.; Lancet, D. PathCards: Multi-source consolidation of human biological pathways. *J. Biol. Databases Curation* **2015**, *2015*. [CrossRef]
50. Harris, S.L.; Levine, A.J. The p53 pathway: Positive and negative feedback loops. *Oncogene* **2005**, *24*, 2899–2908. [CrossRef]
51. Georgakilas, A.G.; Martin, O.A.; Bonner, W.M. p21: A Two-Faced Genome Guardian. *Trends Mol. Med.* **2017**, *23*, 310–319. [CrossRef]
52. Pateras, I.S.; Havaki, S.; Nikitopoulou, X.; Vougas, K.; Townsend, P.A.; Panayiotidis, M.I.; Georgakilas, A.G.; Gorgoulis, V.G. The DNA damage response and immune signaling alliance: Is it good or bad? Nature decides when and where. *Pharmacol. Ther.* **2015**, *154*, 36–56. [CrossRef]
53. Bertoli, C.; Skotheim, J.M.; de Bruin, R.A. Control of cell cycle transcription during G1 and S phases. *Nat. Rev. Mol. Cell Biol.* **2013**, *14*, 518–528. [CrossRef]
54. Shu, K.X.; Li, B.; Wu, L.X. The p53 network: p53 and its downstream genes. *Colloids Surf. Biointerfaces* **2007**, *55*, 10–18. [CrossRef] [PubMed]

55. O'Leary, N.A.; Wright, M.W.; Brister, J.R.; Ciufo, S.; Haddad, D.; McVeigh, R.; Rajput, B.; Robbertse, B.; Smith-White, B.; Ako-Adjei, D.; et al. Reference sequence (RefSeq) database at NCBI: Current status, taxonomic expansion, and functional annotation. *Nucleic Acids Res.* **2016**, *44*, D733–D745. [CrossRef] [PubMed]
56. Christmann, M.; Kaina, B. Transcriptional regulation of human DNA repair genes following genotoxic stress: Trigger mechanisms, inducible responses and genotoxic adaptation. *Nucleic Acids Res.* **2013**, *41*, 8403–8420. [CrossRef] [PubMed]
57. Collin, G.; Huna, A.; Warnier, M.; Flaman, J.M.; Bernard, D. Transcriptional repression of DNA repair genes is a hallmark and a cause of cellular senescence. *Cell Death Dis.* **2018**, *9*, 259. [CrossRef]
58. Kanehisa, M.; Sato, Y.; Furumichi, M.; Morishima, K.; Tanabe, M. New approach for understanding genome variations in KEGG. *Nucleic Acids Res.* **2019**, *47*, D590–D595. [CrossRef]
59. Greene, M.W. Circadian rhythms and tumor growth. *Cancer Lett.* **2012**, *318*, 115–123. [CrossRef]
60. Forssell-Aronsson, E.; Quinlan, R.A. The Impact of Circadian Rhythms on Medical Imaging and Radiotherapy Regimes for the Paediatric Patient. *Radiat. Prot. Dosim.* **2017**, *173*, 16–20. [CrossRef]
61. Jim, H.S.; Lin, H.Y.; Tyrer, J.P.; Lawrenson, K.; Dennis, J.; Chornokur, G.; Chen, Z.; Chen, A.Y.; Permuth-Wey, J.; Aben, K.K.; et al. Common Genetic Variation in Circadian Rhythm Genes and Risk of Epithelial Ovarian Cancer (EOC). *J. Genet. Genome Res.* **2015**, *2*. [CrossRef]
62. Kizaki, T.; Sato, S.; Shirato, K.; Sakurai, T.; Ogasawara, J.; Izawa, T.; Ohira, Y.; Suzuki, K.; Ohno, H. Effect of Circadian Rhythm on Clinical and Pathophysiological Conditions and Inflammation. *Crit. Rev. Immunol.* **2015**, *35*, 261–275. [CrossRef]
63. Silvester, N.; Alako, B.; Amid, C.; Cerdeno-Tarraga, A.; Clarke, L.; Cleland, I.; Harrison, P.W.; Jayathilaka, S.; Kay, S.; Keane, T.; et al. The European Nucleotide Archive in 2017. *Nucleic Acids Res.* **2018**, *46*, D36–D40. [CrossRef]
64. Andrews, S. FastQC: A Quality Control Tool for High Throughput Sequence Data. 2010. Available online: https://www.bioinformatics.babraham.ac.uk/projects/fastqc/ (accessed on 10 March 2020).
65. Ewels, P.; Magnusson, M.; Lundin, S.; Kaller, M. MultiQC: Summarize analysis results for multiple tools and samples in a single report. *Bioinformatics* **2016**, *32*, 3047–3048. [CrossRef]
66. Krueger, F. Trim Galore!: A Wrapper Tool around Cutadapt and Cutadapt and FastQC to Consistently Apply Quality and Adapter Trimming to FastQ Files, with Some Extra Functionality for MspI-Digested RRBS-Type (Reduced Representation Bisufite-Seq) Libraries. 2015. Available online: https://www.bioinformatics.babraham.ac.uk/projects/trim_galore/ (accessed on 10 March 2020).
67. Martin, M. Cutadapt removes adapter sequences from high-throughput sequencing reads. *Embnet. J.* **2011**, *17*, 10–12. [CrossRef]
68. Srivastava, A.; Malik, L.; Sarkar, H.; Zakeri, M.; Almodaresi, F.; Soneson, C.; Love, M.I.; Kingsford, C.; Patro, R. Alignment and mapping methodology influence transcript abundance estimation. *bioRxiv* **2019**. [CrossRef]
69. Huber, W.; Carey, V.J.; Gentleman, R.; Anders, S.; Carlson, M.; Carvalho, B.S.; Bravo, H.C.; Davis, S.; Gatto, L.; Girke, T.; et al. Orchestrating high-throughput genomic analysis with Bioconductor. *Nat. Methods.* **2015**, *12*, 115–121. [CrossRef]
70. Gentleman, R.C.; Carey, V.J.; Bates, D.M.; Bolstad, B.; Dettling, M.; Dudoit, S.; Ellis, B.; Gautier, L.; Ge, Y.; Gentry, J.; et al. Bioconductor: Open software development for computational biology and bioinformatics. *Genome Biol.* **2004**, *5*, R80. [CrossRef]
71. R Core Team. R: A Language and Environment for Statistical Computing. *R Found. Stat. Comput.* **2019**. Available online: https://www.gbif.org/tool/81287/r-a-language-and-environment-for-statistical-computing (accessed on 10 March 2020).
72. RStudio Team. RStudio: Integrated Development for R. *RstudioInc.* **2019**. Available online: https://rstudio.com/ (accessed on 10 March 2020).
73. Love, M.I.; Soneson, C.; Charlotte, H.; Johnson, L.; Pierce, N.; Shepherd, L.; Morgan, M.; Patro, R. Tximeta: Reference sequence checksums for provenance identification in RNA-seq. *PLOS Comput. Biol.* **2019**. [CrossRef]
74. Soneson, C.; Love, M.I.; Robinson, M.D. Differential analyses for RNA-seq: Transcript-level estimates improve gene-level inferences. *F1000Res* **2015**, *4*, 1521. [CrossRef]

75. Benjamini, Y.; Hochberg, Y. Controlling the False Discovery Rate: A Practical and Powerful Approach to Multiple Testing. *J. R. Stat. Soc. Ser. B (Methodol.)* **1995**, *57*, 289–300. [CrossRef]
76. Carlson, M. org.Hs.eg.db: Genome Wide Annotation for Human. 2019. Available online: https://bioconductor.org/packages/release/data/annotation/html/org.Hs.eg.db.html (accessed on 10 March 2020).
77. Yates, B.; Braschi, B.; Gray, K.A.; Seal, R.L.; Tweedie, S.; Bruford, E.A. Genenames.org: The HGNC and VGNC resources in 2017. *Nucleic Acids Res.* **2017**, *45*, D619–D625. [CrossRef]
78. Stouffer, S.A.; Suchman, E.A.; DeVinney, L.C.; Star, S.A.; Williams, R.M. *The American Soldier: Adjustment During Army Life*; Princeton University Press: Princeton, NJ, USA, 1949; p. 1.
79. Mosteller, F.; Bush, R.R. Selected quantitative techniques. In *Handbook of social psychology: Vol. 1. Theory and Method*; Lindzey, G., Ed.; Addison-Wesley: Cambridge, UK, 1954; pp. 289–334.
80. Liao, Y.; Wang, J.; Jaehnig, E.J.; Shi, Z.; Zhang, B. WebGestalt 2019: Gene set analysis toolkit with revamped UIs and APIs. *Nucleic Acids Res.* **2019**, *47*, W199–W205. [CrossRef] [PubMed]
81. Khatri, P.; Sirota, M.; Butte, A.J. Ten years of pathway analysis: Current approaches and outstanding challenges. *Plos Comput. Biol.* **2012**, *8*, e1002375. [CrossRef] [PubMed]
82. Szklarczyk, D.; Gable, A.L.; Lyon, D.; Junge, A.; Wyder, S.; Huerta-Cepas, J.; Simonovic, M.; Doncheva, N.T.; Morris, J.H.; Bork, P.; et al. STRING v11: Protein-protein association networks with increased coverage, supporting functional discovery in genome-wide experimental datasets. *Nucleic Acids Res.* **2019**, *47*, D607–D613. [CrossRef] [PubMed]

International Journal of
Molecular Sciences

MDPI

Article

Genome-Wide DNA Alterations in X-Irradiated Human Gingiva Fibroblasts

Neetika Nath [1,2,†], Lisa Hagenau [1,†], Stefan Weiss [1], Ana Tzvetkova [1,2], Lars R. Jensen [1], Lars Kaderali [2], Matthias Port [3], Harry Scherthan [3] and Andreas W. Kuss [1,*]

1 Human Molecular Genetics Group, Department of Functional Genomics, Interfaculty Institute for Genetics and Functional Genomics, University Medicine Greifswald, 17475 Greifswald, Germany; neetika.nath@uni-greifswald.de (N.N.); lisa.hagenau@uni-greifswald.de (L.H.); stefan.weiss@uni-greifswald.de (S.W.); ana.tzvetkova@uni-greifswald.de (A.T.); jensenl@uni-greifswald.de (L.R.J.)

2 Institute of Bioinformatics, University Medicine Greifswald, 17475 Greifswald, Germany; lars.kaderali@uni-greifswald.de

3 Bundeswehr Institute for Radiobiology Affiliated to the University of Ulm, 80937 München, Germany; MatthiasPort@bundeswehr.org (M.P.); scherth@web.de (H.S.)

* Correspondence: kussa@uni-greifswald.de; Tel.: +49-3834-420-5814

† These authors contributed equally.

Received: 8 July 2020; Accepted: 31 July 2020; Published: 12 August 2020

Abstract: While ionizing radiation (IR) is a powerful tool in medical diagnostics, nuclear medicine, and radiology, it also is a serious threat to the integrity of genetic material. Mutagenic effects of IR to the human genome have long been the subject of research, yet still comparatively little is known about the genome-wide effects of IR exposure on the DNA-sequence level. In this study, we employed high throughput sequencing technologies to investigate IR-induced DNA alterations in human gingiva fibroblasts (HGF) that were acutely exposed to 0.5, 2, and 10 Gy of 240 kV X-radiation followed by repair times of 16 h or 7 days before whole-genome sequencing (WGS). Our analysis of the obtained WGS datasets revealed patterns of IR-induced variant (SNV and InDel) accumulation across the genome, within chromosomes as well as around the borders of topologically associating domains (TADs). Chromosome 19 consistently accumulated the highest SNVs and InDels events. Translocations showed variable patterns but with recurrent chromosomes of origin (e.g., Chr7 and Chr16). IR-induced InDels showed a relative increase in number relative to SNVs and a characteristic signature with respect to the frequency of triplet deletions in areas without repetitive or microhomology features. Overall experimental conditions and datasets the majority of SNVs per genome had no or little predicted functional impact with a maximum of 62, showing damaging potential. A dose-dependent effect of IR was surprisingly not apparent. We also observed a significant reduction in transition/transversion (Ti/Tv) ratios for IR-dependent SNVs, which could point to a contribution of the mismatch repair (MMR) system that strongly favors the repair of transitions over transversions, to the IR-induced DNA-damage response in human cells. Taken together, our results show the presence of distinguishable characteristic patterns of IR-induced DNA-alterations on a genome-wide level and implicate DNA-repair mechanisms in the formation of these signatures.

Keywords: radiation doses; repair mechanism; translocation; transition transversion ratio; IR-induced variants; SNVs; InDels; topological associating domains

1. Introduction

Ionizing radiation (IR) plays an important role in medical diagnostics, nuclear medicine, and radiology. Radiation accidents, on the other hand, represent incidental causes of acute radiation exposures. As repeatedly seen in past events, accidental exposition to IR often involved a single acute dose of IR that led to deterministic damages such as localized injuries and acute radiation sickness [1]. Lower doses can induce stochastic effects that are the consequence of faulty repair of genomic lesions such as DNA double-strand breaks (DSBs). The resulting mutations have the potential to influence cellular functions and can alter the fate of the affected cell(s), which may also contribute to an increased risk for the affected individuals to develop cancer.

On a subcellular level, IR can either act directly on the DNA molecule by breaking one or both strands of the DNA backbone or indirectly by creating free radicals, which in turn damage the genetic material [2]. DSBs are the form of IR-induced damage with the highest biological relevance, since un- or misrepaired DSBs can lead to a loss of genomic integrity and trigger apoptosis or cancer development [3–5]. To prevent this, efficient and accurate cellular DNA repair mechanisms are of vital importance. DSBs are typically repaired by one of two mechanisms, non-homologous end joining (NHEJ) and homologous recombination (HR). During NHEJ, the broken ends of the DNA strands are joined back together without the use of a homologous template, whereas HR uses the homologous chromatid or chromosome as a template for error-free repair [6]. A less common repair mechanism, called microhomology-mediated end joining (MMEJ), uses microhomologous sequences to align the broken strands [7]. DNA repair can occasionally lead to chromosomal rearrangements or translocations when two fragments of different chromosomes are erroneously joined and exchanged [8]. Insertions or deletions of varying size (InDels) and single nucleotide variants (SNV) occur much more frequently than translocations. The former variants contribute substantially to genetic variation, although they can lead to a diverse range of diseases when they occur in functional regions of the DNA [9,10].

The advancement of DNA sequencing technology over the past years now allows the gathering of bulk information at the DNA sequence level [11] and to identify and visualize variants such as SNVs [12], InDels [13], and translocations [14] in the complete human genome sequence. In this study, we have explored a comprehensive dataset from human gingiva fibroblast (HGF) cells that were subjected to increasing X-radiation doses followed by repair intervals of 16 h and 7 days. Using the HiSeq (Illumina) technology, we sequenced the whole genome with high (60-fold) coverage in order to investigate the presence and nature of IR and repair-induced mutational signatures at the chromosomal and nucleotide level.

2. Results

To investigate and gather evidence for IR-induced mutational effects in the genome sequence, we took different aspects of structural and functional features of the genome under consideration. Below we present the mutational patterns observed with respect to translocations (2.1), cytogenetic bands (2.2), and topological domains (2.4), showing specific results for chromosome 19 (2.3) as well as characteristic mutational signatures (2.5). The presented data result from two independent Next-Generation Sequencing (NGS) datasets with 60× (dataset 1) and 20× (dataset2) coverage.

2.1. Translocations

First, we determined the occurrence of inter-chromosomal translocations, i.e., the exchange of chromosomal segments between different chromosomes. Figure 1 shows IR-induced translocations for two different DNA repair intervals (16 h and 7 days) in combination with increasing radiation doses (dataset 1, for dataset 2 results see Appendix A Figure A1A).

Overall, we found 19–24 (dataset 1) and 10–16 (dataset 2) inter-chromosomal translocations after a repair interval of 16 h, and 59–90 (dataset 1) and 20–26 (dataset 2) inter-chromosomal translocations after a 7 days repair interval. This corresponds to an approximately 2- to 3-fold increase in inter-chromosomal translocation counts for all samples after a repair interval of 7 days relative to the respective repair interval of 16 h (Table 1). In general, there was no obvious dose effect. Nonetheless, for the 16 h and 7 days repair interval most translocations involve chromosome 2 in both datasets. In dataset 1, we additionally observed an accumulation of translocations involving chromosome 16 after 7 days, with the results generally being more diverse in both 7 days datasets. Overall, these results show that some chromosomes appear more sensitive to radiation than others and that prolonged time for repair and proliferation after a radiation exposure can alter the aberration spectrum seen early after the insult.

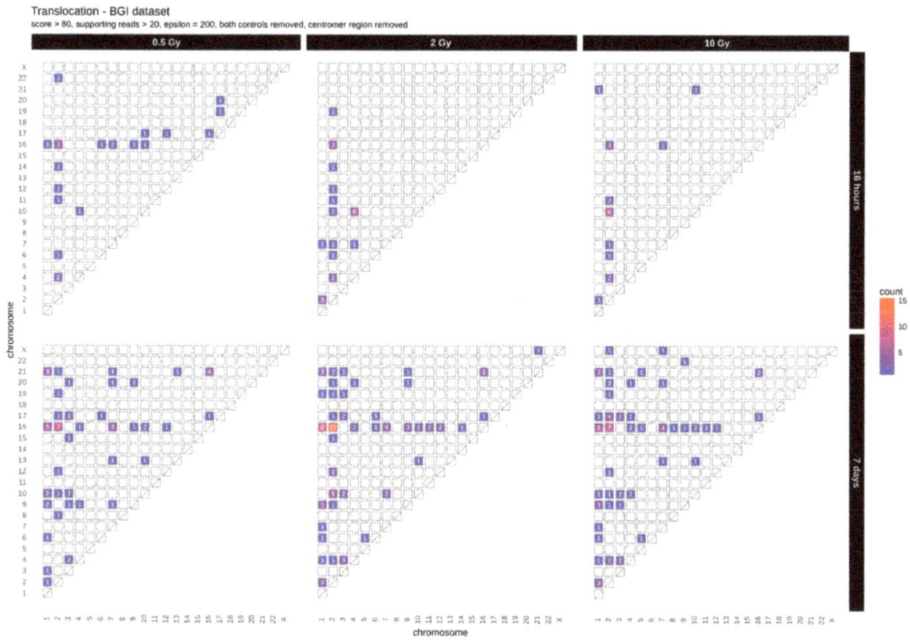

Figure 1. Translocation counts between chromosomes (irrespective of the direction) in genome dataset 1 (generated with BreakDancer).

2.2. Distribution of SNVs and InDels in Genome Dataset

After sample filtering [12], i.e., the removal of sample-specific variants already present prior to IR treatment, approximately 2% of the small variants (SNVs and InDels of insertion length up to 470 nt and deletion length up to 280 nt) remained. These *de novo* IR-induced variants (mutations) were analyzed further (see also Table 1). First, we investigated the genomic accumulation and distribution of SNVs and InDels. We observed few differences in variant accumulation over dose and repair intervals (Figure 2). The maximum increase in variant accumulation was 7.4% for SNVs and 16% for InDels (Table 2) after the 16 h repair interval. There were no consistent dose response effects in our genome sequencing datasets (compare Figure 2 for dataset 1 and Appendix A Figure A2 for dataset 2).

When comparing the occurrence of the two types of variants (SNVs and InDels) before and after IR treatment, we observed a relative increase of InDels as compared to SNVs. The ratio of SNVs:InDels decreased from 5.17–5.37 before sample filtering to 2.04–2.27 after filtering, which represents an approximately 2.5-fold relative increase in InDels among the IR-induced variants (Table 2), a radiation specific effect.

We then looked at the numbers of IR-induced variants per chromosome. The highest variant accumulation was consistently found on chromosome 19 in both datasets and independent of dose and repair interval (Figure 2 for dataset 1 and Appendix A Figure A2 for dataset 2). Chromosome 19 had the highest IR-specific variant accumulation, with this chromosome also showing the highest variant accumulation prior to IR also before sample filtering, i.e., the removal of sample-specific variants already present prior to IR treatment (see Appendix A Figure A1B). However, the differences between the chromosome with the highest (Chr 19) and lowest (Chr 13) number of variants increased 5-fold regarding IR-induced variants. Overall, we saw a spread between highest and lowest variant counts by a factor of about 2 before sample filtering (i.e., in the non-irradiated genome), while this increased to a factor of 10 for the IR-induced variants (Appendix A Figure A1B). In addition to, the ranking of the chromosomes changed with respect to their relative variant load (Compare Figure 2A and Appendix A Figure A1B). For example, while chromosomes 4 and 6 ranked among those with the highest accumulation of variants in control samples (Appendix A Figure A1B), subsequent sample filtering led to a much reduced relative load of IR-induced variants, placing them among the majority of chromosomes with intermediate variant accumulation (Figure 2A,B).

We then aimed to explore IR-induced SNVs and InDels in more detail and focused on the relative occurrence of transitions (Ti, A↔G, and C↔T) and transversions (Tv) on one hand, and the length distribution of the detected InDels on the other. The mean Ti/Tv ratio ranged between 1.11–2.06 (average Ti/Tv = 1.45, for dataset 1) and 1.08–1.65 (average Ti/Tv = 1.25, for dataset 2) among different samples. We failed to note any bias towards any transition or transversion event. After computing individual Ti/Tv values for each chromosome (Figure 2C for dataset 1 and Appendix A Figure A2C for dataset 2), we observed that in dataset 1 chromosome 13 showed the highest mean Ti/Tv ratio with 2.06 at 0.5 Gy after a repair interval of 7 days. Incidentally, chromosome 13 also had the lowest variant accumulation. The lowest Ti/Tv value (dataset 1) was 1.11 on chromosome 20, which showed rather high variant accumulation.

We next analyzed the individual lengths of IR-induced InDels with respect to the reference genome sequence (hg19) by subtracting the nucleotide sequences of our dataset from the reference genome sequence (alt-ref) at the respectively corresponding positions. No significant changes in InDel length distribution were observed after X irradiation as shown in Figure 2D (for dataset 1). In general, deletions (maximum length in dataset 1: 307 nt; in dataset 2: 297 nt) were smaller than insertions (maximum length in dataset 1: 652 nt; in dataset 2: 672 nt). Mononucleotide and dinucleotide InDels were most abundant. Deletions showed a slightly less steep decrease in occurrence with increasing size than insertions (Figure 2D, dataset 1 and Appendix A Figure A2D for dataset 2).

Table 1. Counts of IR-induced variants (sequence annotation with SNPEff annotator) as well as translocations for all samples.

Time	Radiation Dose	Functional Impacts				Inter-Chromosomal Translocations
Dataset 1		High	Moderate	Low	Modifier	
16 h	0.5 Gy	14	189	3785	33,384	22
	2 Gy	21	226	3800	35,714	24
	10 Gy	15	222	3984	35,514	19
7 Days	0.5 Gy	18	216	3408	33,189	59
	2 Gy	12	263	3717	35,336	90
	10 Gy	22	208	3980	36,754	74
Dataset 2						
16 h	0.5 Gy	19	277	4849	50,437	14
	2 Gy	49	262	5119	52,567	10
	10 Gy	30	432	6107	60,561	16
7 Days	0.5 Gy	62	1857	14,227	76,636	23
	2 Gy	36	475	7416	69,550	26
	10 Gy	40	210	11,140	68,032	20

Table 2. Table reporting the depth (DP) and variant counts after and (before) filtering steps.

Time	Radiation Dose	SNVs Depth (DP) After (Before) Filter	InDels Depth (DP) After (Before) Filter	SNVs Counts After (Before) Filter	InDels Counts After (Before) Filter
Dataset 1					
16 h	0.5 Gy	1,029,599 (206,573,706)	334,619 (25,449,747)	34,261 (3,839,076)	15,083 (715,406)
	2 Gy	1,098,772 (221,379,800)	372,440 (27,846,456)	36,674 (3,859,535)	17,982 (739,759)
	10 Gy	1,101,857 (222,284,695)	379,065 (28,064,422)	36,435 (3,854,999)	17,098 (734,831)
7 Days	0.5 Gy	1,030,578 (217,184,675)	352,673 (26,948,643)	34,078 (3,846,812)	15,698 (724,272)
	2 Gy	1,092,867 (217,841,944)	369,254 (27,518,060)	36,256 (3,850,694)	17,188 (734,633)
	10 Gy	1,142,613 (236,100,648)	397,153 (29,973,246)	36,836 (3,778,441)	17,518 (730,883)
Dataset 2					
16 h	0.5 Gy	1,166,694 (130,621,022)	242,653 (6,073,404)	50,516 (3,848,840)	16,987 (252,782)
	2 Gy	1,137,296 (123,863,517)	234,141 (5,764,483)	52,666 (3,852,339)	17,675 (254,521)
	10 Gy	1,474,964 (153,573,569)	351,819 (7,637,043)	60,688 (3,901,398)	25,620 (269,267)
7 Days	0.5 Gy	2,268,357 (187,652,991)	478,893 (9,473,714)	77,818 (3,953,911)	29,045 (274,854)
	2 Gy	1,825,231 (17,5864,812)	432,240 (8,902,319)	69,691 (3,930,167)	27,463 (272,142)
	10 Gy	387,902 (84,432,203)	266,508 (1,973,176)	27,251 (3,510,867)	28,387 (91,240)

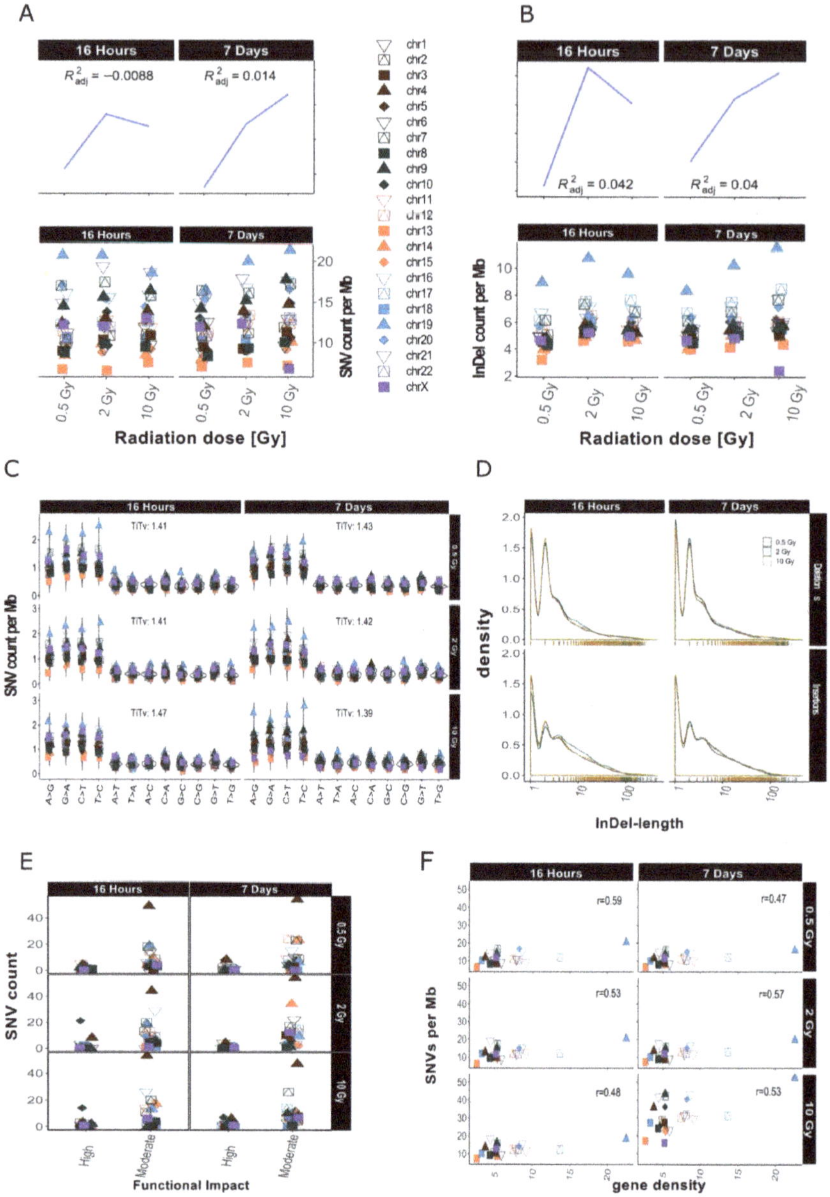

Figure 2. Overview for IR-induced SNV accumulation in genome dataset 1 (for genome dataset 2 see Appendix A Figure A2). Accumulation of SNVs per Mb ((**A**), lower panel) and InDels per Mb ((**B**), lower panel) per chromosome is shown together with the corresponding trends with adjusted root mean square (upper panels). (**C**): SNV types, i.e., as transitions (Ti) and transversions (Tv) per MB per chromosome and Ti/Tv values for each sample. (**D**): Distribution of insertions and deletions with respect to size. (**E**): SNV count per chromosome with respect to functional impact (HIGH and MODERATE functional consequences only). (**F**): SNVs per Mb per chromosome with respect to relative chromosomal gene density. Gene density was computed based on hg19 (GRCh37) using the custom gene set (excluding predicted genes) described b [17].

Next, we explored the functional impact of the IR-induced alterations by annotating these variants. Figure 2E (Appendix A Figure A2E for dataset 2) shows functional impact information based on parameters of gene function as defined by sequence ontologies for dataset 1. Overall, the majority of SNVs were annotated as "modifiers" (which have an even smaller functional impact than the so-called "low-impact" variants) in both datasets (Table 1). High impact was predicted for 12–62 SNVs and the number of variants with moderate impact ranged from 189 to 1857 (Table 1; see also Appendix A Table A1 for more detailed information with respect to chromosome 19). There was, however, no apparent association with dose or repair interval duration.

Previous studies have shown that gene density might play a role in chromosomal susceptibility to radiation damage and/or DNA repair priority [15,16]. Gene density, i.e., the number of genes per number of base pairs in Mb, varies considerably between chromosomes. The chromosome with the highest gene density of 22.53 genes/Mb is chromosome 19, while chromosome 13 has the lowest gene density, with 2.65 genes/Mb. To study whether the accumulation of IR-induced variants per chromosome correlates with gene density, we plotted SNV counts per Mb against gene density [17] and computed correlation scores (see Figure 2F for dataset 1, Appendix A Figure A2F for dataset 2). These were between 0.48 and 0.59, showing a moderately positive linear relationship between gene density and SNV accumulation. This means that gene density alone does not explain the observed variance in SNV accumulation between different chromosomes.

In summary, a clear dose responses relationship could not be obtained for the mutations in our whole genome WGS datasets. We found a relative increase in the number of InDels and observed a rise in variation of variant counts at the chromosomal level among the IR-induced variants. The observed Ti/Tv ratio was lower than the Ti/Tv ratio of the reference genome. Most of the variants had a moderate predicted functional impact and this functional impact did not follow any obvious dose effect. Variant accumulation on the other hand showed a positive correlation with gene density.

2.3. *Intrachromosomal Distribution of SNVs and InDels*

As an example for the intrachromosomal distribution and accumulation of IR-induced sequence alterations, a more detailed analysis of the mutational spectrum of the gene-rich chromosome 19 is presented, which accumulated the highest frequency of IR-induced DNA variants per Mb (see Figure 2A,B). The circular plot in Figure 3A (dataset 1) shows the distribution of SNVs and InDels across the cytogenetic bands of chromosome 19 (for dataset 2 see Appendix A Figure A3A). A large number of variants accumulated in the pericentromeric region p12 with 303 SNVs (0.5 Gy), 295 SNVs (2 Gy), and 183 SNVs (10 Gy) 16 h after IR treatment, and 183 SNVs (0.5 Gy), 251 SNVs (2 Gy) and 240 SNVs (10 Gy) after a 7-day repair interval. For InDels in the pericentromeric region, the numbers were between 49 and 65 for all samples. In addition to that, two other cytogenetic bands (p13.3 and p13.2) on the short arm of chromosome 19 showed increased accumulation of SNVs and InDels. Interestingly, these regions differ markedly in GC-content and gene density with p13.2 being a gene-rich region with high GC content, while p12 is AT-rich and gene-poor.

Subsequently, we focused on topologically associating domains (TADs) as a functional feature of the genome [18]. TADs are self-interacting genomic regions suggesting preferential interaction between loci within these domains relative to interactions to loci outside the domain borders. TADs span from tens to thousands of kilobases and are stable across different cell types and conserved across species (for further information see e.g., [18]). These domains are not evenly distributed throughout individual chromosomes, and their borders differ from those of cytogenetic bands. We explored the distribution of variants across TADs and looked at variant distribution inside and outside the borders of TADs (Figure 3B,C for dataset 1, and Appendix A Figure A3B,C for dataset 2). In order to do that, we binned regions inside and outside of TADs into ten equally long portions. In chromosome 19, we observed that variant accumulation was highest for both SNVs and InDels inside of the largest TAD, which corresponds to cytogenetic band p12. We also found that the variant count in chromosome

19 was always higher inside than outside of TADs with an inside:outside ratio of about 2. Furthermore, most variants were accumulated at the border of TADs.

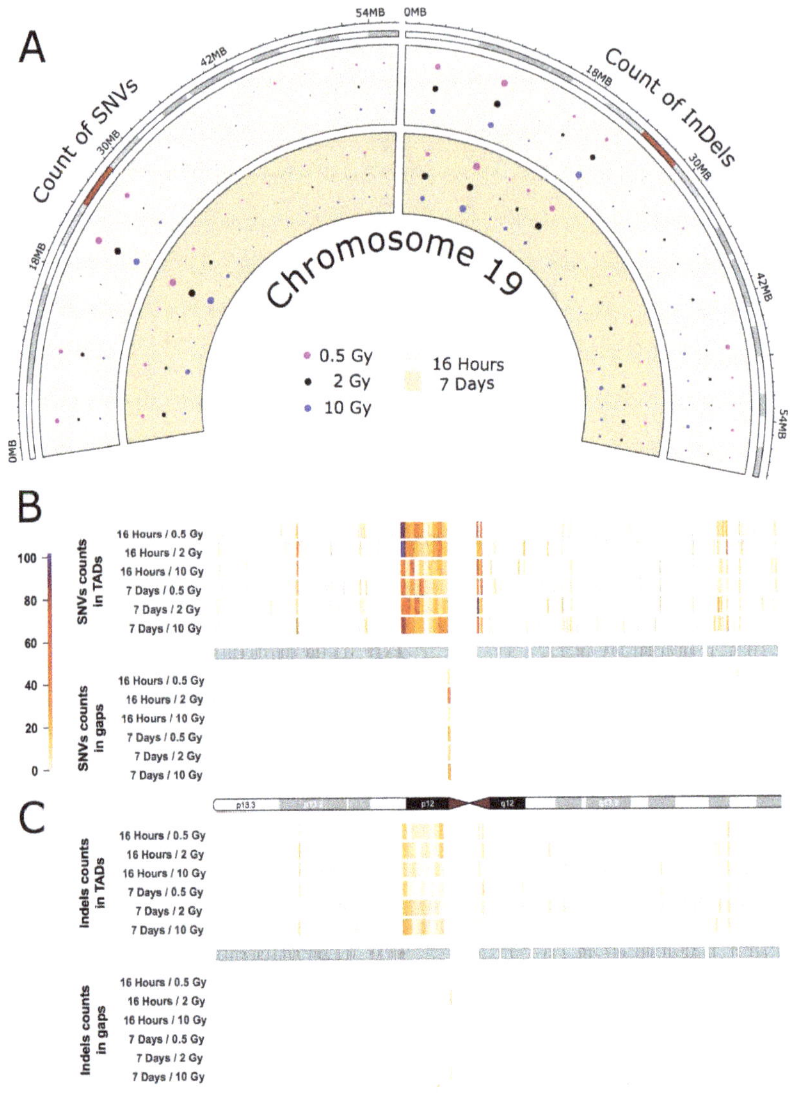

Figure 3. IR-induced variant accumulation across chromosome 19 (dataset 1). (**A**): Distribution of SNVs and InDels over cytogenetic bands. Topologically associating domain (TAD) plot showing the distribution of SNV (**B**) and InDel (**C**) counts inside and outside of TAD coordinates for chromosome 19. The chromosome is represented in the middle of the figure as an ideogram. Each panel is subdivided, showing the SNV and InDel alteration counts inside (above) and outside of TADs (below). The TADs are shown as cyan blocks symmetrically in both panels. Each TAD and each gap (region outside TAD termed as gap) was binned into 10 equally long portions and the counts of the SNVs or InDels were then calculated for each portion. The heatmaps with color gradients from white through orange and red to dark purple represent values ranging from 0 to 104 of variant counts.

The other chromosomes, while having accumulated smaller numbers of IR-induced alterations showed similar results with respect to variant distribution and TAD effects: On all chromosomes, band- and TAD-related patterns of variant accumulation could be observed (for more example see Appendix A Figures A4 and A5).

Next, we investigated putative functional consequences of the IR-induced variants observed on chromosome 19 as defined by sequence ontology [19], again binned over cytogenetic bands (Figure 4, dataset 1, and Appendix A Figure A6, dataset 2). We observed only a few high impact sequence ontologies on 4 different genes on chromosome 19 in the two combined datasets (Appendix A Table A1). Overall, the functional impact of the variants on chromosome 19 is similar to the genome-wide functional impact distribution (Table 1) and we did not find any major difference between chromosome 19 and the other chromosomes with respect to this feature.

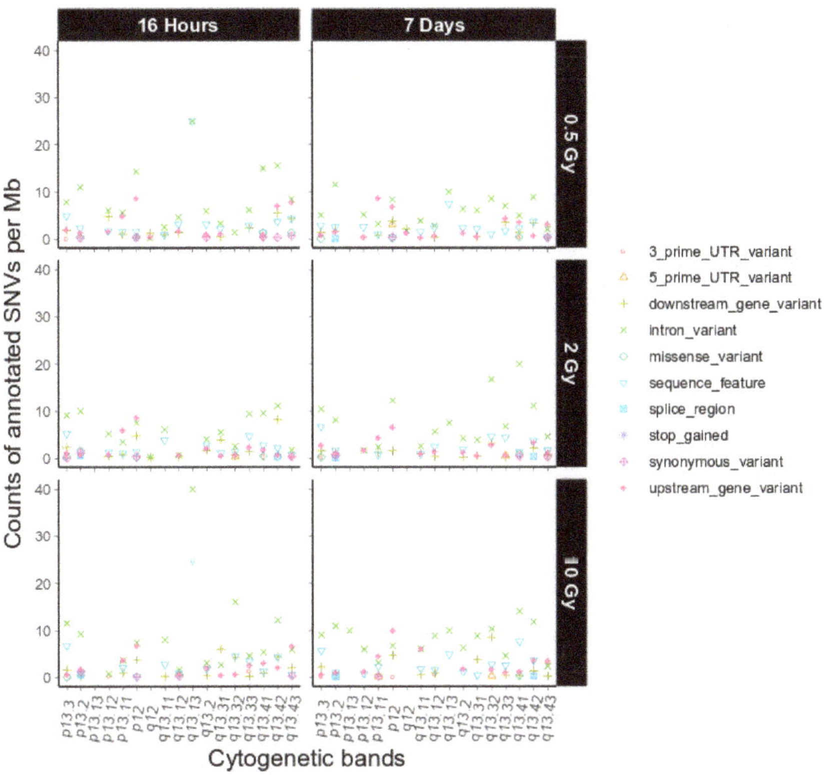

Figure 4. Sequence ontological information for the IR-induced variants observed on chromosome 19 over cytogenetic band, filtered for refseq IDs for dataset 1. For further explanation of the used combination of sequence ontology terms see Section 4.9. Functional Annotation.

2.4. *InDel-specific Signatures*

In order to obtain further clues as to the possible effects of the mechanisms involved in repairing IR-induced DSBs we investigated the occurrence of InDels with respect to their genomic vicinity, i.e., their occurrence in repeat regions, areas of microhomology or regions without any distinguishing features.

Figure 5 shows different short IR-induced InDel (see above) signatures as detected in dataset 1 (for dataset 2 see Appendix A Figure A7) before and after sample filtering (See Table 3 for pertaining terms and definitions). Fold change analysis showed 1.74-fold reduction in insertion signatures after performing filtering steps. For the most part, filtering for IR-induced variants did not have an obvious effect on the distribution patterns. Interestingly, however, we observed a marked increase in the relative abundance of 3 nucleotide deletions in genomic sequence sections without repeat or microhomology, which could be considered characteristic for IR-induced InDel signatures.

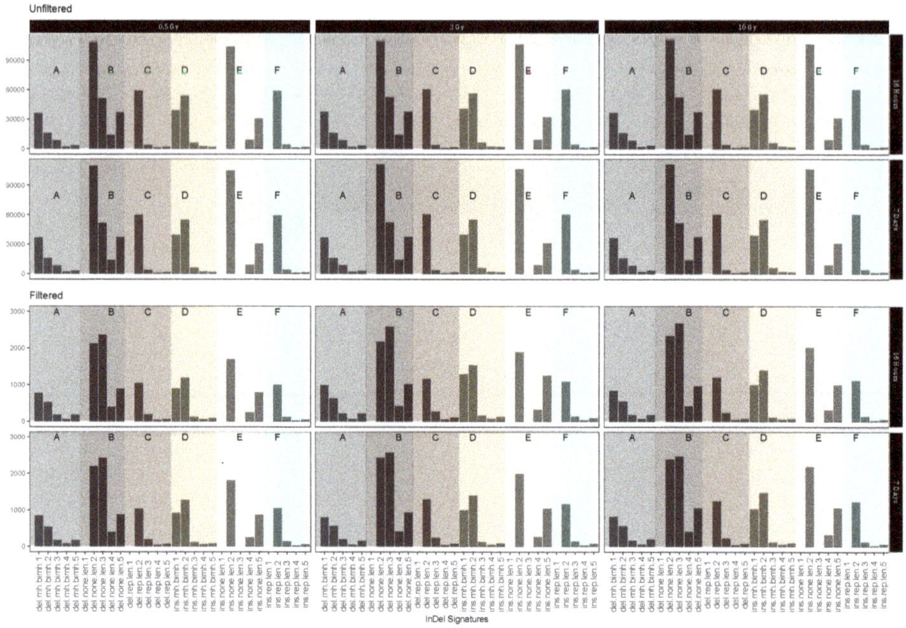

Figure 5. Mutational IR-induced signature plot (dataset 1). On the X-axis, the length and specific vicinity of InDels are represented, while the Y-axis (note differing scales) shows the InDel count. See Table 3 for pertaining terms and definitions.

Table 3. Terms and definitions used for mutational signatures in Figure 5 and Appendix A Figure A7 with average fold change from unfiltered and filtered data.

Terms	Definition	Average Log Fold Change Form Filtered to Unfiltered
Del.mh.bimh.x (A)	Deletion in area of microhomology	−1.66
Del.none.len.x (B)	Deletion in undistinguished area	−1.69
Del.rep.len.x (C)	Deletion in repeat area	−1.72
Ins.mh.bimh.x (D)	Insertion in area of microhomology	−1.70
Ins.none.len.x (E)	Insertion in undistinguished area	−1.74
Ins.rep.len.x (F)	Insertion in repeat area	−1.74

3. Discussion

In this study, we analyzed the IR-induced mutational spectrum of IR-exposed HGFs by whole-genome sequencing. We investigated multiple aspects of IR-induced genomic alterations including translocations and distribution of small variants, and their relationship to structural and functional features of the genome.

3.1. Dose-Response Correlation

Interestingly, after variant calling and filtering, there was no clear dose dependency observed in either dataset both for SNVs and InDels (for dataset 1 Figure 2; Appendix A Figure A2A,B). This contrasts with a previous exome study with a similar experimental setup where we investigated IR-induced variants in exomes of cells exposed to varying doses of X-rays and observed a dose response [12]. Such a dose response may reflect different efficacies of DNA repair pathways on different portions of the genome, of which transcription coupled base excision repair (TCR) is effectively monitoring and repairing the transcribed portion of the genome [20], while homologous recombination and NHEJ contribute to DSB repair in heterochromatin [21]. A dose-independent response on genome level, however, is not inconceivable and might reflect highly efficient DNA repair mechanisms even in cells exposed to a high dose of sparsely ionizing photon radiation. It was found that 99% of all DSBs are correctly repaired in cells after a radiation dose of 2 Gy [22] and that fetal fibroblasts undergo more efficient DNA repair after high-dose photon irradiation [23]. Since human fibroblasts are quite resistant to IR-induced apoptosis and undergo extended G1 checkpoint arrest for days after 10 Gy photon irradiation [24], our 10 Gy X irradiation will predominantly have resulted in reproductive cell death. Furthermore, cells that underwent apoptotic cell death (less than 1.5%; G. Schrock, unpublished observations; [24]) were likely lost from the sequenced population. Additionally, high dose-triggered more efficient DNA repair in fibroblasts [23] could have removed most of the damage after the 10 Gy dose. This may be one way to explain why the number of high-impact, non-lethal mutations were similar for the cells exposed to low or high dose of X radiation (Table 1). Further investigations will have to address DNA repair efficiency to elucidate DNA damage repair and load in irradiated HGFs.

3.2. Mutational Signatures on a Genomic Level

In keeping with a previous study, where IR-induced DNA damage was detected as large-scale rearrangement of the genome, and sensitive chromosome regions were identified [25], we also found evidence that some chromosomes appear more sensitive to radiation than others.

For example, chromosome 2 and 16 contained the highest numbers of translocations (Figure 1). These are likely a result of DSB misrepair where broken DNA strands from different chromosomes get wrongly joined [26]. A recent study correlating chromosomal territories with IR-induced translocations in lymphocytes found that chromosomes 2 and 16 have a high number of interactions and are found in proximally positioned chromosomes [27]. It is possible that the high number of translocation events between chromosome 2 and 16 in our study also reflects spatial proximity in the nucleus. In addition, chromosome 16 has already been shown to be radiation sensitive, which might explain the high number of translocations for its small size [28]. This also correlates with high SNV and InDel rates observed in our study (Figure 2A,B).

We found a lower ratio of SNVs:InDels i.e., a relative increase in InDels, in both datasets when comparing variant counts before and after removing non-IR-induced variants (see Table 2). In a previous study, Adewoye et al. analyzed germline mutation rates induced in mice after parental exposure to IR and found that the rate of induction of *de novo* InDels compared to controls was significantly increased, while the rate of SNVs was not significantly elevated [13]. InDels accumulation after IR treatments is likely caused by additional DSBs induced by IR and has been observed in mice, IR-associated tumor cells, and induced pluripotent stem cells (iPSCs) [13,29–31]. Taken together, this suggests a radiation-specific effect.

When looking at the SNV-type spectrum after filtering at the chromosomal level we did not see any bias towards a specific transition or transversion event (Figure 2C). However, we did observe an IR-induced reduction in Ti/Tv ratios after filtering. In the human genome, the mean Ti/Tv ratio is approximately 2.1, while we calculated a mean Ti/Tv ratio of 1.45 for IR-induced SNVs across all samples. This is in line with different studies in plants and mammals [32,33], who report a conspicuously high proportion of transversions after irradiation. Moreover, it is consistent with our observations in exome data from HGF cells that received an identical treatment as the cells used in this study [34]. A high frequency of transversions are considered a feature of oxidative DNA damage [35] and could here indicate the indirect effect of radiation through hydroxyl radicals by radiolysis of water [6]. On the other hand, mismatch repair (MMR) also leads to an increased number of transversions [36]. While MMR is usually coupled to DNA replication and highly effective in repairing replication errors, several studies have shown cell cycle-independent recruitment of MMR proteins to DNA lesions after treatment with UV light and alkylating agents and propose a mutagenic role for MMR in certain contexts [37,38].

3.3. *Intrachromosomal IR-Related Mutation Patterns*

We showed that the accumulation of both SNVs and InDels after irradiation differs considerably between chromosomes (Figure 2A,B) and while there are differences in the numbers of mutations per chromosome in the non-irradiated controls, the variance there is not as high as in the irradiated samples. This means that some chromosomes are more affected by ionizing radiation than others.

Chromosome 19 showed the highest load of IR-induced alterations under all experimental circumstances, while Chromosome 13 showed substantially less mutational damage. InDels, for example, occur as much as 10 times more often on chromosome 19 as compared to chromosome 13. As these chromosomes are on different ends of the gene density spectrum, we investigated whether the mutational accumulation is correlated to gene density in the whole genome. We observed a moderate positive correlation for gene density and mutation accumulation in our data (Figure 2F), which is in line with previous research, showing that gene-rich chromosomes are more susceptible to DNA damage [16,39].

But also on an intrachromosomal level, patterns of mutation accumulation can be observed in the individual chromosomes. For example, we observed the highest overall rate of SNVs and InDels in Chr19 p12, adjacent to the centromere (Figure 3), where also 1 SNV with putatively deleterious effect is located (affecting ZNF93). This region is relatively gene-poor with a low GC content (40.7%) and a high density of repeat elements and pseudogenes [40] as well as a large cluster of genes encoding zinc finger transcription factors, which are exclusive to primates [41,42]. This contrasts with the finding that gene density is a moderately good predictor for mutation accumulation. Apart from gene density, transcriptional activity and chromatin structure have also been suggested to influence DNA damage and repair [43,44]. It is thus likely that only a combination of various factors will be able to fully explain the differential mutation accumulation patterns/signatures in our data.

Previously, it has been discussed that IR-induced lesions can also have an impact on the 3D organization of the genome and recent results from Hi-C experiments show a decrease in interactions across TAD boundaries after exposure to IR in human fibroblasts [14]. The authors suggest that this could play a role in the ability of cells to maintain genome integrity and prevent deleterious translocations across TAD boundaries [14]. In this study, we evaluated the distribution of IR-induced alterations inside and outside of TADs and found that both SNVs and InDels inside of TADS tend to accumulate towards the TAD boundaries in most chromosomes (see also Appendix A Figures A4 and A5). In mammals, TAD boundaries show enrichment of CCCTC-binding factor (CTCF) binding (for recent review see e.g., [45]), which increases with higher insulation scores [14]. Increased CTCF recruitment to TAD boundaries as a consequence of IR treatment could, therefore, be expected to result in enhanced DNA repair in these areas, as CTCF was found to determine the nanostructure of the γ-H2AX DSB repair domains [46] and to facilitate DNA double-strand break repair by enhancing homologous recombination repair [47,48]. However, it is not yet fully understood in which way CTCF exerts its various functions. It may thus be speculated that CTCF at the bases of TADs will provide a genome-wide distributed pool of chromatin-bound CTCF. If a DSB occurs, CTCF will keep the chromatin loops in its vicinity from flaring out and even facilitate or participate in their repair by its role in HR repair of DSBs.

So, it may be speculated that CTCF binding to the flanking chromatin loops around a DSB will lead to close accumulation of DSB-flanking DNA, thereby facilitating its repair, which is further enhanced given that CTCF is functioning in HR repair. Thus, IR-dependent recruitment of CTCF to TADs may be considered to contribute to the IR-dependent mutation pattern observed.

Interestingly, we consistently observed a relative increase in triplet deletions among IR-induced InDels in areas without repetitive or microhomologous features. This points to a so far unexplored bias of the involved DNA-repair mechanisms in these contexts and might indicate the possibility of an increased tolerance towards in-frame deletions. Taken together, our results reveal distinguishable and characteristic patterns of IR-induced DNA-alterations on a genome-wide level. In keeping with previous findings from studies in *C. elegans* [47] we come to the conclusion that these signatures are the joint product of DNA damage, repair and tolerance.

4. Materials and Methods

4.1. Cell Culture and IR Treatment

Primary HGFs were obtained from Provitro AG (Berlin, Germany), cultured, and treated as previously described [49]. Confluent cell cultures were irradiated with 240 kV X rays at 13 mA (YXLON Maxishot, Hamburg, Germany) filtered with 3 mm beryllium at a dose rate of 1 Gy min^{-1}. Exposure to 0, 0.5, 2 and 10 Gy was combined with two repair intervals (16 h and 7 days, see also Table 2).

4.2. Next-Generation Sequencing Analysis

DNA was prepared using the NucleoSpin® Tissue Kit from Macherey-Nagel (Düren, Germany) and used for NGS library preparation and sequencing. Two independent datasets with 60× (dataset 1) and 20× coverage (dataset 2) were generated, which were sequenced by the BGI Tech Solutions Co. (Hong Kong), and Max Planck Institute for plant breeding research (Cologne, Germany) sequencing facilities, respectively. One Sample (10Gy/7d) from dataset 2 had a 10 × coverage.

4.3. Bioinformatics Analysis

The bioinformatics processing was performed as follows: Paired-end reads obtained through WGS were trimmed using trimmomatric followed by mapping using tophat and CUSHAW to the hg19 (GRCh37) version of the human reference genome. After mapping, we adapted variant calling steps from Genome Analysis Toolkit (GATK) for version 3.3. The corresponding variants were assessed using R version 3.6.1 to perform filtering and further analysis as previously published (for further details see [12,31].

4.4. Data Filtering

In our analysis, we have implemented three filtering steps: (1) sample filtering, (2) coverage filtering and (3) Binomial filtering. For sample filtering, we used variant pools from untreated control samples for both repair intervals to remove IR-independent and cell line-specific variants [41]. For coverage filtering, we used our previously published parameter settings, of a sequencing depth (DP) value of 3 and genotype quality (GQ) value of 20 [31]. For binomial filtering, we have considered an error rate of 0.001 to remove the artifact that is due to sequencing (see again also [31]).

4.5. Translocation Analysis

Translocation analysis was conducted using BreakDancer version 1.4.5 [50] to predict interchromosomal translocations from paired-end sequencing reads for each sample. As suggested by the manual, we used BreakDancer's bam2cfg.pl to create the parameter configuration for each sample following the structural variant analysis using breakdancer-max to determine translocations. For each sample, we removed translocations that were also present in the control samples to keep only translocations that were induced by the repair mechanism after radiation of cells. Translocations were defined as equal if they were present within the same 200-nucleotide window. We then only kept translocations containing at least 20 supported reads as well as a confidence score greater than 80 percent.

4.6. Variant Accumulation Analysis

Variant (V) accumulation analysis was performed for chromosomal bands and TADs Variant counts per Mb were normalized over chromosomal band length using:

$$V/Mb = 1{,}000{,}000 \times (\text{Count of V per cytogenetic band})/(\text{Total length of cytogenetic band}) \quad (1)$$

For TAD-based analyses we used the human TAD coordinates from the Hi-C project [18]. We downloaded the 'TADs in the hg19' folder from the 3D Genome Browser (http://promoter.bx.psu.edu/hi-c/publications.html, accessed on 4 June 2020) that actualizes the TAD coordinates from the Dixon et al. [18] pipeline. To ensure a close match with the cell type used in our study we choose the file "IMR90_Lieberman-raw_TADs", which was generated based on fibroblasts.

4.7. SNV-based analyses

SNVs were subdivided into transitions and transversions to investigate the different substitution types. Transitions refer to a pyrimidine-pyrimidine (C↔T) or purine-purine substitution (A↔G), whereas transversions refer to pyrimidine-purine substitutions or vice versa (A↔C, A↔T, C↔G, G↔T). To compute the transition (Ti) and transversion (Tv) ratio:

$$nTi/nTv = (\text{Transition count}/\text{Transversion count}) \quad (2)$$

4.8. InDel-Based Analyses

InDel counts were further subdivided into numbers of insertions and deletions; nucleotides were counted in the reference genome sequence as "ref" and in the corresponding position in the genome sequence generated in this study as "alt". If count (ref) > count(alt) a deletion was called, while count(ref) < count(alt) was considered an insertion.

In order to identify microhomologous mutational signatures for InDels in our dataset, we use the adapted functions in the R package called "mutSigExtractor" [7]. For InDel signatures, the definitions are dependent on signatures/contexts counts within the data. Basically, InDel signatures were classified first as insertions or deletions, second, according to the length of possible flanking mononucleotide repeats, and third, possible areas of microhomology at InDel boundaries. See Table 3 for the terms used for InDel classification and average log fold changes for the respective alterations.

4.9. Functional Annotation

VCF files were annotated with SnpEff and then processed to summarize the contained sequence ontology information. By default, SnpEff uses the Ensembl transcripts to search for genomic features overlapping the location of the variants. Thus, it is often possible that a specific variant overlaps the same gene ontological feature from different transcripts. In such cases, we counted the feature type for the chromosomal location only once. As the Ensembl transcripts are quite numerous it is also possible to have several different feature types for the same location. To reduce background noise by counting of annotation more than once we used only the protein-coding transcripts included in the "known" Reference Sequence (RefSeq) collection. Finally, we combined some of the features into less specific ones (Table 4).

Table 4. Definitions of combined sequence ontology terms used in this study.

SnpEff Sequence Ontology	Combined Terms
splice_region_variant&intron_variant; splice_region_variant; splice_region_variant&synonymous_variant; splice_region_variant&non_coding_exon_variant	splice_region
splice_acceptor_variant&intron_variant; splice_donor_variant&intron_variant; splice_acceptor_variant&splice_donor_variant&intron_variant	splice_site
missense_variant&splice_region_variant	missense_variant
stop_retained_variant	synonymous_variant
stop_gained&splice_region_variant	stop_gained
5_prime_UTR_premature_start_codon_gain_variant	5_prime_UTR_variant

Author Contributions: Conceptualization, H.S. and A.W.K.; Data curation, N.N. and L.R.J.; Formal analysis, N.N., L.H., S.W. and A.T.; Funding acquisition, A.W.K.; Investigation, N.N., L.H., S.W. and A.T.; Methodology, N.N., L.H., S.W. and A.T.; Writing—original draft, N.N., L.H., A.W.K. and H.S.; Writing—review & editing, S.W., A.T., L.R.J., L.K., H.S., M.P. and A.W.K. All authors have read and agreed to the published version of the manuscript.

Funding: A.W.K. received funding for this project through the German Federal Ministry of Defense (E/U2AD/CF520/DF554).

Acknowledgments: We thank Jessica Müller, Gerrit Schrock, Christian Sperling and Corinna Jensen for excellent technical help. We acknowledge support for the article processing charge from the DFG (German Research Foundation, 393148499) and the Open Access Publication Fund of the University of Greifswald.

Conflicts of Interest: The authors declare no conflict of interest.

Abbreviations

IR	Ionizing radiation
DSB	Double strand break
NHEJ	Non-homologous end joining
HR	Homologous recombination
MMEJ	Microhomology-mediated end joining
InDel	Insertion or deletion
SNV	Single Nucleotide Variants
HGF	Human Gingiva Fibroblasts
NGS	Next generation sequencing
Ti	Transition
Tv	Transversion
WGS	Whole genome sequencing
TAD	Topological Associating Domain
TCR	Transcription-coupled repair
iPSCs	Induced pluripotent stem cells
MMR	Mismatch repair
CTCF	CCCTC-binding factor
GATK	Genome Analysis Toolkit
DP	Sequencing Depth, number of reads during variant calling step
GQ	Genotype Quality based on Phred-Scaled confidence of assigned genotype
RefSeq	Reference Sequence

Appendix A

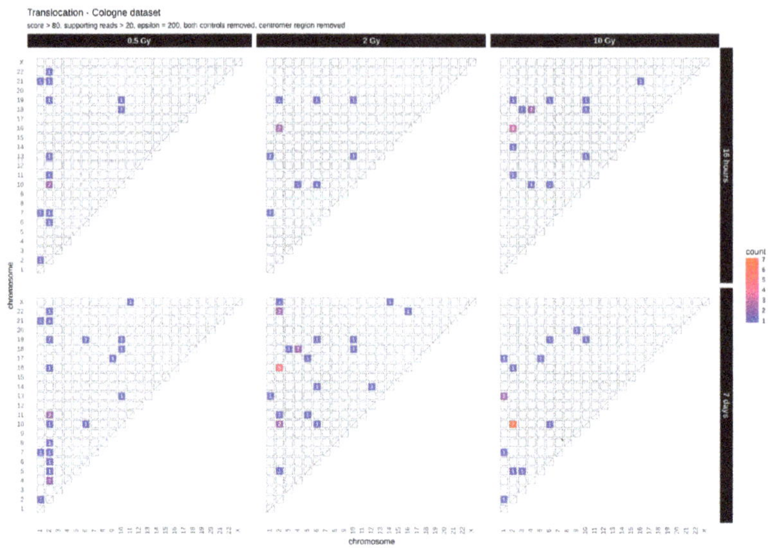

(**A**)

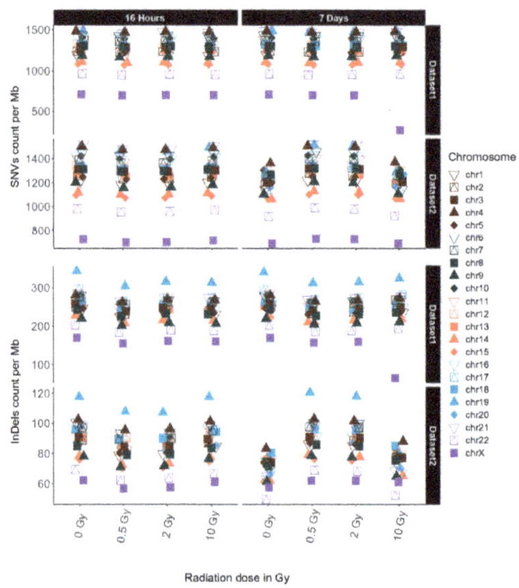

(**B**)

Figure A1. (**A**): Translocation counts between chromosomes (irrespective of the direction) in genome dataset 2 (generated with BreakDancer.). (**B**): Overview for genome dataset 1 and dataset 2 showing accumulation of SNVs per Mb (upper panel) and InDels per Mb (lower panel) per chromosome for unfiltered data.

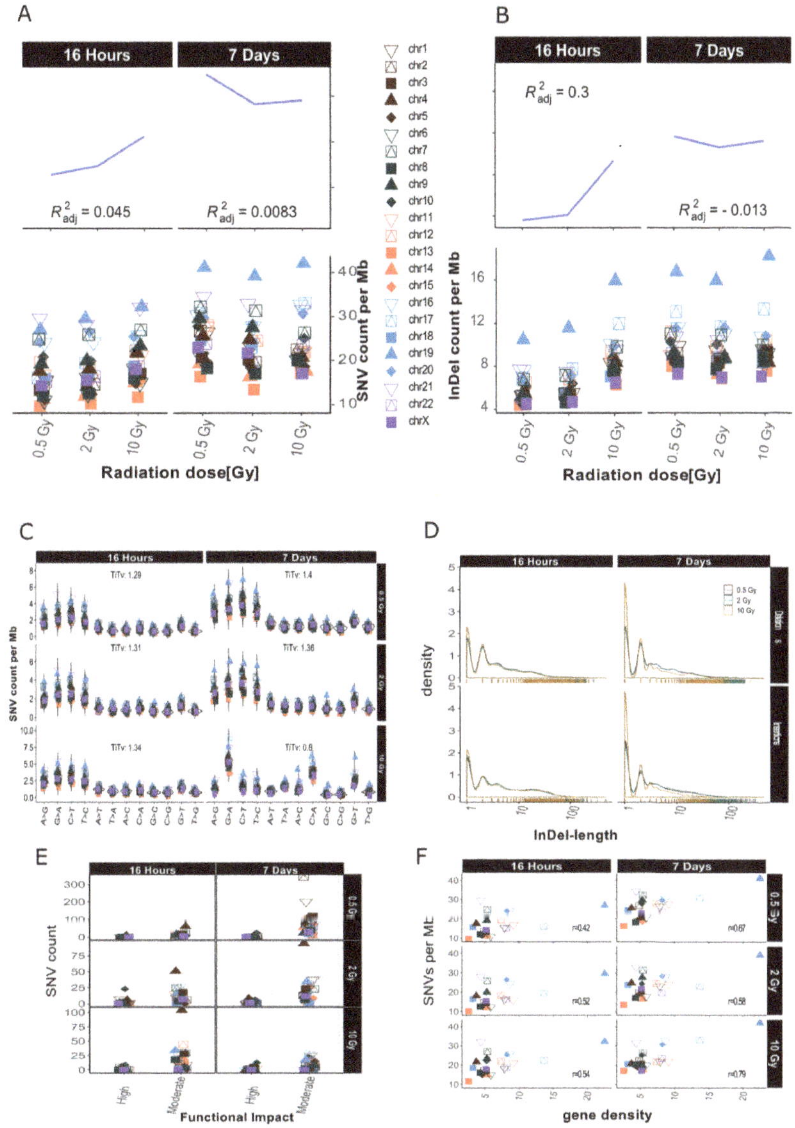

Figure A2. Overview for IR-induced SNV accumulation in genome dataset 2. Accumulation of SNVs per Mb ((**A**), lower panel) and InDels per Mb ((**B**), lower panel) per chromosome is shown together with the corresponding trends with adjusted root mean square (upper panels) for one sample with lower coverage (10Gy/7d) a scaling factor of 2.5 was used for visualization purposes (graphs A and B only). (**C**): SNV types, i.e., as transitions (Ti) and transversions (Tv) per MB per chromosome and Ti/Tv values for each sample. (**D**): Distribution of insertions and deletions with respect to size. (**E**): SNV count per chromosome with respect to functional impact (HIGH and MODERATE functional consequences only). (**F**): SNVs per Mb per chromosome with respect to relative chromosomal gene density. Gene density was computed based on hg19 (GRCh37) using the custom gene set (excluding predicted genes) described b [17].

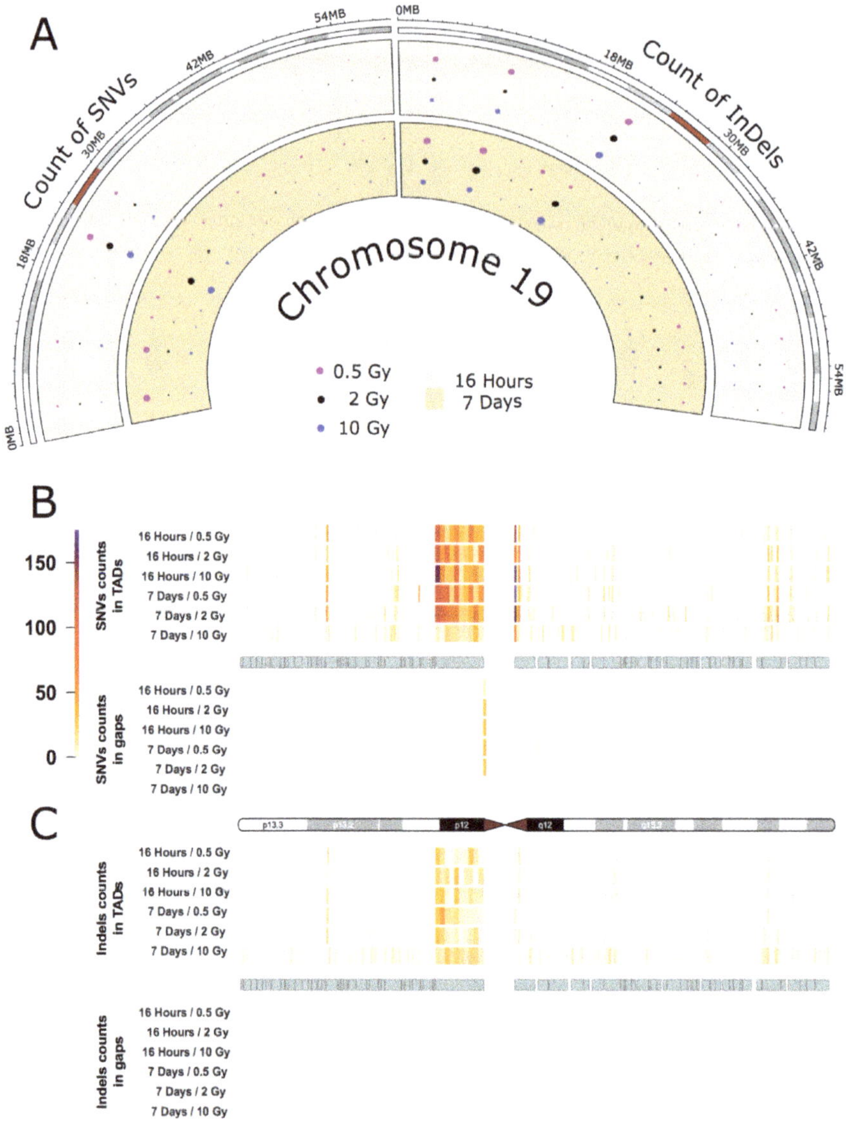

Figure A3. IR-induced variant accumulation across chromosome 19 (dataset 2). (**A**): Distribution of SNVs and InDels over cytogenetic bands. TAD plot showing the distribution of SNV (**B**) and InDel (**C**) counts inside and outside of TAD coordinates for chromosome 19. The chromosome is represented in the middle of the figure as an ideogram. Each panel is subdivided, showing the SNV and InDel alteration counts inside (above) and outside of TADs (below). The TADs are shown as cyan blocks symmetrically in both panels. Each TAD and each gap (region outside TAD termed as gap) was binned into 10 equally long portions and the counts of the SNVs or InDels were then calculated for each portion. The heatmaps with color gradients from white through orange and red to dark purple represent values ranging from 0 to 104 of variant counts.

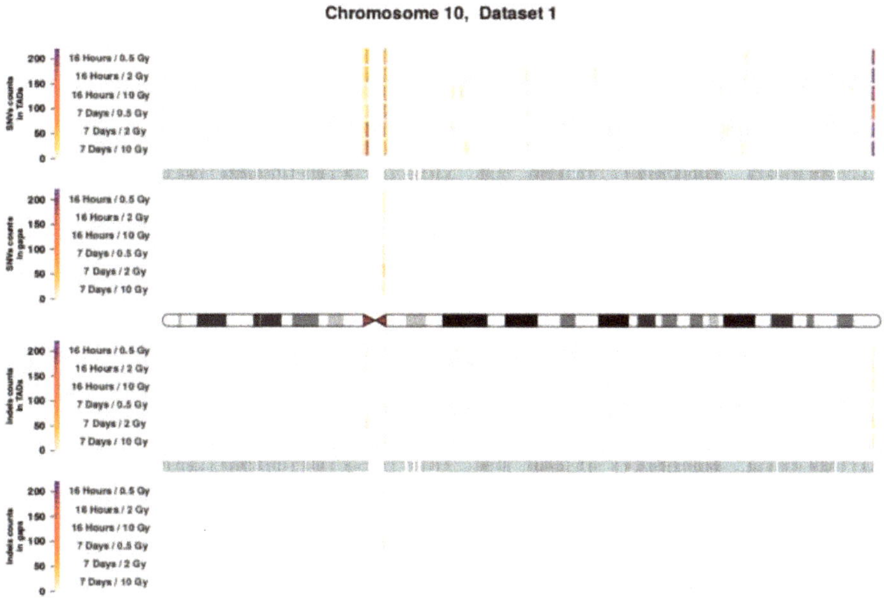

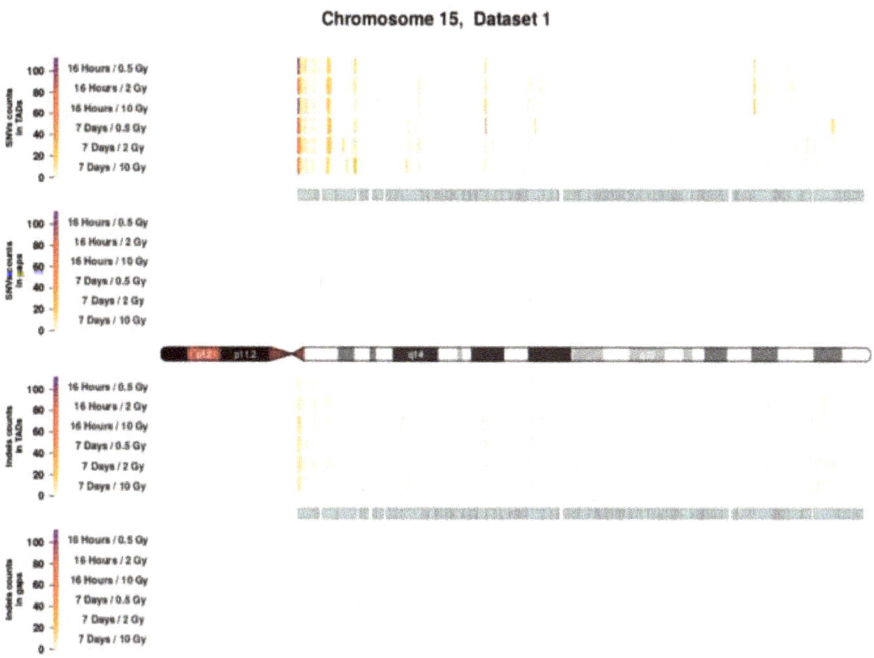

Figure A4. Topological association domains (TADs) plot shows the distribution of SNVs (above panel) and InDels (below panel) counts over within and outside TAD coordinates for Chroosome 10 and 15 in dataset 1.

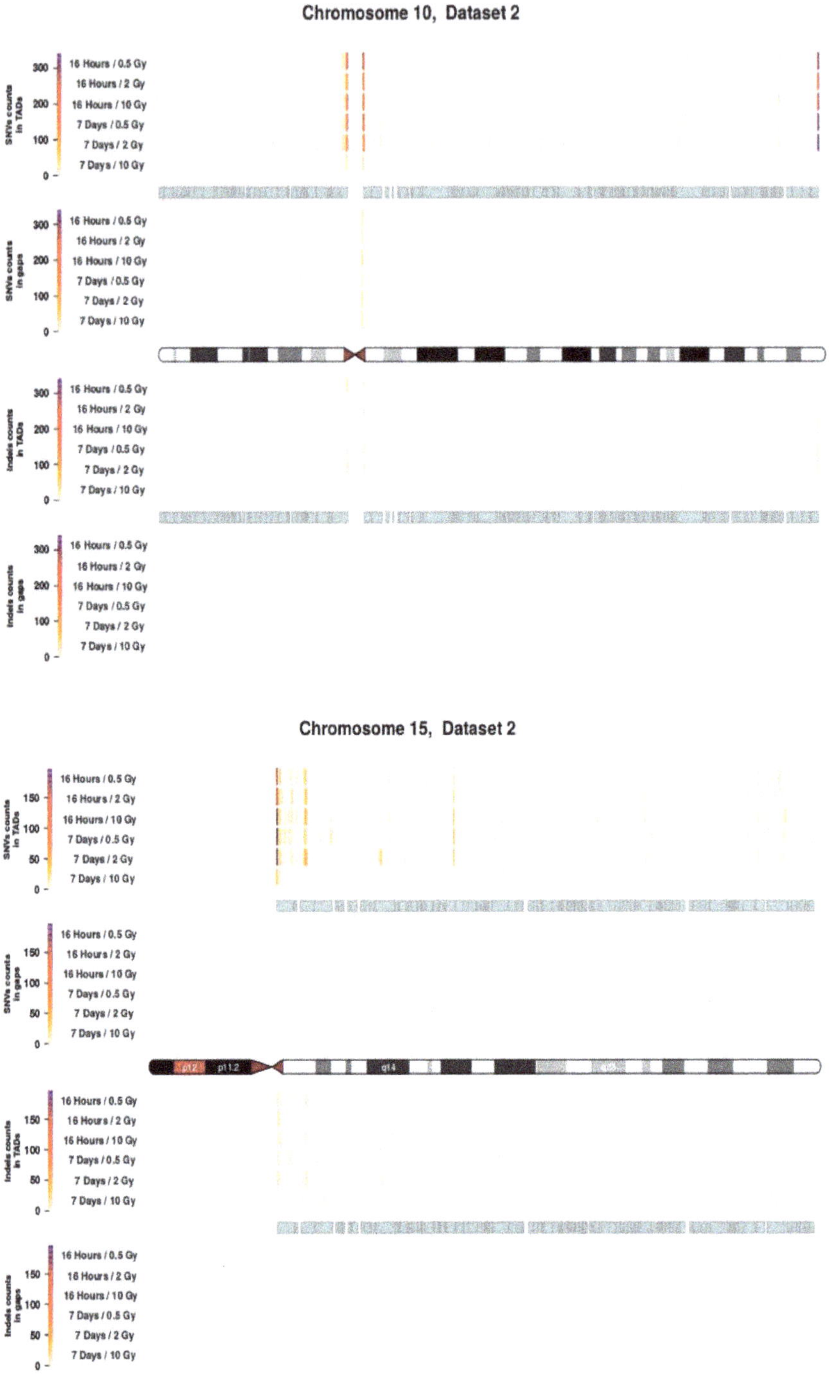

Figure A5. Topological association domains (TADs) plot shows the distribution of SNVs (above panel) and InDels (below panel) counts over within and outside TAD coordinates for Chroosome 10 and 15 in dataset 2.

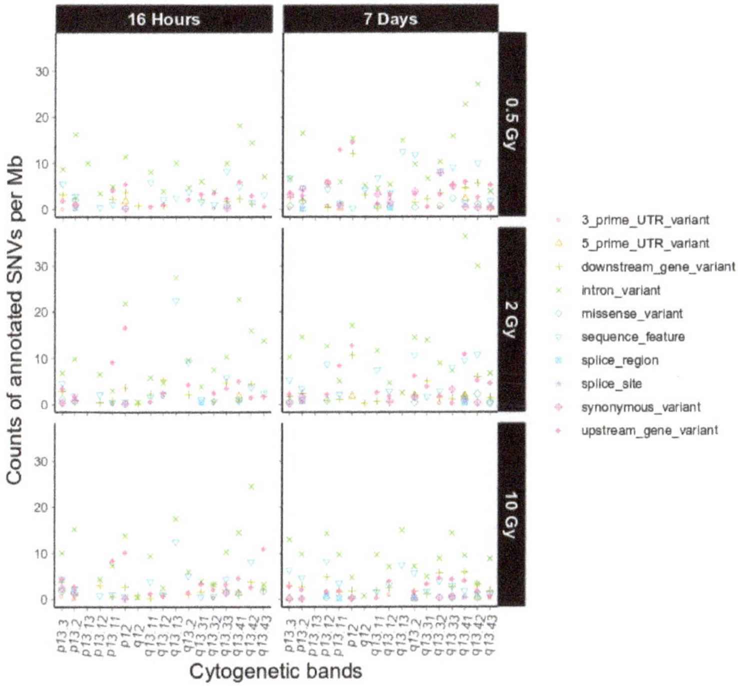

Figure A6. Sequence ontological information for the IR-induced variants observed on chromosome 19 over cytogenetic band, filtered for refseq IDs for dataset 2. For further explanation of the used combination of sequence ontology terms see Table 4.

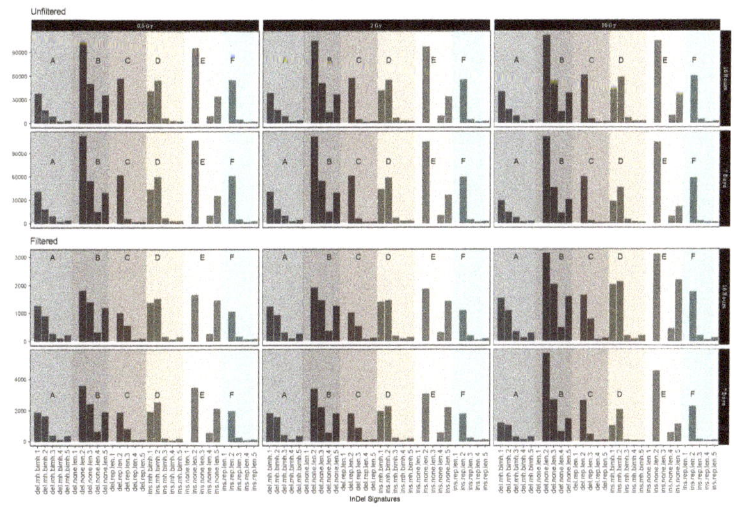

Figure A7. Mutational IR-induced signature plot (dataset 2). On the X-axis the length and specific vicinity of InDels are represented while the Y-axis (note differing scales) shows the InDel count. See Table 3 for pertaining terms and definitions.

Table A1. Table reporting functional consequences for chromosome 19.

ID	Sample ID	Time	Dosage	Functional Effect	Putative Impact	GeneName	Transcript ID	Refseq mrna
chr19:20044988_C/G	393	16 h	0.5 Gy	stop_gained	HIGH	ZNF93	ENST00000343769	NM_031218
chr19:50040366_G/A	448	7 Days	2 Gy	stop_gained	HIGH	RCN3	ENST00000597801	NA
chr19:52904936_A/T	448	7 Days	2 Gy	splice_acceptor_variant&intron_variant	HIGH	ZNF528	ENST00000360465	NM_032423
chr19:52904936_A/T	448	7 Days	2 Gy	splice_acceptor_variant&intron_variant	HIGH	ZNF528	ENST00000436397	NA
chr19:52904936_A/T	448	7 Days	2 Gy	splice_acceptor_variant&intron_variant	HIGH	ZNF528	ENST00000391787	NA
chr19:52904936_A/T	448	7 Days	2 Gy	splice_acceptor_variant&intron_variant	HIGH	ZNF528	ENST00000494167	NA
chr19:52904936_A/T	448	7 Days	2 Gy	splice_acceptor_variant&intron_variant	HIGH	ZNF528	ENST00000493272	NA
chr19:52904936_A/T	448	7 Days	2 Gy	splice_acceptor_variant&intron_variant	HIGH	ZNF528	ENST00000531472	NA
chr19:52904937_G/T	448	7 Days	2 Gy	splice_acceptor_variant&intron_variant	HIGH	ZNF528	ENST00000360465	NM_032423
chr19:52904937_G/T	448	7 Days	2 Gy	splice_acceptor_variant&intron_variant	HIGH	ZNF528	ENST00000436397	NA
chr19:52904937_G/T	448	7 Days	2 Gy	splice_acceptor_variant&intron_variant	HIGH	ZNF528	ENST00000391787	NA
chr19:52904937_G/T	448	7 Days	2 Gy	splice_acceptor_variant&intron_variant	HIGH	ZNF528	ENST00000494167	NA
chr19:52904937_G/T	448	7 Days	2 Gy	splice_acceptor_variant&intron_variant	HIGH	ZNF528	ENST00000493272	NA
chr19:52904937_G/T	448	7 Days	2 Gy	splice_acceptor_variant&intron_variant	HIGH	ZNF528	ENST00000531472	NA
chr19:39229056_A/C	449	7 Days	10 Gy	splice_donor_variant&intron_variant	HIGH	CAPN12	ENST00000328867	NM_144691
chr19:39229056_A/C	449	7 Days	10 Gy	splice_donor_variant&intron_variant	HIGH	CAPN12	ENST00000601953	NA
chr19:52904936_A/T	449	7 Days	10 Gy	splice_acceptor_variant&intron_variant	HIGH	ZNF528	ENST00000360465	NM_032423
chr19:52904936_A/T	449	7 Days	10 Gy	splice_acceptor_variant&intron_variant	HIGH	ZNF528	ENST00000436397	NA
chr19:52904936_A/T	449	7 Days	10 Gy	splice_acceptor_variant&intron_variant	HIGH	ZNF528	ENST00000391787	NA
chr19:52904936_A/T	449	7 Days	10 Gy	splice_acceptor_variant&intron_variant	HIGH	ZNF528	ENST00000494167	NA
chr19:52904936_A/T	449	7 Days	10 Gy	splice_acceptor_variant&intron_variant	HIGH	ZNF528	ENST00000493272	NA
chr19:52904936_A/T	449	7 Days	10 Gy	splice_acceptor_variant&intron_variant	HIGH	ZNF528	ENST00000531472	NA

References

1. Nénot, J.C. Radiation accidents: Lessons learnt for future radiological protection. *Int. J. Radiat Biol.* **1998**, *73*, 435–442. [CrossRef] [PubMed]
2. Ward, J.F. DNA Damage Produced by Ionizing Radiation in Mammalian Cells: Identities, Mechanisms of Formation, and Reparability. In *Progress in Nucleic Acid Research and Molecular Biology*; Cohn, W.E., Moldave, K., Eds.; Academic Press: Cambridge, MA, USA, 1988; pp. 95–125.
3. Van Gent, D.C.; Hoeijmakers, J.H.J.; Kanaar, R. Chromosomal stability and the DNA double-stranded break connection. *Nat. Rev. Genet.* **2001**, *2*, 196–206. [CrossRef] [PubMed]
4. Bree, R.T.; Neary, C.; Samali, A.; Lowndes, N.F. The switch from survival responses to apoptosis after chromosomal breaks. *Dna Repair* **2004**, *3*, 989–995. [CrossRef] [PubMed]
5. Rodgers, K.; McVey, M. Error-Prone Repair of DNA Double-Strand Breaks. *J. Cell Physiol.* **2016**, *231*, 15–24. [CrossRef] [PubMed]
6. Chatterjee, N.; Walker, G.C. Mechanisms of DNA damage, repair, and mutagenesis. *Environ. Mol. Mutagen.* **2017**, *58*, 235–263. [CrossRef] [PubMed]
7. Nik-Zainal, S.; Davies, H.; Staaf, J.; Ramakrishna, M.; Glodzik, D.; Zou, X.; Martincorena, I.; Alexandrov, L.B.; Martin, S.; Wedge, D.C.; et al. Landscape of somatic mutations in 560 breast cancer whole-genome sequences. *Nature* **2016**, *534*, 47–54. [CrossRef] [PubMed]
8. Guirouilh-Barbat, J.; Huck, S.; Bertrand, P.; Pirzio, L.; Desmaze, C.; Sabatier, L.; Lopez, B.S. Impact of the KU80 Pathway on NHEJ-Induced Genome Rearrangements in Mammalian Cells. *Mol. Cell* **2004**, *14*, 611–623. [CrossRef] [PubMed]
9. Lu, Y.; Goldstein, D.B.; Angrist, M.; Cavalleri, G. Personalized Medicine and Human Genetic Diversity. *Cold Spring Harb. Perspect Med.* **2014**, *4*, 1–11. [CrossRef]
10. Bomba, L.; Walter, K.; Soranzo, N. The impact of rare and low-frequency genetic variants in common disease. *Genome Biol.* **2017**, *18*, 1–17. [CrossRef]
11. Bibault, J.E.; Tinhofer, I. The role of Next-Generation Sequencing in tumoral radiosensitivity prediction. *Clin. Transl. Radiat Oncol.* **2017**, *3*, 16–20. [CrossRef]
12. Nath, N.; Esche, J.; Muller, J.; Jensen, L.R.; Port, M.; Stanke, M.; Kaderali, L.; Scherthan, H.; Kuss, A.W. Exome Sequencing Discloses Ionizing-radiation-induced DNA Variants in the Genome of Human Gingiva Fibroblasts. *Health Phys.* **2018**, *115*, 151–160. [CrossRef] [PubMed]
13. Adewoye, A.B.; Lindsay, S.J.; Dubrova, Y.E.; Hurles, M.E. The genome-wide effects of ionizing radiation on mutation induction in the mammalian germline. *Nat. Commun.* **2015**, *6*, 1–8. [CrossRef] [PubMed]
14. Sanders, J.T.; Freeman, T.F.; Xu, Y.; Golloshi, R.; Stallard, M.A.; Martin, R.S.; Balajee, A.S.; McCord, R.P. Radiation-Induced DNA Damage and Repair Effects on 3D Genome Organization. *BioRxiv* **2019**, 740704. [CrossRef]
15. Surrallés, J.; Darroudi, F.; Natarajan, A.T. Low level of DNA repair in human chromosome I heterochromatin. *Genes Chromosom Cancer* **1997**, *20*, 173–184. [CrossRef]
16. Falk, M.; Lukasova, E.; Kozubek, S. Chromatin structure influences the sensitivity of DNA to γ-radiation. *Biochim Biophys Acta Mol Cell Res.* **2008**, *1783*, 2398–2414. [CrossRef]
17. Scherer, S. *Guide to the Human Genome*; Cold Spring Harbor Laboratory Press: Cold Spring Harbor, NY, USA, 2010. [CrossRef]
18. Dixon, J.R.; Selvaraj, S.; Yue, F.; Kim, A.; Li, Y.; Shen, Y.; Hu, M.; Liu, J.S.; Ren, B. Topological domains in mammalian genomes identified by analysis of chromatin interactions. *Nature* **2012**, *485*, 376–380. [CrossRef]
19. Cingolani, P.; Platts, A.; Wang, L.L.; Coon, M.; Nguyen, T.; Wang, L.; Land, S.J.; Lu, X. A program for annotating and predicting the effects of single nucleotide polymorphisms, SnpEff. *Fly* **2012**, *6*, 80–92. [CrossRef]
20. Marteijn, J.A.; Lans, H.; Vermeulen, W.; Hoeijmakers, J.H.J. Understanding nucleotide excision repair and its roles in cancer and ageing. *Nat. Rev. Mol Cell Biol.* **2014**, *15*, 465–481. [CrossRef]
21. Tsouroula, K.; Furst, A.; Rogier, M.; Heyer, V.; Maglott-Roth, A.; Ferrand, A.; Reina-San-Martin, B.; Soutoglou, E. Temporal and Spatial Uncoupling of DNA Double Strand Break Repair Pathways within Mammalian Heterochromatin. *Mol. Cell* **2016**, *63*, 293–305. [CrossRef]
22. Brahme, A. Accurate Description of the Cell Survival and Biological Effect at Low and High Doses and LET's. *J. Radiat Res.* **2011**, *52*, 389–407. [CrossRef]

23. Beck, M.; Moreels, M.; Jacquet, P.; Van Oostveldt, P.; De Vos, W.H.; Baatout, S. X-irradiation induces cell death in fetal fibroblasts. *Int. J. Mol. Med.* **2012**, *30*, 114–118. [CrossRef] [PubMed]
24. Tsuboi, K.; Moritake, T.; Tsuchida, Y.; Tokuuye, K.; Matsumura, A.; Ando, K. Cell cycle checkpoint and apoptosis induction in glioblastoma cells and fibroblasts irradiated with carbon beam. *J. Radiat Res.* **2007**, *48*, 317–325. [CrossRef] [PubMed]
25. Lin, C.-C.; Wu, L.S.-H.; Lee, K.F. The Potential Effect of Different Doses of Ionizing Radiation on Genes and Disease. *Dose-Response* **2019**, *17*. [CrossRef] [PubMed]
26. Cornforth, M.N. Perspectives on the formation of radiation-induced exchange aberrations. *DNA Repair* **2006**, *5*, 1182–1191. [CrossRef] [PubMed]
27. Balajee, A.S.; Sanders, J.T.; Golloshi, R.; Shuryak, I.; McCord, R.P.; Dainiak, N. Investigation of Spatial Organization of Chromosome Territories in Chromosome Exchange Aberrations After Ionizing Radiation Exposure. *Health Phys.* **2018**, *115*, 77–89. [CrossRef] [PubMed]
28. Camparoto, M.L.; Takahashi-Hyodo, S.A.; Dauwerse, J.G.; Natarajan, A.T.; Sakamoto-Hojo, E.T. High susceptibility of chromosome 16 to radiation-induced chromosome rearrangements in human lymphocytes under in vivo and in vitro exposure. *Cytogenet Genome Res.* **2005**, *108*, 287–292. [CrossRef]
29. Behjati, S.; Gundem, G.; Wedge, D.C.; Roberts, N.D.; Tarpey, P.S.; Cooke, S.L.; Loo, P.V.; Alexandrov, L.B.; Ramakrishna, M.; Davies, H.; et al. Mutational signatures of ionizing radiation in second malignancies. *Nat. Commun.* **2016**, *7*, 12605. [CrossRef]
30. Shibata, A.; Jeggo, P.A. Canonical DNA non-homologous end-joining; capacity versus fidelity. *BJR* **2020**, 20190966. [CrossRef]
31. Kucab, J.E.; Zou, X.; Morganella, S.; Joel, M.; Nanda, A.S.; Nagy, E.; Gomez, C.; Degasperi, A.; Harris, R.; Jackson, S.P.; et al. A Compendium of Mutational Signatures of Environmental Agents. *Cell* **2019**, *177*, 821–836.e16. [CrossRef]
32. Yuan, J.; Yeasky, T.M.; Rhee, M.C.; Glazer, P.M. Frequent T:A→G:C transversions in X-irradiated mouse cells. *Carcinogenesis* **1995**, *16*, 83–88. [CrossRef]
33. Hase, Y.; Satoh, K.; Seito, H.; Oono, Y. Genetic Consequences of Acute/Chronic Gamma and Carbon Ion Irradiation of Arabidopsis thaliana. *Front. Plant Sci.* **2020**, *11*, 336. [CrossRef] [PubMed]
34. Nath, N.; Hagenau, L.; Weiss, S.; Tzvetkova, A.; Jensen, L.R.; Kaderali, L.; Port, M.; Scherthan, H.; Kuss, A.W. Ionizing Radiation alters the transition / transversion ratio in the exome of human gingiva fibroblasts. *Health Phys.* **2019**, *119*, 109–117. [CrossRef] [PubMed]
35. Waris, G.; Ahsan, H. Reactive oxygen species: Role in the development of cancer and various chronic conditions. *J. Carcinog. Wolters Kluwer Medknow Publ.* **2006**, 14. [CrossRef]
36. Lujan, S.A.; Williams, J.S.; Pursell, Z.F.; Abdulovic-Cui, A.A.; Clark, A.B.; Nick McElhinny, S.A.; Kunkel, T.A. Mismatch Repair Balances Leading and Lagging Strand DNA Replication Fidelity. *PLoS Genet.* **2012**, *8*. [CrossRef] [PubMed]
37. Tanaka, M.; Takahara, M.; Nukina, K.; Hayashi, A.; Sakai, W.; Sugasawa, K.; Shiomi, Y. Mismatch repair proteins recruited to ultraviolet light-damaged sites lead to degradation of licensing factor Cdt1 in the G1 phase. *Cell Cycle.* **2017**, *16*, 673–684. [CrossRef] [PubMed]
38. Peña-Diaz, J.; Bregenhorn, S.; Ghodgaonkar, M.; Follonier, C.; Artola-Borán, M.; Castor, D.; Lopes, M.; Sartori, A.A.; Jiricny, J. Noncanonical Mismatch Repair as a Source of Genomic Instability in Human Cells. *Mol. Cell.* **2012**, *47*, 669–680. [CrossRef]
39. Mosesso, P.; Palitti, F.; Pepe, G.; Piñero, J.; Bellacima, R.; Ahnstrom, G.; Natarajan, A.T. Relationship between chromatin structure, DNA damage and repair following X-irradiation of human lymphocytes. *Mutat. Res. Genet. Toxicol. Environ. Mutagenesis* **2010**, *701*, 86–91. [CrossRef]
40. Grimwood, J.; Gordon, L.A.; Olsen, A.; Terry, A.; Schmutz, J.; Lamerdin, J.; Hellsten, U.; Goodstein, D.; Couronne, O.; Tran-Gyamfi, M.; et al. The DNA sequence and analysis of human chromosome 19. *Nature* **2004**, *428*, 522–528. [CrossRef]
41. Bellefroid, E.J.; Marine, J.C.; Matera, A.G.; Bourguignon, C.; Desai, T.; Healy, K.C.; Bray-Ward, P.; Martial, J.A.; Ihle, J.N.; Ward, D.A. Emergence of the ZNF91 Krüppel-associated box-containing zinc finger gene family in the last common ancestor of anthropoidea. *Proc. Natl. Acad. Sci. USA* **1995**, *92*, 10757. [CrossRef]
42. Eichler, E.E.; Hoffman, S.M.; Adamson, A.A.; Gordon, L.A.; McCready, P.; Lamerdin, J.E.; Mohrenweiser, H.W. Complex β-Satellite Repeat Structures and the Expansion of the Zinc Finger Gene Cluster in 19p12. *Genome Res.* **1998**, *8*, 791–808. [CrossRef]

43. Jinks-Robertson, S.; Bhagwat, A.S. Transcription-Associated Mutagenesis. *Annu Rev. Genet.* **2014**, *48*, 341–359. [CrossRef] [PubMed]
44. Lavelle, C.; Foray, N. Chromatin structure and radiation-induced DNA damage: From structural biology to radiobiology. *Int. J. Biochem. Cell Biol.* **2014**, *49*, 84–97. [CrossRef] [PubMed]
45. Szabo, Q.; Bantignies, F.; Cavalli, G. Principles of genome folding into topologically associating domains. *Sci. Adv.* **2019**, *5*. [CrossRef]
46. Natale, F.; Rapp, A.; Yu, W.; Maiser, A.; Harz, H.; Scholl, A.; Grulich, S.; Anton, T.; Hörl, D.; Chen, W.; et al. Identification of the elementary structural units of the DNA damage response. *Nat. Commun.* **2017**, *8*. [CrossRef]
47. Lang, F.; Li, X.; Zheng, W.; Li, Z.; Lu, D.; Chen, G.; Gong, D.; Yang, L.; Fu, J.; Shi, P.; et al. CTCF prevents genomic instability by promoting homologous recombination-directed DNA double-strand break repair. *Proc. Natl. Acad. Sci. USA* **2017**, *114*, 10912–10917. [CrossRef] [PubMed]
48. Hilmi, K.; Jangal, M.; Marques, M.; Zhao, T.; Saad, A.; Zhang, C.; Luo, V.M.; Syme, A.; Rejon, C.; Yu, Z.; et al. CTCF facilitates DNA double-strand break repair by enhancing homologous recombination repair. *Sci. Adv.* **2017**, *3*, 1–14. [CrossRef]
49. Weissmann, R.; Kacprowski, T.; Peper, M.; Esche, J.; Jensen, L.R.; Van Diepen, L.; Port, M.; Kuss, A.W.; Scherthan, H. Transcriptome alterations in X-irradiated human gingiva fibroblasts. *Health Phys.* **2016**, *111*, 75–84. [CrossRef]
50. McPherson, A.; Chen, K.; Wu, C.; Wallis, J.W.; Wyatt, A.W.; McLellan, M.D.; McGrath, S.D.; Wendl, M.C.; Zhang, Q.; Locke, D.P.; et al. BreakDancer: An algorithm for high-resolution mapping of genomic structural variation: Article: Nature Methods. *Nat. Methods.* **2009**, *6*, 677–681. [CrossRef]

International Journal of
Molecular Sciences

MDPI

Article

SRC Tyrosine Kinase Inhibitor and X-rays Combined Effect on Glioblastoma Cell Lines

Filippo Torrisi [1,2,†], Luigi Minafra [2,3,†], Francesco P. Cammarata [2,3,*], Gaetano Savoca [2,3], Marco Calvaruso [2,3], Nunzio Vicario [1], Laura Maccari [4], Elodie A. Pérès [5], Hayriye Özçelik [5], Myriam Bernaudin [5], Lorenzo Botta [4], Giorgio Russo [2,3], Rosalba Parenti [1,*] and Samuel Valable [5]

1 Department of Biomedical and Biotechnological Sciences (BIOMETEC), University of Catania, 95123 Catania, Italy; filippo.torrisi@unict.it (F.T.); nunziovicario@unict.it (N.V.)
2 National Institute for Nuclear Physics, Laboratori Nazionali del Sud, INFN-LNS, 95123 Catania, Italy; luigi.minafra@ibfm.cnr.it (L.M.); savoca.gaetano@gmail.com (G.S.); marco.calvaruso@ibfm.cnr.it (M.C.); giorgio.russo@ibfm.cnr.it (G.R.)
3 Institute of Molecular Bioimaging and Physiology, National Research Council, IBFM-CNR, 90015 Cefalù, Italy
4 Lead Discovery Siena s.r.l. (LDS), via Vittorio Alfieri, 31, Castelnuovo Berardenga, 53019 Siena, Italy; l.maccari@leaddiscoverysiena.it (L.M.); l.botta@leaddiscoverysiena.it (L.B.)
5 UNICAEN, CEA, CNRS, ISTCT/CERVOxy Group, GIP Cyceron, Normandie University, 14074 Caen, France; peres@cyceron.fr (E.A.P.); ozcelik@cyceron.fr (H.Ö.); bernaudin@cyceron.fr (M.B.); samuel.valable@cnrs.fr (S.V.)
* Correspondence: francesco.cammarata@ibfm.cnr.it (F.P.C.); parenti@unict.it (R.P.)
† These authors contributed equally to this work.

Received: 12 May 2020; Accepted: 28 May 2020; Published: 30 May 2020

Abstract: Glioblastoma (GBM) is one of the most lethal types of tumor due to its high recurrence level in spite of aggressive treatment regimens involving surgery, radiotherapy and chemotherapy. Hypoxia is a feature of GBM, involved in radioresistance, and is known to be at the origin of treatment failure. The aim of this work was to assess the therapeutic potential of a new targeted c-SRC inhibitor molecule, named Si306, in combination with X-rays on the human glioblastoma cell lines, comparing normoxia and hypoxia conditions. For this purpose, the dose modifying factor and oxygen enhancement ratio were calculated to evaluate the Si306 radiosensitizing effect. DNA damage and the repair capability were also studied from the kinetic of γ-H2AX immunodetection. Furthermore, motility processes being supposed to be triggered by hypoxia and irradiation, the role of c-SRC inhibition was also analyzed to evaluate the migration blockage by wound healing assay. Our results showed that inhibition of the c-SRC protein enhances the radiotherapy efficacy both in normoxic and hypoxic conditions. These data open new opportunities for GBM treatment combining radiotherapy with molecularly targeted drugs to overcome radioresistance.

Keywords: Glioblastoma; ionizing radiation; hypoxia; DNA damage; combined treatments

1. Introduction

Radiotherapy (RT) represents a gold standard in the treatment of glioblastoma (GBM) that remains one of the most aggressive primary brain tumors with a high rate of recurrence [1]. Clinical data reported that RT is a positive prognostic factor on the survival of patients, as compared to patients that receive surgery or chemotherapy alone [2]. However, there are no clinical studies demonstrating an overall survival improvement with RT dosing above the standard of 60 Gy for 30 fractions (2 Gy/day), showing that there are two main issues to overcome: (i) Avoiding radiation side effects and (ii) reducing GBM radioresistance. The hypoxic pattern of GBM has been widely described and represents one of the main factors inducing radioresistance [3]. Hypoxic microenvironment reduces non-repairable DNA damage mediated by RT, as described by the hypothesis of oxygen fixation. Indeed, under normoxic conditions, molecular oxygen permanently fixes the DNA damage induced by free radicals produced in water radiolysis (indirect effects of ionizing radiation), being very genotoxic. Such a role, under hypoxic conditions, is proportionally reduced, thus affecting indirect damage induced by RT and establishing so called GBM radioresistance, leading to non-repairable DNA double strand breaks [4,5]. Therefore, hypoxic microenvironment, particularly pronounced in GBM, represents a poor prognosis factor, as shown both in preclinical models [6] and in human GBM patients [3,7]. Moreover, hypoxia mediates a favourable microenvironment to the growth and renewal of GBM stem cells and to the activation of specific proteins, involved in cell proliferation, angiogenesis, migration and invasion, that are the biological basis of GBM recurrence [8,9]. Among these proteins, SRC proto-oncogene non-receptor tyrosine kinase (c-SRC), a member of non-receptor SRC family kinases (SFKs), drives GBM carcinogenesis and progression, and is involved in intracellular signalling pathways related with hypoxia [10]. Several factors are involved in the activation of c-SRC, including focal adhesion kinase (FAK), integrins or tyrosine kinase growth factor receptors, like epidermal growth factor receptor (EGFR) [11]. Hypoxia stimulates the interaction of vIIIEGFR with the integrin β3 in GBM cells, activating a signalling pathways c-SRC-dependent resulting in the up-regulation of the cancer cell invasion markers, like matrix metalloproteinase-2 (MMP-2) and matrix metalloproteinase-9 (MMP-9) [12]. Therefore, c-SRC and its related network represent a key protein for targeted therapy.

Si306 (Lead Discovery Siena, Siena, Italy) is a molecule of the pyrazolo[3,4-d]pyrimidines family, which has been shown to inhibit c-SRC kinase protein activity [13]. Previous preclinical studies confirmed that Si306 was able to cross the intact blood–brain barrier and to progressively accumulate into the brain for 24 h after the post-intravenous injection. Moreover, it has been demonstrated that Si306 in combination with X-rays showed a synergic anti-proliferative effect in both in vitro and in vivo GBM models [14].

Herein, we aimed at investigating the Si306 capability to increase the radiotherapy efficacy both in normoxic and hypoxic conditions on the GBM cells, increasing the current knowledge on radiosensitizing effects of the novel c-SRC inhibitor Si306. For this purpose, we investigated the radiosensitizing effect of Si306 on two GBM cell lines, U251-MG and U87-MG, irradiated with X-rays in both normoxic (21% of oxygen) and hypoxic (1% of oxygen) conditions, and evaluated the degree of proliferation and migration. In addition, γH2AX foci detection by immunofluorescence was performed to quantify the radiation-induced DNA double-strand break formation and the DNA damage repair ability. Our results showed that c-SRC inhibition acted synergistically with radiation treatment, reducing clonogenic and migration ability and increasing DNA damage in GBM cells, in both normoxic and hypoxic conditions.

2. Results

2.1. c-SRC Inhibition Improves the Efficacy of Radiotherapy on U251-MG Cell Line

2.1.1. Evaluation of Cell Survival from Clonogenic Assay

In order to compare the effects of increasing doses of X-rays (0, 2, 4, 6, 8 Gy) on U251-MG cell survival in normoxic (21% O_2) and hypoxic (1% O_2) conditions and in combination with 10 μM and 20 μM of the Si306 molecule, we performed clonogenic assays on the U251-MG cell line. The surviving fraction (SF) values were plotted against the dose to obtain dose-response curves. Dose modifying factor (DMF) and oxygen enhancement ratio (OER) were also calculated to evaluate treatment efficiency. The results showed a radiation dose dependent decrease in clone number with a significant effect with the exposition concomitant to Si306 (Figure 1a,b). Of note, U251-MG cells exhibited hypoxia-induced radioresistance with an OER of 1.27 (Figure 2a,b and Table 1). In normoxic conditions, the exposure to Si306 combined with RT induced a decrease in SF values with a DMF of 1.38 at the concentration of

20 μM (Figure 3a,b and Table 1). In hypoxic conditions, the effect of combined treatment was increased in culture exposed to Si306 versus control. The synergistic effect of Si306 and RT was further confirmed by the OER reduction of about 11%, demonstrating that c-SRC inhibition had a significant role as radiosensitizer in hypoxic conditions (Figure 4a,b and Table 1).

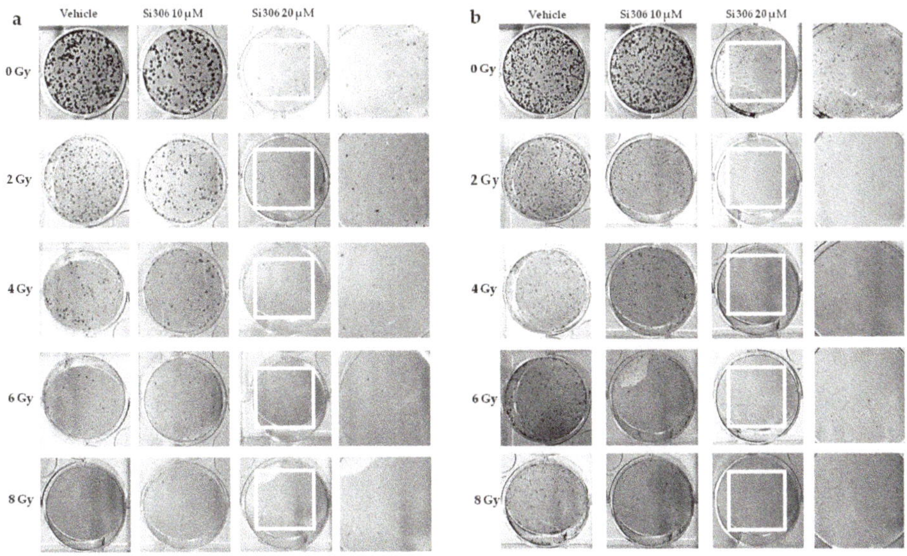

Figure 1. U251 clones after X-ray irradiation combined with Si306 in normoxia (21% oxygen) (**a**) and hypoxia (1% oxygen) (**b**).

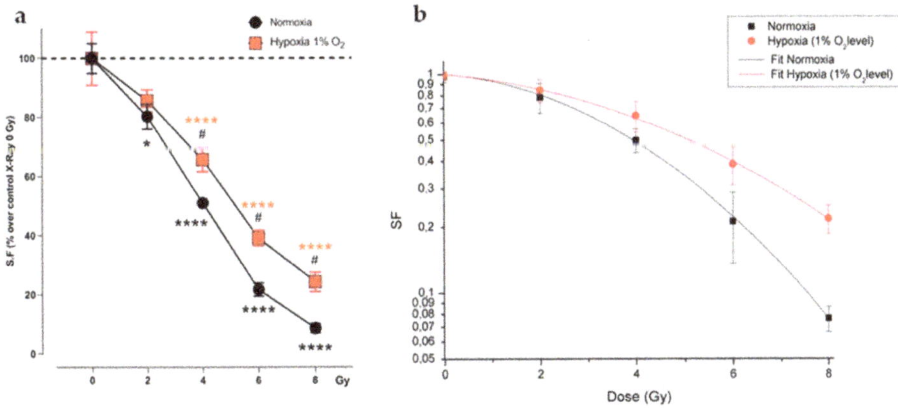

Figure 2. U251-MG irradiated cells in normoxia and hypoxia. (**a**) Surviving fraction (SF) plot of normoxic and hypoxic U251-MG cells exposed to 0, 2, 4, 6 and 8 Gy. Data are mean ± SEM of $n = 3$ independent experiments. * p-value < 0.05 and **** p-value < 0.0001 versus normoxia 0 Gy; # p-value < 0.05 versus each dose in normoxia ($F_{Si306conc.} = 133.8$, p-value < 0.0001; $F_{Gy} = 15.49$, p-value $= 0.0003$; $F_{Si306conc.\ x\ Gy} = 1.568$, p-value $= 0.1973$. Two-way ANOVA with Holm–Šídák post-hoc test). (**b**) Linear-quadratic adjustment of the data of U251 cell survival curves treated with X-rays in hypoxia and normoxia.

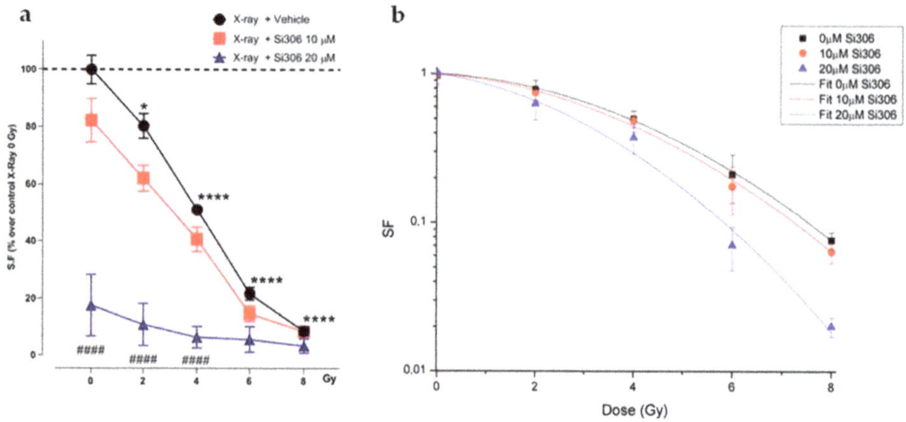

Figure 3. Cell survival of irradiated cells in normoxia with Si306 exposure. (**a**) SF plot of normoxic U251-MG cells exposed to 0, 2, 4, 6 and 8 Gy and treated with vehicle, 10 or 20 μM Si306. Data are mean ± SEM of $n = 3$ independent experiments. * p-value < 0.05 and **** p-value < 0.0001 versus 0 Gy in normoxia; #### p-value < 0.0001 versus only irradiated cells with 0, 2 and 4 Gy in normoxia ($F_{Si306conc.} = 89.17$, p-value < 0.0001; $F_{Gy} = 124.5$, p-value < 0.0001; $F_{Si306conc. \times Gy} = 14.64$, p-value < 0.0001. Two-way ANOVA with Holm–Šídák post-hoc test Two-way ANOVA with Holm–Šídák post-hoc test). (**b**) Linear-quadratic adjustment of the data of U251 cell survival curves treated with X-rays only and combined with Si306 in normoxia.

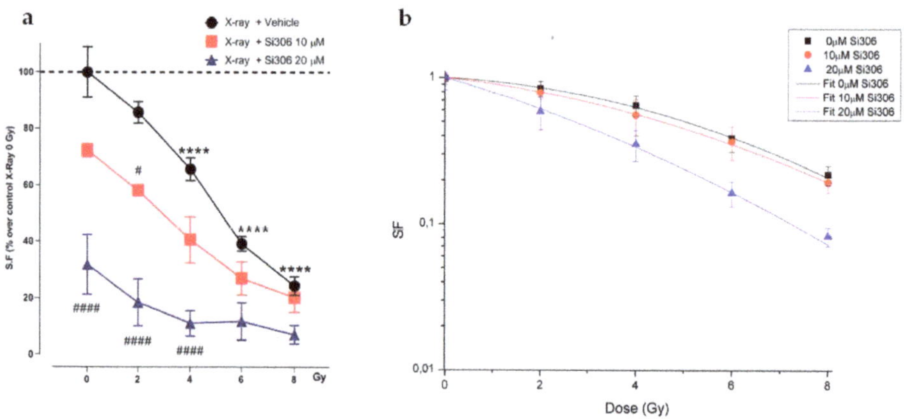

Figure 4. SF of irradiated cells with Si306 exposure in hypoxia. (**a**) Mean ± SEM, three independent experiments; **** p-value < 0.0001 versus 0 Gy in normoxia; # p-value < 0.05 and #### p-value < 0.0001 versus X-rays + vehicle at the same dose ($F_{Si306conc.} = 34.09$, p-value < 0.0001; $F_{Gy} = 77.95$, p-value < 0.0001; $F_{Si306conc. \times Gy} = 3.929$, p-value = 0.0012. Two-way ANOVA with Holm–Šídák post-hoc test). (**b**) Linear-quadratic adjustment of the data of U251 cell survival curves treated with X-rays only and combined with Si306 in hypoxia.

Table 1. Dose modifying factor (DMF) and oxygen enhancement ratio (OER) values calculated as isoeffective dose at surviving fraction of 0.5.

Treatment	Normoxia SF50% (Gy)	Hypoxia SF50% (Gy)	Normoxia DMF	Hypoxia DMF	OER
X-rays + vehicle	4.09	5.18	1	1	1.27
X-rays + 10 μM Si306	3.86	4.53	1.05	1.15	1.17
X-rays+ 20 μM Si306	2.54	2.67	1.38	1.94	1.05

2.1.2. Radiobiological Meaning of A, B and A/B Ratio Parameters

DMF and OER changes were also related to the α and β parameters analysis. These values displayed differences between groups (normoxia versus hypoxia) and treatment (vehicle versus Si306) (Tables 2 and 3). The Si306 treatment combined with X-rays induced an α value increase in both conditions, in particular in the hypoxic one. Indeed, 10 and 20 μM Si306 showed α values of 0.092 ± 0.010 and 0.219 ± 0.025, respectively, as compared to control cultures (α value = 0.037 ± 0.024). This means that, in hypoxia, the linear contribution to damage is higher than in normoxia. The increase in β value is greater in normoxia rather than in hypoxia after exposure to Si306 in combination with irradiation, maybe due to ROS decrease in hypoxic condition. However, the DNA direct damage associated with α component may contribute to the OER decrease. Moreover, our data provided important evidence on the α/β value meaning that is an inverse reflection of a tissue sensitivity to dose fractionation. According to the α/β ratio, tissues are classified as early (low α/β) or late (high α/β) responding [15]. Therefore, the significant increase in the α/β ratio observed in hypoxia may represent a change in cellular radiobiological response leading to tissue patterns with a reduced ability to repair damage and with a greater accumulation of lethal lesions.

Table 2. α and β parameters by fitting the cell survival to the linear-quadratic (LQ) model in normoxia. Values correspond to mean ± SEM; three independent experiments.

Treatment Normoxia	α (Gy-1)	β (Gy-2)	α/β (Gy)
X-rays + vehicle	0.037 ± 0.011	0.036 ± 0.009	1.03
X-ray s+ 10 μM Si306	0.060 ± 0.039	0.035 ± 0.009	1.71
X-rays+ 20 μM Si306	0.077 ± 0.009	0.052 ± 0.005	1.48

Table 3. α and β parameters estimated by fitting the cell survival to the linear-quadratic in normoxia (LQ) model in hypoxia. Values correspond to mean ± SEM; three independent experiments.

Treatment Hypoxia	α (Gy-1)	β (Gy-2)	α/β (Gy)
X-rays + vehicle	0.037 ± 0.024	0.020 ± 0.005	1.85
X-rays + 10 μM Si306	0.092 ± 0.010	0.013 ± 0.002	7.07
X-rays + 20 μM Si306	0.219 ± 0.025	0.014 ± 0.005	15.64

2.2. c-SRC Inhibition Sustains Radiation-Induced DNA Damage over Time

The DNA damage was evaluated by γH2AX immunofluorescence during the maximum of foci formation and also damage repair capacity (2 and 24 h after X-ray radiation, respectively) [9]. Immunofluorescence analyses showed that in normoxia and hypoxia, the exposure to Si306 in combination with irradiation led to a signal increase that was not significant 2 h after irradiation compared to X-rays only (Figure 5a,b). The synergistic effect of the Si306 molecule with IR became significant 24 h after treatment, where the foci signal was maintained at high levels in the case of combined treatment, compared to irradiation alone: 48 % and 41% of U251 cells, exposed, respectively, with 10 μM and 20 μM of Si306, were still positive compared to 10% of only irradiated U251 cells in normoxia. More interestingly, in hypoxia we showed a persistence of 21% and 27% positive U251-MG cells, both irradiated and exposed to 10 μM Si306 and 20 μM of Si306, respectively, compared to 5% of only irradiated cells (Figure 6a,b). To further confirm this observation, the immunofluorescence assay was repeated on the U87-MG GBM cell line. The results obtained were similar, since the differences in the foci γH2AX signal between the treatment conditions with vehicle and with Si306 were not significant 2 h after irradiation (Figure 7a,b). The increase in foci γH2AX expression was statistically significant only 24 h after irradiation in the combined treatments: In normoxia, after irradiation and Si306 pre-treatment, 35% (10 μM) and 31% (20 μM) of U87 cells were positive versus 15% of only irradiated U87 cells; similar results were obtained in hypoxia, since 18% and 28% of irradiated and Si306 pre-treated U87 cells, respectively, with 10 μM and 20 μM, were positive compared to 10% of only irradiated U87 cells (Figure 8a,b).

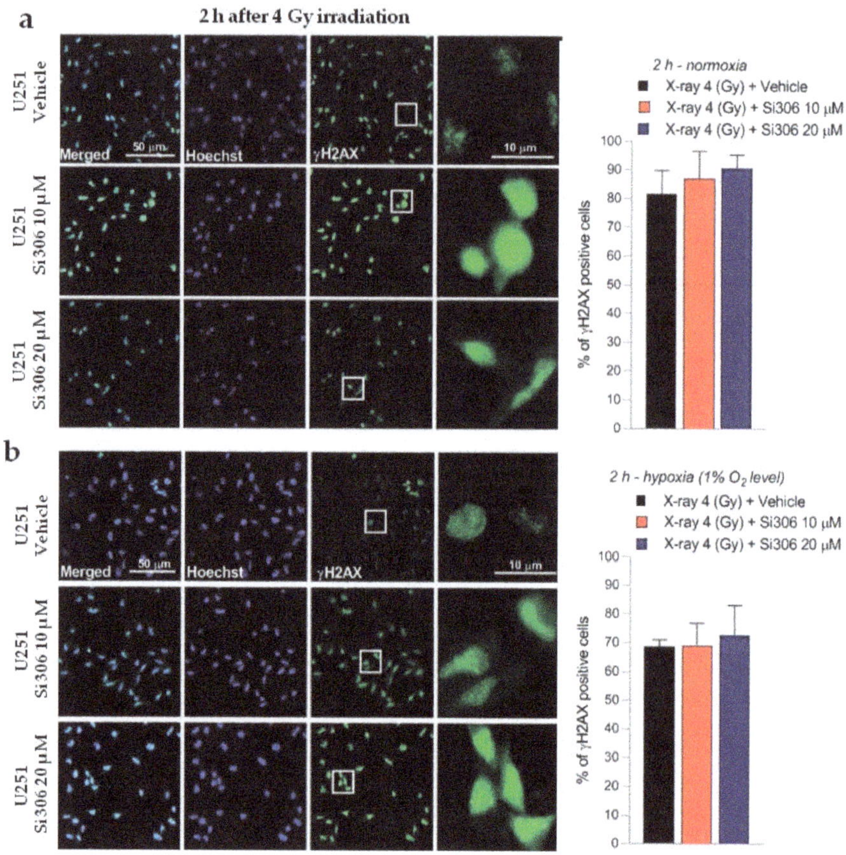

Figure 5. Representative pictures with inserts (white squares) and quantification of U251-MG positive cells for γH2AX performed 2 h after 4 Gy irradiation in normoxia (**a**) and hypoxia (**b**). Data are mean ± SD of n = 3 independent experiments; 4 Gy + vehicle; ($F_{normoxia}$ = 2.030, p-value = 0.1564; $F_{hypoxia}$ = 0.5685, p-value = 0.5798. One-way ANOVA with Holm–Šídák post-hoc test).

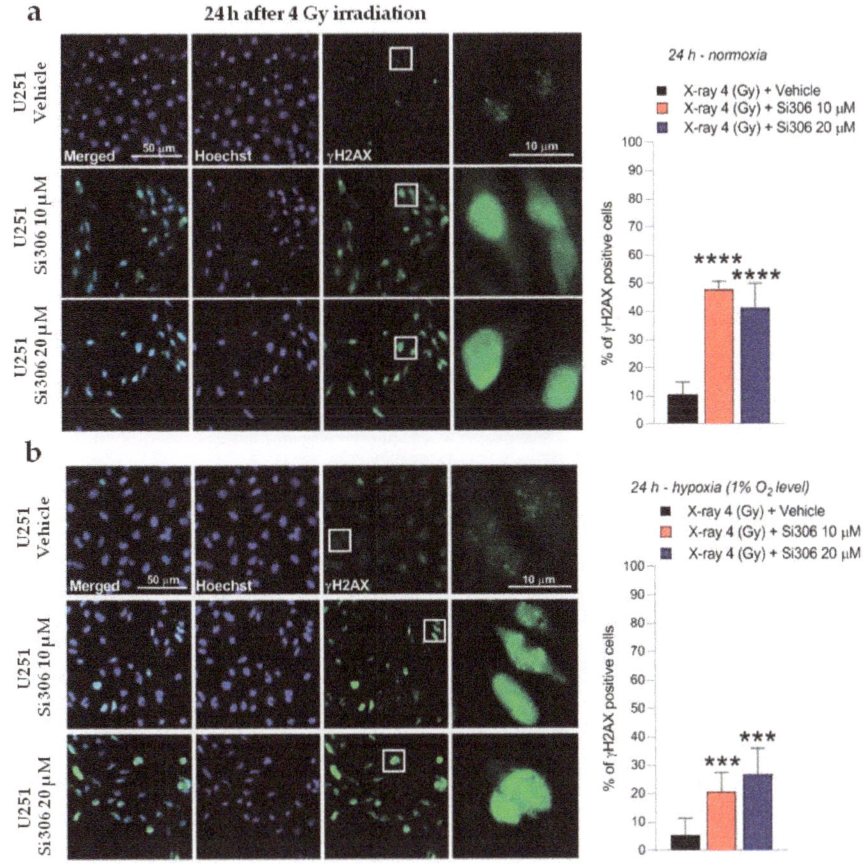

Figure 6. Representative pictures with inserts (white squares) and quantification of U251-MG positive cells for γH2AX realized 24 h after 4 Gy irradiation normoxia (**a**) and in hypoxia (**b**). Data are mean ± SD of $n = 3$ independent experiments; *** p-value < 0.001 and **** p-value < 0.0001 versus 4 Gy + vehicle ($F_{normoxia} = 87.81$, p-value < 0.0001; $F_{hypoxia} = 18.87$, p-value < 0.0001. One-way ANOVA with Holm–Šidák post-hoc test).

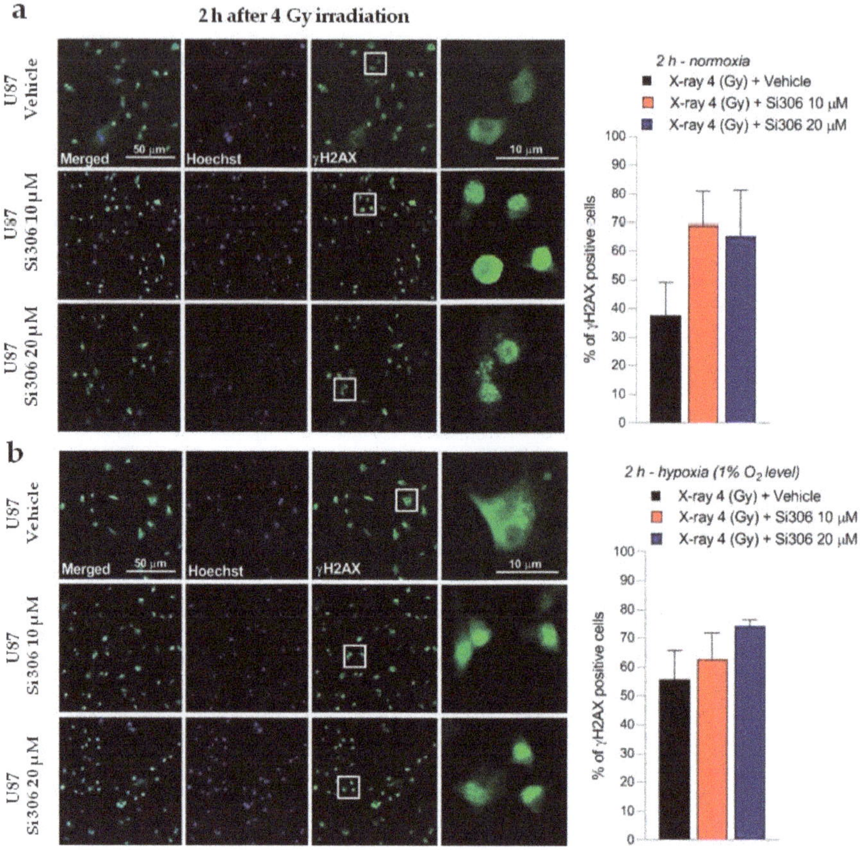

Figure 7. Representative pictures with inserts (white squares) and quantification of U87 positive cells for γH2AX performed 2 h after 4 Gy irradiation in normoxia (**a**) and hypoxia (**b**) Data are mean ± SD of $n = 3$ independent experiments ($F_{normoxia} = 5.787$, p-value < 0.0329; $F_{hypoxia} = 4.048$, p-value < 0.0557. One-way ANOVA with Holm–Šídák post-hoc test).

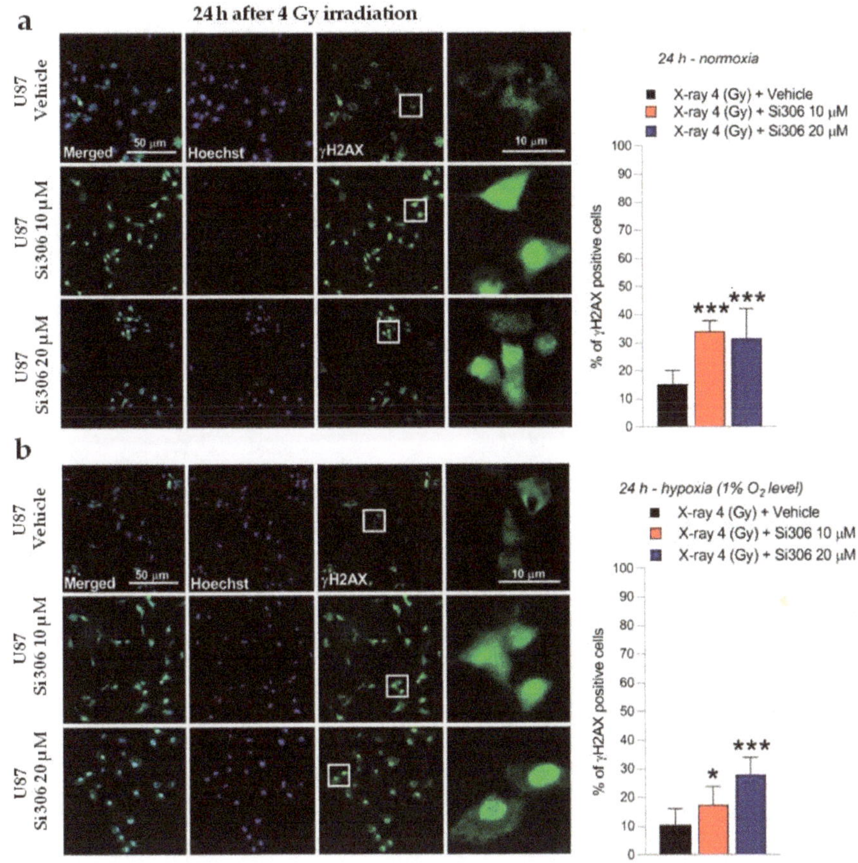

Figure 8. Representative pictures with inserts (white squares) and quantification of U87 positive cells for γH2AX realized 24 h after 4 Gy irradiation in normoxia (**a**) and hypoxia (**b**). Data are mean ± SD of $n = 3$ independent experiments; * p-value < 0.05 and *** p-value < 0.001 versus 4 Gy + vehicle ($F_{normoxia} = 16.82$, p-value < 0.0001; $F_{hypoxia} = 12.77$, p-value = 0.0004. One-way ANOVA with Holm–Šídák post-hoc test).

2.3. c-SRC Inhibition Reduces Cell Migration

Migration and invasion of malignant glioma play a key role in GBM progression. Therefore, we examined, by wound healing assay, the effect of c-SRC inhibition on migration in irradiated U251-MG cells, being highly invasive, as reported in previous studies [16]. The results of wound healing assay showed an inhibitory effect of the Si306 molecule on the migration of the U251-MG cells. The addition of the Si306 molecule at both concentrations of 10 µM and 20 µM reduced the migration index of cells compared to those not irradiated and irradiated with a vehicle, in both normoxic and hypoxic conditions (Figure 9a,b).

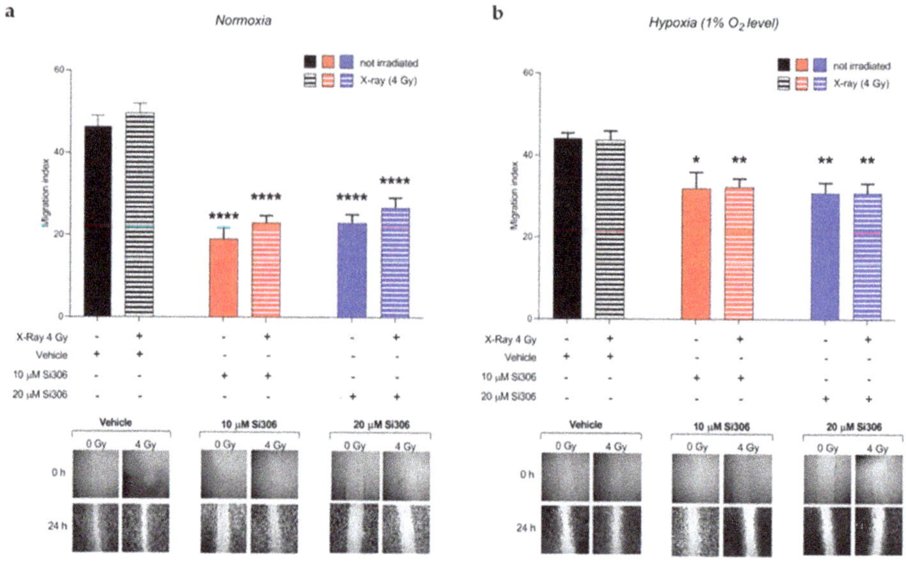

Figure 9. Effects of Si306 on migration of U251-MG cells in normoxia (**a**) and hypoxia (**b**). Data are mean ± SEM of $n = 3$ independent experiments. * p-value < 0.05, ** p-value < 0.01 and **** p-value < 0.0001 versus vehicle or vehicle + irradiation at 24 h after scratch ($F_{normoxia} = 32.59$, p-value < 0.0001; $F_{hypoxia} = 6.907$, p-value < 0.0001. One-way ANOVA with Holm–Šídák post-hoc test).

3. Discussion

The poor prognosis of GBM represents an urgent clinical need and reinforces the necessity to explore and to develop novel therapeutic approaches. According to the clinical guidelines for the treatment of newly diagnosed GBM, only concomitant temozolomide with fractionated radiotherapy is indicated to significantly improve median survival (14.6 versus 12.1 months) and progression free survival (6.9 versus 5 months) as compared to RT alone, but high recurrences are still observed [17]. Therefore, specific cancer molecular targets are expected to have a synergistic effect to increase the efficacy of RT, overcoming radioresistance and modulating the irradiation dose delivered to enhance RT intrinsic sensitivity. During the last decade, molecular investigation on pathobiological mechanisms of GBM promoted research to develop molecularly targeted drugs (i.e., targeted therapy), including monoclonal antibodies (mAb) and tyrosine-kinase inhibitors (TKi), but their efficacy in the clinical practice is still limited as compared to conventional chemotherapy regimen [18]. c-SRC is a non-receptor tyrosine kinase (nRTK), interacting with many intracellular proteins, involved in GBM proliferation, invasion, motility and angiogenesis [10]. Previous evidence showed that hypoxia enhanced phosphorylation of tyrosine 416 in c-SRC, thus leading to protein-tyrosine kinase domain activation and to the downstream induction of VEGF expression, promoting angiogenesis [19]. Hypoxia may promote GBM progression and invasion throughout the integrin β3/FAK/SRC/EGFRvIII signalling axis, linking tumor cells and their surrounding environment [12]. Moreover, c-SRC activates HIF-1α and glucose uptake, thus fostering GBM proliferation rate [20].

Recently, we investigated Si306 molecule, a member of the pyrazolo[3,4-d] pyrimidines family, which is able to selectively bind and inactivate the ATP site of c-SRC protein, acting as ATP competitive inhibitor type I/II [13]. Combined approaches with X-rays irradiation showed that Si306 is able to reduce proliferation, survival and clonogenic ability of GBM cell lines, also promoting carcinoma-associated fibroblasts throughout TGFβ [14]. We previously showed that a combination of Si306 and proton irradiation holds great potential to induce synergic cytotoxic effects and modulate the complex gene network in in vitro models of GBM [21]. Given the pronounced hypoxia observed during GBM development and progression, we aimed at studying the role of the Si306 and X-ray combination in hypoxic conditions, generating dose/response curves and calculating OER in addition to DMF to evaluate the relationship of these two parameters.

We first confirmed that Si306 was able to reduce cell survival in normoxia and, importantly, whether such an effect was preserved in hypoxic conditions. Notably, clonogenic assay revealed that c-SRC inactivation had a significant impact in hypoxic cells, leading to a higher DMF and a lower OER. The α and β values also support these data, showing that the robust increase in the α/β ratio in hypoxic conditions was related to an increase in α value, thus indicating improved non-repairable DNA damage [22]. A potential explanation of such a significant effect of Si306 in hypoxic GBM cells may be related to the intrinsic biological response to low oxygen levels [23]. Hypoxia induces radioresistance promoting GBM invasion and activating specific intracellular machinery that also relies on c-SRC activation [12,24]. Previous studies showed that RT itself may positively relate to activation of invasion and migration mechanisms involving c-SRC proteins [25,26]. Our evidence suggests that Si306 contributes to reducing efficacy of endogenous self-protective mechanisms that took place in hypoxic conditions, particularly sensitizing cell populations relying on c-SRC activation [27]. Furthermore, the analysis of the γH2AX foci showed the c-SRC inhibition increases radiation-induced DNA damage and slows down the DNA repair abilities in both normoxic and hypoxic conditions. Importantly, Si306 treatment was also able to dramatically reduce cell migration in both normoxic and hypoxic conditions, thus indicating a substantial role of c-SRC pathway inhibition in GBM invasiveness.

Altogether, our data support the hypothesis that c-SRC inhibition may represent a promising approach to improve RT efficacy. Our evidences are in accordance with previous observations with the reference compound of c-SRC-family inhibitor PP2 [28] and with Si306 [14,29,30]. To date, the most important nTKI is the dual inhibitor c-SRC/Abl (Dasatinib) that was tested alone and in combination with mAb anti-VEGF (Bevacizumab), TKi of EGFR (Erlotinib) and alkylating agent (Lomustine) in clinical trials for recurrent GBM [31–34]. Results from randomized phase I/II trial of Dasatinib combined with Temozolomide and radiotherapy for newly diagnosed GBM does not show increased survival as compared to standard therapy alone [35]. The limitations of Dasatinib were associated to pharmacokinetics aspects due to efflux transporters P-glycoprotein, which are highly expressed in the blood–brain barrier and GBM cells [36]. On this aspect, recent evidence showed that Si306 hold higher cell growth inhibitory potential as compared to Dasatinib, and it was found to reduce P-gp activity in GBM cells with multidrug resistance phenotype in addition to an optimal brain penetration and accumulation on mice [37].

This work provided addition data supporting the benefit of c-SRC inhibition to enhance RT and, for the first time, investigated the efficacy of radiotherapy combined with c-SRC inhibition comparing normoxic and hypoxic conditions on GBM cell lines. Interestingly, our results indicated that Si306 molecule has a radiosensitizing effect on GBM cells both in normoxia and hypoxia, showing that it could be considered in a targeted strategy for GBM treatment.

4. Materials and Methods

4.1. Cell Culture and Hypoxia Experiments

The U251-MG and U87-MG human GBM cell lines were purchased from American Type Culture Collections (ATCC, Manassas, VA, USA) and cultured as previously described [21]. Cells were maintained in an exponentially growing culture condition, at 37 °C in a humidified atmosphere with 21% O_2 and 5% CO_2 (normoxic condition) and were subcultured in 75 cm^2 standard tissue culture flasks. The U87-MG cells were used as additional cell line only for γ-H2AX immunofluorescence analyses.

For hypoxic experiments, 15 h after seeding, cells were transferred in the hypoxic workstation (IN VIVO2 1000, Ruskinn; Awel International, Blain, France), balanced with 94% N_2 and 5% CO_2 to maintain a gas concentration of 1% O_2 at 37 °C (hypoxia). During experiment, cells were refilled with fresh medium previously equilibrated with the gas mixture containing 1% O_2 in order to maintain this concentration from the beginning of the treatment with the drug.

4.2. Irradiation and Drug Treatments

Irradiation was performed in a biological irradiator (CellRad®, Faxitron, Edimex Le Plessis Grammoire, France) with a dose rate of 2 Gy/min, 130 kV and 5.0 mA. GBM cell irradiation was carried out using dose values of 2, 4, 6 and 8 Gy for clonogenic assay. 4 Gy dose was used for γ-H2AX immunofluorescence and migration assay.

The compound Si306 was provided by Lead Discovery Siena (Siena, Italy) as a stock powder and was dissolved in Dimethylsulfoxide (DMSO, Saint Quentin Fallavier, France). The Si306 molecule was diluted at a final concentration of 10 μM and 20 μM with fresh medium, in which GBM cells were maintained for 24 h. After irradiation, cells were replaced with fresh medium in order to remove the Si306 and manteined in normoxia or hypoxia up to the end of the experiment. The control samples for all biological tests were supplemented with vehicle (i.e., 0.5% DMSO).

4.3. Clonogenic Assay

Cells were seeded in a 6-well plates in triplicate at a density of 80–420 cells/cm^2, according to the dose delivered and to the vehicle or drug concentration. Then, irradiation was performed using the dose values of 2, 4, 6 and 8 Gy. After irradiation, cells were incubated for 7–10 days in normoxia and hypoxia condition until the colony formation. The colonies were incubated with 0.05% crystal violet diluted in 20% ethanol (Saint Quentin Fallavier, France) for 30 min at room temperature. SF was determined according to the plating efficiency (PE) as we previously described [9]. Briefly, we calculated the PE, dividing the counted colony by the total plated cells. We then calculated the SF as a ratio of sample PE over control PE. For each experiment, the effect of each dose of radiation alone and combined with Si306 was evaluated on three individual wells of cell culture and each experiment was performed in triplicate.

4.4. Radiobiological Parameters Calculation

Surviving fraction values were adjusted according to the LQ model, which utilizes a multi-parameter equation for each individual experimental curve, the form of which is: $S(D)/S(0)=e^{(-\alpha D-\beta D^{\wedge}2)}$, where S(D) is the fraction of cells that survive at a given dose (D) and S(0) is the fraction of cells at 0 Gy; so we get $\alpha[Gy^{-1}]$ and $\beta[Gy^{-2}]$ with their own standard deviation [21,38]. The DMF, which represents the dose of irradiation required to obtain the isoeffect, was calculated as previously described [21]. The OER, which is defined as the ratio of dose given under hypoxic conditions to the dose resulting in the same effect when given under normoxia [39], was also calculated. For both values of DMF and OER, the surviving fraction of 50% was considered a biological isoeffect at 0 μM, 10 μM and 20 μM of Si306.

4.5. γ-H2AX Immunofluorescence Analysis

Cells were seeded on sterile cover-glasses on 24 multiwell plates. After 8 h, cells were exposed to Si306 treatment for 24 h. Cells were then irradiated with 4 Gy and fixed in paraformaldehyde 4% at 2 and 24 h post-irradiation. Samples were then incubated with bovine serum albumin (BSA) 3% (Saint Quentin Fallavier, France), Tween 0.1% in PBS (Saint Quentin Fallavier, France) as blocking solution and to permeate cells for 30 min at room temperature. Indirect staining was performed using a primary antibody anti-γH2AX (1/1000; Abcam, ab26350, Paris, France) dissolved in BSA 1%, Tween 0.1% in PBS overnight at 4 °C. Then, samples were washed three times with Tween 0.1% in PBS for 5 min. Samples were incubated with Alexa-488-conjugated anti-mouse secondary antibody (1/500;

Thermofisher Scientific, A-21202, Montigny Le Bretonneux, France) for 1 h. Nuclei were counterstained adding Hoechst 33342 stain (10 µg/mL; Saint Quentin Fallavier, France) for 1 h at room temperature. After three washes in PBS, samples were coverslipped and images were acquired using a Leica DM6000 microscope with a 20× objective. FITC and DAPI filter were used to detect foci γ-H2AX (in green) and nuclear signals (in blue), respectively. Quantifications were performed as previously described [40–42]. Briefly, images were analyzed using FIJI application software (version 2.0.0-rc-69/1.52p). Each region of interest was analyzed applying the iso-data threshold on immunofluorescence images of Hoechst and γ-H2AX and data are expressed as percentage of γ-H2AX positive nuclei over total Hoechst positive cells. Investigators blinded to the treatment groups performed all quantifications.

4.6. Migration Assay

Cells were seeded in 24 multiwell plates and incubated at both normoxic and hypoxic conditions. Following cell adhesion, Si306 molecule was added for 24 h. Mitomycin C (3 µL/mL, Saint Quentin Fallavier, France) was used to block cell proliferation. Samples were irradiated with 4 Gy, and immediately after the irradiation a horizontal scratch was created using a sterile tip in the center of the cell monolayer. After 24 h samples were washed with PBS to remove floating cells and were stained with crystal violet solution as mentioned above. Images were acquired at 0 hand 24 h post-scratch and the area between scratch edges was quantified. The scratch wound closure percentage was calculated as follows: The scratch area 0 h − the scratch area 24 h / (the scratch area 0 h) × 100%.

4.7. Statistical Analyses

All tests were performed in GraphPad Prism (version 5.00, GraphPad Software, San Diego, CA, USA). Data were tested for normality using a D'Agostino and Pearson omnibus normality test and subsequently assessed for homogeneity of variance. For comparison of $n > 3$ groups, one-way or two-way ANOVA was used where appropriate, followed by Holm–Šídák post-hoc test.

5. Conclusions

Further studies will help to better characterize the biological effects of Si306 in terms of cell toxicity and potential side effects. Taken together, the cell survival reduction, supported by DMF and LQ model, the DNA damage increase and the migration inhibition are all effects induced by the combination of a Si306 molecule and X-rays in both conditions of normoxia and hypoxia. For this reason, Si306 is a potential candidate as a new radiosensitizer in targeted therapy to overcome radioresistance in GBM disease.

Author Contributions: Conceptualization, F.T., L.M. (Luigi Minafra), F.P.C., S.V.; methodology, F.T., L.M. (Luigi Minafra), F.P.C., G.S., M.C., E.A.P., H.Ö., S.V.; investigation, F.T., L.M. (Luigi Minafra), S.V.; data curation and formal analysis, F.T., L.M. (Luigi Minafra), N.V. and S.V.; resources, L.M. (Laura Maccari), M.B., L.B., G.R., R.P., S.V.; writing—original draft preparation, F.T.; writing—review and editing, F.T., L.M. (Luigi Minafra), F.P.C., N.V., R.P., S.V.; supervision, F.T., L.M. (Luigi Minafra), F.P.C., M.B., G.R., R.P., S.V.; project administration, F.P.C., M.B., L.B., G.R., R.P., S.V.; funding acquisition, M.B., G.R., R.P., S.V. All authors have read and agreed to the published version of the manuscript.

Funding: This work was partially supported by Institut National du Cancer (INCA 11699) and HABIONOR European project, co-funded by the Normandy County Council, the French State in the framework of the interregional development Contract "Vallée de la Seine" 2015-2020. This work was partially supported by the National Institute for Nuclear Physics (INFN) Commissione Scientifica Nazionale 5 (CSN5) Call 'MoVe-IT'. This work was also supported by PBCT PRIN: Progetti di Ricerca di Rilevante Interesse Nazionale – PRIN 2017 – Prot. 2017XKWWK9. F.T. was supported by the ERASMUS+ Programme, Key Action 1, 2018/2019 – Student Mobility for Traineeship (Sapienza, University of Rome, Italy) and by the PhD programme in Biotechnology (Biometec, University of Catania, Italy). N.V. was supported by the PON AIM R&I 2014-2020 - E66C18001240007.

Conflicts of Interest: The authors declare no conflict of interest. The funders had no role in the design of the study; in the collection, analyses, or interpretation of data; in the writing of the manuscript, or in the decision to publish the results.

Abbreviations

GBM	Glioblastoma
RT	Radiotherapy
MMP-2	Matrix metalloproteinase-2
MMP-9	Matrix metalloproteinase-9
SFKs	SRC family kinases
FAK	Focal adhesion kinase
EGFR	Epidermal growth factor receptor
SF	Surviving fraction
PE	Plating efficiency
DMF	Dose modifying factor
OER	Oxygen enhancement ratio
LQ	Linear-quadratic
mAb	Monoclonal antibodies
TKi	Tyrosine-kinase inhibitors
nRTK	Non receptor tyrosine kinase
ECM	Extracellular matrix
DMSO	Dimethylsulfoxide
BSA	Bovine serum albumin

References

1. Wen, P.Y.; Weller, M.; Lee, E.Q.; Alexander, B.A.; Barnholtz-Sloan, J.S.; Barthel, F.P.; Batchelor, T.T.; Bindra, R.S.; Chang, S.M.; Chiocca, E.A.; et al. Glioblastoma in Adults: A Society for Neuro-Oncology (SNO) and European Society of Neuro-Oncology (EANO) Consensus Review on Current Management and Future Directions. *Neuro-Oncology* **2020**. [CrossRef] [PubMed]
2. Cabrera, A.R.; Kirkpatrick, J.P.; Fiveash, J.B.; Shih, H.A.; Koay, E.J.; Lutz, S.; Petit, J.; Chao, S.T.; Brown, P.D.; Vogelbaum, M.; et al. Radiation therapy for glioblastoma: Executive summary of an American Society for Radiation Oncology Evidence-Based Clinical Practice Guideline. *Pract. Radiat. Oncol.* **2016**, *6*, 217–225. [CrossRef] [PubMed]
3. Gerstner, E.R.; Zhang, Z.; Fink, J.R.; Muzi, M.; Hanna, L.; Greco, E.; Prah, M.; Schmainda, K.M.; Mintz, A.; Kostakoglu, L.; et al. ACRIN 6684: Assessment of Tumor Hypoxia in Newly Diagnosed Glioblastoma Using 18F-FMISO PET and MRI. *Clin. Cancer Res.* **2016**, *22*, 5079–5086. [CrossRef] [PubMed]
4. Liu, C.; Lin, Q.; Yun, Z. Cellular and molecular mechanisms underlying oxygen-dependent radiosensitivity. *Radiat. Res.* **2015**, *183*, 487–496. [CrossRef] [PubMed]
5. Grimes, D.R.; Partridge, M. A mechanistic investigation of the oxygen fixation hypothesis and oxygen enhancement ratio. *Biomed. Phys. Eng. Express* **2015**, *1*, 045209. [CrossRef] [PubMed]
6. Valable, S.; Corroyer-Dulmont, A.; Chakhoyan, A.; Durand, L.; Toutain, J.; Divoux, D.; Barre, L.; MacKenzie, E.T.; Petit, E.; Bernaudin, M.; et al. Imaging of brain oxygenation with magnetic resonance imaging: A validation with positron emission tomography in the healthy and tumoural brain. *J. Cereb Blood Flow Metab.* **2017**, *37*, 2584–2597. [CrossRef]
7. Ponte, K.F.; Berro, D.H.; Collet, S.; Constans, J.M.; Emery, E.; Valable, S.; Guillamo, J.S. In Vivo Relationship Between Hypoxia and Angiogenesis in Human Glioblastoma: A Multimodal Imaging Study. *J. Nucl. Med.* **2017**, *58*, 1574–1579. [CrossRef]
8. Persano, L.; Rampazzo, E.; Della Puppa, A.; Pistollato, F.; Basso, G. The three-layer concentric model of glioblastoma: Cancer stem cells, microenvironmental regulation, and therapeutic implications. *Sci. World J.* **2011**, *11*, 1829–1841. [CrossRef]
9. Peres, E.A.; Gerault, A.N.; Valable, S.; Roussel, S.; Toutain, J.; Divoux, D.; Guillamo, J.S.; Sanson, M.; Bernaudin, M.; Petit, E. Silencing erythropoietin receptor on glioma cells reinforces efficacy of temozolomide and X-rays through senescence and mitotic catastrophe. *Oncotarget* **2015**, *6*, 2101–2119. [CrossRef]
10. Ahluwalia, M.S.; de Groot, J.; Liu, W.M.; Gladson, C.L. Targeting SRC in glioblastoma tumors and brain metastases: Rationale and preclinical studies. *Cancer Lett.* **2010**, *298*, 139–149. [CrossRef]

11. Keller, S.; Schmidt, M.H.H. EGFR and EGFRvIII Promote Angiogenesis and Cell Invasion in Glioblastoma: Combination Therapies for an Effective Treatment. *Int. J. Mol. Sci.* **2017**, *18*. [CrossRef] [PubMed]
12. Liu, Z.; Han, L.; Dong, Y.; Tan, Y.; Li, Y.; Zhao, M.; Xie, H.; Ju, H.; Wang, H.; Zhao, Y.; et al. EGFRvIII/integrin beta3 interaction in hypoxic and vitronectinenriching microenvironment promote GBM progression and metastasis. *Oncotarget* **2016**, *7*, 4680–4694. [CrossRef] [PubMed]
13. Schenone, S.; Radi, M.; Musumeci, F.; Brullo, C.; Botta, M. Biologically driven synthesis of pyrazolo[3,4-d]pyrimidines as protein kinase inhibitors: An old scaffold as a new tool for medicinal chemistry and chemical biology studies. *Chem. Rev.* **2014**, *114*, 7189–7238. [CrossRef] [PubMed]
14. Calgani, A.; Vignaroli, G.; Zamperini, C.; Coniglio, F.; Festuccia, C.; Di Cesare, E.; Gravina, G.L.; Mattei, C.; Vitale, F.; Schenone, S.; et al. Suppression of SRC Signaling Is Effective in Reducing Synergy between Glioblastoma and Stromal Cells. *Mol. Cancer Ther.* **2016**, *15*, 1535–1544. [CrossRef] [PubMed]
15. O'Rourke, S.F.; McAneney, H.; Hillen, T. Linear quadratic and tumour control probability modelling in external beam radiotherapy. *J. Math. Biol.* **2009**, *58*, 799–817. [CrossRef]
16. Hu, G.; Fang, W.; Liu, N.; Li, C. Effects of mir-128a on the invasion and proliferation of glioma U251 cells. *Oncol. Lett.* **2019**, *17*, 891–896. [CrossRef]
17. Stupp, R.; Mason, W.P.; van den Bent, M.J.; Weller, M.; Fisher, B.; Taphoorn, M.J.; Belanger, K.; Brandes, A.A.; Marosi, C.; Bogdahn, U.; et al. Radiotherapy plus concomitant and adjuvant temozolomide for glioblastoma. *N. Engl. J. Med.* **2005**, *352*, 987–996. [CrossRef]
18. Buglione, M.; Triggiani, L.; Borghetti, P.; Pedretti, S.; Pasinetti, N.; Magrini, S.M. The "Radioresistance" of Glioblastoma in the Clinical Setting, and the Present Therapeutic Options. In *Radiobiology of Glioblastoma*; Pirtoli, L., Gravina, G., Giordano, A., Eds.; Springer: Jersey City, NJ, USA, 2016; pp. 15–27.
19. Mukhopadhyay, D.; Tsiokas, L.; Zhou, X.M.; Foster, D.; Brugge, J.S.; Sukhatme, V.P. Hypoxic induction of human vascular endothelial growth factor expression through c-Src activation. *Nature* **1995**, *375*, 577–581. [CrossRef]
20. Valle-Casuso, J.C.; Gonzalez-Sanchez, A.; Medina, J.M.; Tabernero, A. Hif-1 and C-Src Mediate Increased Glucose Uptake Induced by Endothelin-1 and Connexin43 In Astrocytes. *PLoS ONE* **2012**, *7*, e32448. [CrossRef]
21. Cammarata, F.P.; Torrisi, F.; Forte, G.I.; Minafra, L.; Bravata, V.; Pisciotta, P.; Savoca, G.; Calvaruso, M.; Petringa, G.; Cirrone, G.A.P.; et al. Proton Therapy and Src Family Kinase Inhibitor Combined Treatments on U87 Human Glioblastoma Multiforme Cell Line. *Int. J. Mol. Sci.* **2019**, *20*, 4745. [CrossRef]
22. Choi, J.; Kang, J.O. Basics of Particle Therapy Ii: Relative Biological Effectiveness. *Radiat. Oncol. J.* **2012**, *30*, 1–13. [CrossRef] [PubMed]
23. Plaks, V.; Kong, N.; Werb, Z. The Cancer Stem Cell Niche: How Essential Is The Niche in Regulating Stemness of Tumor Cells? *Cell Stem Cell* **2015**, *16*, 225–238. [CrossRef] [PubMed]
24. Skuli, N.; Monferran, S.; Delmas, C.; Favre, G.; Bonnet, J.; Toulas, C.; Cohen-Jonathan Moyal, E. Alphavbeta3/Alphavbeta5 Integrins-Fak-Rhob: A Novel Pathway for Hypoxia Regulation in Glioblastoma. *Cancer Res.* **2009**, *69*, 3308–3316. [CrossRef] [PubMed]
25. Park, C.M.; Park, M.J.; Kwak, H.J.; Lee, H.C.; Kim, M.S.; Lee, S.H.; Park, I.C.; Rhee, C.H.; Hong, S.I. Ionizing Radiation Enhances Matrix Metalloproteinase-2 Secretion and Invasion of Glioma Cells Through Src/Epidermal Growth Factor Receptor-Mediated P38/Akt And Phosphatidylinositol 3-Kinase/Akt Signaling Pathways. *Cancer Res.* **2006**, *66*, 8511–8519. [CrossRef] [PubMed]
26. Roos, A.; Ding, Z.; Loftus, J.C.; Tran, N.L. Molecular and Microenvironmental Determinants of Glioma Stem-Like Cell Survival and Invasion. *Front. Oncol.* **2017**, *7*, 120. [CrossRef]
27. Brown, J.M.; Wilson, W.R. Exploiting Tumour Hypoxia in Cancer Treatment. *Nat. Rev. Cancer* **2004**, *4*, 437–447. [CrossRef]
28. Eom, K.Y.; Cho, B.J.; Choi, E.J.; Kim, J.H.; Chie, E.K.; Wu, H.G.; Kim, I.H.; Paek, S.H.; Kim, J.S.; Kim, I.A. The Effect of Chemoradiotherapy with SRC Tyrosine Kinase Inhibitor, PP2 and Temozolomide on Malignant Glioma Cells In Vitro and In Vivo. *Cancer Res. Treat.* **2016**, *48*, 687–697. [CrossRef]
29. Carraro, F.; Naldini, A.; Pucci, A.; Locatelli, G.A.; Maga, G.; Schenone, S.; Bruno, O.; Ranise, A.; Bondavalli, F.; Brullo, C.; et al. Pyrazolo[3,4-D]Pyrimidines as Potent Antiproliferative and Proapoptotic Agents Toward A431 And 8701-Bc Cells in Culture Via Inhibition Of C-Src Phosphorylation. *J. Med. Chem.* **2006**, *49*, 1549–1561. [CrossRef]

30. Tintori, C.; Fallacara, A.L.; Radi, M.; Zamperini, C.; Dreassi, E.; Crespan, E.; Maga, G.; Schenone, S.; Musumeci, F.; Brullo, C.; et al. Combining X-Ray Crystallography and Molecular Modeling Toward the Optimization of Pyrazolo[3,4-D]Pyrimidines as Potent C-Src Inhibitors Active in Vivo Against Neuroblastoma. *J. Med. Chem.* **2015**, *58*, 347–361. [CrossRef]
31. Lassman, A.B.; Pugh, S.L.; Gilbert, M.R.; Aldape, K.D.; Geinoz, S.; Beumer, J.H.; Christner, S.M.; Komaki, R.; DeAngelis, L.M.; Gaur, R.; et al. Phase 2 Trial of Dasatinib in Target-Selected Patients With Recurrent Glioblastoma (Rtog 0627). *Neuro-Oncology* **2015**, *17*, 992–998. [CrossRef]
32. Galanis, E.; Anderson, S.K.; Twohy, E.L.; Carrero, X.W.; Dixon, J.G.; Tran, D.D.; Jeyapalan, S.A.; Anderson, D.M.; Kaufmann, T.J.; Feathers, R.W.; et al. A phase 1 and randomized, placebo-controlled phase 2 trial of bevacizumab plus dasatinib in patients with recurrent glioblastoma: Alliance/North Central Cancer Treatment Group N0872. *Cancer* **2019**, *125*, 3790–3800. [CrossRef] [PubMed]
33. Reardon, D.A.; Vredenburgh, J.J.; Desjardins, A.; Peters, K.B.; Sathornsumetee, S.; Threatt, S.; Sampson, J.H.; Herndon, J.E., 2nd; Coan, A.; McSherry, F.; et al. Phase 1 trial of dasatinib plus erlotinib in adults with recurrent malignant glioma. *J. Neuro-Oncol.* **2012**, *108*, 499–506. [CrossRef]
34. Franceschi, E.; Stupp, R.; van den Bent, M.J.; van Herpen, C.; Laigle Donadey, F.; Gorlia, T.; Hegi, M.; Lhermitte, B.; Strauss, L.C.; Allgeier, A.; et al. EORTC 26083 phase I/II trial of dasatinib in combination with CCNU in patients with recurrent glioblastoma. *Neuro-Oncology* **2012**, *14*, 1503–1510. [CrossRef] [PubMed]
35. Laack, N.N.; Galanis, E.; Anderson, S.K.; Leinweber, C.; Buckner, J.C.; Giannini, C.; Geoffroy, F.J.; Johnson, D.R.; Lesser, G.J.; Jaeckle, K.A.; et al. Randomized, placebo-controlled, phase II study of dasatinib with standard chemo-radiotherapy for newly diagnosed glioblastoma (GBM), NCCTG N0877 (Alliance). *J. Clin. Oncol.* **2017**, *33*, 2013. [CrossRef]
36. Agarwal, S.; Mittapalli, R.K.; Zellmer, D.M.; Gallardo, J.L.; Donelson, R.; Seiler, C.; Decker, S.A.; Santacruz, K.S.; Pokorny, J.L.; Sarkaria, J.N.; et al. Active efflux of Dasatinib from the brain limits efficacy against murine glioblastoma: Broad implications for the clinical use of molecularly targeted agents. *Mol. Cancer Ther.* **2012**, *11*, 2183–2192. [CrossRef]
37. Fallacara, A.L.; Zamperini, C.; Podolski-Renic, A.; Dinic, J.; Stankovic, T.; Stepanovic, M.; Mancini, A.; Rango, E.; Iovenitti, G.; Molinari, A.; et al. A New Strategy for Glioblastoma Treatment: In Vitro and In Vivo Preclinical Characterization of Si306, a Pyrazolo[3,4-d]Pyrimidine Dual Src/P-Glycoprotein Inhibitor. *Cancers* **2019**, *11*, 848. [CrossRef]
38. Chapman, J.D. Can the Two Mechanisms of Tumor Cell Killing by Radiation Be Exploited for Therapeutic Gain? *J. Radiat. Res.* **2014**, *55*, 2–9. [CrossRef]
39. Joiner, M.; van der Kogel, A. *Basic Clinical Radiobiology*, 4th ed.; Taylor & Fransis Group: London, UK, 2009; p. 375.
40. Vicario, N.; Bernstock, J.D.; Spitale, F.M.; Giallongo, C.; Giunta, M.A.S.; Li Volti, G.; Gulisano, M.; Leanza, G.; Tibullo, D.; Parenti, R.; et al. Clobetasol Modulates Adult Neural Stem Cell Growth via Canonical Hedgehog Pathway Activation. *Int. J. Mol. Sci.* **2019**, *20*. [CrossRef]
41. Mauri, E.; Sacchetti, A.; Vicario, N.; Peruzzotti-Jametti, L.; Rossi, F.; Pluchino, S. Evaluation of RGD functionalization in hybrid hydrogels as 3D neural stem cell culture systems. *Biomater. Sci.* **2018**, *6*, 501–510. [CrossRef]
42. Gulino, R.; Vicario, N.; Giunta, M.A.S.; Spoto, G.; Calabrese, G.; Vecchio, M.; Gulisano, M.; Leanza, G.; Parenti, R. Neuromuscular Plasticity in a Mouse Neurotoxic Model of Spinal Motoneuronal Loss. *Int. J. Mol. Sci.* **2019**, *20*. [CrossRef]

International Journal of
Molecular Sciences

Article

Different Mechanisms Underlie the Metabolic Response of GBM Stem-Like Cells to Ionizing Radiation: Biological and MRS Studies on Effects of Photons and Carbon Ions

Alessandra Palma [1,†], Sveva Grande [1,†], Lucia Ricci-Vitiani [2], Anna Maria Luciani [1], Mariachiara Buccarelli [2], Mauro Biffoni [2], Valentina Dini [1,3], Giuseppe A. P. Cirrone [4], Mario Ciocca [5], Laura Guidoni [1,‡], Roberto Pallini [6], Vincenza Viti [1,§] and Antonella Rosi [1,*]

1 National Centre for Innovative Technologies in Public Health, Istituto Superiore di Sanità, 00161 Rome, Italy; alessandra.palma@iss.it (A.P.); sveva.grande@iss.it (S.G.); annamaria.luciani@iss.it (A.M.L.); valentina.dini@iss.it (V.D.); laguidoni@gmail.com (L.G.); vincefl.viti@gmail.com (V.V.)

2 Department of Oncology and Molecular Medicine, Istituto Superiore di Sanità, 00161 Rome, Italy; lucia.riccivitiani@iss.it (L.R.-V.); mariachiara.buccarelli@iss.it (M.B.); mauro.biffoni@iss.it (M.B.)

3 Istituto Nazionale di Fisica Nucleare INFN Sez. di Roma, 00185 Rome, Italy

4 National Institute for Nuclear Physics, Laboratori Nazionali del Sud, INFN-LNS, 95123 Catania, Italy; pablo.cirrone@lns.infn.it

5 Centro Nazionale di Adroterapia Oncologica (CNAO)-National Center for Oncological Hadrontherapy, 27100 Pavia, Italy; mario.ciocca@cnao.it

6 Department of Neuroscience, Fondazione Policlinico Universitario A. Gemelli, Università Cattolica del Sacro Cuore, 00168 Rome, Italy; roberto.pallini@unicatt.it

* Correspondence: antonella.rosi@iss.it; Tel.: +39-06-49903159

† These authors equally contributed to the manuscript.

‡ Retired since 1 January 2013.

§ Retired since 1 August 2010.

Received: 22 May 2020; Accepted: 20 July 2020; Published: 21 July 2020

Abstract: Glioblastoma multiforme (GBM) is a malignant primary brain tumor with very poor prognosis, high recurrence rate, and failure of chemo-radiotherapy, mainly due to a small fraction of cells with stem-like properties (GSCs). To study the mechanisms of GSCs resistance to radiation, two GSC lines, named line #1 and line #83, with different metabolic patterns and clinical outcome, were irradiated with photon beams and carbon ions and assessed by ^{1}H Magnetic Resonance Spectroscopy (MRS). Both irradiation modalities induced early cytotoxic effects in line #1 with small effects on cell cycle, whereas a proliferative G2/M cytostatic block was observed in line #83. MR spectroscopy signals from mobile lipids (ML) increased in spectra of line #1 after photon and C-ion irradiation with effects on lipid unsaturation level, whereas no effects were detected in line #83 spectra. Gamma-Aminobutyric Acid (GABA), glutamic acid (glu) and Phosphocreatine (pCr) signals showed a significant variation only for line #1 after carbon ion irradiation. Glucose (glc) level and lactate (Lac) extrusion behaved differently in the two lines. Our findings suggest that the differences in irradiation response of GSCs #1 and #83 lines are likely attributable to their different metabolic fingerprint rather than to the different radiation types.

Keywords: glioblastoma; stem cells; metabolism; MRS; photon beams; carbon ions

1. Introduction

Glioblastoma multiforme (GBM) is the most common malignant glioma in adults. Despite innovative research efforts for targeted therapies, the outcome remains poor [1]. To enhance sensitivity to radiotherapy (RT), the mechanisms underlying the different cell response to radiation need to be further elucidated and new strategies developed [2,3].

The cancer stem cell (CSC) hypothesis attributes treatment failure to a small fraction of self-renewing cells with stem-like properties and high resistance to radiation [4,5] that seems to be responsible for tumor recurrence in GBM. Because of the high resistance to photon beam irradiation [6], several studies have examined the response of CSCs to irradiation with proton beams and charged particles. In particular, heavy ion radiotherapy has been shown to have potential advantages over photons in treating many radioresistant human colon cancers [7]; a recent paper shows that the combination of proton therapy and a radiosensitive compound could overcome GBM resistance to conventional treatments [8]. However, results from clinical trials conducted through a comparative analysis between protons, carbon ions and photon beam (Cinderella and Cleopatra), are still pending [9–11]. A Phase II clinical trial on selected GBM patients used a boost of C-ion RT prior to the initiation of the standard treatment course [12]. Using a 3D model for investigating GSC radiosensitivity to proton beam and carbon ion irradiation, Chiblack showed an enhanced biological effectiveness of C-ion RT in vivo, due to its potent antiangiogenic effects and eradication of radioresistant hypoxic tumor cells [13,14]. As other cancers, GBM displays high heterogeneity among patients, with relevant differences in genomic, transcriptomic, proteomic, and metabolomic features, and quite different cell populations seem to be present in the same tumor [15,16]. Both inter- and intra-individual heterogeneity may then be responsible for treatment failure and tumor relapse, suggesting the necessity of tailored therapies. In previous studies, subtypes of high-grade glioma have been identified on the basis of molecular gene expression [17] which includes the proneural, proliferative, and mesenchymal subtypes. The hypothesis that the ability to escape or to mitigate radiation damage could be due to different subtype characteristics deserves to be investigated. Further profiling studies revealed the presence of two distinct subsets of GSCs based on gene expression analysis: (a) a subset, displaying a Glioblastoma full stem-like phenotype (GSf-like), highly tumorigenic and invasive in vivo, and an enriched proneural gene expression signature; (b) the other with a Glioblastoma restricted stem-like phenotype (GSr-like) showing expression signatures more similar to stabilized cell lines than to the original patient tumors [18,19]. Moreover, it has been shown that stratification of GSCs according to their metabolic MRS signals parallels the genetic profiles typical of GSf/GSr cell lines [18]. Finally, GSCs were characterized by high heterogeneity [20] since cells with the same brain origin and similar clinical parameters belonged to different metabolic clusters and showed different genetic signatures [18,19,21]. These results led to investigate if the response to irradiation in different GSC lines could be due to different genetic and metabolic fingerprints rather than to the radiation type.

Metabolomics and metabolic profiling, as biochemical picture of a particular phenotype, can be explored by MRS as a tool to further understand metabolic changes occurring during tumor initiation, progression, and radiotherapy resistance [15]. Improved knowledge of the metabolism of brain tumors may contribute to discover new diagnostic or prognostic biomarkers, overcoming the problem of poor prognosis. In particular, metabolic reprogramming is a hallmark of GSCs [22] used to identify malignance signatures and open new ways to develop tailored therapeutic approaches. Stratifying patients according to molecular biomarker profiles is a key step to manage patient heterogeneity.

To study the mechanisms involved in GSCs radiation resistance and evaluate whether their response to irradiation is due to different radiation type or to intrinsic GSCs characteristics, two GSC lines, named #1 and #83, were selected on the basis of their metabolic profiles and genetic signatures, deeply analyzed in previous studies [18,21]. Particularly a mixed neural–astrocyte metabolic phenotype with a strong neuronal fingerprint was found in line #1, while a glioma-like metabolism prevailed in line #83. We performed MRS experiments on these cells after irradiation with ^{137}Cs photons and C-ions (CNAO, Pavia, Italy and INFN-LNS, Catania, Italy) at different Linear Energy Transfer (LET) value at a dose of 20 Gy. Effects on cell survival and cell cycle were also evaluated.

2. Results

GSC lines #1 and #83 were selected from a panel of GSCs on the basis of their biological and metabolic characteristics [20,21]. Although they both derived from tumors localized close to the temporal subventricular zone, they have different overall survival (12.5 and 8.5 months for line #1 and line #83, respectively), different CD133 positivity (more than 80% in line #1 and less than 1% in line #83); moreover line #1 cells grow as neurospheres, while line #83 cells grow mostly as monolayer. Finally, they belong to different metabolic and genetic clusters [18–21]. Cell growth of both control and 20 Gy photon irradiated lines are shown in Figure 1A,B. A cytotoxic effect was observed for line #1 at short and late times (A), while only a cytostatic effect was observed in early times of growth for line #83 cells (B). These latter, in late times, maintain their growth in culture, and resume proliferation once they have been re-seeded, (see Inset in Figure 1B). A similar behavior was observed after irradiation with carbon ions (Figure 1C,D). Cell growth data were confirmed by the trend of the cell cycle in of the two lines. The line #1 cells after carbon ion irradiation accumulated in G0/G1 phase due to a proliferation arrest (Figure 1G). Moreover, the percentage of cells in S phase decreased in all carbon ion irradiated samples and at late time after photon irradiation (Figure 1E,G). Differently, cells from line #83 were significantly blocked after both photon and carbon ion irradiation in G2 phase where cell damage repair is known to occur (Figure 1F,H).

Apoptosis for both lines is shown in Supplementary Figure S1. Apoptosis onset was observed in line #1 by 2 days after cell irradiation, then apoptosis increased at 3 days, becoming relevant at 4 days after irradiation. For cell line #1 mainly necrosis gave a relevant contribution to cell death. On the contrary in line #83 the percentage of dead cells was smaller in the time range studied (Figure S1).

Metabolomics and metabolic profiling of the two cell lines have been explored by MRS. As already reported, the analysis of metabolic profiles for lines #1 and #83 indicated that these two lines strongly differ in several metabolites as shown in box plots and Fold Change evaluations shown in Figure S2.

The whole MR spectra of the two lines, the analysed metabolic signals (lipid, glutathione and energy metabolism), selected region of 2D COSY spectra and the corresponding intensity values are shown in Figures 2 and 3 for line #1 and line #83, respectively. The MRS spectral signals were assigned according to [21].

In particular, 1D and 2D spectra of photon beam irradiated samples (green trace) and their controls (blue trace) of lines #1 and #83 together with the spectral regions of interest are shown in A and B panels of Figures 2 and 3, respectively, where changes induced by irradiation on the analysed metabolites can be observed. Similarly, 1D and 2D spectra of carbon ion irradiated samples (green trace) and their controls (blue trace) of lines #1 and #83 and the regions of interest where metabolite changes can be detected are shown in C and D panels of Figures 2 and 3, respectively. Spectra were acquired 48 h after irradiation.

Mean values of lipid, glutathione and energy metabolites signal intensities, standard deviations and p values, calculated on at least 5 samples, are shown in Figure 2E (photon beams) and Figure 2F (carbon ions) for line #1 and Figure 3E (photon beams) and Figure 3F (carbon ions) for line #83, respectively.

The intensities of signals from lipids (ML) increased in spectra of line #1 after both photon and C ion irradiation (Figure 2A,C,E,F), while remain unaffected in spectra from line #83 (Figure 3A,C,E,F). The behavior of lipid signals was also monitored as cross-peak A in 2D spectra from lipid chains giving the same results for both lines (Figure 2B,D–F; Figure 3B,D–F). It is worth noting the intrinsic high level of ML in line #83 as reported in Figure S2A,B. Clone analysis for line #83 is shown in Figure S2D to testify the higher heterogeneity of this line with respect to line #1.

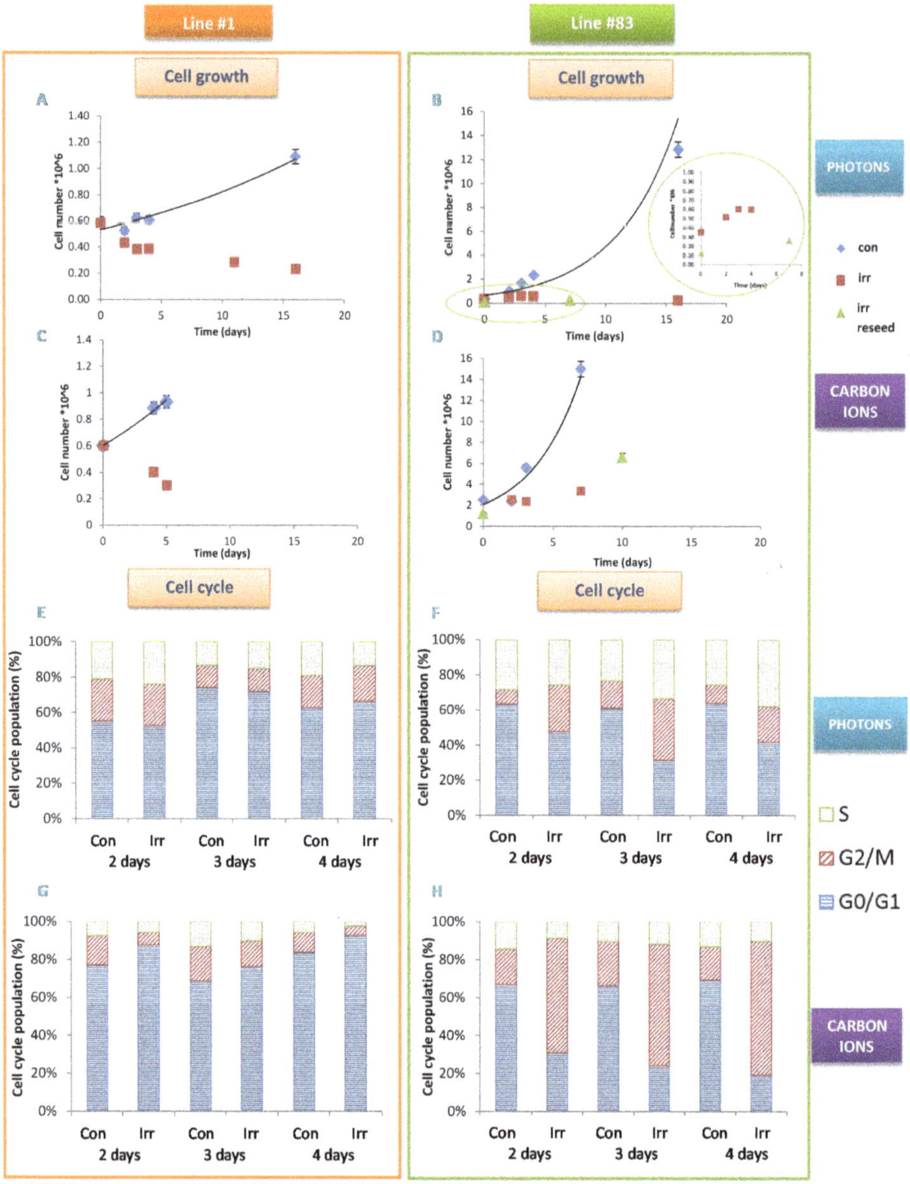

Figure 1. Irradiation affects growth and cell cycle of line #1 and #83 in a different way. Growth curves of line #1 (**A**,**C**) and of line #83 (**B**,**D**) after photon and carbon ion irradiation at 20 Gy. For line #83 values of cell growth after reseeding ("irr reseed") are also reported. In panel B the detailed expanded region of cell growth for line #83 is shown. Exemplificative experiments are reported; error bars correspond to measurement error (5%). Distribution of cell cycle phases for line #1 (**E**,**G**) and for line #83 (**F**,**H**) after photon and carbon ion irradiation at 20 Gy as a function of time after irradiation (Con: control and Irr: irradiated sample).

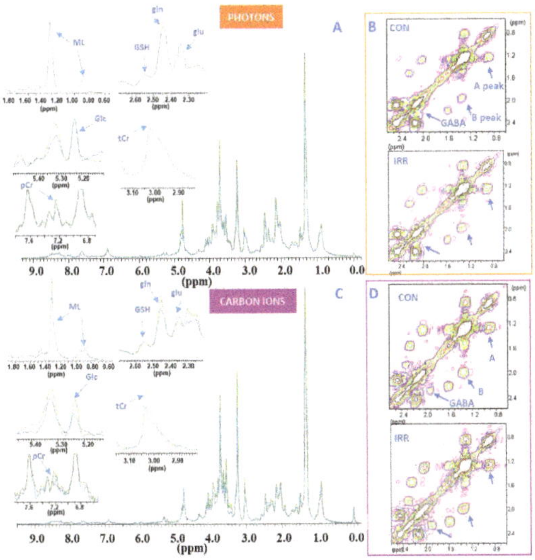

Line #1

E — PHOTONS

Metabolite	Mean	St. dev.	p value
LIPID METABOLISM			
ML			
Con	3.05	0.51	0.02*
Irr	4.27	0.17	
A peak			
Con	2.59	0.29	0.03*
Irr	3.72	0.58	
GSH AND RELATED METABOLITES			
GSH			
Con	0.98	0.16	0.33
Irr	1.03	0.12	
Gln			
Con	4.57	2.11	0.48
Irr	4.48	0.96	
Glu			
Con	1.19	0.13	0.40
Irr	1.15	0.17	
GABA			
Con	0.84	0.70	0.16
Irr	1.50	0.72	
ENERGY METABOLISM			
Glucose			
Con	0.14	0.03	0.04*
Irr	0.08	0.02	
pCr			
Con	0.54	0.13	0.37
Irr	0.59	0.20	

F — CARBON IONS

Metabolite	Mean	St. dev.	p value
LIPID METABOLISM			
ML			
Con	2.60	0.57	0.03*
Irr	4.20	0.80	
A peak			
Con	1.86	0.36	0.02*
Irr	2.99	0.51	
GSH AND RELATED METABOLITES			
GSH			
Con	1.28	0.16	0.06
Irr	1.04	0.15	
Gln			
Con	9.23	1.03	0.36
Irr	9.48	0.43	
Glu			
Con	0.95	0.07	0.01*
Irr	0.73	0.08	
GABA			
Con	0.33	0.08	0.02*
Irr	0.83	0.19	
ENERGY METABOLISM			
Glucose			
Con	0.13	0.02	0.01*
Irr	0.08	0.01	
pCr			
Con	0.49	0.04	0.02*
Irr	0.75	0.10	

Figure 2. ^{1}H NMR spectra of line #1 cells, signals of interest and metabolic modifications induced by radiation treatment. (**A**,**B**) 1D and 2D spectra of photon beam irradiated line #1 cell samples (green trace) and corresponding controls (blue trace) together with the spectral regions of interest for the analysis: Mobile Lipids (ML, 1.28 ppm), glutathione and related metabolites (GSH 2.55 ppm, gln 2.45 ppm, glu 2.35 ppm), glucose (glc 5.22 ppm), total creatine (tCr, 3.02 ppm) and phosphocreatine (pCr, 7.25 ppm). In panel B cross peak A, arising from the correlation of CH_3-terminal methyl group of Fatty Acid (FA) chains with the bulk of FA (excluding omega-3 FAs), at (0.89–1.28) ppm, cross peak B, representing both mono and polyunsaturated FA at (2.02–1.28) ppm and GABA cross peak at (2.30–1.90) ppm are labelled. (**C**,**D**) 1D and 2D spectra of carbon ion irradiated samples (green trace) at 20 Gy and corresponding controls (blue trace) of lines #1 and #83. Region of interest were expanded. All spectra were acquired 48 h after irradiation at a dose of 20 Gy with both photons and carbon ions. Assignments were performed as reported in [21]. Tables (**E**) (photon beams) and (**F**) (carbon ions) show mean values of all analyzed metabolites signal intensities, standard deviations and p values, calculated on at least 5 experiments performed 48 h after irradiation and corresponding controls. Stars indicate statistically significant variations (p value < 0.05).

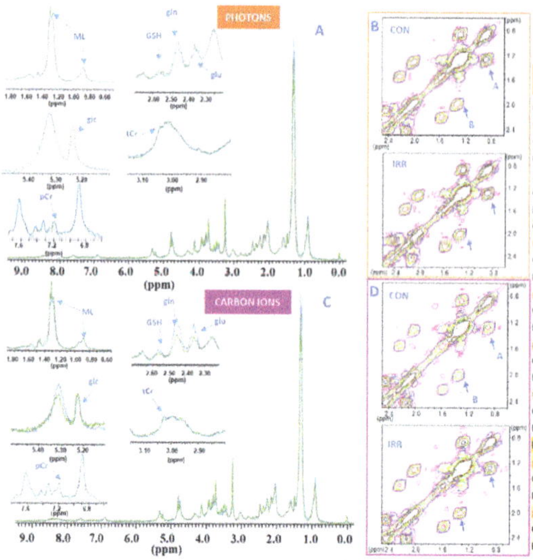

Line #83

E — PHOTONS

Metabolite	Mean	St. dev.	p value
LIPID METABOLISM			
ML			
Con	13.60	2.00	0.12
Irr	11.71	0.70	
A peak			
Con	14.03	5.03	0.48
Irr	13.89	2.79	
GSH AND RELATED METABOLITES			
GSH			
Con	0.89	0.14	0.13
Irr	0.75	0.12	
gln			
Con	5.54	0.90	0.29
Irr	5.10	0.90	
glu			
Con	1.40	0.19	0.39
Irr	1.43	0.05	
GABA			
Con	n.d.	--	--
Irr	n.d.	--	
ENERGY METABOLISM			
glucose			
Con	0.24	0.01	0.17
Irr	0.19	0.04	
pCr			
Con	0.40	0.06	0.35
Irr	0.43	0.07	

F — CARBON IONS

Metabolite	Mean	St. dev.	p value
LIPID METABOLISM			
ML			
Con	12.55	5.08	0.47
Irr	12.27	2.81	
A peak			
Con	10.92	2.99	0.46
Irr	10.74	1.55	
GSH AND RELATED METABOLITES			
GSH			
Con	1.21	0.20	0.07
Irr	0.03	0.10	
gln			
Con	6.23	1.24	0.19
Irr	5.41	0.62	
glu			
Con	1.41	0.15	0.18
Irr	1.28	0.17	
GABA			
Con	n.d	--	--
Irr	n.d.	--	
ENERGY METABOLISM			
glucose			
Con	0.26	0.04	0.18
Irr	0.22	0.06	
pCr			
Con	0.42	0.05	0.48
Irr	0.46	0.08	

Figure 3. ^{1}H NMR spectra of line #83 cells, signals of interest and metabolic modifications induced by radiation (**A**,**B**) 1D and 2D spectra of photon beam irradiated line #83 cell samples (green trace) and corresponding controls (blue trace) together with the spectral regions of interest for the analysis: Mobile Lipids (ML, 1.28 ppm), glutathione and related metabolites (GSH 2.55 ppm, glu 2.35 ppm, gln 2.45 ppm), glucose (glc 5.22 ppm), total creatine (tCr, 3.02 ppm) and phosphocreatine (pCr, 7.25 ppm). In panel B the cross peak A, arising from the correlation of CH_3-terminal methyl group of Fatty Acid (FA) chains with the bulk of FA (excluding omega-3 FAs), at (0.89–1.28) ppm, and cross peak B, representing both mono and polyunsaturated FA at (2.02–1.28) are labelled. Cross peak diagnostic for GABA is barely detectable (not shown). (**C**,**D**) 1D and 2D spectra of carbon ions irradiated samples (green trace) at 20 Gy and corresponding controls (blue trace) of lines #1 and #83. Region of interest were expanded. All spectra were acquired 48 h after irradiation at a dose of 20 Gy with both photons and carbon ions. Assignments were performed as reported in [21]. Tables (**E**) (photon beams) and (**F**) (carbon ions) show mean values of all analyzed metabolites signal intensities, standard deviations and *p* values, calculated on at least 5 experiments performed 48 h after irradiation and corresponding controls.

It is interesting to note that the level of lipid unsaturation, measured as the B/A ratio of cross peak intensities (Figures 2B and 3B), was strongly influenced in line #1 with respect to line #83 (Figure 4A). Changes in the B/A ratio are in fact indicative of the increasing level of unsaturated fatty acids compared to total FA. Principal component analysis (PCA) shows that the control values of line #1 are always separated from those irradiated, independently of the radiation type, while in line #83 there is no statistically significant difference between control and irradiated samples (Figure 4A).

The signals of GSH, gln, glu and GABA were influenced in line #1 and line #83 after irradiation with photons and carbon ions, though to a different extent (Figures 2 and 3, respectively). In particular, in line #1 the intensity of the glu signal decreases, while the GABA signal increases in intensity in a statistically significant way only for carbon ion irradiation, whereas no statistically significant effects were observed for these signals in line #83, where the GABA signal was not detectable (Figures 2 and 3).

Unsupervised Cluster analysis of GSH, gln, glu signal intensities from spectra was also performed. The resulting dendrogram is shown in Figure 4B. As expected, the two lines were grouped in different cluster cutting the tree at appropriate level. Control and irradiated samples of lines #1 and #83 are grouped together independently of the type of irradiation.

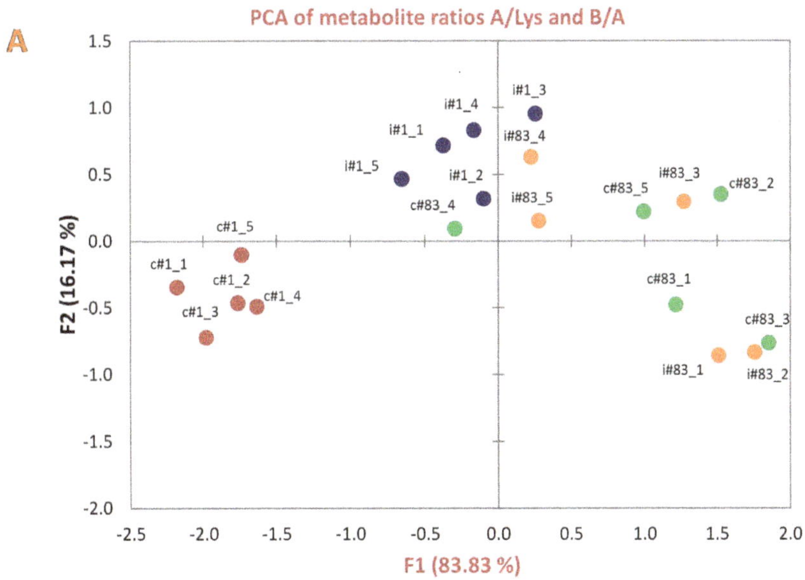

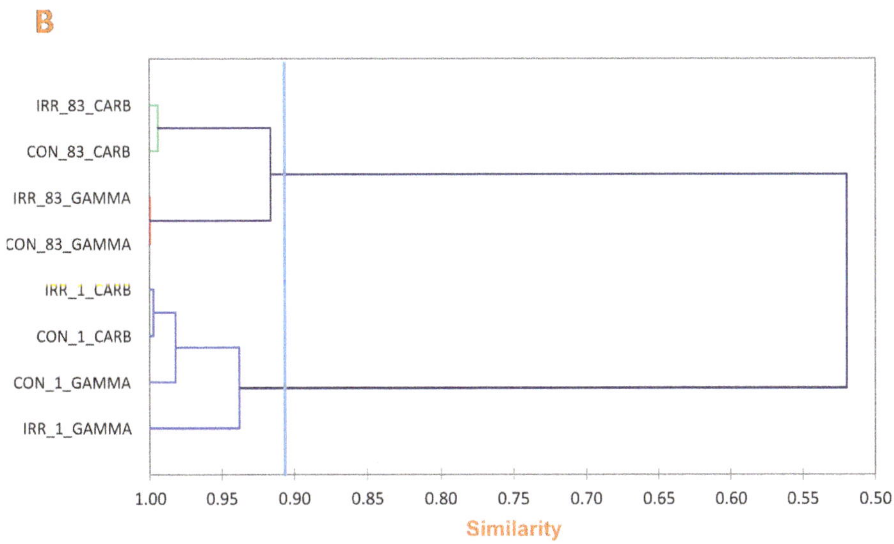

Figure 4. Statistical analysis confirms a different metabolic behaviour of line #1 and #83 cells after irradiation (**A**) F1 vs. F2 plot from Principal Component Analysis of observations of ML cross peak A and cross peak ratio B/A from 2D NMR spectra, for line #1 and #83. Peak A at (0.89–1.28) ppm arises from the correlation of CH3-terminal methyl group of FA chains with the bulk of FA (excluding omega-3 FAs). Peak B at (2.02–1.28) ppm represents both mono and polyunsaturated FA with a different contribution from n-9 and n-3 chains. F1 and F2 are the first and the second components in the PCA space, that account for 83.83% and 16.17%, respectively representing 100% of the total variances existing along the obtained data. Each point in the plot corresponds to the mean value of at least three experiments at different times after irradiation. (**B**) Dendrogram resulting from Unsupervised Cluster Analysis of GSH, gln, glu and GABA mean signal intensities from experiments at different times after irradiation.

In addition, the signals of phosphocreatine (pCr) and total creatine (tCr) in line #1 spectra were strongly influenced by irradiation with carbon ions (Figure 2C,F), while no statistically significant effects were observed in the spectra of line #83 (Figure 3A,C,E,F).

Finally, the intensity of the glucose signal (glc) clearly decreased in the spectra of line #1 after irradiation with photon and carbon ions (Figure 2A,C,E,F), while the decrease of glc signal intensity in line #83 spectra after irradiation was not statistically significant (Figure 3A,C,E,F). The lactate signal from spectra of culture media of the two cell lines increased in time, but only minor changes were observed after photon and carbon ion irradiation (Figure 5A). Lac_{con}/Lac_{irr} ratio for both types of irradiation in both lines increased as a function of time, (Figure 5B), being always greater than 100% for photons and less than 100% for carbon ions. This ratio, that can be used as an indication of the Warburg effect entity according to what reported in [23], is significantly higher in line #1 than in line #83, especially taking into account the lower cell number of line #1 with respect to line #83. This indicated an increased anaerobic glycolysis as confirmed by increased Warburg effect (Figure 5B,C, where a snapshot of the effect at 4 days after irradiation is shown).

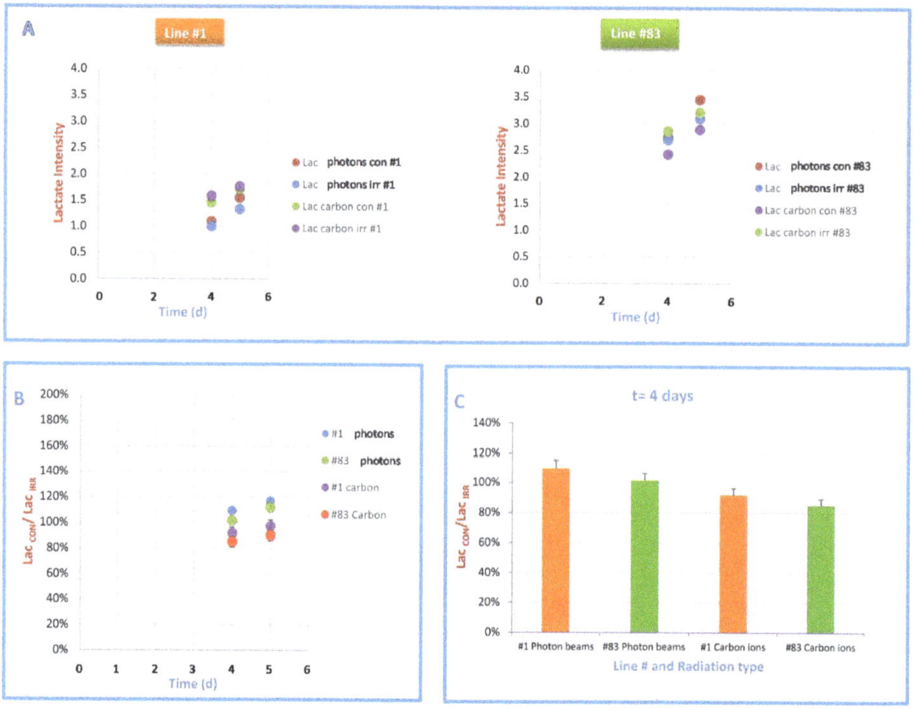

Figure 5. Lactate signal and cell number after photon and carbon ion irradiation. (**A**) lactate extrusion in culture medium, after photon and carbon ion irradiation at 20 Gy, as a function of time, for control and irradiated samples of line #1 and of line #83. Lactate doublet signal at 1.44 ppm was assigned and valuated in growth media spectra according to (Grande 2016) (**B**) Lac_{con}/Lac_{irr} ratio from lines #1 and #83 culture media spectra after irradiation with photons and carbon ions at 20 Gy as a function of time. (**C**) Lac_{con}/Lac_{irr} values calculated at 4 days after irradiation, indicative of the entity of Warburg effect for both lines.

Effects on the selected metabolites after irradiation with both photon beams and carbon ions as a function of time are reported in Figure S3 only for line #1, where statistically significant differences were observed. These effects remained stable until 4 days after irradiation (Figure S3).

Finally, a synthesis of the metabolic response of the two GSC cell lines to photon and C-ion irradiation is reported in Table 1 together with their genetic and metabolic different profiles. Line #83 metabolism is not affected by irradiation with photons and Carbon ions. These cells undergo a G2 block more intense after Carbon ion irradiation, but the ability to recover is similar in both cases. Differently line #1 seems to be more responsive to Carbon ions with respect to photons, but in both cases, cells die and are not able to recover in any way.

Table 1. GSC lines metabolic and biological response to photon and carbon ion irradiation at 20 Gy.

	GSC#1		GSC#83	
Genetic Profile [18]	**GSf-Like**		**GSr-Like**	
Metabolic Cluster [21]	**Proneural Like**		**Glioma Like**	
	Photon	**Carbon ions**	**Photon**	**Carbon ions**
Cell growth	early cytotoxic effect	early cytotoxic effect	cytostatic effect	cytostatic effect
Cell cycle	No effect	G0/G1 accumulation	G2/M block	G2/M block
	Cell metabolite signals			
Lipid	Increase	Increase	No effect	No effect
Glutamic acid	No effect	Decrease	No effect	No effect
GABA	No effect	Increase	N.D.	N.D.
Glucose	Decrease	Decrease	No effect	No effect
Phosphocreatine	No effect	Increase	No effect	No effect

3. Discussion

In GBM the self-renewing, tumorigenic GSC population likely contributes to tumor initiation and therapeutic resistance. In addition, as normal stem and progenitor cells participate in tissue development and repair, these developmental strategies re-emerge in GSCs to support the progressive growth of tumors [5,17,24–33]. Several strategies have been proposed to investigate the mechanisms leading to GSC escape from therapeutic treatments. Charged ion therapy was proposed [13,14,34,35] to overcome GSC radiation resistance because of its potential greater effectiveness than photon therapy, but no definitive results have been obtained. Emerging evidence shows that metabolic reprogramming of glioblastoma mediates resistance to standard therapies [36]. Glioblastoma is associated with a significant increase in glycolysis and knockdown of glycolytic genes strongly inhibits GBM growth [37]. The high utilization of glucose suggests that inhibiting glycolysis may have a therapeutic value. Also, isocitrate dehydrogenases (IDHs) metabolism plays a crucial role in the tricarboxylic acid cycle; in IDH wild type (wt) GBM, the upregulation of wtIDH1 (wild type isocitrate dehydrogenase 1) fuels therapy resistance [36]. It is known that glutamine levels are higher in GBM tissue than in the surrounding normal brain and that the ability to metabolize glutamine is critical for GBM proliferation and survival. Elevated glutaminase and glutamate levels following mTOR (mammalian target of rapamycin) kinase inhibitor treatment promoted GBM survival while combined inhibition of mTOR kinase and glutaminase resulted in massive synergistic tumor cell death and growth inhibition. These results highlight a critical role for compensatory glutamine metabolism [38].

High GBM heterogeneity seems to play a relevant role in GSC ability to escape therapies. A recent study that used single-cell RNA sequencing of primary GBMs revealed extraordinary diversity within each tumor [39]. The high level of heterogeneity in GBM has been recently invoked [39,40] as one of the reasons for treatment failure and tumor relapse. Stratifying patients according to molecular biomarker profiles is a key step to manage patient heterogeneity.

Metabolomics may help in clarifying stem cell fate and reprogramming function [41,42]. In the present study two GSC lines, #83 and #1, characterized by different metabolic profiles [21] and clone heterogeneity [20] were analyzed as response to photons and carbon ions in terms of effects on cell survival, cell cycle and metabolism. Highly heterogeneous clones of line #83 [20] may be responsible for different responses to irradiation if compared to homogeneous clones of line #1. Indeed, while #83 clones 1 and 2 resulted similar to the origin line #83, significant differences in clone 3 were observed for

NAA and Myo metabolites and in clones 4 and 5 for lipid concentration. Finally, in clone 5 the alfa amino-adipic-acid signal, previously correlated to poor prognosis [43], was observed.

A previous study on 52 GSC lines analyzed in terms of LD50 after photon irradiation reported a similar response to a 20 Gy irradiation for both lines [6]. However, in the present study the two cell lines behave differently after irradiation with both photons and carbon ions; only line #83 cells are able to recover after carbon ions and photon beam irradiation, in agreement with preliminary data on cell survival. The presence of a plateau in the survival levels after charged particles and photon irradiation at doses higher than 10 Gy, suggested the presence of radio-resistant subpopulations in this GSC line [44]. According to this hypothesis, some of the heterogeneous clones from line #83 [20] may be able to recover and to reestablish the original pattern, differently from line #1 cells that do not show such heterogeneity. The different response to radiation treatments of the two lines here examined in terms of cell cycle and metabolic changes may be attributable mainly to the intrinsic characteristics of the lines, rather than to radiation type.

Cell cycle checkpoints and mechanisms involved in DNA repair are clearly interdependent, with the choice of repair mechanism consistently adjusted throughout the cell cycle [45,46]. Existing models established in different cell types have shown that radio-induced double-strand breaks (DSB) undergo faster repair in the G2 phase during the first hours following irradiation [47]. Additionally, particle therapy is effective at cell killing irrespective of the cell cycle phase, unlike photons [48,49], where cells are more resistant in late S and G2 phase [46,50]. It is still unclear if cells preferentially select specific pathway to repair DSBs generated by high LET radiation, unlike low LET radiation-induced DSBs. The arrest of the line #83 cells in the G2/M phase, where DNA repair processes are known to occur, may explain the higher radioresistance of this cell line, differently from cell line #1, that undergo cytotoxic effect without relevant G2/M arrest. The effect on line #83 cells seems more relevant after carbon ion compared to photon irradiation. This may be due to higher Relative Biological Effectiveness (RBE) value able to induce less reparable damages in agreement with the role played by DSBs in high LET particle irradiation. In this situation a higher permanence of cells in G2 phase may allow them to better repair and to recover at late times after irradiation.

Lipids detected as Mobile Lipids play a relevant role in GSC escaping mechanisms to radiotherapy. The lipid signal increased in line #1 after irradiation, while only small fluctuations could be detected in line #83. This increase in MLs in line #1 can be attributed to the accumulation of lipids in the cytoplasm as a consequence of metabolism impairment [51,52]. In contrast, in line #83, MLs are segregated and stored mainly in cytoplasmic lipid droplets where they are protected from oxidation and from the formation of unstable lipid peroxides, which are toxic to cells [53]. It has been recently reported [54] that lipolytic inhibitor G0/G1 switch gene 2 (G0S2) is upregulated in radioresistant GSCs and elevated in clinical GBM. GBM patients with high G0S2 expression had significantly shorter overall survival compared with those with low expression of G0S2. Moreover, G0S2-induced radioresistance is related with G0S2-mediated lipid droplet stability. The intrinsic notable difference in lipid content (high in line #83, low in line #1) of the two lines is consistent with this hypothesis and with the different patient survival.

The two lines belong to distinct metabolic clusters, where lipid content is one of the important metabolic parameters differentiating mesenchymal from neuronal fingerprint [20,21]. The different effects after photon or carbon ion irradiation in the two lines support the hypothesis of a metabolic switch of line #1 cells towards the mesenchymal fingerprint and not a mechanism to protect themselves from irradiation. This interpretation is supported by the observation of the increase in unsaturation of lipid chain in irradiated line #1 towards more typical value of line #83. Our results highlight the physiological relevance of Lipid Droplets in lipid metabolic reprogramming of cancer cells [55].

As GSH is concerned it is well known the role of this metabolite as the most important active oligopeptide in the maintenance of redox homeostasis, with a radiation protection role against oxidative stress. The absence, after irradiation, of relevant changes of this metabolite in both cell lines shows that GSH is not a key metabolite in protecting GSCs from irradiation. On the contrary, the GABA increase, concomitantly with a glu decrease in irradiated cells from line #1, is particularly relevant after carbon ion irradiation and suggests a GABA synthesis increase. Indeed, GABA is synthesized via enzymatic decarboxylation of L-glutamate involving glutamic acid decarboxylase (GAD) and pyridoxal-L-phosphate (PLP) as cofactor [56]. Unfortunately, changes of oxygenation do not clearly differentiate conditions for GABA increase/decrease, as its increase was observed after continuous hypoxia in neurons, but not in glia while intermittent hypoxia would inhibit the GAD enzyme activity [56]. Noteworthy, a powerful eradication of radioresistant hypoxic tumor cells was attributed to carbon ion irradiation [57]. The GABA increase in line #1 cells, evident only after carbon irradiation, could be related to such effect, confirming a non-complete eradication of radioresistant hypoxic cells also due to the absence of the protecting agent against irradiation GSH.

Cancer cells have significant heterogeneity in glucose metabolism. Dysfunction of mitochondria, loss of tumor suppressors, the hypoxic microenvironment, and oncogene-driven metabolic reprogramming are initiating events of abnormal energy metabolism in cancer cells [58]. At present a widely accepted hypotheses suggests that energy metabolism is strictly involved taking into account that one of the established features of cancer is the deregulation of cellular energetics [59]. Glycolysis represents an important tool for tumor cells driving the malignant progression [60]. Particularly GBM, like most malignant solid tumors, is characterized by high levels of glycolysis with large amounts of lactic acid production and many reports have identified the Warburg effect to be implicated in resistance to cytotoxic stress induced by radiotherapy. In line with this hypothesis, in a previous study [20] it has been shown that in line #83, where aerobic glycolysis is predominant, the glucose signal remains unchanged after irradiation. On the contrary, glucose decrease after both types of irradiation for line #1 is likely due to a shift from OXPHOS to aerobic glycolysis, suggested as typical cell response to external challenge [58] and the consequent rise of the Warburg effect in this line, confirmed by the higher decrease of the ratio Lac_{con}/Lac_{irr}.

In tumor cells, glycolysis is the main supplier of intracellular ATP and cytosolic phosphocreatine-creatine kinase (CK) was found associated with enzymes of glycolysis [61–63] though the role of CK in glycolysis has not been fully elucidated. The CK shuttle system is a key metabolite in the brain and central nervous system networks. In line #1 pCr signal increased by irradiation with carbon ions for a possible impairment of the CK shuttle system, likely impairing cytosolic CK by oxidative and radical damage [64]. As pCr is involved in energy exchange, it is possible to hypothesize a contribution of the pCr metabolic pathway in line #1, differently to what observed for line #83. All these effects on energetic metabolism could be hypothesized as a further mechanism contributing to the higher radiosensitivity of line #1 compared to line #83.

In conclusion metabolism of line #83, analyzed by MRS, is unresponsive to irradiation, while in line #1 cells alterations of metabolic patterns of FA, GABA, pCr, and glc confirm the involvement of cell metabolism, including energy pathways, in cell radiosensitivity. Differences in radiation response may, therefore, be likely attributed to genetic and metabolic differences between the two lines, representative of mesenchymal and proneural profiles, regardless of radiation type. Further experiments are foreseen to have more insight in the observed effects. In this respect, proton MRS may contribute to study cell metabolism in GSCs with different metabolic and genetic signatures after different radiotherapeutic approaches.

4. Materials and Methods

4.1. Patients and Tumor Characterization

Tumor tissue samples were collected from adult patients with GBM tumors (WHO grade IV) undergoing complete or partial surgical resection at the Institute of Neurosurgery, Catholic University School of Medicine in Rome. Informed consent was obtained from the patients before surgery. All patients provided written informed consent according to the research proposals approved by the Ethical Committee of the Catholic University School of Medicine, UCSC (Prot. 4720/17). Patients were eligible for the study if a diagnosis of glioblastoma multiforme was established histologically according to the WHO classification [65]. Details about patient's treatments and GSCs isolation were given elsewhere [21,66]. Experimental protocols were approved by the ethical committee of the Catholic University of Rome.

The expression of the proliferation marker Ki-67 and of Phosphatase and Tensin Homolog were characterized on tumor specimen by immunohistochemistry on deparaffinized sections using the avidin-biotin-peroxidase complex methods (ABC-Elite Kit, Vector Laboratories, Inc., Burlingame, CA, USA), anti-Ki67 monoclonal antibody (MIB-1, Dako, Agilent, Santa Clara, CA, USA) and anti-Phosphatase and Tensin Homolog mouse monoclonal antibody (clone 28H6; Novo Castra, Newcastle, UK). O6-methylguanine-DNA methyltransferase promoter methylation patterns were assessed on genomic DNA extracted from paraffin-embedded tissue by methylation-specific PCR as previously described [66,67]. Levels of VEGF and EGFRvIII were assessed as previously described [66].

4.2. Glioblastoma Stem-Like Cell Cultures

GSCs were isolated through mechanical dissociation of the tumor tissue and cultured in a serum-free medium supplemented with epidermal growth factor and basic fibroblast growth factor as previously described [66]. The in vivo tumorigenic potential of GBM neurospheres was assessed by intracranial or subcutaneous cell injection in immunocompromised mice. GSC lines were validated by Short Tandem repeat DNA fingerprinting as previously described [68,69].

Cell number in biological experiments has been evaluated by a Scepter 2.0 Coulter Counter (Merck, Darmstad, Germany).

Clones from GSCs were obtained by plating single cells into 96 wells plate. After 4 weeks, single clones were mechanically dissociated and replated to expand the culture.

For irradiation experiments GSCs were seeded to be expanded. At 3th passage cells were pooled and stored as 1×10^6 cells/vial in nitrogen and 50 vials were prepared to be used for all the experiments at identical starting conditions.

Seven days before the experiments, the GSC #1 and GSC #83 cell lines were seeded in T175 cm^2 flasks at a density of about 4×10^4/mL and 2×10^4/mL in 50 mL total volume, respectively. Just before irradiation, cells were collected and placed in different sterilized sample holders, depending on the irradiation facility used: commercial 1 mL polystyrene cuvettes (INFN-LNS and ISS) and commercial UVette (CNAO). These sample holders were then centrifuged at low speed to obtain cell pellets 4 mm and 2 mm thick, respectively.

4.3. Irradiation Conditions

Irradiation with C-ions was carried out at the multidisciplinary irradiation beamline facility of the INFN-LNS in Catania and at the CNAO hadrontherapy facility in Pavia.

At the INFN-LNS, 62 AMeV in energy carbon ions, exited in air and reached the entrance of the cell suspensions at 52 MeV/u, corresponding to a Linear Energy Transfer (LET) value of 42 keV/μm. However, the thickness of the 4-mm cell pellet does affect the LET of C-ions, with a significant variation in the average LET from 42 to 56 keV/μm. Absolute dosimetry was performed following the IAEA TRS 398 Code of Practice using a plane parallel PTW Markus ionisation chamber [70]. At the CNAO facility, a 22 mm homogeneous Spread-Out-Bragg-Peak (SOBP) was obtained by C-ion beams of initial energies ranging from 187 to 214 MeV/u. Samples were placed in sealed vessels at the depth of 82 mm in water, in order to achieve an average LET value of 83 keV/μm, with no significant variation between the entrance and the exit of the 2 mm cell pellet (LET of 80 and 86 keV/μm, respectively).

^{137}Cs γ-rays were obtained from a Gammacell 40 Exactor (Nordion) at the ISS, Rome, with a dose rate of about 0.8 Gy min^{-1}.

All irradiations were performed at room temperature, with a single high radiation dose of 20 Gy comparable with: total doses delivered during radiation therapy in a fractionated regimen, single high doses used in other therapeutic modalities (i.e., intraoperative radiotherapy and stereotactic radiosurgery), and the boost before the chemo-radiotherapic treatments used for the treatment of patient with Carbon ions [12]. The same experimental protocols were used for charged particles and photon irradiations. Cell suspensions were centrifuged in 1-mL polystyrene cuvettes to obtain a 4-mm thick pellet, and irradiated; after irradiation, samples were prepared as in [20] for biological end points and MRS experiments.

4.4. ^1H MRS Cell Sample Preparation

Cells were removed, washed in Phosphate-buffered saline (PBS) and centrifuged at 162 rcf for 3 min. The pellet was suspended in PBS with 20% D_2O and 2 mM Sodium 3-(trimethylsilyl) propionate-2,2,3,3-d4 (TMSP) used as a frequency standard. A 15 μL aliquot of the suspension was inserted into a 1 mm MRS microtube and centrifuged to obtain a packed cell volume.

Conditioned growth media, collected from cell cultures, were added with 20% D_2O and 2 mM TMSP and transferred into a 1 mm MRS microtube. All MRS reagents were purchased from Cambridge Isotope Laboratories, Inc., Tewksbury, MA, USA.

4.5. ^1H MRS Measurements

^1H MRS experiments were run on a digital Bruker Avance spectrometer (Bruker Scientific LLC, Billerica, MA, USA) at 400.14 MHz, equipped with a 1mm microprobe. Both one dimensional (1D) and two-dimensional Correlation SpectroscopY (2D COSY) experiments were performed, at T = 298 K.

1D ^1H MRS spectra of GSCs and culture media were acquired with a 90° RF pulse, the number of scans (ns) was equal to 1000 (sufficient to obtain a good signal-to-noise ratio) for cell spectra while ns = 4000 was used for culture media spectra. When indicated, a Lorentzian-Gaussian function was applied in the time domain, before Fourier transformation.

2D COSY spectra were acquired with a 90°-t1–90°-t2 pulse sequence and ns = 32 for cell or ns = 128 for culture media samples. Spectra were acquired as a matrix of 512 × 128 data points in time domain.

NMR parameters were obtained in at least three independent experiments and data are expressed as mean ± standard deviation (SD) values. WINNMR software (Bruker Scientific LLC, Billerica, MA, USA) was used to perform 1D signal deconvolution and 2D cross peak integration as reported in [21]. Macromolecule signal at 0.89 ppm intensity was used as internal reference for 1D measurement, while 2D signal integrals were normalized to the intensity of the lysine (Lys) cross-peak at 1.70–3.00 ppm. This peak was considered representative of the cellular mass, as it was found to be constant in a number of cell models and tissue samples [21].

4.6. Statistical Analysis

Unsupervised agglomerative hierarchical Clustering, Principal Component Analysis and Student's t test were performed utilizing XLSTAT software (Addinsoft^{TM} Paris, France) version 2012.2.02.

Supplementary Materials: Supplementary Materials can be found at http://www.mdpi.com/1422-0067/21/14/5167/s1. Figure S1: Percentage of line #1 (A) and line #83 (C) photon beam irradiated cells in different phases as obtained by the Annexin V apoptosis assay (Q1: prenecrotic cells, Q2: late apoptotic + necrotic cells; Q3: living cells; Q4: early apoptotic cells). Data from both irradiated and control samples are compared as a function of time after irradiation. Percentage of living #1 (B) and #83 (D) cells, calcuated as Q3/(Q1 + Q2 + Q3 + Q4) * 100. Data from both irradiated and control samples are compared as a function of time after irradiation. Figure S2: Metabolic characteristics of cell line #1 and #83 and their clones Panel A: Box and whisker plots describing the distribution of intensity values of metabolic signals calculated in spectra from line #1 (red boxes) and line #83 cells (green boxes). Reported values are from mobile lipids (ML, A, B- see main text for description), total Creatine (tCr), and Glutathione (GSH). Bar graphs in panels B and C show fold changes between line #1 and line #83 (calculated as log2(FC)). The above mentioned metabolites were evaluated together with glutammine (Gln) and glutammate (Glu). All variations are statistically significant. For Mobile lipids signals a more detailed analysis (panel C) was performed on cross peaks from 2D spectra (A, B, E, F) given their higher statistical significance ($p < 0.01$). Panel D: Box and whiskers plots describing the distribution of intensity values of metabolic signals calculated in spectra from line #83 (green boxes) and its clones (yellow boxes). Reported values are from mobile lipids (ML, A). Bar graph in panel E shows fold changes between line #83 and its clones. Mobile lipid cross peaks from 2D spectra (A, B, E, F) were evaluated. All variations are statistically significant ($p < 0.05$). Clone analysis for line #1 vs line #1 is not shown because no (statistically) significant differences were found. Figure S3: Metabolic changes observed by mr spectroscopy in line #1 cell spectra. Intensity of metabolic signals from 1D and 2D spectra as a function of time after photon (A, C) and carbon ion (B, D) irradiation. Both irradiated and control values are reported. A, B: mobile lipid signals (ML, from 1D spectra and A peak from 2D spectra); C: Glucose signal; D: Glutammic acid, GABA, Phosphocreatine and Glucose signals.

Author Contributions: Conceptualization, A.P., S.G., L.R.-V., M.B. (Mauro Biffoni), R.P., and A.R.; Data curation, A.P., S.G., M.B. (Mariachiara Buccarelli), and V.D.; Formal analysis, A.P., S.G., V.D., and V.V.; Investigation, A.P., S.G., L.R.-V., A.M.L., L.G., V.V. and A.R.; Methodology, A.M.L., M.B. (Mariachiara Buccarelli), G.A.P.C., and M.C.; Resources, L.R.-V., G.A.P.C., M.C., and R.P.; Supervision, A.R.; Writing—original draft, A.P., S.G., A.M.L., and A.R.; Writing—review & editing, A.P., S.G., L.R.-V., A.M.L., M.B. (Mariachiara Buccarelli), M.B. (Mauro Biffoni), V.D., G.A.P.C., M.C., L.G., R.P., V.V. and A.R. All authors have read and agreed to the published version of the manuscript.

Funding: This research was funded by National Institute of Nuclear Physics (INFN) in the framework of INFN Group V "RADIOSTEM" Experiment.

Acknowledgments: We thank Maria Antonella Tabocchini for fruitful discussion on irradiation set up and on biological results.

Conflicts of Interest: The authors declare no conflict of interest.

Abbreviations

COSY	COrrelation SpectroscopY
CSC	Cancer Stem Cells
DSB	Double-Strand Breaks
FA	Fatty Acid
GABA	Gamma-Aminobutyric Acid
GBM	Glioblastoma multiforme
glc	glucose
gln	glutamine
glu	glutamic acid
GSC	Glioblastoma Stem-like Cells
GSf-like	Glioblastoma full stem-like phenotype,
GSH	Glutathione
GSr-like	Glioblastoma restricted stem-like phenotype
IDHs	Isocitrate DeHydrogenases
Lac	Lactate
LET	Linear Energy Transfer
ML	Mobile Lipids
MRS	Magnetic Resonance Spectroscopy
mTOR	Mammalian Target Of Rapamycin
PBS	Phosphate-buffered saline
PCA	Principal Component Analysis
pCr	Phosphocreatine
RBE	Relative Biological Effectiveness
RT	Radiotherapy
tCr	total Creatine

References

1. Bryukhovetskiy, I.; Ponomarenko, A.; Lyakhova, I.; Zaitsev, S.; Zayats, Y.; Korneyko, M.; Eliseikina, M.; Mischenko, P.; Shevchenko, V.; Shanker Sharma, H.; et al. Personalized regulation of glioblastoma cancer stem cells based on biomedical technologies: From theory to experiment (Review). *Int. J. Mol. Med.* **2018**, *42*, 691–702. [CrossRef] [PubMed]
2. Han, X.; Xue, X.; Zhou, H.; Zhang, G. A molecular view of the radioresistance of gliomas. *Oncotarget* **2017**, *8*, 100931–100941. [CrossRef] [PubMed]
3. Shen, H.; Hau, E.; Joshi, S.; Dilda, P.J.; McDonald, K.L. Sensitization of glioblastoma cells to irradiation by modulating the glucose metabolism. *Mol. Cancer Ther.* **2015**, *14*, 1794–1804. [CrossRef] [PubMed]

4. Singh, S.K.; Hawkins, C.; Clark, I.D.; Squire, J.A.; Bayani, J.; Hide, T.; Henkelman, R.M.; Cusimano, M.D.; Dirks, P.B. Identification of human brain tumour initiating cells. *Nature* **2004**, *432*, 396–401. [CrossRef] [PubMed]
5. Krause, M.; Dubrivska, A.; Linge, A.; Baumann, M. Cancer stem cells: Radioresistance, prediction of radiotherapy outcome and specific targets for combined treatments. *Adv. Drug Deliv. Rev.* **2017**, *109*, 63–73. [CrossRef] [PubMed]
6. D'Alessandris, Q.G.; Biffoni, M.; Martini, M.; Runci, D.; Buccarelli, M.; Cenci, T.; Signore, M.; Stancato, L.; Olivi, A.; De Maria, R.; et al. The clinical value of patient-derived glioblastoma tumorspheres in predicting treatment response. *Neuro Oncol.* **2017**, *19*, 1097–1108. [CrossRef]
7. Cui, X.; Oonishi, K.; Tsujii, H.; Yasuda, T.; Matsumoto, Y.; Furusawa, Y.; Akashi, M.; Kamada, T.; Okayasu, R. Effects of carbon ion beam on putative colon cancer stem cells and its comparison with X-rays. *Cancer Res.* **2011**, *71*, 3676–3687. [CrossRef]
8. Cammarata, F.; Torrisi, F.; Forte, G.; Minafra, L.; Bravatà, V.; Pisciotta, P.; Savoca, G.; Calvaruso, M.; Petringa, G.; Cirrone, G.; et al. Proton Therapy and Src family kinase inhibitor combined treatments on u87 human glioblastoma multiforme cell line. *Int. J. Mol. Sci.* **2019**, *20*, 4745. [CrossRef]
9. Combs, S.E.; Burkholder, I.; Edler, L. Randomised phase I/II study to evaluate carbon ion radiotherapy versus fractionated stereotactic radiotherapy in patients with recurrent or progressive gliomas: The Cinderella trial. *BMC Cancer* **2010**, *10*, 53. [CrossRef]
10. Combs, S.E.; Kieser, M.; Rieken, S. Randomized phase II study evaluating a carbon ion boost applied after combined radiochemotherapy with temozolomide versus a proton boost after radiochemotherapy with temozolomide in patients with primary glioblastoma: The Cleopatra trial. *BMC Cancer* **2010**, *10*, 478. [CrossRef]
11. Malouff, T.D.; Peterson, J.L.; Mahajan, A.; Trifiletti, D.M. Carbon ion radiotherapy in the treatment of gliomas: A review. *J. Neuro Oncol.* **2019**, *145*, 191–199. [CrossRef] [PubMed]
12. Kong, L.; Gao, J.; Hu, J.; Lu, R.; Yang, J.; Qiu, X.; Hu, W.; Lu, J.J. Carbon ion radiotherapy boost in the treatment of glioblastoma: A randomized phase I/III clinical trial. *Cancer Commun.* **2019**, *39*, 5. [CrossRef] [PubMed]
13. Chiblak, S.; Tang, Z.; Campos, B.; Gal, Z.; Unterberg, A.; Debus, J.; Herold-Mende, C.; Abdollahi, A. Radiosensitivity of patient-derived glioma stem cell 3-dimensional cultures to photon, proton, and carbon irradiation. *Int. J. Radiat. Oncol. Biol. Phys.* **2016**, *95*, 112–119. [CrossRef] [PubMed]
14. Chiblak, S.; Tang, Z.; Lemke, D.; Knoll, M.; Dokic, I.; Warta, R.; Moustafa, M.; Mier, W.; Brons, S.; Carmen Rapp, C.; et al. Carbon irradiation overcomes glioma radioresistance by eradicating stem cells and forming an antiangiogenic and immunopermissive niche. *JCI Insight* **2019**, *4*, e123837. [CrossRef]
15. Aboud, O.A.; Weiss, R.H. New opportunities from the cancer metabolome. *Clin. Chem.* **2013**, *59*, 138–146. [CrossRef]
16. Boesch, M.; Sopper, S.; Zeimet, A.G.; Reimer, D.; Gastl, G.; Ludewig, B.; Wolf, D. Heterogeneity of cancer stem cells: Rationale for targeting the stem cell niche. *Biochim. Biophys. Acta* **2016**, *1866*, 276–289. [CrossRef]
17. Phillips, H.S.; Kharbanda, S.; Chen, R.; Forrest, W.F.; Soriano, R.H.; Wu, T.D.; Misra, A.; Nigro, J.M.; Colman, H.; Soroceanu, L.; et al. Molecular subclasses of high-grade glioma predict prognosis, delineate a pattern of disease progression, and resemble stages in neurogenesis. *Cancer Cell* **2006**, *9*, 157–173. [CrossRef]
18. Marziali, G.; Signore, M.; Buccarelli, M.; Grande, S.; Palma, A.; Biffoni, M.; Rosi, A.; D'Alessandris, Q.G.; Martini, M.; Larocca, L.M.; et al. Metabolic/proteomic signature defines two glioblastoma subtypes with different clinical outcome. *Sci. Rep.* **2016**, *6*, 21557. [CrossRef]
19. Marziali, G.; Buccarelli, M.; Giuliani, A.; Ilari, R.; Grande, S.; Palma, A.; D'Alessandris, Q.G.; Martini, M.; Biffoni, M.; Pallini, R.; et al. A three-microRNA signature identifies two subtypes of glioblastoma patients with different clinical outcomes. *Mol. Oncol.* **2017**, *11*, 1115–1129. [CrossRef]
20. Grande, S.; Palma, A.; Ricci-Vitiani, L.; Luciani, A.M.; Buccarelli, M.; Biffoni, M.; Molinari, A.; Calcabrini, A.; D'Amore, E.; Guidoni, L.; et al. Metabolic heterogeneity evidenced by mrs among patient-derived glioblastoma multiforme stem-like cells accounts for cell clustering and different responses to drugs. *Stem Cells Int.* **2018**, *3292704*, 1–16. [CrossRef]
21. Guidoni, L.; Ricci-Vitiani, L.; Rosi, A.; Palma, A.; Grande, S.; Luciani, A.M.; Pelacchi, F.; di Martino, S.; Colosimo, C.; Biffoni, M.; et al. ^{1}H NMR detects different metabolic profiles in glioblastoma stem-like cells. *NMR Biomed.* **2014**, *27*, 129–145. [CrossRef] [PubMed]

22. Ryall, J.G.; Cliff, T.; Dalton, S.; Sartorelli, V. Metabolic reprogramming of stem cell epigenetics. *Cell Stem Cell* **2015**, *17*, 651–662. [CrossRef] [PubMed]
23. Hao, W.; Chang, C.P.B.; Tsao, C.C.; Xu, J. Oligomycin-induced bioenergetic adaptation in cancer cells with heterogeneous bioenergetic organization. *J. Biol. Chem.* **2010**, *285*, 12647–12654. [CrossRef] [PubMed]
24. Krause, M.; Yaromina, A.; Eicheler, W.; Koch, U.; Baumann, M. Cancer stem cells: Targets and potential biomarkers for radiotherapy. *Clin. Cancer Res.* **2011**, *17*, 7224–7229. [CrossRef] [PubMed]
25. Wang, J.; Wakeman, T.P.; Lathia, J.D.; Hjelmeland, A.B.; Wang, X.F.; White, R.R.; Rich, J.N.; Sullenger, B.A. Notch promotes radioresistance of glioma stem cells. *Stem Cells* **2010**, *28*, 17–28. [CrossRef]
26. Bao, S.; Wu, Q.; McLendon, R.E.; Hao, Y.; Shi, Q.; Hjelmeland, A.B.; Dewhirst, M.W.; Bigner, D.D.; Rich, J.N. Glioma stem cells promote radioresistance by preferential activation of the DNA damage response. *Nature* **2006**, *444*, 756–760. [CrossRef] [PubMed]
27. Baralis, E.; Bertotti, A.; Baumann, M.; Krause, M.; Hill, R. Exploring the role of cancer stem cells in radioresistance. *Nat. Rev. Cancer* **2008**, *8*, 545–554.
28. Diehn, M.; Cho, R.W.; Lobo, N.A.; Kalisky, T.; Dorie, M.J.; Kulp, A.N.; Qian, D.; Lam, J.S.; Ailles, L.E.; Wong, M.; et al. Association of reactive oxygen species levels and radioresistance in cancer stem cells. *Nature* **2009**, *458*, 780–783. [CrossRef]
29. Pajonk, F.; Vlashi, E.; McBride, W.H. Radiation resistance of cancer stem cells: The 4 R's of radiobiology revisited. *Stem Cells* **2010**, *28*, 639–648. [CrossRef]
30. Cheng, L.; Wu, Q.; Huang, Z.; Guryanova, O.A.; Huang, Q.; Shou, W.; Rich, J.N.; Bao, S. L1CAM regulates DNA damage checkpoint response of glioblastoma stem cells through NBS1. *EMBO J.* **2011**, *30*, 800–813. [CrossRef]
31. Vlashi, E.; Pajonk, F. Cancer stem cells, cancer cell plasticity and radiation therapy. *Semin. Cancer Biol.* **2014**, *31*, 28–35. [CrossRef] [PubMed]
32. Ahmed, S.U.; Carruthers, R.; Gilmour, L.; Yildirim, S.; Watts, C.; Chalmers, A.J. Selective inhibition of parallel DNA damage response pathways optimizes radiosensitization of glioblastoma stem-like cells. *Cancer Res.* **2015**, *75*, 4416–4428. [CrossRef]
33. De Bacco, F.; D'Ambrosio, A.; Casanova, E.; Orzan, F.; Neggia, R.; Albano, R.; Verginelli, F.; Cominelli, M.; Poliani, P.L.; Luraghi, P.; et al. MET inhibition overcomes radiation resistance of glioblastoma stem-like cells. *EMBO Mol. Med.* **2016**, *8*, 550–556. [CrossRef] [PubMed]
34. Chew, M.T.; Nisbet, A.; Suzuki, M.; Matsufuji, N.; Murakami, T.; Jones, B.; Bradley, D.A. Potential lethal damage repair in glioblastoma cells irradiated with ion beams of various types and levels of linear energy transfer. *J. Radiat. Res.* **2018**, *60*, 59–68. [CrossRef] [PubMed]
35. Oike, T.; Sato, H.; Noda, S.; Nakano, T. Translational research to improve the efficacy of carbon ion radiotherapy: Experience of gunma university. *Front. Oncol.* **2016**, *6*, 139. [CrossRef] [PubMed]
36. Zhou, W.; Wahl, D.R. Metabolic abnormalities in glioblastoma and metabolic strategies to overcome treatment resistance. *Cancers* **2019**, *11*, 1231. [CrossRef]
37. Sanzey, M.; Aminah, S.; Rahim, A.; Oudin, A.; Dirkse, A.; Kaoma, T.; Vallar, L.; Herold-Mende, C.; Bjerkvig, R.; Golebiewska, A.; et al. Comprehensive analysis of glycolytic enzymes as therapeutic targets in the treatment of glioblastoma. *PLoS ONE* **2015**, *10*, e0123544. [CrossRef]
38. Tanaka, K.; Sasayama, T.; Irino, Y.; Takata, K.; Nagashima, H.; Satoh, N.; Kyotani, K.; Mizowaki, T.; Imahori, T.; Ejima, Y.; et al. Compensatory glutamine metabolism promotes glioblastoma resistance to mTOR inhibitor treatment. *J. Clin. Investig.* **2015**, *125*, 1591–1602. [CrossRef]
39. Patel, A.P.; Tirosh, I.; Trombetta, J.J.; Shalek, A.K.; Gillespie, S.M.; Wakimoto, H.; Cahill, D.P.; Nahed, B.V.; Curry, W.T.; Martuza, R.L.; et al. Single-cell RNA-seq highlights intratumoral heterogeneity in primary glioblastoma. *Science* **2014**, *344*, 1396–1401. [CrossRef]
40. Yoshida, G.J. Metabolic reprogramming: The emerging concept and associated therapeutic strategies. *J. Exp. Clin. Cancer Res.* **2015**, *34*, 111. [CrossRef]
41. Li, C.; Wang, J. Quantifying cell fate decisions for differentiation and reprogramming of a human stem cell network: Landscape and biological paths. *PLoS Comput. Biol.* **2013**, *9*, e1003165. [CrossRef] [PubMed]
42. Agnihotri, S.; Zadeh, G. Metabolic reprogramming in glioblastoma: The influence of cancer metabolism on epigenetics and unanswered questions. *Neuro Oncol.* **2016**, *18*, 160–172. [CrossRef]

43. Rosi, A.; Ricci-Vitiani, L.; Biffoni, M.; Grande, S.; Luciani, A.M.; Palma, A.; Runci, D.; Cappellari, M.; De Maria, R.; Guidoni, L.; et al. ^{1}H NMR spectroscopy of glioblastoma stem-like cells identifies alpha-aminoadipate as a marker of tumor aggressiveness. *NMR Biomed.* **2015**, *28*, 317–326. [CrossRef] [PubMed]
44. Pecchia, I.; Dini, V.; Ricci-Vitiani, L.; Biffoni, M.; Balduzzi, M.; Fratini, E.; Belli, M.; Campa, A.; Esposito, G.; Cirrone, G.; et al. Gliobastoma Stem cells: Radiobiological response to ionising radiation of different qualities. *Radiat. Prot. Dosim.* **2015**, *166*, 374–378. [CrossRef] [PubMed]
45. Tachon, G.; Cortes, U.; Guichet, P.O.; Rivet, P.; Balbous, A.; Masliantsev, K.; Karayan-Tapon, L.; Berger, A.; Boissonnade, O.; Wager, M. Cell cycle changes after glioblastoma stem cell irradiation: The major role of RAD51. *Int. J. Mol. Sci.* **2018**, *19*, 3018. [CrossRef]
46. Karanam, K.; Kafri, R.; Loewer, A.; Lahav, G. Quantitative live cell imaging reveals a gradual shift between DNA repair mechanisms and a maximal use of HR in mid S phase. *Mol. Cell* **2012**, *47*, 320–329. [CrossRef]
47. Schulz, A.; Meyer, F.; Dubrovska, A.; Borgmann, K. Cancer stem cells and radioresistance: DNA repair and beyond. *Cancers* **2019**, *11*, 862. [CrossRef]
48. Lopez Perez, R.; Nicolay, N.H.; Wolf, J.C.; Frister, M.; Schmezer, P.; Weber, K.J.; Huber, P.E. DNA damage response of clinical carbon ion versus photon radiation in human glioblastoma cells. *Radiother. Oncol.* **2019**, *133*, 77–86. [CrossRef]
49. Magee, J.A.; Piskounova, E.; Morrison, S.J. Cancer stem cells: Impact, heterogeneity, and uncertainty. *Cancer Cell* **2012**, *21*, 283–296. [CrossRef]
50. Wang, Q.E. DNA damage responses in cancer stem cells: Implications for cancer therapeutic strategies. *World J. Biol. Chem.* **2015**, *6*, 57. [CrossRef]
51. Griffin, J.L.; Lehtimäki, K.K.; Valonen, P.; Gröhn, O.; Kettunen, M.I.; Ylä-Herttuala, S.; Pitkänen, A.; Nicholson, J.K.; Kauppinen, R.A. Assignment of ^{1}H nuclear magnetic resonance visible polyunsaturated fatty acids in BT4C gliomas undergoing ganciclovir-thymidine kinase gene therapy-induced programmed cell death. *Cancer Res.* **2003**, *63*, 3195–3201. [PubMed]
52. Langelier, B.; Linard, A.; Bordat, C.; Lavialle, M.; Heberden, C. Long chain-polyunsaturated fatty acids modulate membrane phospholipid composition and protein localization in lipid rafts of neural stem cell cultures. *J. Cell Biochem.* **2010**, *110*, 1356–1364. [CrossRef] [PubMed]
53. Antal, O.; Hackler, L.J.; Shen, J.; Mán, I.; Hideghéty, K.; Kitajka, K.; Puskás, L.G. Combination of unsaturated fatty acids and ionizing radiation on human glioma cells: Cellular, biochemical and gene expression analysis. *Lipids Health Dis.* **2014**, *13*, 142. [CrossRef] [PubMed]
54. Wang, Y.; Hou, Y.; Zhang, W.; Alvarez, A.A.; Bai, Y.; Hu, B.; Cheng, S.Y.; Yang, K.; Li, Y.; Feng, H. Lipolytic inhibitor G0S2 modulates glioma stem-like cell radiation response. *J. Exp. Clin. Cancer Res.* **2019**, *38*, 1–13. [CrossRef]
55. Chen, Y.; Huang, W.; Zheng, M.; Xue, Y.; Yang, J.; Liu, W.; Han, S. The mTOR inhibitor rapamycin synergizes with a fatty acid synthase inhibitor to induce cytotoxicity in ER/HER2-positive breast cancer cells. *PLoS ONE* **2014**, *9*, e97697. [CrossRef]
56. Kumar, G.K. Hypoxia. 3. Hypoxia and neurotransmitter synthesis. *Am. J. Physiol. Cell Physiol.* **2011**, *300*, C743–C751. [CrossRef]
57. Konings, K.; Vandevoorde, C.; Baselet, B.; Baatout, S.; Moreels, M. Combination therapy with charged particles and molecular targeting: A promising avenue to overcome radioresistance. *Front. Oncol.* **2020**, *10*, 128. [CrossRef]
58. Yang, W.; Liu, Y.; Gao, R.; Yu, H.; Sun, T. HDAC6 inhibition induces glioma stem cells differentiation and enhances cellular radiation sensitivity through the SHH/Gli1 signaling pathway. *Cancer Lett.* **2018**, *415*, 164–176. [CrossRef]
59. May, J.L.; Kouri, F.M.; Hurley, L.A.; Liu, J.; Tommasini-Ghelfi, S.; Ji, Y.; Gao, P.; Calvert, A.R.; Lee, A.; Chandel, N.S.; et al. IDH3α regulates one-Carbon metabolism in glioblastoma. *Sci. Adv.* **2019**, *5*, eaat0456. [CrossRef]
60. Icard, P.; Shulman, S.; Farhat, D.; Steyaert, J.M.; Alifano, M.; Lincet, H. How the Warburg effect supports aggressiveness and drug resistance of cancer cells? *Drug Resist. Updates* **2018**, *38*, 1–113. [CrossRef]
61. Gerlach, G.; Hofer, H.W. Interaction of immobilized phosphofructokinase with soluble muscle proteins. *Biochim. Biophys. Acta (BBA) Gen. Subj.* **1986**, *881*, 398–404. [CrossRef]

62. Kraft, T.; Hornemann, T.; Stolz, M.; Nier, V.; Wallimann, T. Coupling of creatine kinase to glycolytic enzymes at the sarcomeric I-band of skeletal muscle: A biochemical study in situ. *J. Muscle Res. Cell Motil.* **2000**, *21*, 691–703. [CrossRef] [PubMed]
63. Storey, K.B.; Hochachka, P.W. Enzymes of energy metabolism from a vertebrate facultative anaerobe, Pseudemys scripta. Turtle heart Phosphofructokinase. *J. Biol. Chem.* **1974**, *249*, 1423–1427. [PubMed]
64. Schlattner, U.; Tokarska-Schlattner, M.; Wallimann, T. Mitochondrial creatine kinase in human health and disease. *Biochim. Biophys. Acta* **2006**, *1762*, 164–180. [CrossRef]
65. Kleihues, P.; Cavenee, W. (Eds.) Pathology and genetics of tumours of the nervous system. In *World Health Organization Classification of Tumours*, 2nd ed.; France IARC Press: Lyon, France, 2000.
66. Pallini, R.; Ricci-Vitiani, L.; Banna, G.L.; Signore, M.; Lombardi, D.; Todaro, M.; Stassi, G.; Martini, M.; Maira, G.; Larocca, L.M.; et al. Cancer stem cell analysis and clinical outcome in patients with glioblastoma multiforme. *Clin. Cancer Res.* **2008**, *14*, 8205–8212. [CrossRef] [PubMed]
67. Esteller, M.; Hamilton, S.R.; Burger, P.C.; Baylin, S.B.; Herman, J.G. Inactivation of the DNA repair gene O6-methylguanine-DNA methyltransferase by promoterhypermethylation is a common event in primary hu-man neoplasia. *Cancer Res.* **1999**, *59*, 793–797. [PubMed]
68. Martini, M.; Cenci, T.; D'Alessandris, G.Q.; Cesarini, V.; Cocomazzi, A.; Ricci-Vitiani, L.; De Maria, R.; Pallini, R.; Larocca, L.M. Epigenetic silencing of Id4 identifies a glioblastoma subgroup with a better prognosis as a consequence of an inhibition of angiogenesis. *Cancer* **2013**, *119*, 1004–1012. [CrossRef]
69. Lulli, V.; Buccarelli, M.; Martini, M.; Signore, M.; Biffoni, M.; Giannetti, S.; Morgante, L.; Marziali, G.; Ilari, R.; Pagliuca, A.; et al. miR-135b suppresses tumorigenesis in glioblastoma stem-like cells impairing proliferation, migration and self-renewal. *Oncotarget* **2015**, *6*, 37241–37256. [CrossRef]
70. IAEA TRS-398 Absorbed Dose Determination in External Beam Radiotherapy: An International Code of Practice for Dosimetry Based on Standards of Absorbed Dose to Water. Available online: http://www-naweb.iaea.org/nahu/DMRP/documents/CoP_V12_2006-06-05.pdf (accessed on 22 May 2020).

International Journal of
Molecular Sciences

MDPI

Review

Ionizing Radiation as a Source of Oxidative Stress—The Protective Role of Melatonin and Vitamin D

Jarosław Nuszkiewicz *, Alina Woźniak and Karolina Szewczyk-Golec *

Department of Medical Biology and Biochemistry, Faculty of Medicine, Ludwik Rydygier Collegium Medicum in Bydgoszcz, Nicolaus Copernicus University in Toruń, 24 Karłowicza St, 85-092 Bydgoszcz, Poland; al1103@cm.umk.pl

* Correspondence: jnuszkiewicz@cm.umk.pl (J.N.); karosz@cm.umk.pl (K.S.-G.)

Received: 29 June 2020; Accepted: 11 August 2020; Published: 13 August 2020

Abstract: Ionizing radiation (IR) has found widespread application in modern medicine, including medical imaging and radiotherapy. As a result, both patients and healthcare professionals are exposed to various IR doses. To minimize the negative side effects of radiation associated with oxidative imbalance, antioxidant therapy has been considered. In this review, studies on the effects of melatonin and vitamin D on radiation-induced oxidative stress are discussed. According to the research data, both substances meet the conditions for use as agents that protect humans against IR-induced tissue damage. Numerous studies have confirmed that melatonin, a hydro- and lipophilic hormone with strong antioxidant properties, can potentially be used as a radioprotectant in humans. Less is known about the radioprotective effects of vitamin D, but the results to date have been promising. Deficiencies in melatonin and vitamin D are common in modern societies and may contribute to the severity of adverse side effects of medical IR exposure. Hence, supporting supplementation with both substances seems to be of first importance. Interestingly, both melatonin and vitamin D have been found to selectively radiosensitise cancer cells, which makes them promising adjuvants in radiotherapy. More research is needed in this area, especially in humans.

Keywords: ionizing radiation; melatonin; oxidative stress; radioprotection; reactive oxygen species; vitamin D

1. Introduction

The field of radiology dates back to 1895 when the German scientist Wilhelm Konrad Roentgen discovered X-rays [1]. Since then, ionizing radiation (IR) has found wide application in medicine, both in diagnostics and in therapies [2–5]. The use of medical imaging, especially roentgenodiagnostics and computed tomography, and radiotherapy exposes both patients and medical professionals to the harmful side effects of radiation [2,3]. IR occurs naturally in the environment, having accompanied humanity since its dawn. Its sources are natural radioisotopes found in soil and cosmic rays reaching the Earth's surface [6–8]. This radiation is called background radiation and its value changes with natural conditions [7,9]. The highest background radiation dose values of circa 0.26 Gy/year are observed in Ramsar (Iran) [10]. This dose is 10–100 times higher than the average one, but no greater incidence of cancer or other IR-related diseases is observed in this region [10]. This is due to radiation hormesis, which is an evolutionary adaptation to the presence of background radiation and the development of appropriate repair systems [11].

The mechanism of deleterious IR action is strongly associated with increasing oxidative stress in irradiated tissues [12]. IR is capable of penetrating the cells of living organisms, where it induces the ionization of both organic and inorganic compounds [13,14]. Due to the high water content in cells, radiolysis of water molecules by IR is the main process contributing to the increased formation of reactive oxygen species (ROS) [15,16]. ROS rapidly react with macromolecules, including proteins, nucleic acids and lipids, leading to cell dysfunction and apoptotic cell death [12]. As a result of augmented oxidative stress, not only direct negative side effects, but also ROS-related diseases may develop. Therefore, it is especially important to identify effective and safe prophylactic compounds to protect people from IR damage [4]. Undoubtedly, the substances considered in this type of supporting therapy should demonstrate an ability to counteract excessive oxidative stress.

Recently, attention has been paid to radioprotective properties of two hormones, whose synthesis depends on the specific light wavelengths, namely melatonin and vitamin D [17,18]. Both substances are endogenous compounds, but their deficiencies have been widely described in modern societies [19,20]. Melatonin as a strong direct and indirect antioxidant has been considered a radioprotector since the beginning of the 21st century [21–23]. However, animal model and in vitro studies have not been translated into human use yet [23]. Vitamin D, originally associated with bone homeostasis, has been found to perform many regulatory functions, affecting, among other things, the oxidative–antioxidant balance of the body [24–26]. Thus, its use to prevent irradiation side effects has also been taken under consideration, but data are limited and require more research.

Taking into account the relevance of the problem, the aim of the current review was to provide new scientific data on the protective effects of melatonin and vitamin D against oxidative damage caused by ionizing radiation. The current state of knowledge, including possible molecular mechanisms of action, is discussed. We hope that our review will be an impetus for further research on the use of both hormones in preventing deleterious side effects of ionizing radiation, especially in the field of human studies.

2. Ionizing Radiation as a Source of Reactive Oxygen Species

IR is a form of energy transfer that is able to cause ionization of a material medium while interacting with it [7]. This energy can be transferred by means of electromagnetic waves, including X radiation, gamma radiation and a small range of ultraviolet (UV) radiation with short wavelength and high energy, or through alpha and beta particles [27,28]. Each type of radiation differs in its energy, penetration and biological effects of the exposure. Alpha particles, consisting of two protons and two neutrons, have a short range due to their high mass [29]. There are two types of beta radiation. Beta minus radiation consists of electrons, while beta plus radiation consists of positrons, which are the antimatter counterpart of the electron [30]. Both X and gamma rays are characterized by high penetration and a plate made of lead is needed as an effective shield against them [31]. UV radiation capable of causing ionization has a wavelength in the range of 100–280 nm (UVC) and is absorbed by the atmosphere [28,32].

An important parameter used in dosimetry characterizing IR is linear energy transfer (LET), which determines the average amount of energy lost per unit of length transferred by radiation quanta [33]. High LET values are characteristic of alpha particles, neutrons and cosmic rays (heavy ions) [34]. Alpha particles, compared to other types of radiation, are characterized by shallow penetration, so the radiation energy is deposited at a shorter distance [35]. Neutron radiation and heavy ions, characteristic of cosmic radiation, have a greater range and penetrate deeper than alpha particles [34]. Low LET, typical of beta and gamma types of radiation, involves deposition of energy over a longer distance, causing less damage per distance unit [36].

High LET alpha radiation interacts mainly with molecules on the surface of the tissue by destroying its structure [36]. The most common source of alpha radiation in the environment is one of the natural radon isotopes, namely radon-222 [37]. Despite limited tissue penetration, alpha particles have high relative biological effectiveness. They can cause significant damage, especially in tissues sensitive to alpha particles due to their shallowness, such as bronchial epithelium. This makes radon, as an inhaled residential gas, a significant cause of lung cancer [37]. Characterized by higher penetration, low LET radiation is mainly responsible for the generation of ROS by ionization of atoms [35,38].

It should be noticed that most environmental, occupational and medical IR sources expose people to simultaneous action of different types of radiation. The interaction of low and high LET radiation may lead to increased and more complex biological damage [35].

IR, absorbed by tissues and cells, affects their functioning and structure to various extents, depending on the dose and type of radiation [13,14]. In affected cells, ROS are generated mainly through the radiolysis of water molecules (decay by the action of radiation quanta) or the excitation of water molecules and their decay [15,16,39]. IR can also indirectly influence the oxidative–antioxidant homeostasis by damaging different biomolecules [12]. The altered molecules, such as DNA or proteins responsible for stabilizing the DNA structure, become more susceptible to damage caused by ROS [40,41]. In addition, antioxidants or genes encoding for enzymatic antioxidants can be damaged, which directly increases the oxidative stress [40,41]. A meta-analysis carried out by Einor et al. [42], based on 41 studies concerning various biological matrices, proved that IR, even at low doses, generates ROS.

In biological systems, the state in which the amount of ROS and reactive nitrogen species (RNS) exceeds the physiological ability to maintain homeostasis is called oxidative stress [43,44]. ROS, which are products of excitation and one-, two- and three-electron reduction of the oxygen molecule, are characterized by much greater reactivity than the oxygen in the ground state [45,46]. ROS are a broad concept, including ions, atoms, as well as molecules and radicals such as hydrogen peroxide (H_2O_2), singlet oxygen ($^1O_2$), superoxide anion radical ($O_2^{\bullet-}$) and hydroxyl radical ($OH^{\bullet}$) [46,47]. The hydroxyl radical is the most dangerous for tissues due to its high reactivity and the ability to oxidize many cell components, such as lipids, proteins, carbohydrates and deoxyribonucleic acids [48,49]. As a result of lipid peroxidation, reactive lipid derivatives are formed, which are capable of oxidative damage to other biomolecules [50]. Depending on the fatty acid that undergoes oxidation, trans-4-hydroxy-2-nonenal (4-HNE) and/or malondialdehyde (MDA) are formed as one of the end products, used as markers of the lipid peroxidation level [44]. Oxidative modifications of the protein structure have been observed in many pathophysiological conditions, including the ageing process, as well as apoptosis [51,52]. They lead to a loss of spatial conformation and biological properties, impeded degradation and accumulation of modified protein products, such as protein carbonyl derivatives [51,52]. Oxidative stress causes damage to both mitochondrial and nuclear DNA, which may result in mutation and carcinogenesis. The marker of DNA damage is 8-oxoguanine, a chemical derivative of guanine [53,54]. Oxidative stress is associated with many diseases, including epidemiologically significant diseases of affluence such as cancer [55], cardiovascular disease [56], obesity [57], neurodegenerative diseases [58] and allergic diseases [59].

Oxygen metabolism and the prevalence of ROS have forced living organisms to develop appropriate counteraction mechanisms to minimize the negative effects of oxidative stress [60]. The antioxidant defense system consists of endogenous and exogenous elements. Antioxidant enzymes, which include superoxide dismutases (SODs), catalase (CAT), glutathione peroxidases (GPxs) and glutathione reductase (GR), the enzyme necessary for the proper functioning of GPx, are a part of the endogenous primary enzymatic defense [61,62]. In addition to antioxidant enzymes, reduced glutathione (GSH), a cofactor for GPx, proteins (ferritin, transferrin, ceruloplasmin, albumin), uric acid, melatonin and vitamin D take part in the prevention of excessive oxidative stress [17,63–65]. Carotenoids, vitamins A, C, and E, selenium, and polyphenols are the main exogenous antioxidants [66,67]. The cooperation of both endogenous and exogenous antioxidants maintains the oxidative and antioxidant balance, preventing the negative effects of oxidative stress but enabling ROS to perform physiologically important functions as mediators of intercellular communication [68].

In numerous studies, the effect of ionizing radiation on the oxidative stress level has been examined [39,69–71]. Different radiation qualities and doses have been used in the experiments during recent years [72–75]. According to Kang et al. [72], a dose of 2 Gy γ-irradiation at a dose rate of 1.1 Gy/min affected ROS generation in murine splenocyte cell culture. The level of oxidative stress was determined by a method using 2′,7′-dichlorofluorescin diacetate (DCFH-DA), which penetrates inside the cells and is hydrolysed by intracellular esterase into 2′,7′-dichlorofluorescin (DCFH). DCFH reacts with ROS and is converted to highly fluorescent 2′,7′-dichlorodihydrofluorescein (DCF). Fluorescence was assessed 24 h after irradiation and a significant increase in ROS levels was observed as a result of radiation. Similar observations were made by Shaban et al. [75] in a study whose purpose was to investigate the effect of gamma radiation at a dose of 2, 4, 6, 8 and 10 Gy (delivered in four fractions at one-day intervals at a dose rate of 0.5 Gy/min) in male Albino Sprague-Dawley rat testis. The authors examined blood samples and histopathologically evaluated the irradiated tissues. After exposure to IR, increases in MDA, nitric oxide and calcium ion levels were observed, while SOD and CAT activities and GSH concentration decreased. Karimi et al. [73] also described the relationship between gamma radiation at a dose of 15 Gy (at a dose rate of 0.985 Gy/min) and oxidative stress after irradiation of rat lenses. Two days after the exposure to IR, the animals were sacrificed and an increase in MDA concentrations and a decrease in GSH levels were detected in the tested lenses. Rezaeyan et al. [74] irradiated the adult male Sprague-Dawley rat chest area. The applied X-ray at a dose of 18 Gy in one fraction increased oxidative stress 24 h after the exposure through increased MDA levels and decreased SOD activities. It can be summarized that exposure to high doses of IR leads to increased ROS production, enhanced lipid peroxidation and reduced enzymatic antioxidant defense in a dose-, dose-rate- and LET-dependent manner, while low doses of low LET radiation may upregulate antioxidant defense, including the stimulation of GSH synthesis [39]. It has been proven that IR affects ROS and RNS cell metabolism, activating different signaling pathways and disrupting the normal redox system [69,71]. These changes lead to the dysregulation of the activities of cyclooxygenases, lipoxygenases, nitric oxide synthases, and nicotinamide adenine dinucleotide phosphate oxidases, accompanied by mitochondrial dysfunction [69,71]. It is also worth noting that the response to IR is tissue-dependent, with acute damage but fast regeneration for tissues with rapid turnover [70].

3. Melatonin—A Circadian Rhythm Regulator with Antioxidant Properties

Melatonin (N-acetyl-5-methoxytryptamine) is a hormone synthesized and secreted mainly by the pineal gland present in the brain of vertebrates [76]. Extrapineal sources of melatonin are localised in bone marrow, skin, platelets, lymphocytes, retina, the gastrointestinal tract, and the Harderian gland [77,78]. It was first isolated from the bovine pineal gland by Aaron Lerner in 1958 [79] and since then researchers have explored new aspects of this hormone.

The pineal gland is an unpaired structure localized between thalamic bodies in the quadrigeminal cistern [80]. In an adult human, this small neuroendocrine gland reaches 5–9 mm in length, 1–5 mm in width, and 3–5 mm in thickness and weighs about 100–180 mg [81]. The substrate for melatonin biosynthesis in pinealocytes is the amino acid containing an indole ring, tryptophan [76,80]. With the tryptophan hydroxylase enzyme (TPH), the tryptophan molecule is converted to 5-hydroxytryptophan (oxitriptan). Then aromatic L-amino acid decarboxylase (AAAD), using pyridoxal phosphate (PLP) as a coenzyme, catalyzes the reaction in which serotonin is formed [82]. 5-hydroxytryptamine (serotonin), a neurotransmitter colloquially called the happiness hormone, is an intermediate for aralkylamine *N*-acetyltransferase (AANAT), which in the presence of acetyl coenzyme A (acetyl CoA) leads to the biosynthesis of N-acetylserotonin (normelatonin) [83]. The last stage of melatonin biosynthesis takes place with the participation of the enzyme acetylserotonin O-methyltransferase (ASMT) and S-adenosyl methionine (SAM), a coenzyme in methylation reactions [84].

Biosynthesis and secretion of melatonin by pinealocytes are regulated by the presence of electromagnetic radiation in the visible light range, especially light with a wavelength of 460–480 nm, which is perceived as blue light [85]. The highest secretion of melatonin is observed between 3:00 a.m. and 4:00 a.m. (with normal circadian rhythms) [86]. The plasma melatonin concentrations during these hours range from 18.5 to 180 pg/mL [87]. Night work and the use of computer screens or smartphones at night, typical of the modern society, lead to reduced melatonin synthesis [20]. In humans, the endogenous master clock, which controls many physiological processes and behavior patterns, is located in the hypothalamic suprachiasmatic nucleus (SCN) [88]. Light reaching intrinsically photosensitive retinal ganglion cells is received by a photopigment sensitive to blue light called melanopsin [88,89]. The signal is transmitted via the retinohypothalamic tract to SCN located above the optic chiasm [80,89]. The SCN has direct connections to other hypothalamic nuclei and the pineal gland [90]. In this way, information sent by SCN regulates melatonin synthesis. Melatonin secreted into the circulatory system affects SCN by feedback, and other tissues by regulating their chronobiology [91]. The *Clock*, *Bmal1*, *Cry1-2*, *Per1-2* genes, whose expression is modulated by melatonin, play an important role in regulating SCN [88,92,93]. The *Bmal1* and *Clock* gene transcription products combine together to form heterodimers, which attach to the promoter region of the *Per* and *Cry* genes to initiate their transcription [94]. In the absence of light, greater transcription of the *Bmal1* and *Clock* genes is observed [94]. In the cytoplasm, PER and CRY proteins combine into a heterodimer, which inhibits further transcription of the genes responsible for their synthesis [92]. Additionally, melatonin is known to attenuate *Cry1* gene expression [93]. The combination of molecular clocks based on the promotion and inhibition of specific gene transcription, and regulation based on external stimuli, namely the presence or lack of blue light, allow the circadian rhythms to function properly. Melatonin acts as a regulator and synchronizer of these processes.

Melatonin is an endocrine, paracrine and autocrine hormone, so it has an effect on tissues distant from the synthesis site, on neighbouring cells, and directly on the cells that synthesize it [82,95]. The action of melatonin occurs through membrane G protein-coupled receptors (MT1, MT2, MT3), but also through nuclear receptors (RZR/RORα) and calmodulin [96,97]. The number of tissues in which MT1 and MT2 receptors have been detected demonstrates the broad spectrum of the compound's activity, including the liver, kidneys, retina, ovaries, testes, mammary glands, gallbladder, immune cells, cardiovascular system, exocrine pancreas, duodenal enterocytes, brain (hypothalamus, SCN, pituitary), blood vessels, gastrointestinal tract, adipocytes, and skin [98–100]. MT1 (MTNR1A), consisting of 350 amino acid residues, couples to pertussis toxin-sensitive G_i and toxin-insensitive $G_{q/11}$ proteins, inhibits cAMP response element-binding protein (CREB) phosphorylation, forskolin-stimulated cAMP and protein kinase A signaling, and increases potassium conductance through Kir internally rectifying channels [98,99]. MT2 (MTNR1B), consisting of 362 amino acid residues, inhibits cGMP formation and forskolin-stimulated cAMP production, reduces calcium-dependent dopamine release in the retina and activates protein kinase C (PKC) in the SCN [98,101].

The effect of melatonin is not limited to regulating circadian and seasonal rhythms. Melatonin also modulates the functioning of the immune system [102] and has anti-inflammatory properties [100,103–105]. Reduced concentration of melatonin is observed in many pathophysiological conditions and its supplementation may affect the course of disorders, such as neurodegenerative diseases, including Alzheimer's disease [106,107], primary headache disorders [108], obesity [105,109], diabetes mellitus type 2 [110,111] and hypertension [105,112].

Numerous studies indicate strong antioxidant properties of melatonin [17,77,113–116]. The molecule can cross the blood–brain barrier and its activity is not limited to the central nervous system (CNS) but it also affects other tissues distant from the site of synthesis [117]. The melatonin is soluble in both water and lipid environments, so it can act as an antioxidant in the aqueous environment inside the cells and in body fluids, as well as in plasma membranes of cells and cell organelles [118]. Research into the antioxidant properties of melatonin has confirmed that this hormone and its metabolites neutralize numerous ROS and RNS molecules, including H_2O_2, $^1O_2$, $O^{2\bullet-}$, peroxynitrite (ONOO-), as well as very reactive $OH^\bullet$ [17,119]. Melatonin metabolism products such as N1-acetyl-N2-formyl-5-methoxykynuramine (AFMK), N1-acetyl-5-methoxykynuramine (AMK) and 3-hydroxymelatonin (3-OHM) are also ROS and RNS scavengers [77,120,121]. The antioxidant properties of melatonin are due to its chemical structure, specifically the aromatic indole ring rich in delocalized electrons, a source of electrons in ROS and RNS neutralization reactions [122]. Melatonin may also indirectly affect the oxidative–antioxidant balance, stimulating the expression of genes encoding for some antioxidant enzymes. This effect is observed in the case of SODs, GPxs and GR [123,124]. The effect of melatonin and its chemical derivatives on the oxidoreductive balance is shown in Figure 1.

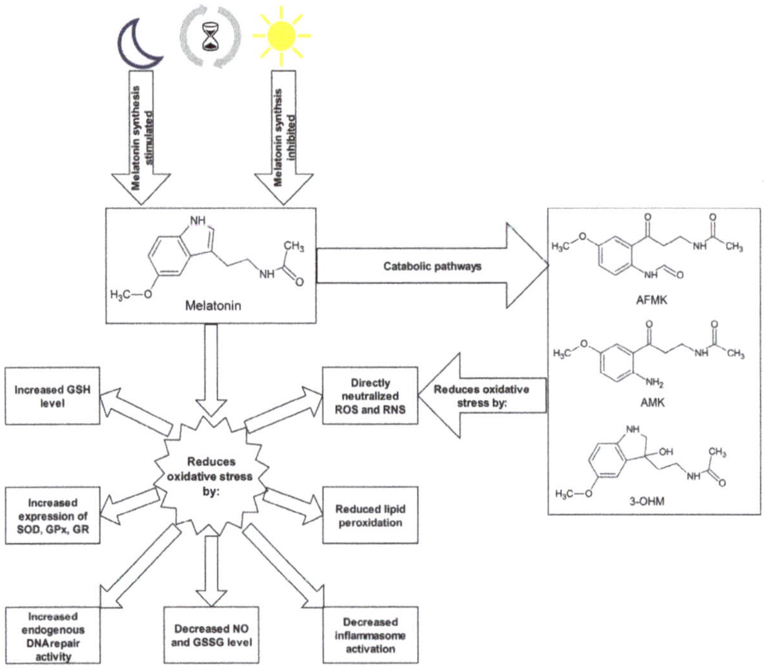

Figure 1. Melatonin and its metabolites as antioxidants. Abbreviations used: 3-OHM—3- hydroxymelatonin, AFMK—N1-acetyl-N2-formyl-5-methoxykynuramine, AMK—N1-acetyl-5-methoxykynuramine, GPx—glutathione peroxidase, GR—glutathione reductase, GSH—glutathione, GSSG—glutathione disulphide, NO—nitric oxide, RNS—reactive nitrogen species, ROS—reactive oxygen species, SOD—superoxide dismutase.

The study of the effect of melatonin on organisms exposed to IR has been described mainly in animal models [23]. Numerous studies have been conducted on the effect of melatonin pretreatment in the irradiation of specific parts of the body [125–130], indicating a reduction in lipid peroxidation, an improvement in enzymatic and non-enzymatic antioxidant defense, stimulation of the DNA damage response, and a reduction in the inflammatory state and histopathological changes. According to Fernandez-Gil et al. [125], melatonin may protect small intestine cells from toxic products formed during radiation therapy used in the oral mucosa. In this experiment, the researchers used adult male Wistar rats, which were given 3% melatonin gel (the total melatonin dose was 45 mg/day for 21 days). The rats were anesthetized prior to irradiation with a dose of 7.5 Gy/day (X-ray) for five consecutive days, where only the oral cavity was irradiated. Melatonin was applied in the oral cavity, starting 48 h before the first irradiation. After the sacrifice of animals, small intestine samples were taken for further analysis. Oral cavity irradiation resulted in small intestinal damage associated with oxidative stress. An increase in lipid peroxidation and nitrite/nitrate was observed, compared to non-irradiated controls. In the melatonin treated group, a significant increase in the activities and protein levels of GPx, GR, SOD2 and a substantial decrease in inflammasome activation in the small intestine were described, compared to the irradiated but non melatonin-treated group. In the Gurses et al. [126] study, Wistar rats were given a 50 mg/kg dose of melatonin (injected intraperitoneally) 15 min prior to irradiation of the anatomical area surrounding the heart. A dose of 18 Gy was used in one fraction. Six months after exposure to radiation, the rats were sacrificed, and histopathological preparations were performed to assess changes in the study and control groups. The use of melatonin prevented the development of vasculitis, and also reduced myocyte necrosis and cardiac fibrosis.

According to the experiments concerning whole body irradiation, melatonin administered both before and after IR exposure increased the survival rate of examined animals, reduced symptoms of acute irradiation disease, decreased histopathological changes, and improved the oxidative–antioxidant balance in the organism [130–136]. In a mouse model study, Vasin et al. [136] examined the effect of the whole body exposure to IR at a dose of 9.5–10 Gy at a power of 0.077–0.171 Gy/min given in one fraction. Melatonin was dissolved in water at a concentration of 5 mg/L and was administered from 3 to 30 days after the irradiation. The daily dosage of melatonin was changed with the onset of acute radiation sickness to 0.9–1.0 mg/kg and 1.2 mg/kg during recovery. The group of mice treated with melatonin showed less severe symptoms of acute radiation sickness and significantly higher survival was observed within 30 days of irradiation, compared to the control group. At the peak of radiation sickness (12 days after the irradiation), the average number of leukocytes in the group of mice supplemented with melatonin was higher than in the control group by 40%. Similar results were noted in a study conducted by Haddadi et al. [127] on adult male Wistar rats that received melatonin (100 mg/kg b.w.) intraperitoneally 30 min before irradiation and 5 mg/kg per day after irradiation for a maximum of 22 weeks. The animals were anesthetized and sacrificed at 4 and 24 h of irradiation, as well as 1, 3, 8, 16, 20, and 22 weeks after the treatment. The total radiation dose was 22 Gy at a dose rate of 1.8 Gy/min. The authors of the study indicate that the survival of animals from the melatonin-treated group was higher than the control group. In addition, in the melatonin group, lower expression of vascular endothelial growth factor (VEGF) and fewer histopathological changes were shown.

Research data on human studies is very limited. Vijayalaxmi et al. [137] performed an in vitro study on the effect of melatonin on radiated human peripheral blood samples and obtained very promising results. Approximately 15 min before the administration of 300 mg melatonin orally, volunteers gave a blood sample. The next blood collection took place 1 and 2 h after melatonin supplementation. Every blood sample was exposed in vitro to 1 Gy of gamma radiation at a dose rate of 1.087 Gy/min. The lymphocytes were examined to determine the amount of primary DNA damage. A significant increase in melatonin concentration in both serum and leukocytes was observed as early as 1 h after the administration of melatonin. The extent of primary DNA damage was reduced in both blood samples taken 1 and 2 h after melatonin administration, compared to the blood taken before melatonin supplementation. It is worth emphasizing that no negative effects of such a high dose of melatonin (300 mg) were observed. The dose of melatonin given to the subjects in the study is 30 times higher than the safe dose of the substance (10 mg/day) recommended in the treatment of sleep disorders, so further research is crucial to determining the appropriate amount of melatonin needed to protect people from the side effects of irradiation. Table 1 presents the summary of studies on the effects of melatonin on organisms exposed to ionizing radiation.

Table 1. Research on the impact of ionizing radiation on the generation of reactive oxygen species and the radioprotective role of melatonin.

Subjects	Melatonin Dosage (Route of Administration)	Time of Melatonin Administration	Radiation Dosage (Irradiation Area)	Outcomes	Reference
Adult male Sprague-Dawley rats	10 and 20 mg/kg (IP injection)	Immediately before and after irradiation	X-ray radiation of 8 Gy (whole body)	Melatonin reduced the levels of MDA and increased the GSH concentration.	[134]
Adult female Sprague-Dawley rats	30 and 5 mg/kg (IP injection)	30 min prior to irradiation and on the following days of experiment	Gamma radiation of 5 and 8 Gy (total cranial)	Melatonin decreased the formation of late side effects of radiation. Melatonin administration during radiotherapy protected ocular lenses against radiation-induced oxidative injuries.	[129]

Table 1. *Cont.*

Subjects	Melatonin Dosage (Route of Administration)	Time of Melatonin Administration	Radiation Dosage (Irradiation Area)	Outcomes	Reference
Adult male Wistar rats	100 and 5 mg/kg (IP injection)	30 min before irradiation and once a day per after irradiation	Gamma radiation 22 Gy (cervical segment of the spinal cord)	Melatonin increased survival rate and decreased histopathological changes.	[127]
Adult male Wistar rats	50 mg/kg (IP injection)	15 min prior to irradiation	18 Gy (anatomical area of the heart position)	Melatonin prevented the development of vasculitis, reduced myocyte necrosis and cardiac fibrosis.	[126]
Adult male Wistar rats	10, 20, and 10 mg/kg (IP injection)	Before irradiation, just after irradiation and 24h after irradiation	Gamma radiation 8 Gy, twice (whole body and abdominopelvic)	Melatonin administration inhibited primary spermatocyte degeneration.	[130]
Adult male Wistar rats	0.2 mg/day (IP injection)	Once a day for 14 days before irradiation	Gamma radiation 8 Gy (whole body)	Melatonin had a protective effect on suprarenal gland.	[131]
Adult male Wistar rats	5 and 10 mg/kg (IP injection)	30 min before irradiation	Gamma radiation 6 Gy (whole body)	Melatonin decreased hepatic MDA and nitric oxide (NO) levels.	[135]
Adult male Wistar rats	45 mg/day (PO)	Once a day for 21 days before irradiation	X-ray 7.5 Gy/day for five consecutive days (oral cavity)	Melatonin increased the activities and protein levels of GPx, GR, SOD2 and strongly decreased inflammasome activation.	[125]
Adult both sexes Wistar rats	100 mg/kg (IP injection)	For 5 days post radiation	Total dose of 7.2 Gy in two fractions (whole body)	Melatonin reduced MDA level, rates of oedema, necrosis, neuronal degeneration, and vasodilatation.	[133]
Adult male mice	From 0.9–1.0 to 1.2 mg/kg (PO)	From the third day after irradiation	Gamma radiation 9.5–10 Gy (whole body)	Melatonin reduced symptoms of acute radiation sickness, increased survival rate and leukocyte level.	[136]
Adult male Swiss albino mice	0.1 mg/kg/day (PO)	15 consecutive days prior to radiation	Gamma radiation 6, 8 and 10 Gy (whole body)	Melatonin reduced lipid peroxidation, glutathione disulphide (GSSG) level, deficit in the body and organ weight. Melatonin increased GSH level and survival rate.	[132]
Young adult male squirrels	250 mg/kg (SC injection)	Once a day for four weeks before irradiation	X-ray radiation of 2.06 Gy (abdominal, near the splenic region)	Long term melatonin treatment protected the splenocytes and modulated endogenous DNA repair activity.	[128]
In vitro, human blood	300 mg (PO)	1 h before irradiation of blood sample	Gamma radiation 1 Gy (blood sample)	Melatonin reduced primary DNA damage.	[137]

Abbreviations used: IP injection—intraperitoneal injection, PO—oral administration, SC injection—subcutaneous injection.

To complete the picture of the relationship between melatonin and IR, it should be emphasised that melatonin has been found to radiosensitize cancer cells in a selective manner [69,138–140]. Melatonin's ability to sensitize cancer cells to irradiation, along with its radioprotective properties, makes it an ideal adjuvant in radiotherapy [69]. In the case of neck squamous cell carcinoma (HNSCC cell lines), melatonin (0.1, 0.5, 1.0, and 1.5 mM melatonin combined with 8 Gy irradiation) was described to enhance radiation cytotoxicity by stimulating mitochondrial ROS generation, apoptosis and autophagy [138]. It was also observed that melatonin (pretreatment with 1 mmol/L melatonin for 2 h) effectively inhibited cellular proliferation of the human colorectal carcinoma cell line HCT 116, and decreased colony formation rate and cell migration counts following IR exposure (gamma rays, 0–8 Gy) [140]. This effect was associated with activation of the caspase-dependent apoptotic pathway, cell cycle arrest in G2/M, and an impaired DNA double-strand break repair. Moreover, in the study, it was shown that melatonin in combination with IR treatment significantly suppressed tumor cell growth in colorectal tumor xenografts. Analogous results have been confirmed in breast cancer [139]. What remains unestablished is the use of melatonin combined with IR in patients, including the effects of the treatment, the time-lapse between melatonin administration and radiotherapy, as well as the optimal dosage of melatonin in humans exposed to IR. Further patient-based studies, such as pre-clinical and randomized control trials are needed to explain all uncertainties.

4. Vitamin D—Function and Antioxidant Effect

Vitamin D is a group of organic chemical compounds belonging to the group secosteroids, among which calcitriol (1,25-dihydroxycholecalciferol) performs the highest biological (hormonal) activity [26,141]. Vitamin D is currently at the center of research interest for many scientists due to its widespread deficiency, reaching about 30–50% on a global scale, especially in older age groups [19,142]. Many scientists have been involved in the research on the discovery and description of vitamin D properties. The largest contribution was made by Sir Edward Mellanby [143], Elmer McCollum [144] and Adolf Windaus [145], who in 1928 received the Nobel Prize for their work on vitamin D [146]. Vitamin D comes from both external sources and from the body's own synthesis [147]. Two forms of vitamin D are taken with food, namely cholecalciferol (vitamin D_3) and ergocalciferol (vitamin D_2) [148]. Fatty fish (such as salmon, mackerel, herring), meat, egg yolks, milk and butter are sources of cholecalciferol, while fungi, yeast and some plants contain ergocalciferol [149,150]. Vitamin D taken from food sources is only a fraction of the daily requirement for this compound [151]. The first stage of calcitriol biosynthesis is the transformation of 7-dehydrocholesterol in the skin under the influence of UV radiation at a wavelength of approximately 290–315 nm (UVB) [152,153]. For that reason, vitamin D is sometimes called the "sunshine vitamin". Excessive exposure to UV radiation does not cause the formation of toxic amounts of previtamin D because it photoisomerises into two biologically inert products, lumisterol and tachysterol [154]. Previtamin D undergoes spontaneous isomerisation to provitamin D (cholecalciferol) under the influence of body temperature [155]. Then, cholecalciferol both formed in the skin and originating from dietary sources binds to a specific transport protein, vitamin D-binding protein (DBP), and is transported to the liver [156]. Hydroxylation with cytochrome P450 CYP2R1 enzymes occurs in the liver. The product of this reaction, 25-hydroxyvitamin D, binds to DBP and is transported to the kidney for subsequent hydroxylation by the enzyme CYP27B1 [157]. The end product of this pathway is the hormonally active form of vitamin D, calcitriol, which is stored mainly in adipose tissue [158,159]. The vitamin D receptor (VDR) belongs to a subfamily of nuclear receptors that act as transcription factors [160]. VDR is heterodimerized with the retinoid-X receptor (RXR), which causes a change in its spatial conformation. The resulting heterodimer binds to appropriate promoter sites of vitamin D-dependent genes [161]. VDR occurs in almost all cells and tissues, including the skeletal system, cells involved in immune modulation, brain, heart, skin, gonads, prostate, breast and gut [162]. Originally, calcitriol was considered to be associated only with calcium-phosphate metabolism by cooperating with parathyroid hormone and the skeletal system, stimulating the absorption of dietary calcium from the gastrointestinal tract, promoting renal tubular reabsorption of calcium, and inducing the release of calcium from bones [163]. However, the role of vitamin D is known to be much greater and its deficiency is associated not only with diseases of the skeletal system, such as osteomalacia or osteoporosis in adults and rickets in children [164], but also with depression [165], cancer [152], adverse cardiovascular risk profile [166], obesity [24], type 2 diabetes mellitus [25] and autoimmune thyroid disease [167]. The reference vitamin D concentration range is 30–50 ng/mL (75–125 nmol/L) [168,169]. It should be added that this is the level of 25-hydroxyvitamin D, not calcitriol, that is tested because of lower test costs, higher analyte stability and good correlation with the concentration of the hormonally active form in the organism [168,169].

Vitamin D is thought to have antioxidant properties although involved mechanisms have not been fully described yet and further research is required [18,170]. Vitamin D, acting through its nuclear receptors, can stimulate the expression of genes coding for antioxidant enzymes such as SODs and GPxs [171]. It has also been confirmed that after exposure of the skin to UV radiation, calcitriol and its precursors increase p53 levels, which reduces intracellular ROS [172]. In addition, calcitriol has been shown to induce the synthesis of metallothioneins, which are ROS scavengers [172]. Tang et al. [173] reported that MDA levels negatively correlated with serum vitamin D levels in patients with non-segmental vitiligo. Furthermore, the researchers pointed out that vitamin D protected human melanocytes against ROS by activation of Wnt/β-catenin signaling. In addition, Jain et al. [174] showed a positive link between vitamin D and GSH concentrations, as well as a reduction in the levels of pro-inflammatory cytokines (monocyte chemoattractant protein 1 and interleukin 8), which lead to reduced ROS generation. In this study, U937 monocyte cells were treated with calcitriol at the concentration of 0, 10, and 25 nM for 24 h. Similar results were described by Dzik et al. [175]. In their study, patients, qualified for lumbar spine surgery utilizing static or dynamic implants, were supplemented with 25-hydroxyvitamin D at a daily dose of 3200 IU (equal to 80 μg) for 5 weeks. Vitamin D supplementation to appropriate serum levels reduced oxidative stress in skeletal muscle. The patients with prior vitamin D deficiency showed increases in Cu/ZnSOD and GPx activities in paraspinal muscles after supplementation. Chen et al. [176] tested 10α-hydroxylase knockout mice ($1\alpha(OH)ase^{-/-}$) supplemented with calcitriol at a dose of 1 μg/kg. The authors noted that low calcitriol levels were associated with higher oxidative stress. In addition, calcitriol regulated the expression of nuclear factor-erythroid-2-related factor 2 (Nrf2), which controls antioxidant and detoxification enzymes. In response to reduced ROS levels, SOD2 activity decreased. Sepehrmanesh et al. [177] confirmed that vitamin D supplementation led to a significant increase in GSH concentrations. Patients with major depressive disorder were supplemented with 25-hydroxyvitamin D at a weekly dose of 50,000 IU (equal to 1.25 mg) for 8 weeks. In addition to the increase in the GSH level, there was also a significant increase in total antioxidant capacity (TAC). On the other hand, Barzegari et al. [178] did not observe significant changes in SOD, CAT, and GPx activities, as well as in the MDA and TAC levels, despite a 8-week calcitriol supplementation at 50,000 IU once a week. The study was based on a double-blind, randomized, placebo-controlled clinical trial, involving 50 patients with type 2 diabetes and nephropathy. Undoubtedly, further studies on the antioxidant function of vitamin D are required. The main mechanisms of vitamin D action as an antioxidant are shown in Figure 2.

Considering the broad spectrum of vitamin D action in the organism, it has been identified as a potential protective agent against radiation-induced damage [179]. However, the analysis of the available literature indicates very limited research data on the radioprotective role of vitamin D in the context of IR action. It was observed that the administration of vitamin D_3 (0.7 μg of vitamin D_3 or 28 IU/100 g body mass) to chronically irradiated Wistar rats (0.01 Gy per day for 30 days) induced the normalization of carbohydrate metabolism by improving the activities of glycolytic enzymes in erythroid and myeloid bone marrow cells [180]. In the in vitro study of Müller et al. [181], it was found that the cell growth and clonogenic survival of irradiated keratinocytes (cell line HaCaT), pretreated with calcitriol, were significantly increased when compared to the untreated cells after IR exposure. In the experiment, exponentially growing HaCaT cells were irradiated with X-rays (total dose of 0 to 7.5 Gy was delivered with a dose rate of 1 Gy/min). To assess the vitamin D effect, the HaCaT cells were incubated with 10 nmol/L $1\alpha,25(OH)_2D_3$ for 48 h before irradiation. It was demonstrated that vitamin D improved cell growth and survival, as well as inhibited the radiation-induced up-regulation of adhesion molecule expression on human keratinocytes. These results were confirmed by Langberg et al. [182], who proved that treatment with calcitriol (100 nmol/L 24 h before and for 24–48 h after IR) inhibited caspase-dependent and -independent programmed cell death occurring within 48 h of irradiation, increased the colony formation capacity, attenuated radiation-induced increase in matrix metalloproteinase-9 and mRNA levels in irradiated (4 Gy with a dose rate 2 Gy/min) HaCaT keratinocytes. The same cell line was also the subject of the Trémezaygues et al. [183] experiment. It was observed that the pretreatment of HaCaT-keratinocytes with $1,25(OH)_2D_3$ (100 nmol/L) over 48 h differentially modulated harmful effects of IR (1–5 Gy) in a dose- and time-dependent manner, indicating a protective effect of vitamin D against relatively low IR (1–2 Gy). A study on the cell culture of human umbilical vein endothelial cells (HUVEC), conducted by Marampon et al. [184], showed the protective effect of vitamin D against damage caused by ROS generated under the influence of IR. Cell cultures were preincubated in a solution with the addition of vitamin D in concentrations of 25, 50, 75 and 100 nmol/L for 24 h, then transferred to the growth medium and irradiated with X-rays (total dose of 0 to 8 Gy was delivered with a dose rate of 1.3 Gy/min). Vitamin D preincubation reduced the amount of ROS by the protection of proliferating and quiescent cells via the regulation of the mitogen-activated protein kinase (MAPK) pathway, prevented apoptosis by activating signal-regulated kinases (ERKs) in proliferating HUVEC, and inhibited p38, associated with ageing in quiescent cells.

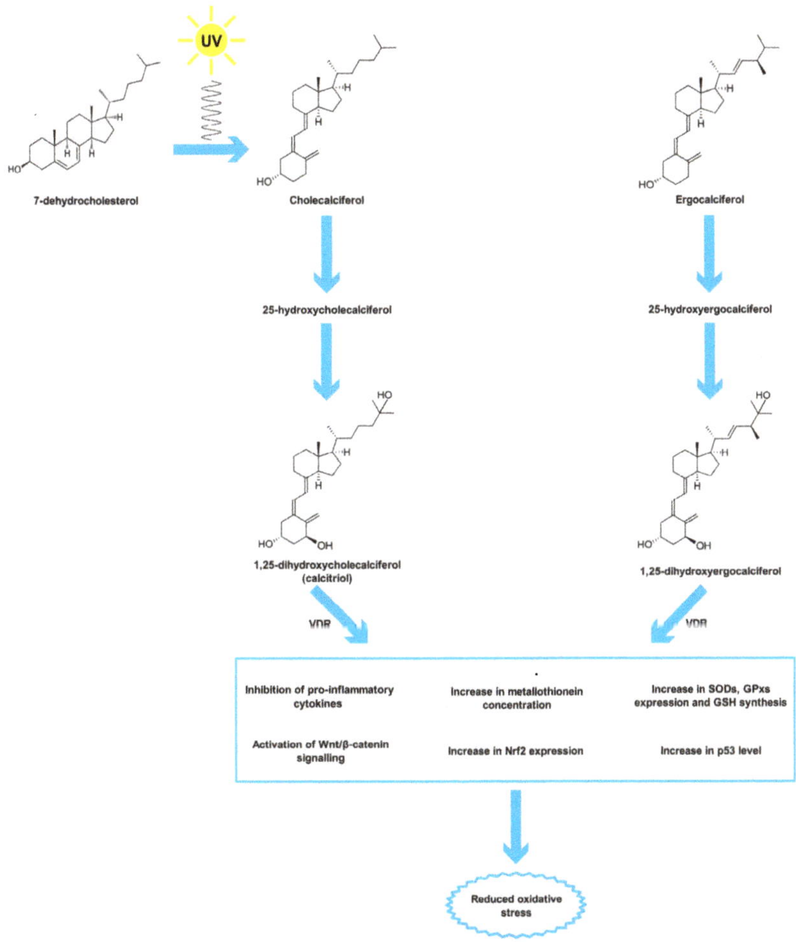

Figure 2. Antioxidant properties of vitamin D. Abbreviations used: GPxs—glutathione peroxidases, GSH—glutathione, Nrf2—nuclear factor-erythroid-2-related factor 2, SODs—superoxide dismutases, VDR—vitamin D receptor.

As with melatonin, vitamin D and its analogues have been found to selectively radiosensitize cancer cells, including breast and non-small cell lung tumor cells [185–189], which makes it a promising adjuvant in radiotherapy, enhancing the treatment effect and reducing side effects. In irradiated (5 times of 2 Gy administered over a period of 3 days) MCF-7 breast tumor cells, pretreatment with a hormonally active form of vitamin D (100 nmol/L 1,25$(OH)_2D_3$ for 72 h) promoted autophagy, sensitized the cells to IR and suppressed the proliferative recovery occurring after radiation alone [185]. This effect was not observed in the BT474 breast tumor cell line with low-level expression of VDR, suggesting a receptor-mediated action of calcitriol. Moreover, similar responses were not detected in a model of normal human fibroblasts [187]. The promotion of an enhanced response to radiation by 1,25-D_3 in non-small cell lung cancer cells has been found to be mediated by VDR, tumour protein p53 and AMPK pathways [188]. Normal human bronchial cells and cardiomyocytes were not radiosensitized by vitamin D in this study [188].

Interestingly, it was found that chronic exposure to IR affected the vitamin D_3 active form level and caused modifications of enzymes involved in vitamin D metabolism [190]. In accordance with this study, Kaminskyi et al. [191] described significantly lower vitamin D concentrations among the populations of radiologically contaminated regions of Chernivtsi oblast due to the Chornobyl catastrophe, compared to those in the uncontaminated Ukrainian cities of Charnivtsi and Vyzhnytsia. Therefore, the deficiency of vitamin D in patients during radiotherapy or in medical professionals chronically exposed to low IR doses should be taken into consideration in further research on the supplementary treatment.

5. Conclusions

This review points to the important role of ionizing radiation as an inducer of oxidative stress, which occurs in the pathogenesis and the course of many diseases. ROS are not only generated during medical procedures that require the use of IR but also when the organism is exposed to sunlight and background radiation present in the environment. The endogenous synthesis of two compounds with antioxidant potential described in this paper, namely melatonin and vitamin D, depends on the presence of light (visible or UV). Numerous studies emphasize the role of melatonin as an antioxidant and its protective effects against IR damage. This hormone both directly and indirectly neutralizes ROS. In the case of vitamin D, further experiments are required that could align its antioxidant mechanisms and protection against IR, as previous publications show conflicting findings. As a result of ever-growing use of IR in medicine, more and more people are being exposed to IR at different doses, including several dozen Gy during radiotherapy. Thus, supportive therapies for both patients and medical professionals are of first importance. Synthetic radioprotective compounds have a limited use because they often induce some undesirable side effects, especially at doses required to achieve maximal radioprotection. According to the research data presented in the review, melatonin could be the best candidate for a radioprotectant in people. Less is known about vitamin D. However, the results have been promising so far. The supporting supplementation with both substances seems to be also important in the context of common deficiencies in melatonin and vitamin D in modern societies, which may contribute to the severity of adverse side effects of medical IR exposure. Moreover, both substances have been found to selectively radiosensitize cancer cells, which makes them promising adjuvants for enhancing the anticancer effect of radiotherapy and improving therapeutic outcomes. Thus, in light of existing studies, melatonin and vitamin D are worth considering as agents for protecting professionals exposed to radiation and patients diagnosed or treated with radiation. Nevertheless, more research is needed in this area, especially in humans. Most importantly, appropriate doses of melatonin and vitamin D, effective in protecting against radiation and safe for people, should be established and tested in clinical trials.

Author Contributions: Conceptualization, J.N. and K.S.-G.; original draft preparation J.N. and K.S.-G.; formal analysis J.N., A.W. and K.S.-G.; visualisation, J.N. and K.S.-G.; review and editing, J.N., A.W. and K.S.-G.; editing and final approval, J.N., A.W. and K.S.-G. All authors have read and agreed to the published version of the manuscript.

Funding: This work received no external funding.

Conflicts of Interest: The authors declare no conflict of interest.

Abbreviations

$^1O_2$	singlet oxygen
3-OHM	3-hydroxymelatonin
4-HNE	trans-4-hydroxy-2-nonenal
AAAD	aromatic L-amino acid decarboxylase
AANAT	aralkylamine *N*-acetyltransferase
AFMK	N1-acetyl-N2-formyl-5-methoxykynuramine
AMK	N1-acetyl-5-methoxykynuramine
ASMT	acetylserotonin O-methyltransferase
CAT	catalase
CNS	central nervous system
CREB	cAMP response element-binding protein
DBP	vitamin D-binding protein
DCF	2′,7′-dichlorodihydrofluorescein
DCFH	2′,7′-dichlorofluorescin
DCFH-DA	2′,7′-dichlorofluorescin diacetate
ERKs	signal-regulated kinases
GPx	glutathione peroxidase
GR	glutathione reductase
GSH	glutathione
GSSG	glutathione disulfide
HUVEC	human umbilical vein endothelial cells
IP	intraperitoneal injection
IR	ionizing radiation
LET	linear energy transfer
MAPKs	mitogen-activated protein kinase pathway
MDA	malondialdehyde
Nrf2	nuclear factor-erythroid-2-related factor 2
$O_2^{\bullet-}$	superoxide anion radical
$OH^{\bullet}$	hydroxyl radical
ONOO-	peroxynitrite
PKC	protein kinase C
PLP	pyridoxal phosphate
PO	oral administration
RNS	reactive nitrogen species
ROS	reactive oxygen species
RXR	retinoid-X receptor
SAM	S-adenosyl methionine
SC	subcutaneous injection
SCN	suprachiasmatic nucleus
SOD	superoxide dismutase
TAC	total antioxidant capacity
TPH	tryptophan hydroxylase
VDR	vitamin D receptor

References

1. Bamgbose, B.O.; Suwaid, M.A.; Kaura, M.A.; Sugianto, I.; Hisatomi, M.; Asaumi, J. Current status of oral and maxillofacial radiology in West Africa. *Oral Radiol.* **2018**, *34*, 105–112. [CrossRef]
2. Hickling, S.; Xiang, L.; Jones, K.C.; Parodi, K.; Assmann, W.; Avery, S.; Hobson, M.; El Naqa, I. Ionizing radiation-induced acoustics for radiotherapy and diagnostic radiology applications. *Med. Phys.* **2018**, *45*, 707–721. [CrossRef] [PubMed]
3. Do, K.H. General Principles of Radiation Protection in Fields of Diagnostic Medical Exposure. *J. Korean Med. Sci.* **2016**, *31*, 6–9. [CrossRef] [PubMed]

4. Burgio, E.; Piscitelli, P.; Migliore, L. Ionizing Radiation and Human Health: Reviewing Models of Exposure and Mechanisms of Cellular Damage. An Epigenetic Perspective. *Int. J. Environ. Res. Public Health* **2018**, *15*, 1971. [CrossRef] [PubMed]
5. Abuelhia, E. Awareness of ionizing radiation exposure among junior doctors and senior medical students in radiological investigations. *J. Radiol. Prot.* **2017**, *37*, 59–67. [CrossRef]
6. Indriolo, N.; Neufeld, D.A.; Gerin, M.; Schilke, P.; Benz, A.O.; Winkel, B.; Menten, K.M.; Chambers, E.T.; Black, J.H.; Bruderer, S.; et al. Herschelsurvey of Galactic Oh+, H2O+, and H3O+: Probing the Molecular Hydrogen Fraction and Cosmic-Ray Ionization Rate. *Astrophys. J.* **2015**, *800*, 1–26. [CrossRef]
7. Zdrojewicz, Z.; Szlagor, A.; Wielogórska, M.; Nowakowska, D.; Nowakowski, J. Influence of ionizing radiation on human body. *Fam. Med. Prim. Care Rev.* **2016**, *2*, 174–179. [CrossRef]
8. Bassez, M.P. Water, air, Earth and cosmic radiation. *Orig. Life Evol. Biosph.* **2015**, *45*, 5–13. [CrossRef]
9. Baldwin, J.; Grantham, V. Radiation Hormesis: Historical and Current Perspectives. *J. Nucl. Med. Technol.* **2015**, *43*, 242–246. [CrossRef]
10. Jargin, S.V. Hormesis and radiation safety norms: Comments for an update. *Hum. Exp. Toxicol.* **2018**, *37*, 1233–1243. [CrossRef]
11. Shibamoto, Y.; Nakamura, H. Overview of Biological, Epidemiological, and Clinical Evidence of Radiation Hormesis. *Int. J. Mol. Sci.* **2018**, *19*, 2387. [CrossRef] [PubMed]
12. Buonanno, M.; de Toledo, S.M.; Pain, D.; Azzam, E.I. Long-term consequences of radiation-induced bystander effects depend on radiation quality and dose and correlate with oxidative stress. *Radiat. Res.* **2011**, *175*, 405–415. [CrossRef] [PubMed]
13. Thiagarajan, A.; Yamada, Y. Radiobiology and radiotherapy of brain metastases. *Clin. Exp. Metastasis* **2017**, *34*, 411–419. [CrossRef] [PubMed]
14. Kirsch, D.G.; Diehn, M.; Kesarwala, A.H.; Maity, A.; Morgan, M.A.; Schwarz, J.K.; Bristow, R.; Demaria, S.; Eke, I.; Griffin, R.J.; et al. The Future of Radiobiology. *J. Natl. Cancer Inst.* **2018**, *110*, 329–340. [CrossRef]
15. Cui, F.; Ma, N.; Han, X.; Chen, N.; Xi, Y.; Yuan, W.; Xu, Y.; Han, J.; Xu, X.; Tu, Y. Effects of 60 Co γ gamma Irradiation on the Reproductive Function of Caenorhabditis elegans. *Dose-Response* **2019**, *17*, 1–6. [CrossRef]
16. Santacruz-Gomez, K.; Sarabia-Sainz, A.; Acosta-Elias, M.; Sarabia-Sainz, M.; Janetanakit, W.; Khosla, N.; Melendrez, R.; Montero, M.P.; Lal, R. Antioxidant activity of hydrated carboxylated nanodiamonds and its influence on water gamma-radiolysis. *Nanotechnology* **2018**, *29*, 1–9. [CrossRef]
17. Reiter, R.J.; Mayo, J.C.; Tan, D.X.; Sainz, R.M.; Alatorre-Jimenez, M.; Qin, L. Melatonin as an antioxidant: Under promises but over delivers. *J. Pineal Res.* **2016**, *61*, 253–278. [CrossRef]
18. Sepidarkish, M.; Farsi, F.; Akbari-Fakhrabadi, M.; Namazi, N.; Almasi-Hashiani, A.; Maleki Hagiagha, A.; Heshmati, J. The effect of vitamin D supplementation on oxidative stress parameters: A systematic review and meta-analysis of clinical trials. *Pharmacol. Res.* **2019**, *139*, 141–152. [CrossRef]
19. Podzolkov, V.I.; Pokrovskaya, A.E.; Panasenko, O.I. Vitamin D deficiency and cardiovascular pathology. *Ter. Arkhiv* **2018**, *90*, 144–150. [CrossRef]
20. Touitou, Y.; Reinberg, A.; Touitou, D. Association between light at night, melatonin secretion, sleep deprivation, and the internal clock: Health impacts and mechanisms of circadian disruption. *Life Sci.* **2017**, *173*, 94–106. [CrossRef]
21. Shirazi, A.; Ghobadi, G.; Ghazi-Khansari, M. A radiobiological review on melatonin: A novel radioprotector. *J. Radiat. Res.* **2007**, *48*, 263–272. [CrossRef] [PubMed]
22. Vijayalaxmi; Reiter, R.J.; Tan, D.X.; Herman, T.S.; Thomas, C.R., Jr. Melatonin as a radioprotective agent: A review. *Int. J. Radiat. Oncol. Biol. Phys.* **2004**, *59*, 639–653. [CrossRef] [PubMed]
23. Zetner, D.; Andersen, L.P.; Rosenberg, J. Melatonin as Protection Against Radiation Injury: A Systematic Review. *Drug Res.* **2016**, *66*, 281–296. [CrossRef] [PubMed]
24. Savastano, S.; Barrea, L.; Savanelli, M.C.; Nappi, F.; Di Somma, C.; Orio, F.; Colao, A. Low vitamin D status and obesity: Role of nutritionist. *Rev. Endocr. Metab. Disord.* **2017**, *18*, 215–225. [CrossRef] [PubMed]
25. Issa, C.M. Vitamin D and Type 2 Diabetes Mellitus. *Adv. Exp. Med. Biol.* **2017**, *996*, 193–205. [CrossRef]
26. Christakos, S.; Dhawan, P.; Verstuyf, A.; Verlinden, L.; Carmeliet, G. Vitamin D: Metabolism, Molecular Mechanism of Action, and Pleiotropic Effects. *Physiol. Rev.* **2016**, *96*, 365–408. [CrossRef]
27. Alizadeh, E.; Orlando, T.M.; Sanche, L. Biomolecular damage induced by ionizing radiation: The direct and indirect effects of low-energy electrons on DNA. *Annu. Rev. Phys. Chem.* **2015**, *66*, 379–398. [CrossRef]

28. Castronuovo, D.; Sofo, A.; Lovelli, S.; Candido, V.; Scopa, A. Effects of UV-C radiation on common dandelion and purple coneflower: First results. *Int. J. Plant Biol.* **2017**, *8*, 61–64. [CrossRef]
29. Sgouros, G.; Hobbs, R.; Josefsson, A. Dosimetry and Radiobiology of Alpha-Particle Emitting Radionuclides. *Curr. Radiopharm.* **2018**, *11*, 209–214. [CrossRef]
30. Dell'Oro, S.; Marcocci, S.; Viel, M.; Vissani, F. Neutrinoless Double Beta Decay: 2015 Review. *Adv. High Energy Phys.* **2016**, *2016*, 1–37. [CrossRef]
31. Kozlovska, M.; Cerny, R.; Otahal, P. Attenuation of X and Gamma Rays in Personal Radiation Shielding Protective Clothing. *Health Phys.* **2015**, *109*, 205–211. [CrossRef] [PubMed]
32. Demasters, G.; Di, X.; Newsham, I.; Shiu, R.; Gewirtz, D.A. Potentiation of radiation sensitivity in breast tumor cells by the vitamin D3 analogue, EB 1089, through promotion of autophagy and interference with proliferative recovery. *Mol. Cancer Ther.* **2006**, *5*, 2786–2797. [CrossRef] [PubMed]
33. Hubenak, J.R.; Zhang, Q.; Branch, C.D.; Kronowitz, S.J. Mechanisms of injury to normal tissue after radiotherapy: A review. *Plast. Reconstr. Surg.* **2014**, *133*, 49–56. [CrossRef] [PubMed]
34. Takahashi, A.; Ikeda, H.; Yoshida, Y. Role of High-Linear Energy Transfer Radiobiology in Space Radiation Exposure Risks. *Int. J. Part Ther.* **2018**, *5*, 151–159. [CrossRef]
35. Sollazzo, A.; Shakeri-Manesh, S.; Fotouhi, A.; Czub, J.; Haghdoost, S.; Wojcik, A. Interaction of low and high LET radiation in TK6 cells-mechanistic aspects and significance for radiation protection. *J. Radiol. Prot.* **2016**, *36*, 721–735. [CrossRef]
36. Tharmalingam, S.; Sreetharan, S.; Kulesza, A.V.; Boreham, D.R.; Tai, T.C. Low-Dose Ionizing Radiation Exposure, Oxidative Stress and Epigenetic Programing of Health and Disease. *Radiat. Res.* **2017**, *188*, 525–538. [CrossRef]
37. Lorenzo-Gonzalez, M.; Torres-Duran, M.; Barbosa-Lorenzo, R.; Provencio-Pulla, M.; Barros-Dios, J.M.; Ruano-Ravina, A. Radon exposure: A major cause of lung cancer. *Expert. Rev. Respir. Med.* **2019**, *13*, 839–850. [CrossRef]
38. Acheva, A.; Haghdoost, S.; Sollazzo, A.; Launonen, V.; Kamarainen, M. Presence of Stromal Cells Enhances Epithelial-to-Mesenchymal Transition (EMT) Induction in Lung Bronchial Epithelium after Protracted Exposure to Oxidative Stress of Gamma Radiation. *Oxidative Med. Cell. Longev.* **2019**, *2019*, 1–14. [CrossRef]
39. Azzam, E.I.; Jay-Gerin, J.P.; Pain, D. Ionizing radiation-induced metabolic oxidative stress and prolonged cell injury. *Cancer Lett.* **2012**, *327*, 48–60. [CrossRef]
40. Sage, E.; Shikazono, N. Radiation-induced clustered DNA lesions: Repair and mutagenesis. *Free Radic. Biol. Med.* **2017**, *107*, 125–135. [CrossRef]
41. Sylvester, C.B.; Abe, J.I.; Patel, Z.S.; Grande-Allen, K.J. Radiation-Induced Cardiovascular Disease: Mechanisms and Importance of Linear Energy Transfer. *Front. Cardiovasc. Med.* **2018**, *5*, 1–9. [CrossRef]
42. Einor, D.; Bonisoli-Alquati, A.; Costantini, D.; Mousseau, T.A.; Moller, A.P. Ionizing radiation, antioxidant response and oxidative damage: A meta-analysis. *Sci. Total Environ.* **2016**, *548–549*, 463–471. [CrossRef] [PubMed]
43. Sies, H. Oxidative stress: A concept in redox biology and medicine. *Redox Biol.* **2015**, *4*, 180–183. [CrossRef] [PubMed]
44. Breitenbach, M.; Eckl, P. Introduction to Oxidative Stress in Biomedical and Biological Research. *Biomolecules* **2015**, *5*, 1169–1177. [CrossRef] [PubMed]
45. Filetti, F.M.; Vassallo, D.V.; Fioresi, M.; Simoes, M.R. Reactive oxygen species impair the excitation-contraction coupling of papillary muscles after acute exposure to a high copper concentration. *Toxicol. Vitro* **2018**, *51*, 106–113. [CrossRef] [PubMed]
46. Pospisil, P.; Prasad, A.; Rac, M. Mechanism of the Formation of Electronically Excited Species by Oxidative Metabolic Processes: Role of Reactive Oxygen Species. *Biomolecules* **2019**, *9*, 258. [CrossRef]
47. Georgiou, C.D.; Zisimopoulos, D.; Kalaitzopoulou, E.; Quinn, R.C. Radiation-Driven Formation of Reactive Oxygen Species in Oxychlorine-Containing Mars Surface Analogues. *Astrobiology* **2017**, *17*, 319–336. [CrossRef]
48. Leser, M.; Chapman, J.R.; Khine, M.; Pegan, J.; Law, M.; Makkaoui, M.E.; Ueberheide, B.M.; Brenowitz, M. Chemical Generation of Hydroxyl Radical for Oxidative 'Footprinting'. *Protein Pept. Lett.* **2019**, *26*, 61–69. [CrossRef]
49. Cheignon, C.; Tomas, M.; Bonnefont-Rousselot, D.; Faller, P.; Hureau, C.; Collin, F. Oxidative stress and the amyloid beta peptide in Alzheimer's disease. *Redox Biol.* **2018**, *14*, 450–464. [CrossRef]

50. Ahotupa, M. Oxidized lipoprotein lipids and atherosclerosis. *Free Radic. Res.* **2017**, *51*, 439–447. [CrossRef]
51. Gebicki, J.M. Oxidative stress, free radicals and protein peroxides. *Arch. Biochem. Biophys.* **2016**, *595*, 33–39. [CrossRef] [PubMed]
52. Hauck, A.K.; Huang, Y.; Hertzel, A.V.; Bernlohr, D.A. Adipose oxidative stress and protein carbonylation. *J. Biol. Chem.* **2019**, *294*, 1083–1088. [CrossRef] [PubMed]
53. Czarny, P.; Wigner, P.; Galecki, P.; Sliwinski, T. The interplay between inflammation, oxidative stress, DNA damage, DNA repair and mitochondrial dysfunction in depression. *Prog. Neuropsychopharmacol. Biol. Psychiatry* **2018**, *80*, 309–321. [CrossRef] [PubMed]
54. Bisht, S.; Faiq, M.; Tolahunase, M.; Dada, R. Oxidative stress and male infertility. *Nat. Rev. Urol.* **2017**, *14*, 470–485. [CrossRef] [PubMed]
55. Klaunig, J.E. Oxidative Stress and Cancer. *Curr. Pharm. Des.* **2018**, *24*, 4771–4778. [CrossRef]
56. Takahashi, K.; Okumura, H.; Guo, R.; Naruse, K. Effect of Oxidative Stress on Cardiovascular System in Response to Gravity. *Int. J. Mol. Sci.* **2017**, *18*, 1426. [CrossRef]
57. Manna, P.; Jain, S.K. Obesity, Oxidative Stress, Adipose Tissue Dysfunction, and the Associated Health Risks: Causes and Therapeutic Strategies. *Metab. Syndr. Relat. Disord.* **2015**, *13*, 423–444. [CrossRef]
58. Yaribeygi, H.; Panahi, Y.; Javadi, B.; Sahebkar, A. The Underlying Role of Oxidative Stress in Neurodegeneration: A Mechanistic Review. *CNS Neurol. Disord. Drug Targets* **2018**, *17*, 207–215. [CrossRef]
59. Sahiner, U.M.; Birben, E.; Erzurum, S.; Sackesen, C.; Kalayci, O. Oxidative stress in asthma: Part of the puzzle. *Pediatr. Allergy Immunol.* **2018**, *29*, 789–800. [CrossRef]
60. Torres-Cuevas, I.; Parra-Llorca, A.; Sanchez-Illana, A.; Nunez-Ramiro, A.; Kuligowski, J.; Chafer-Pericas, C.; Cernada, M.; Escobar, J.; Vento, M. Oxygen and oxidative stress in the perinatal period. *Redox Biol.* **2017**, *12*, 674–681. [CrossRef]
61. Wang, S.; He, G.; Chen, M.; Zuo, T.; Xu, W.; Liu, X. The Role of Antioxidant Enzymes in the Ovaries. *Oxidative Med. Cell. Longev.* **2017**, *2017*, 1–14. [CrossRef] [PubMed]
62. Veal, E.; Jackson, T.; Latimer, H. Role/s of 'Antioxidant' Enzymes in Ageing. *Subcell. Biochem.* **2018**, *90*, 425–450. [CrossRef] [PubMed]
63. Pisoschi, A.M.; Pop, A. The role of antioxidants in the chemistry of oxidative stress: A review. *Eur. J. Med. Chem.* **2015**, *97*, 55–74. [CrossRef] [PubMed]
64. Pingitore, A.; Lima, G.P.; Mastorci, F.; Quinones, A.; Iervasi, G.; Vassalle, C. Exercise and oxidative stress: Potential effects of antioxidant dietary strategies in sports. *Nutrition* **2015**, *31*, 916–922. [CrossRef]
65. Wimalawansa, S.J. Vitamin D Deficiency: Effects on Oxidative Stress, Epigenetics, Gene Regulation, and Aging. *Biology (Basel)* **2019**, *8*, 30. [CrossRef]
66. Siti, H.N.; Kamisah, Y.; Kamsiah, J. The role of oxidative stress, antioxidants and vascular inflammation in cardiovascular disease (a review). *Vasc. Pharm.* **2015**, *71*, 40–56. [CrossRef]
67. Prauchner, C.A. Oxidative stress in sepsis: Pathophysiological implications justifying antioxidant co-therapy. *Burns* **2017**, *43*, 471–485. [CrossRef]
68. Forrester, S.J.; Kikuchi, D.S.; Hernandes, M.S.; Xu, Q.; Griendling, K.K. Reactive Oxygen Species in Metabolic and Inflammatory Signaling. *Circ. Res.* **2018**, *122*, 877–902. [CrossRef]
69. Farhood, B.; Goradel, N.H.; Mortezaee, K.; Khanlarkhani, N.; Najafi, M.; Sahebkar, A. Melatonin and cancer: From the promotion of genomic stability to use in cancer treatment. *J. Cell. Physiol.* **2018**, *234*, 5613–5627. [CrossRef]
70. McBride, W.H.; Schaue, D. Radiation-induced tissue damage and response. *J. Pathol.* **2020**, *250*, 647–655. [CrossRef]
71. Yahyapour, R.; Motevaseli, E.; Rezaeyan, A.; Abdollahi, H.; Farhood, B.; Cheki, M.; Rezapoor, S.; Shabeeb, D.; Musa, A.E.; Najafi, M.; et al. Reduction-oxidation (redox) system in radiation-induced normal tissue injury: Molecular mechanisms and implications in radiation therapeutics. *Clin. Transl. Oncol.* **2018**, *20*, 975–988. [CrossRef] [PubMed]
72. Kang, J.A.; Yoon, S.H.; Rho, J.K.; Jang, B.S.; Choi, D.S.; Lee, D.E.; Byun, E.B.; Jeon, J.; Park, S.H. Radioprotective effect of hesperetin against gamma-irradiation-induced DNA damage and immune dysfunction in murine splenocytes. *Food Sci. Biotechnol.* **2016**, *25*, 163–168. [CrossRef]
73. Karimi, N.; Monfared, A.S.; Haddadi, G.H.; Soleymani, A.; Mohammadi, E.; Hajian-Tilaki, K.; Borzoueisileh, S. Radioprotective effect of hesperidin on reducing oxidative stress in the lens tissue of rats. *Int. J. Pharm. Investig.* **2017**, *7*, 149–154. [CrossRef] [PubMed]

74. Rezaeyan, A.; Haddadi, G.H.; Hosseinzadeh, M.; Moradi, M.; Najafi, M. Radioprotective effects of hesperidin on oxidative damages and histopathological changes induced by X-irradiation in rats heart tissue. *J. Med. Phys.* **2016**, *41*, 182–191. [CrossRef] [PubMed]
75. Shaban, N.Z.; Ahmed Zahran, A.M.; El-Rashidy, F.H.; Abdo Kodous, A.S. Protective role of hesperidin against gamma-radiation-induced oxidative stress and apoptosis in rat testis. *J. Biol. Res.-Thessal.* **2017**, *24*, 1–11. [CrossRef]
76. Tordjman, S.; Chokron, S.; Delorme, R.; Charrier, A.; Bellissant, E.; Jaafari, N.; Fougerou, C. Melatonin: Pharmacology, Functions and Therapeutic Benefits. *Curr. Neuropharmacol.* **2017**, *15*, 434–443. [CrossRef]
77. Zhang, H.M.; Zhang, Y. Melatonin: A well-documented antioxidant with conditional pro-oxidant actions. *J. Pineal Res.* **2014**, *57*, 131–146. [CrossRef]
78. Acuna-Castroviejo, D.; Escames, G.; Venegas, C.; Diaz-Casado, M.E.; Lima-Cabello, E.; Lopez, L.C.; Rosales-Corral, S.; Tan, D.X.; Reiter, R.J. Extrapineal melatonin: Sources, regulation, and potential functions. *Cell. Mol. Life Sci.* **2014**, *71*, 2997–3025. [CrossRef]
79. Lerner, A.B.; Case, J.D.; Takahashi, Y.; Lee, T.H.; Mori, W. Isolation of Melatonin, the Pineal Gland Factor That Lightens Melanocytes. *J. Am. Chem. Soc.* **1958**, *80*, 2587. [CrossRef]
80. Amaral, F.G.D.; Cipolla-Neto, J. A brief review about melatonin, a pineal hormone. *Arch. Endocrinol. Metab.* **2018**, *62*, 472–479. [CrossRef]
81. Nichols, D.E. N,N-dimethyltryptamine and the pineal gland: Separating fact from myth. *J. Psychopharmacol.* **2018**, *32*, 30–36. [CrossRef] [PubMed]
82. Cipolla-Neto, J.; Amaral, F.G.D. Melatonin as a Hormone: New Physiological and Clinical Insights. *Endocr. Rev.* **2018**, *39*, 990–1028. [CrossRef] [PubMed]
83. Talib, W.H. Melatonin and Cancer Hallmarks. *Molecules* **2018**, *23*, 518. [CrossRef] [PubMed]
84. Claustrat, B.; Leston, J. Melatonin: Physiological effects in humans. *Neurochirurgie* **2015**, *61*, 77–84. [CrossRef] [PubMed]
85. Reiter, R.J.; Tan, D.X.; Galano, A. Melatonin: Exceeding expectations. *Physiology* **2014**, *29*, 325–333. [CrossRef] [PubMed]
86. Waller, K.L.; Mortensen, E.L.; Avlund, K.; Fagerlund, B.; Lauritzen, M.; Gammeltoft, S.; Jennum, P. Melatonin and cortisol profiles in late midlife and their association with age-related changes in cognition. *Nat. Sci. Sleep* **2016**, *8*, 47–53. [CrossRef]
87. Kennaway, D.J. A critical review of melatonin assays: Past and present. *J. Pineal Res.* **2019**, *67*, 1–14. [CrossRef]
88. Pfeffer, M.; Korf, H.W.; Wicht, H. Synchronizing effects of melatonin on diurnal and circadian rhythms. *Gen. Comp. Endocrinol.* **2018**, *258*, 215–221. [CrossRef]
89. Giudice, A.; Crispo, A.; Grimaldi, M.; Polo, A.; Bimonte, S.; Capunzo, M.; Amore, A.; D'Arena, G.; Cerino, P.; Budillon, A.; et al. The Effect of Light Exposure at Night (LAN) on Carcinogenesis via Decreased Nocturnal Melatonin Synthesis. *Molecules* **2018**, *23*, 1308. [CrossRef]
90. Zisapel, N. New perspectives on the role of melatonin in human sleep, circadian rhythms and their regulation. *Br. J. Pharmacol.* **2018**, *175*, 3190–3199. [CrossRef]
91. Hardeland, R. Melatonin and the pathologies of weakened or dysregulated circadian oscillators. *J. Pineal Res.* **2017**, *62*, 1–16. [CrossRef]
92. Perez, S.; Murias, L.; Fernandez-Plaza, C.; Diaz, I.; Gonzalez, C.; Otero, J.; Diaz, E. Evidence for clock genes circadian rhythms in human full-term placenta. *Syst. Biol. Reprod. Med.* **2015**, *61*, 360–366. [CrossRef] [PubMed]
93. Jahanban-Esfahlan, R.; Mehrzadi, S.; Reiter, R.J.; Seidi, K.; Majidinia, M.; Baghi, H.B.; Khatami, N.; Yousefi, B.; Sadeghpour, A. Melatonin in regulation of inflammatory pathways in rheumatoid arthritis and osteoarthritis: Involvement of circadian clock genes. *Br. J. Pharmacol.* **2018**, *175*, 3230–3238. [CrossRef] [PubMed]
94. Vriend, J.; Reiter, R.J. Melatonin feedback on clock genes: A theory involving the proteasome. *J. Pineal Res.* **2015**, *58*, 1–11. [CrossRef]
95. Ma, N.; Zhang, J.; Reiter, R.J.; Ma, X. Melatonin mediates mucosal immune cells, microbial metabolism, and rhythm crosstalk: A therapeutic target to reduce intestinal inflammation. *Med. Res. Rev.* **2019**, *40*, 606–632. [CrossRef]
96. Emens, J.S.; Burgess, H.J. Effect of Light and Melatonin and Other Melatonin Receptor Agonists on Human Circadian Physiology. *Sleep Med. Clin.* **2015**, *10*, 435–453. [CrossRef]

97. Emet, M.; Ozcan, H.; Ozel, L.; Yayla, M.; Halici, Z.; Hacimuftuoglu, A. A Review of Melatonin, Its Receptors and Drugs. *Eurasian J. Med.* **2016**, *48*, 135–141. [CrossRef] [PubMed]
98. Liu, J.; Clough, S.J.; Hutchinson, A.J.; Adamah-Biassi, E.B.; Popovska-Gorevski, M.; Dubocovich, M.L. MT1 and MT2 Melatonin Receptors: A Therapeutic Perspective. *Annu. Rev. Pharmacol. Toxicol.* **2016**, *56*, 361–383. [CrossRef] [PubMed]
99. Ng, K.Y.; Leong, M.K.; Liang, H.; Paxinos, G. Melatonin receptors: Distribution in mammalian brain and their respective putative functions. *Brain Struct. Funct.* **2017**, *222*, 2921–2939. [CrossRef] [PubMed]
100. Favero, G.; Franceschetti, L.; Bonomini, F.; Rodella, L.F.; Rezzani, R. Melatonin as an Anti-Inflammatory Agent Modulating Inflammasome Activation. *Int. J. Endocrinol.* **2017**, *2017*, 1–13. [CrossRef]
101. Oishi, A.; Cecon, E.; Jockers, R. Melatonin Receptor Signaling: Impact of Receptor Oligomerization on Receptor Function. *Int. Rev. Cell. Mol. Biol.* **2018**, *338*, 59–77. [CrossRef]
102. Mortezaee, K.; Potes, Y.; Mirtavoos-Mahyari, H.; Motevaseli, E.; Shabeeb, D.; Musa, A.E.; Najafi, M.; Farhood, B. Boosting immune system against cancer by melatonin: A mechanistic viewpoint. *Life Sci.* **2019**, *238*, 1–8. [CrossRef] [PubMed]
103. Carrascal, L.; Nunez-Abades, P.; Ayala, A.; Cano, M. Role of Melatonin in the Inflammatory Process and its Therapeutic Potential. *Curr. Pharm. Des.* **2018**, *24*, 1563–1588. [CrossRef] [PubMed]
104. Hardeland, R. Aging, Melatonin, and the Pro- and Anti-Inflammatory Networks. *Int. J. Mol. Sci.* **2019**, *20*, 1223. [CrossRef] [PubMed]
105. Prado, N.J.; Ferder, L.; Manucha, W.; Diez, E.R. Anti-Inflammatory Effects of Melatonin in Obesity and Hypertension. *Curr. Hypertens. Rep.* **2018**, *20*, 1–12. [CrossRef]
106. Alghamdi, B.S. The neuroprotective role of melatonin in neurological disorders. *J. Neurosci. Res.* **2018**, *96*, 1136–1149. [CrossRef]
107. Shukla, M.; Govitrapong, P.; Boontem, P.; Reiter, R.J.; Satayavivad, J. Mechanisms of Melatonin in Alleviating Alzheimer's Disease. *Curr. Neuropharmacol.* **2017**, *15*, 1010–1031. [CrossRef]
108. Gelfand, A.A.; Goadsby, P.J. The Role of Melatonin in the Treatment of Primary Headache Disorders. *Headache* **2016**, *56*, 1257–1266. [CrossRef]
109. Mostafavi, S.A.; Akhondzadeh, S.; Mohammadi, M.R.; Keshtkar, A.A.; Hosseini, S.; Eshraghian, M.R.; Motlagh, T.A.; Alipour, R.; Keshavarz, S.A. Role of Melatonin in Body Weight: A Systematic Review and Meta-Analysis. *Curr. Pharm. Des.* **2017**, *23*, 3445–3452. [CrossRef]
110. Sharma, S.; Singh, H.; Ahmad, N.; Mishra, P.; Tiwari, A. The role of melatonin in diabetes: Therapeutic implications. *Arch. Endocrinol. Metab.* **2015**, *59*, 391–399. [CrossRef]
111. Karamitri, A.; Jockers, R. Melatonin in type 2 diabetes mellitus and obesity. *Nat. Rev. Endocrinol.* **2019**, *15*, 105–125. [CrossRef] [PubMed]
112. Nduhirabandi, F.; Maarman, G.J. Melatonin in Heart Failure: A Promising Therapeutic Strategy? *Molecules* **2018**, *23*, 1819. [CrossRef] [PubMed]
113. Manchester, L.C.; Coto-Montes, A.; Boga, J.A.; Andersen, L.P.; Zhou, Z.; Galano, A.; Vriend, J.; Tan, D.X.; Reiter, R.J. Melatonin: An ancient molecule that makes oxygen metabolically tolerable. *J. Pineal Res.* **2015**, *59*, 403–419. [CrossRef]
114. Reiter, R.J.; Rosales-Corral, S.; Tan, D.X.; Jou, M.J.; Galano, A.; Xu, B. Melatonin as a mitochondria-targeted antioxidant: One of evolution's best ideas. *Cell. Mol. Life Sci.* **2017**, *74*, 3863–3881. [CrossRef] [PubMed]
115. Tan, D.X.; Manchester, L.C.; Esteban-Zubero, E.; Zhou, Z.; Reiter, R.J. Melatonin as a Potent and Inducible Endogenous Antioxidant: Synthesis and Metabolism. *Molecules* **2015**, *20*, 18886–18906. [CrossRef] [PubMed]
116. Karaaslan, C.; Suzen, S. Antioxidant properties of melatonin and its potential action in diseases. *Curr. Top. Med. Chem.* **2015**, *15*, 894–903. [CrossRef] [PubMed]
117. Osier, N.; McGreevy, E.; Pham, L.; Puccio, A.; Ren, D.; Conley, Y.P.; Alexander, S.; Dixon, C.E. Melatonin as a Therapy for Traumatic Brain Injury: A Review of Published Evidence. *Int. J. Mol. Sci.* **2018**, *19*, 1539. [CrossRef]
118. Asghari, M.H.; Moloudizargari, M.; Bahadar, H.; Abdollahi, M. A review of the protective effect of melatonin in pesticide-induced toxicity. *Expert Opin. Drug Metab. Toxicol.* **2017**, *13*, 545–554. [CrossRef]
119. Vishnoi, S.; Raisuddin, S.; Parvez, S. Glutamate Excitotoxicity and Oxidative Stress in Epilepsy: Modulatory Role of Melatonin. *J. Environ. Pathol. Toxicol. Oncol.* **2016**, *35*, 365–374. [CrossRef]

120. Jaworek, J.; Szklarczyk, J.; Bonior, J.; Kot, M.; Goralska, M.; Pierzchalski, P.; Reiter, R.J.; Czech, U.; Tomaszewska, R. Melatonin metabolite, N(1)-acetyl-N(1)-formyl-5-methoxykynuramine (AFMK), attenuates acute pancreatitis in the rat: In vivo and in vitro studies. *J. Physiol. Pharm. Off. J. Pol. Physiol. Soc.* **2016**, *67*, 411–421.
121. Galano, A.; Tan, D.X.; Reiter, R.J. On the free radical scavenging activities of melatonin's metabolites, AFMK and AMK. *J. Pineal Res.* **2013**, *54*, 245–257. [CrossRef] [PubMed]
122. Martinez, G.R.; Almeida, E.A.; Klitzke, C.F.; Onuki, J.; Prado, F.M.; Medeiros, M.H.; Di Mascio, P. Measurement of melatonin and its metabolites: Importance for the evaluation of their biological roles. *Endocrine* **2005**, *27*, 111–118. [CrossRef]
123. Hardeland, R. Melatonin and the electron transport chain. *Cell. Mol. Life Sci.* **2017**, *74*, 3883–3896. [CrossRef] [PubMed]
124. He, R.; Cui, M.; Lin, H.; Zhao, L.; Wang, J.; Chen, S.; Shao, Z. Melatonin resists oxidative stress-induced apoptosis in nucleus pulposus cells. *Life Sci.* **2018**, *199*, 122–130. [CrossRef] [PubMed]
125. Fernandez-Gil, B.; Moneim, A.E.; Ortiz, F.; Shen, Y.Q.; Soto-Mercado, V.; Mendivil-Perez, M.; Guerra-Librero, A.; Acuna-Castroviejo, D.; Molina-Navarro, M.M.; Garcia-Verdugo, J.M.; et al. Melatonin protects rats from radiotherapy-induced small intestine toxicity. *PLoS ONE* **2017**, *12*, e0174474. [CrossRef] [PubMed]
126. Gurses, I.; Ozeren, M.; Serin, M.; Yucel, N.; Erkal, H.S. Histopathological evaluation of melatonin as a protective agent in heart injury induced by radiation in a rat model. *Pathol. Res. Pract.* **2014**, *210*, 863–871. [CrossRef]
127. Haddadi, G.; Shirazi, A.; Sepehrizadeh, Z.; Mahdavi, S.R.; Haddadi, M. Radioprotective effect of melatonin on the cervical spinal cord in irradiated rats. *Cell J.* **2013**, *14*, 246–253.
128. Sharma, S.; Haldar, C.; Chaube, S.K. Effect of exogenous melatonin on X-ray induced cellular toxicity in lymphatic tissue of Indian tropical male squirrel, Funambulus pennanti. *Int. J. Radiat. Biol.* **2008**, *84*, 363–374. [CrossRef]
129. Shirazi, A.; Haddadi, G.H.; Asadi-Amoli, F.; Sakhaee, S.; Ghazi-Khansari, M.; Avand, A. Radioprotective effect of melatonin in reducing oxidative stress in rat lenses. *Cell J.* **2011**, *13*, 79–82. [CrossRef]
130. Take, G.; Erdogan, D.; Helvacioglu, F.; Goktas, G.; Ozbey, G.; Uluoglu, C.; Yucel, B.; Guney, Y.; Hicsonmez, A.; Ozkan, S. Effect of melatonin and time of administration on irradiation-induced damage to rat testes. *Braz. J. Med. Biol. Res.* **2009**, *42*, 621–628. [CrossRef]
131. Alicelebic, S.; Mornjakovic, Z.; Susko, I.; Cosovic, E.; Beganovic-Petrovic, A. The role of pineal gland and exogenous melatonin on the irradiation stress response of suprarenal gland. *Bosn. J. Basic Med. Sci.* **2006**, *6*, 18–21. [CrossRef] [PubMed]
132. Bhatia, A.L.; Manda, K. Study on pre-treatment of melatonin against radiation-induced oxidative stress in mice. *Environ. Toxicol. Pharmacol.* **2004**, *18*, 13–20. [CrossRef] [PubMed]
133. Erol, F.S.; Topsakal, C.; Ozveren, M.F.; Kaplan, M.; Ilhan, N.; Ozercan, I.H.; Yildiz, O.G. Protective effects of melatonin and vitamin E in brain damage due to gamma radiation: An experimental study. *Neurosurg. Rev.* **2004**, *27*, 65–69. [CrossRef]
134. Sener, G.; Jahovic, N.; Tosun, O.; Atasoy, B.M.; Yegen, B.C. Melatonin ameliorates ionizing radiation-induced oxidative organ damage in rats. *Life Sci.* **2003**, *74*, 563–572. [CrossRef] [PubMed]
135. Taysi, S.; Koc, M.; Buyukokuroglu, M.E.; Altinkaynak, K.; Sahin, Y.N. Melatonin reduces lipid peroxidation and nitric oxide during irradiation-induced oxidative injury in the rat liver. *J. Pineal Res.* **2003**, *34*, 173–177. [CrossRef]
136. Vasin, M.V.; Ushakov, I.B.; Kovtun, V.Y.; Semenova, L.A.; Koroleva, L.V.; Galkin, A.A.; Afanas'ev, R.V. Therapeutic effect of long-term melatonin treatment on the course and fatal outcome of modeled acute radiation sickness. *Bull. Exp. Biol. Med.* **2014**, *156*, 776–777. [CrossRef]
137. Vijayalaxmi; Reiter, R.J.; Herman, T.S.; Meltz, M.L. Melatonin reduces gamma radiation-induced primary DNA damage in human blood lymphocytes. *Mutat. Res.* **1998**, *397*, 203–208. [CrossRef]
138. Fernandez-Gil, B.I.; Guerra-Librero, A.; Shen, Y.Q.; Florido, J.; Martinez-Ruiz, L.; Garcia-Lopez, S.; Adan, C.; Rodriguez-Santana, C.; Acuna-Castroviejo, D.; Quinones-Hinojosa, A.; et al. Melatonin Enhances Cisplatin and Radiation Cytotoxicity in Head and Neck Squamous Cell Carcinoma by Stimulating Mitochondrial ROS Generation, Apoptosis, and Autophagy. *Oxidative Med. Cell. Longev.* **2019**, *2019*, 1–13. [CrossRef]
139. Griffin, F.; Marignol, L. Therapeutic potential of melatonin for breast cancer radiation therapy patients. *Int. J. Radiat. Biol.* **2018**, *94*, 472–477. [CrossRef]

140. Wang, Q.; Sun, Z.; Du, L.; Xu, C.; Wang, Y.; Yang, B.; He, N.; Wang, J.; Ji, K.; Liu, Y.; et al. Melatonin Sensitizes Human Colorectal Cancer Cells to gamma-ray Ionizing Radiation In Vitro and In Vivo. *Int. J. Mol. Sci.* **2018**, *19*, 3974. [CrossRef]
141. Ratheesh, V.; Subramanian, S.; Prakash, P.S.G.; Victor, D.J. Evaluation of Association of Vitamin D Receptor Genetic Polymorphism with Severe Chronic Periodontitis in an Ethnic Tamilian Population. *Genet. Test. Mol. Biomark.* **2018**, *22*, 615–621. [CrossRef] [PubMed]
142. Holick, M.F. The vitamin D deficiency pandemic: Approaches for diagnosis, treatment and prevention. *Rev. Endocr. Metab. Disord.* **2017**, *18*, 153–165. [CrossRef] [PubMed]
143. Mellanby, E. An Experimental Investigation On Rickets. *Lancet* **1919**, *193*, 407–412. [CrossRef] [PubMed]
144. McCollum, E.V.; Simmonds, N.; Becker, J.E.; Shipley, P. Studies on experimental rickets XXI. An experimental demonstration of the existence of a vitamin which promotes calcium deposition. *J. Biol. Chem.* **1922**, *53*, 293–312.
145. Windaus, A.; Schenck, F.; Werder, F. Über das antirachitisch wirksame Bestrahlungsprodukt ans 7-Dehydro-cholesterin. *Hoppe-Seyler's Zeitschrift für physiologische Chemie* **1936**, *241*, 100–103. [CrossRef]
146. Jones, G. The discovery and synthesis of the nutritional factor vitamin D. *Int. J. Paleopathol.* **2018**, *23*, 96–99. [CrossRef]
147. Wilson, L.R.; Tripkovic, L.; Hart, K.H.; Lanham-New, S.A. Vitamin D deficiency as a public health issue: Using vitamin D2 or vitamin D3 in future fortification strategies. *Proc. Nutr. Soc.* **2017**, *76*, 392–399. [CrossRef]
148. Borel, P.; Caillaud, D.; Cano, N.J. Vitamin D bioavailability: State of the art. *Crit. Rev. Food Sci. Nutr.* **2015**, *55*, 1193–1205. [CrossRef]
149. Cardwell, G.; Bornman, J.F.; James, A.P.; Black, L.J. A Review of Mushrooms as a Potential Source of Dietary Vitamin D. *Nutrients* **2018**, *10*, 1498. [CrossRef]
150. Duffy, S.K.; O'Doherty, J.V.; Rajauria, G.; Clarke, L.C.; Hayes, A.; Dowling, K.G.; O'Grady, M.N.; Kerry, J.P.; Jakobsen, J.; Cashman, K.D.; et al. Vitamin D-biofortified beef: A comparison of cholecalciferol with synthetic versus UVB-mushroom-derived ergosterol as feed source. *Food Chem.* **2018**, *256*, 18–24. [CrossRef]
151. Vaes, A.M.M.; Brouwer-Brolsma, E.M.; van der Zwaluw, N.L.; van Wijngaarden, J.P.; Berendsen, A.A.M.; van Schoor, N.; van der Velde, N.; Uitterlinden, A.; Lips, P.; Dhonukshe-Rutten, R.A.M.; et al. Food sources of vitamin D and their association with 25-hydroxyvitamin D status in Dutch older adults. *J. Steroid Biochem. Mol. Biol.* **2017**, *173*, 228–234. [CrossRef] [PubMed]
152. Jeon, S.M.; Shin, E.A. Exploring vitamin D metabolism and function in cancer. *Exp. Mol. Med.* **2018**, *50*, 1–14. [CrossRef] [PubMed]
153. Smolensky, M.H.; Sackett-Lundeen, L.L.; Portaluppi, F. Nocturnal light pollution and underexposure to daytime sunlight: Complementary mechanisms of circadian disruption and related diseases. *Chronobiol. Int.* **2015**, *32*, 1029–1048. [CrossRef] [PubMed]
154. Juzeniene, A.; Grigalavicius, M.; Juraleviciute, M.; Grant, W.B. Phototherapy and vitamin D. *Clin. Dermatol.* **2016**, *34*, 548–555. [CrossRef] [PubMed]
155. Duchow, E.G.; Cooke, N.E.; Seeman, J.; Plum, L.A.; DeLuca, H.F. Vitamin D binding protein is required to utilize skin-generated vitamin D. *Proc. Natl. Acad. Sci. USA* **2019**, *116*, 24527–24532. [CrossRef]
156. Denburg, M.R.; Bhan, I. Vitamin D-Binding Protein in Health and Chronic Kidney Disease. *Semin. Dial.* **2015**, *28*, 636–644. [CrossRef]
157. Bahrami, A.; Sadeghnia, H.R.; Tabatabaeizadeh, S.A.; Bahrami-Taghanaki, H.; Behboodi, N.; Esmaeili, H.; Ferns, G.A.; Mobarhan, M.G.; Avan, A. Genetic and epigenetic factors influencing vitamin D status. *J. Cell. Physiol.* **2018**, *233*, 4033–4043. [CrossRef]
158. Jean, G.; Souberbielle, J.C.; Chazot, C. Vitamin D in Chronic Kidney Disease and Dialysis Patients. *Nutrients* **2017**, *9*, 328. [CrossRef]
159. Abbas, M.A. Physiological functions of Vitamin D in adipose tissue. *J. Steroid Biochem. Mol. Biol.* **2017**, *165*, 369–381. [CrossRef]
160. Gil, A.; Plaza-Diaz, J.; Mesa, M.D. Vitamin D: Classic and Novel Actions. *Ann. Nutr. Metab.* **2018**, *72*, 87–95. [CrossRef]
161. Deuster, E.; Jeschke, U.; Ye, Y.; Mahner, S.; Czogalla, B. Vitamin D and VDR in Gynecological Cancers-A Systematic Review. *Int. J. Mol. Sci.* **2017**, *18*, 2328. [CrossRef]
162. Marino, R.; Misra, M. Extra-Skeletal Effects of Vitamin D. *Nutrients* **2019**, *11*, 1460. [CrossRef] [PubMed]

163. Christakos, S.; Veldurthy, V.; Patel, N.; Wei, R. Intestinal Regulation of Calcium: Vitamin D and Bone Physiology. *Adv. Exp. Med. Biol.* **2017**, *1033*, 3–12. [CrossRef] [PubMed]
164. Bouillon, R.; Carmeliet, G. Vitamin D insufficiency: Definition, diagnosis and management. *Best Pract. Res. Clin. Endocrinol. Metab.* **2018**, *32*, 669–684. [CrossRef] [PubMed]
165. Parker, G.B.; Brotchie, H.; Graham, R.K. Vitamin D and depression. *J. Affect. Disord.* **2017**, *208*, 56–61. [CrossRef] [PubMed]
166. Pilz, S.; Verheyen, N.; Grubler, M.R.; Tomaschitz, A.; Marz, W. Vitamin D and cardiovascular disease prevention. *Nat. Rev. Cardiol.* **2016**, *13*, 404–417. [CrossRef] [PubMed]
167. Wang, J.; Lv, S.; Chen, G.; Gao, C.; He, J.; Zhong, H.; Xu, Y. Meta-analysis of the association between vitamin D and autoimmune thyroid disease. *Nutrients* **2015**, *7*, 2485–2498. [CrossRef]
168. Ferrari, D.; Lombardi, G.; Banfi, G. Concerning the vitamin D reference range: Pre-analytical and analytical variability of vitamin D measurement. *Biochem. Med.* **2017**, *27*, 1–14. [CrossRef]
169. Pludowski, P.; Holick, M.F.; Grant, W.B.; Konstantynowicz, J.; Mascarenhas, M.R.; Haq, A.; Povoroznyuk, V.; Balatska, N.; Barbosa, A.P.; Karonova, T.; et al. Vitamin D supplementation guidelines. *J. Steroid Biochem. Mol. Biol.* **2018**, *175*, 125–135. [CrossRef]
170. Tagliaferri, S.; Porri, D.; De Giuseppe, R.; Manuelli, M.; Alessio, F.; Cena, H. The controversial role of vitamin D as an antioxidant: Results from randomised controlled trials. *Nutr. Res. Rev.* **2019**, *32*, 99–105. [CrossRef]
171. Hajiluian, G.; Abbasalizad Farhangi, M.; Nameni, G.; Shahabi, P.; Megari-Abbasi, M. Oxidative stress-induced cognitive impairment in obesity can be reversed by vitamin D administration in rats. *Nutr. Neurosci.* **2018**, *21*, 744–752. [CrossRef] [PubMed]
172. Jagoda, S.V.; Dixon, K.M. Protective effects of 1,25 dihydroxyvitamin D3 and its analogs on ultraviolet radiation-induced oxidative stress: A review. *Redox Rep.* **2020**, *25*, 11–16. [CrossRef] [PubMed]
173. Tang, L.; Fang, W.; Lin, J.; Li, J.; Wu, W.; Xu, J. Vitamin D protects human melanocytes against oxidative damage by activation of Wnt/beta-catenin signaling. *Lab. Investig.* **2018**, *98*, 1527–1537. [CrossRef] [PubMed]
174. Jain, S.K.; Micinski, D. Vitamin D upregulates glutamate cysteine ligase and glutathione reductase, and GSH formation, and decreases ROS and MCP-1 and IL-8 secretion in high-glucose exposed U937 monocytes. *Biochem. Biophys. Res. Commun.* **2013**, *437*, 7–11. [CrossRef]
175. Dzik, K.; Skrobot, W.; Flis, D.J.; Karnia, M.; Libionka, W.; Kloc, W.; Kaczor, J.J. Vitamin D supplementation attenuates oxidative stress in paraspinal skeletal muscles in patients with low back pain. *Eur. J. Appl. Physiol.* **2018**, *118*, 143–151. [CrossRef]
176. Chen, L., Yang, R ; Qiao, W.; Yuan, X.; Wang, S.; Goltzman, D.; Miao, D. 1,25-Dihydroxy vitamin D prevents tumorigenesis by inhibiting oxidative stress and inducing tumor cellular senescence in mice. *Int. J. Cancer* **2018**, *143*, 368–382. [CrossRef]
177. Sepehrmanesh, Z.; Kolahdooz, F.; Abedi, F.; Mazroii, N.; Assarian, A.; Asemi, Z.; Esmaillzadeh, A. Vitamin D Supplementation Affects the Beck Depression Inventory, Insulin Resistance, and Biomarkers of Oxidative Stress in Patients with Major Depressive Disorder: A Randomized, Controlled Clinical Trial. *J. Nutr.* **2016**, *146*, 243–248. [CrossRef]
178. Barzegari, M.; Sarbakhsh, P.; Mobasseri, M.; Noshad, H.; Esfandiari, A.; Khodadadi, B.; Gargari, B.P. The effects of vitamin D supplementation on lipid profiles and oxidative indices among diabetic nephropathy patients with marginal vitamin D status. *Diabetes Metab. Syndr. Clin. Res. Rev.* **2019**, *13*, 542–547. [CrossRef]
179. Hayes, D.P. The protection afforded by vitamin D against low radiation damage. *Int. J. Low Radiat.* **2008**, *5*, 368–394. [CrossRef]
180. Starikovich, L.S.; Aragon, G.A.; Vernikovska, Y.I.; Vigovska, T.V.; Veliky, M.M. Effect of a vitamin D3-based nutritional supplement ('Videchol') on carbohydrate metabolism of rats following chronic low dose-rate irradiation. *J. Radiol. Prot.* **2001**, *21*, 269–276. [CrossRef]
181. Muller, K.; Schinn, M.; Reichrath, J.; Meineke, V. 1alpha,25-Dihydroxyvitamin D3 modulates the response of human keratinocytes to ionizing radiation exposure. *Anticancer Res.* **2006**, *26*, 2735–2741. [PubMed]
182. Langberg, M.; Rotem, C.; Fenig, E.; Koren, R.; Ravid, A. Vitamin D protects keratinocytes from deleterious effects of ionizing radiation. *Br. J. Dermatol.* **2009**, *160*, 151–161. [CrossRef] [PubMed]
183. Tremezaygues, L.; Seifert, M.; Vogt, T.; Tilgen, W.; Reichrath, J. 1,25-Dihydroxyvitamin D3 modulates effects of ionizing radiation (IR) on human keratinocytes: In vitro analysis of cell viability/proliferation, DNA-damage and -repair. *J. Steroid Biochem. Mol. Biol.* **2010**, *121*, 324–327. [CrossRef]

184. Marampon, F.; Gravina, G.L.; Festuccia, C.; Popov, V.M.; Colapietro, E.A.; Sanita, P.; Musio, D.; De Felice, F.; Lenzi, A.; Jannini, E.A.; et al. Vitamin D protects endothelial cells from irradiation-induced senescence and apoptosis by modulating MAPK/SirT1 axis. *J. Endocrinol. Investig.* **2016**, *39*, 411–422. [CrossRef] [PubMed]
185. Bristol, M.L.; Di, X.; Beckman, M.J.; Wilson, E.N.; Henderson, S.C.; Maiti, A.; Fan, Z.; Gewirtz, D.A. Dual functions of autophagy in the response of breast tumor cells to radiation: Cytoprotective autophagy with radiation alone and cytotoxic autophagy in radiosensitization by vitamin D 3. *Autophagy* **2012**, *8*, 739–753. [CrossRef]
186. Jeong, Y.; Swami, S.; Krishnan, A.V.; Williams, J.D.; Martin, S.; Horst, R.L.; Albertelli, M.A.; Feldman, B.J.; Feldman, D.; Diehn, M. Inhibition of Mouse Breast Tumor-Initiating Cells by Calcitriol and Dietary Vitamin D. *Mol. Cancer Ther.* **2015**, *14*, 1951–1961. [CrossRef]
187. Polar, M.K.; Gennings, C.; Park, M.; Gupta, M.S.; Gewirtz, D.A. Effect of the vitamin D3 analog ILX 23-7553 on apoptosis and sensitivity to fractionated radiation in breast tumor cells and normal human fibroblasts. *Cancer Chemother. Pharmacol.* **2003**, *51*, 415–421. [CrossRef]
188. Sharma, K.; Goehe, R.W.; Di, X.; Hicks, M.A., 2nd; Torti, S.V.; Torti, F.M.; Harada, H.; Gewirtz, D.A. A novel cytostatic form of autophagy in sensitization of non-small cell lung cancer cells to radiation by vitamin D and the vitamin D analog, EB 1089. *Autophagy* **2014**, *10*, 2346–2361. [CrossRef]
189. Wilson, E.N.; Bristol, M.L.; Di, X.; Maltese, W.A.; Koterba, K.; Beckman, M.J.; Gewirtz, D.A. A switch between cytoprotective and cytotoxic autophagy in the radiosensitization of breast tumor cells by chloroquine and vitamin D. *Horm. Cancer* **2011**, *2*, 272–285. [CrossRef]
190. Tissandie, E.; Gueguen, Y.; Lobaccaro, J.M.; Aigueperse, J.; Gourmelon, P.; Paquet, F.; Souidi, M. Chronic contamination with 137Cesium affects Vitamin D3 metabolism in rats. *Toxicology* **2006**, *225*, 75–80. [CrossRef]
191. Kaminskyi, O.V.; Pankiv, V.I.; Pankiv, I.V.; Afanasyev, D.E. Vitamin D content in population of radiologically contaminated areas in chernivtsi oblast (pilot project). *Probl. Radiac. Med. Radiobiol.* **2018**, *23*, 442–451. [CrossRef] [PubMed]

International Journal of
Molecular Sciences

MDPI

Article

Molecular Investigation on a Triple Negative Breast Cancer Xenograft Model Exposed to Proton Beams

Francesco P. Cammarata [1,2,†], Giusi I. Forte [1,2,†], Giuseppe Broggi [3], Valentina Bravatà [1,2,*], Luigi Minafra [1,2], Pietro Pisciotta [2,4], Marco Calvaruso [1,2], Roberta Tringali [5], Barbara Tomasello [5], Filippo Torrisi [2,6], Giada Petringa [2], Giuseppe A. P. Cirrone [2], Giacomo Cuttone [2], Rosaria Acquaviva [5], Rosario Caltabiano [3] and Giorgio Russo [1,2]

1 Institute of Molecular Bioimaging and Physiology (IBFM-CNR), 90015 Cefalù (Palermo), Italy; francesco.cammarata@ibfm.cnr.it (F.P.C.); giusi.forte@ibfm.cnr.it (G.I.F.); luigi.minafra@ibfm.cnr.it (L.M.); marco.calvaruso@ibfm.cnr.it (M.C.); giorgio.russo@ibfm.cnr.it (G.R.)

2 National Laboratory of South, National Institute for Nuclear Physics (LNS-INFN), 95123 Catania, Italy; pietro.pisciotta@lns.infn.it (P.P.); filippo.torrisi@unict.it (F.T.); giada.petringa@lns.infn.it (G.P.); pablo.cirrone@lns.infn.it (G.A.P.C.); cuttone@lns.infn.it (G.C.)

3 Department of Medical, Surgical and Advanced Technological Sciences "Gian Filippo Ingrassia", Section of Anatomic Pathology, University of Catania, 95123 Catania, Italy; giuseppe.broggi@gmail.com (G.B.); rosario.caltabiano@unict.it (R.C.)

4 Department of Radiation Oncology, University Medical Center Groningen, 9713 Groningen, The Netherlands

5 Department of Drug Science, Section of Biochemistry, University of Catania, 95125 Catania, Italy; roberta.tringali@live.it (R.T.); btomase@unict.it (B.T.); racquavi@unict.it (R.A.)

6 Department of Biomedical and Biotechnological Sciences (BIOMETEC), University of Catania, 95124 Catania, Italy

* Correspondence: valentina.bravata@ibfm.cnr.it

† These authors contributed equally to this work.

Received: 23 June 2020; Accepted: 29 August 2020; Published: 1 September 2020

Abstract: Specific breast cancer (BC) subtypes are associated with bad prognoses due to the absence of successful treatment plans. The triple-negative breast cancer (TNBC) subtype, with estrogen (ER), progesterone (PR) and human epidermal growth factor-2 (HER2) negative receptor status, is a clinical challenge for oncologists, because of its aggressiveness and the absence of effective therapies. In addition, proton therapy (PT) represents an effective treatment against both inaccessible area located or conventional radiotherapy (RT)-resistant cancers, becoming a promising therapeutic choice for TNBC. Our study aimed to analyze the in vivo molecular response to PT and its efficacy in a MDA-MB-231 TNBC xenograft model. TNBC xenograft models were irradiated with 2, 6 and 9 Gy of PT. Gene expression profile (GEP) analyses and immunohistochemical assay (IHC) were performed to highlight specific pathways and key molecules involved in cell response to the radiation. GEP analysis revealed in depth the molecular response to PT, showing a considerable immune response, cell cycle and stem cell process regulation. Only the dose of 9 Gy shifted the balance toward pro-death signaling as a dose escalation which can be easily performed using proton beams, which permit targeting tumors while avoiding damage to the surrounding healthy tissue.

Keywords: triple-negative breast cancer (TNBC); proton therapy; xenograft mice; microarray

1. Introduction

Breast cancer (BC) is a heterogeneous disease classified in several subgroups based on molecular and genomic profiles, associated with different treatment responses [1,2]. So, following a BC diagnosis, the most immediate challenge in patient management is the prognosis determination and identification of the most appropriate therapeutic approach. Moreover, specific BC subtypes are associated with a worse prognosis due to the absence of successful treatment plans. In this sense, the triple-negative breast cancer (TNBC) subtype, which represents 15–20% of BC incidence, is a clinical challenge for oncologists, because of its aggressiveness and absence of successful therapies. Considering that TNBC is featured by estrogen (ER), progesterone (PR) and human epidermal growth factor-2 (HER2) negative receptor status, patients cannot be treated with specific hormonal or targeted therapy, and a gold standard chemotherapy has not been established yet [3]. In addition, considering the molecular differences among BC subtypes, the choice of a unique treatment plan common to all BC patients, including radiotherapy (RT), may not be the best option. Moreover, taking into account a relapse rate ranging from 25% to 40% after treatment of TNBC, the acquisition of chemo and radioresistance is taken into consideration. However, several authors have demonstrated that breast conserving surgery in tandem with RT was less likely to develop locoregional recurrence compared with mastectomy [4–6].

Over the last decade, the technological development of RT has led to more performing and innovative technologies which can deliver, with high precision, increasing doses saving the organ at risk, and high-dose radiation on small-sized tumor targets [1,7,8]. This property of the proton beam is due to its typical curve of energy deposition through the matter, releasing the well-known Bragg peak, which can represent a better conformational option in respect to conventional photon beams [3].

Then, proton therapy (PT) provides a substantial physical advantage compared to conventional RT by using X or Gamma radiation rays, even if its biological advantages still remain understudied.

However, in BC treatment, the prospective use of PT in place of conventional RT would result in a lower ionizing radiation (IR) dose to the heart and lungs, especially if the tumor is located in the left mammary gland [9–14]. Furthermore, several authors have debated the use of conventional RT for TNBC patients [11,15,16]. Because of such advantages and due to clinically encouraging results, PT is currently used for different aggressive forms of cancers and its use is rapidly growing, especially for those sited in the proximity of organs at risk. However, few data are available regarding proton-induced molecular changes, particularly in mammary gland breast cells, a topic that therefore deserves to be accurately described.

In this scenario, we have recently described the dose response effects on cell survival induced by proton beam irradiation by using an in vitro approach, and, for the first time to our knowledge, the radiation-induced gene expression profiles (GEPs) and immunological profiles produced by the BC cell lines with different aggressive phenotypes including the well-known TNBC model, the MDA-MB-231 BC cells (metastatic, basal, triple-negative) [7].

Here we decided to generate a TNBC xenograft model using MDA-MB-231, with the aim to test the in vivo PT efficacy and to search for specific PT molecular signature, using increasing doses of proton beam (2, 6 and 9 Gy) [17]. The dose range of 2–9 Gy was chosen with the aim of understanding the different molecular response observable with the dose escalation. Furthermore, the dose of 2 and 9 Gy were precedingly used in our in vitro experiments, allowing us to perform a GEP analysis comparison between in vitro and in vivo molecular responses induced by the same PT dose. In addition, an immunohistochemical (IHC) characterization was performed to search for particular markers of specific biological processes.

The results of this comparison show the activation of specific pathways associated with the dose delivered and the time window analyzed in the MDA-MB-231 xenograft model, which sustains the control of key cellular processes, such as stem cell proliferation, cell cycle and cell death, and is thus able to control cell fate. In particular, the major PT efficacy is evident with the dose of 9 Gy, in which the survival/cell death balance is clearly moved toward tumor cell killing. These findings encourage the use of proton beams for the treatment of more aggressive tumor forms because its specific energy deposition curve allows the use of dose escalation while saving organs at risk.

2. Results

2.1. Monte Carlo Simulation Depth Dose Profile and LET (Linear Energy Transfer) Assessment

As above described, specific treatment plans were developed for each animal used in this project. In turn, the best irradiation configuration and dose distributions were applied using the GEANT4-based application (GEometry ANd Tracking-4). DICOM (Digital Imaging and COmmunications in Medicine) CT images were employed to define the animal composition and geometry inside the GEANT4 framework. The medium LET (linear energy transfer) value calculated within the tumor was equal to 6.68 keV/μm. In Figure 1 an example of dose distribution calculated by our application is provided.

(a)

(b)

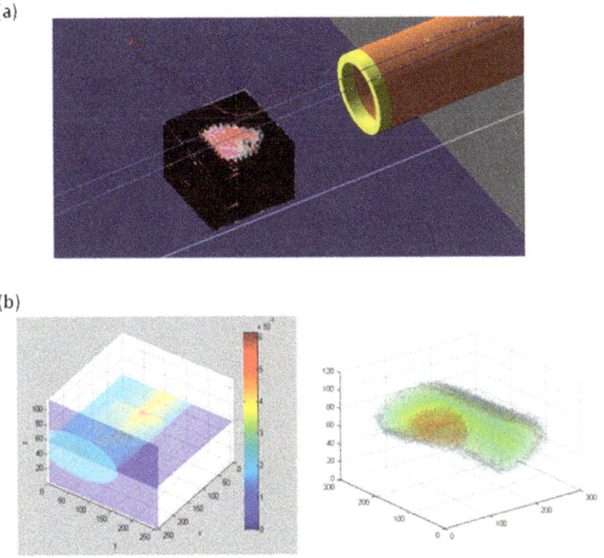

Figure 1. (**a**) Screenshot of DICOM volume in GEANT4. (**b**) Example of dose distribution calculated by Monte Carlo application.

2.2. Immunohistochemical Evaluation of Key Markers

Staining intensity of key regulator molecules involved in crucial cellular processes has been evaluated in order to better highlight the PT efficacy and understand which biological processes "are involved in the successful treatment plan" or "in the success of RT treatment" considering these results together with the whole genome GEP analysis, performed at the same doses. These markers' expression has been correlated to the dose delivered (2, 6 e 9 Gy) in both the two-time window analyzed of 72 h and 10 days after PT. As shown in Table 1 these were: CD133 (Cluster of Differentiation 133) as cancer stem cells marker, Cyclin D1 as cell cycle positive regulator marker, ki67 as proliferation rate marker, cleaved Caspase 3 as apoptotic marker, CD68 (Cluster of Differentiation 68) as macrophage marker. In addition, Table 1 reports the evaluation of necrosis extension and the apoptotic cell count, in order to better elucidate the death mechanism induced by PT.

Table 1. Immunohistochemical tumor characterization.

Samples	Number of Necrotic Foci	Extension of Necrosis (%)	Caspase-3			CD133			Ki67	CD68	Cyclin D1	Apoptotic Count (10 High Power Fields)
			IS	ES	IRS	IS	ES	IRS				
Untreated (72 h)	0	0	1	1	1	2	1	2	60%	Negative	Negative	10
Untreated (10 days)	4	0.45	2	4	8	3	3	9	70%	Negative	Negative	11
MDA-MB-231 Xenograft mice exposed to 2 Gy (72 h)	1	10%	1	2	2	2	2	4	50%	Negative	Positive (+IS 2; ES 2; IRS 4)	12
MDA-MB-231 Xenograft mice exposed to 2 Gy (10 days)	3	20%	2	3	6	3	3	9	65%	Positive (+5–20%)	Positive (+IS 1; ES 2; IRS 2)	16
MDA-MB-231 Xenograft mice exposed to 6 Gy (72 h)	3	25%	2	4	8	3	3	9	50%	Negative	Positive (+IS 2; ES 2; IRS 4)	14
MDA-MB-231 Xenograft mice exposed to 6 Gy (10 days)	2	15%	2	2	4	2	3	6	70%	Positive (+5–20%)	Positive (+IS 2; ES 2; IRS 4)	14
MDA-MB-231 Xenograft mice exposed to 9 Gy (72 h)	2	50%	2	4	8	3	3	9	60%	Negative	Positive (+IS 2; ES 3; IRS 6)	15
MDA-MB-231 Xenograft mice exposed to 9 Gy (10 days)	1	5%	1	2	2	2	2	4	60%	Positive (++20–50%)	Positive (+IS 1; ES 3; IRS 3)	14

IS: intensity score (0–3); ES: extent score (0–4); IRS: intensity reactivity score (0–12). CD68 positivity was expressed using (+) symbol scale: lower positivity was expressed with one (+), while its increment with multiple (+).

High Power Fields (HPF)

Figure 2A,B display representative H&E-stained slices deriving from samples analyzed 10 days and 72 h, respectively, after PT with 9 Gy of IR dose. In particular, after irradiation with 9 Gy, small foci of necrosis were observed 10 days post-irradiation (Figure 2A), while extensive necrosis was highlighted after 72 h (Figure 2B). In particular, the necrosis extension strongly increases with the dose escalation in the time window of 72 h post-PT, whereas 10 days after PT, we can observe an increase with the dose of 2 Gy with respect to untreated sample, and, then, a reduced necrotic extension at 6 and 9 Gy respect to the 2 Gy sample. Regarding apoptosis, a strong increase of cleaved CASP3 was highlighted in a dose- and time-dependent manners, except for samples treated with 9 Gy at the time window of 10 days post-PT. Figure 2C shows the high cleaved CASP3 expression after 72 h post-PT with the dose of 9 Gy.

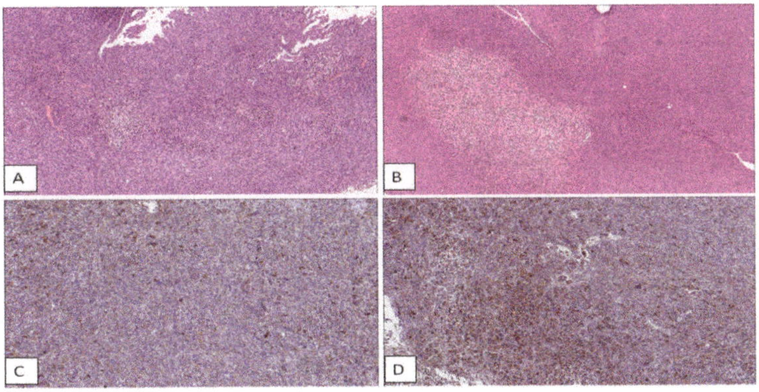

Figure 2. Representative slides of hematoxylin/eosin (H&E) and immunohistochemical staining. (**A**) Small necrosis foci after 10 days post-PT (proton therapy) with the dose of 9 Gy; (**B**) extensive necrosis after 72 h post-PT with the dose of 9 Gy; high immunohistochemical expression of caspase-3 (**C**) and CD133 (**D**) in tumor after 72 h post-PT with the dose of 9 Gy.

In addition, the apoptotic count showed a slight increase with dose escalation in both the time window of 72 h and 10 days post-treatment. In order to understand the reduction of necrotic foci and cleaved CASP3 expression with higher doses in samples observed 10 days post-PT, we assessed the CD68, as a macrophage marker. Indeed, its positivity was found only in samples analyzed 10 days after PT, with higher scores just in the 9 Gy treated sample, rendering evident the scavenger effect by these cells, able to eliminate dead tumor cells and debris.

On the other hand, survival signals are activated by the treatment, as suggested by a mild increase of Cyclin D1 in samples treated with all the three doses, both after 72 h or 10 days from PT exposures, with the only exception of the 9 Gy sample analyzed after 10 days, most likely due to the above-mentioned scavenger effect by macrophages. Finally, the stem cell marker CD133 was found up-regulated after PT using all the doses of 2, 6 and 9 Gy in the early time-window of 72 h. Instead, after 10 days post-irradiation, a CD133 strong increase is observed with respect to the untreated sample, and then a progressive reduction with 6 and 9 Gy in respect to the 2 Gy sample. Figure 2D shows the CD133 expression in a tumor treated with 9 Gy observed 72 h post-PT. Finally, the Ki67 marker expression did not show significant changes.

2.3. Overview of cDNA Microarray Gene Expression Analyses

As above described, in this work we analyzed the GEPs induced by PT irradiation using 2, 6 and 9 Gy of IR doses, on BC xenograft mice models in order to highlight genes and cellular processes involved in radioresistance/radiosensitivity balance regulation.

In detail, we analyzed the three following configurations:

(i) MDA-MB-231 xenograft mice_2Gy_PT
(ii) MDA-MB-231 xenograft mice_6Gy_PT
(iii) MDA-MB-231 xenograft mice_9Gy_PT

Comparative differential gene expression analyses revealed that a conspicuous number of genes had significantly altered expression levels by 2-fold or greater, compared to the untreated reference group as follows: MDA-MB-231 xenograft mice_2Gy_PT, 1256 differentially expressed genes (DEGs; 457 down-regulated and 799 up-regulated); MDA-MB-231 xenograft mice_6Gy_PT, 848 DEGs (279 down-regulated and 569 up-regulated); and MDA-MB-231 xenograft mice_9Gy_PT, 1279 DEGs (407 down-regulated and 872-up regulated).

Moreover, up- and down-regulated transcripts for each configuration analyzed in this study were selected and grouped according to their involvement in specific biological pathways using the DAVID tool, as displayed in Tables 2–5.

Table 2. GEPs (gene expression profiles) DAVID analysis of MDA-MB-231 xenograft mice exposed to 2 Gy of proton irradiation.

Top 5 Molecular Pathways of Differentially Expressed Genes of MDA-MB-231 Xenograft Mice Exposed to 2 Gy of Proton Irradiations				
	Term	**Count**	***p* Value**	**Genes**
1	Graft-versus-host disease	12	8.1×10^{-7}	*HLA-DQB1, HLA-DRB1, HLA-A, IL1B, FASLG, HLA-C, HLA-DPA1, HLA-B, HLA-DPB1, HLA-DOA, HLA-E, HLA-DRA*
2	Allograft rejection	11	2.1×10^{-5}	*HLA-DQB1, HLA-DRB1, HLA-A, FASLG, HLA-C, HLA-DPA1, HLA-B, HLA-DPB1, HLA-DOA, HLA-E, HLA-DRA*
3	Antigen processing and presentation	15	5.8×10^{-5}	*CIITA, HLA-DQB1, HLA-DRB1, HLA-A, HSPA1A, HLA-C, HLA-B, HLA-E, CD74, HSPA6, KIR3DL3, HLA-DPA1, HLA-DPB1, HLA-DOA, HLA-DRA*
4	Phagosome	21	2.0×10^{-4}	*HLA-DQB1, NOS1, HLA-DRB1, HLA-A, HLA-C, HLA-B, SFTPA1, ITGB3, HLA-E, CLEC4M, FCAR, CD209, COMP, TUBAL3, HLA-DPA1, SCARB1, HLA-DPB1, HLA-DOA, ATP6V0D2, TUBB4A, HLA-DRA*
5	Cell adhesion molecules (CAMs)	19	7.7×10^{-4}	*PVR, HLA-DQB1, HLA-DRB1, CLDN5, HLA-A, NLGN1, HLA-C, HLA-B, HLA-E, CLDN15, ALCAM, NCAM2, SDC1, CD2, MADCAM1, HLA-DPA1, HLA-DPB1, HLA-DOA, HLA-DRA*

Table 3. GEPs DAVID analysis of MDA-MB-231 xenograft mice exposed to 6 Gy of proton irradiation.

Top 5 Molecular Pathways of Differentially Expressed Genes of MDA-MB-231 Xenograft Mice Exposed to 6 Gy of Proton Irradiations				
	Term	**Count**	***p* Value**	**Genes**
1	Proteoglycans in cancer	22	1.2×10^{-5}	*NANOG, ERBB4, ROCK2, HCLS1, ERBB2, FASLG, IGF2, FZD3, HGF, DCN, ITGB3, MMP2, PXN, KDR, CTNNB1, SMO, MAPK13, HPSE, PLCG2, HSPB2, PRKACA, TWIST1*
2	Leukocyte transendothelial migration	11	9.5×10^{-3}	*ITGAL, ROCK2, MAPK13, PLCG2, CLDN5, CTNND1, MYLPF, JAM2, MMP2, PXN, CTNNB1*
3	Graft-versus-host disease	5	3.2×10^{-2}	*HLA-DQB1, HLA-DRB1, FASLG, HLA-DPA1, HLA-DPB1*
4	Allograft rejection	5	4.6×10^{-2}	*HLA-DQB1, HLA-DRB1, FASLG, HLA-DPA1, HLA-DPB1*
5	Phagosome	11	4.9×10^{-2}	*HLA-DQB1, TUBA8, CD36, HLA-DRB1, TUBAL3, HLA-DPA1, SFTPA1, COLEC11, HLA-DPB1, ITGB3, TUBB4A*

Table 4. GEPs DAVID analysis of MDA-MB-231 xenograft mice exposed to 9 Gy of proton irradiation.

Top 15 Molecular Pathways of Differentially Expressed Genes of MDA-MB-231 Xenograft Mice Exposed to 9 Gy of Proton Irradiations				
	Term	**Count**	***p* Value**	**Genes**
1	Proteoglycans in cancer	26	8.4×10^{-5}	*ERBB2, ITGB3, DCN, MMP2, GPC3, ANK2, HPSE, PPP1R12A, PIK3R5, PIK3R3, PIK3CG, NANOG, WNT10B, ROCK2, ITGA2, IGF2, FZD3, PRKCG, FZD5, FLNB, KDR, WNT2B, EIF4B, MAPK13, VEGFA, WNT11*
2	Rap1 signaling pathway	26	0.0002	*PRKCZ, FGFR3, RAP1GAP, TLN2, CTNND1, LPAR3, FGF13, ITGB3, ITGAM, RAC3, RASGRP2, RAPGEF4, PIK3R5, PIK3R3, ANGPT2, PLCB2, PIK3CG, FYB, GNAO1, GRIN1, PRKCG, KDR, DOCK4, MAPK13, VEGFA, PDGFRA*
3	Pathways in cancer	40	0.0002	*FGFR3, APC2, PTGS2, ERBB2, GNG13, CXCL8, LPAR3, FGF13, MMP2, SUFU, AGTR1, CDKN2A, RAC3, CASP8, RASGRP2, PIK3R5, HHIP, PIK3R3, PLCB2, PIK3CG, CEBPA, PTGER1, COL4A3, WNT10B, HSP90AA1, BCR, VHL, RALBP1, ROCK2, TGFBR2, BRCA2, ITGA2, PRKCG, FZD3, FZD5, STAT1, WNT2B, VEGFA, PDGFRA, WNT11*
4	Signaling pathways regulating pluripotency of stem cells	19	0.0006	*PIK3CG, NANOG, WNT10B, FGFR3, ONECUT1, APC2, PAX6, FZD3, FZD5, WNT2B, RIF1, HAND1, MAPK13, PIK3R5, WNT11, JAK3, SKIL, BMPR1B, PIK3R3*
5	VEGF signaling pathway	11	0.002	*PIK3CG, PTGS2, RAC3, MAPK13, VEGFA, PPP3R2, PRKCG, NOS3, PIK3R5, PIK3R3, KDR*
6	Platelet activation	16	0.005	*PIK3CG, PRKCZ, ROCK2, TLN2, COL3A1, ITGA2, ITGB3, PRKG1, MAPK13, RASGRP2, PPP1R12A, PIK3R5, NOS3, PIK3R3, PLCB2, COL11A1*
7	Phagosome	17	0.008	*HLA-DQB1, NOS1, HLA-DRB1, ITGA2, C1R, SFTPA1, ITGB3, HLA-DQA2, ITGAM, TUBA8, ATP6V0E2, CD209, COMP, TUBAL3, HLA-DPA1, HLA-DPB1, HLA-DRA*
8	Antigen processing and presentation	11	0.008	*HLA-DQB1, HSP90AA1, HSPA2, HLA-DRB1, CD8A, KIR3DL3, HLA-DPA1, HLA-DPB1, HLA-DQA2, CD74, HLA-DRA*
9	Graft-versus-host disease	7	0.008	*HLA-DQB1, CD86, HLA-DRB1, HLA-DPA1, HLA-DPB1, HLA-DQA2, HLA-DRA*
10	Focal adhesion	21	0.009	*PIK3CG, COL4A3, ROCK2, TLN2, ERBB2, COL3A1, ITGA10, ITGA2, PRKCG, ITGB3, FLNB, KDR, RAC3, COMP, VEGFA, PPP1R12A, PDGFRA, PIK3R5, PIK3R3, COL11A1, SHC4*
11	Ras signaling pathway	22	0.01	*PIK3CG, PLD1, PLA2G16, FGFR3, RALBP1, GRIN1, GNG13, PRKCG, FGF13, RGL1, KDR, RAC3, RASGRP2, VEGFA, PDGFRA, ZAP70, PIK3R5, SYNGAP1, PIK3R3, PLA2G2D, ANGPT2, SHC4*
12	Allograft rejection	7	0.01	*HLA-DQB1, CD86, HLA-DRB1, HLA-DPA1, HLA-DPB1, HLA-DQA2, HLA-DRA*
13	Chemokine signaling pathway	18	0.03	*CXCL1, PIK3CG, PRKCZ, FGR, ROCK2, GNG13, CXCL8, CCL4L2, STAT1, CCL15, CCL26, CCR3, RASGRP2, PIK3R5, JAK3, PIK3R3, PLCB2, SHC4*
14	HIF-1 signaling pathway	11	0.04	*PIK3CG, VHL, PFKFB3, HMOX1, ERBB2, VEGFA, PRKCG, NOS3, PIK3R5, PIK3R3, ANGPT2*
15	Wnt signaling pathway	14	0.04	*WNT10B, APC2, ROCK2, PPP3R2, FZD3, PRKCG, FZD5, WNT2B, CSNK2A1, SOST, SFRP1, RAC3, WNT11, PLCB2*

Table 5. Gene expression analyses of key genes of TNBC (triple-negative breast cancer) xenograft mice exposed to Proton RadioTherapy PRT.

Gene Expression Analyses of Key Genes of TNBC Xenograft Mice Exposed to PT							
Gene Symbol	**Description**	**2 Gy PRT**		**6 Gy PRT**		**9 Gy PRT**	
		Array	**qRT-PCR**	**Array**	**qRT-PCR**	**Array**	**qRT-PCR**
BAX	BCL2 associated X, apoptosis regulator	−1.25	0.51	−1.12	0.32	1.14	2.26
BCL2	BCL2 apoptosis regulator	−1.20	0.56	−1.20	0.28	1.18	1.87
CCNA2	Cyclin A2	−1.24	0.62	−2.73	0.30	Und.	2.24
CD24	CD24 molecule	1.43	1.24	1.23	Und.	Und.	2.92
CD44	CD44 molecule	1.37	0.97	1.07	Und.	1.24	3.16
CDC20	Cell division cycle 20	−1.11	0.48	−1.11	0.83	1.04	3.83
CDC25	Cell division cycle 25	−1.19	0.38	Und	0.51	1.11	1.51
FOS	Fos proto-oncogene, AP-1 transcription factor subunit	4.95	3.35	1.12	Und.	1.91	30.56
FOSB	FosB proto-oncogene, AP-1 transcription factor subunit	2.51	1.07	1.34	0.57	1.30	6.35
GDF15	Growth differentiation factor 15	1.86	1.64	1.72	1.23	2.50	10.27
MMP9	Matrix metallopeptidase 9	−1.16	0.21	Und	0.38	1.33	2.31
RRAD	Ras related glycolysis inhibitor and calcium channel regulator	1.50	3.78	1.54	1.95	1.71	6.51
TAF7L	TATA-box binding protein associated factor 7 like	1.14	1.27	Und	Und.	1.18	1.80
TNF	Tumor necrosis factor	1.34	4.14	Und	1.31	1.17	5.02
TP53	Tumor protein p53	n.a.	3.11	1.33	1.09	1.25	3.28
TP53INP1	Tumor protein p53 inducible nuclear protein 1	1.04	1.40	1.23	1.76	Und.	4.53
WNT5A	Wnt family member 5A	1.17	2.12	1.26	2.64	1.14	9.28

Triple negative breast cancer (TNBC); undetected (Und.); proton therapy (PT).

In detail, as shown in Tables 2–4, graft-versus-host disease, allograft rejection and phagosome cellular pathways were deregulated in all the three configuration assayed in this work, underling their key role strictly linked to proton irradiation rather than to the dose delivered.

The first two pathways are known to be related to immune reactivity of the recipient against the transplanted allograft. On the other hand, a phagosome is able to maintain host homeostasis because it is involved in the degradation of pathogens and cellular death as well as in other processes linked to antigen-presenting processes and the recovery of inflammatory mediators [18].

In addition, after 2 Gy of PT, our preclinical tumor model was able to deregulate the antigen processing and presentation pathway, once again underlying the immune system activation after PT, as well as the cell adhesion molecules (CAMs) involved in binding and communication with other cells.

On the other hand, after 6 Gy of PT irradiation, leukocyte transendothelial migration process and proteoglycans in the cancer pathway were deregulated. Precisely, leukocyte migration from the blood into tissues is vital for immune surveillance and inflammation while proteoglycans are key molecules and effectors of cell surface and microenvironments, known to have multiple functions in cancer and angiogenesis due their ability to regulate neoplastic growth and neovascularization [19].

Interestingly, after 2 and 6 Gy of PT irradiation, we selected only the top five statistical and biological relevant pathways. In contrast, after 9 Gy of PT irradiation we selected more numerous deregulated cellular networks and, thus, in Table 4 we reported the top 15 statistical and biological relevant pathways. In summary, some of these are involved in cell–cell communication and/or immune system activation (i.e., antigen processing and presentation; Rap1 signaling pathway; graft-versus-host disease; chemokine signaling pathway; focal adhesion; allograft rejection, platelet activation, phagosome), others are related to tumor progression, angiogenesis and invasiveness (i.e., pathways in cancer; VEGF signaling pathway, proteoglycans in cancer; Ras signaling pathway, signaling pathways regulating pluripotency of stem cells; HIF-1 signaling pathway; Wnt signaling pathway). Candidate genes were selected and used to validate microarray datasets by qRT-PCR analyses that confirm gene-expression trends (Table 5). Interestingly, as displayed, MDA-MB-231 xenograft mice exposed to PT with 2 and 6 Gy have shown overall comparable gene expression trends, unlike samples treated with the high dose of 9 Gy.

2.4. Commonly Deregulated Genes and Pathways among the PT Schedules

Moreover, in order to study the number of unique and shared differentially expressed genes between MDA-MB-231 xenograft mice exposed to IR doses of 2, 6 and 9 Gy, we performed Venn diagrams (Figure 3). As shown, a great amount of genes were commonly deregulated in the configurations assayed: 489 genes were deregulated after 2 and 6 Gy of proton irradiation; 469 genes changed their expression levels after 2 and 9 Gy; and 407 were deregulated both after 6 and 9 Gy of PT, respectively shown in Figure 3A–C. Finally, as shown in Figure 3D, we decided to highlight the deregulated gene set shared between all the configurations assayed (2, 6 and 9 Gy) and, thus, we selected the 290-gene signature linked to proton cell response rather than to the dose delivered (Figure 3D). The respective DAVID analysis of this gene list results in the following pathways: antigen processing and presentation; graft-versus-host disease; allograft rejection; phagosome and complement and coagulation cascades signaling (Table 6), confirming once again the crucial role of immune system activation, in response to irradiation by proton beams of a xenograft model.

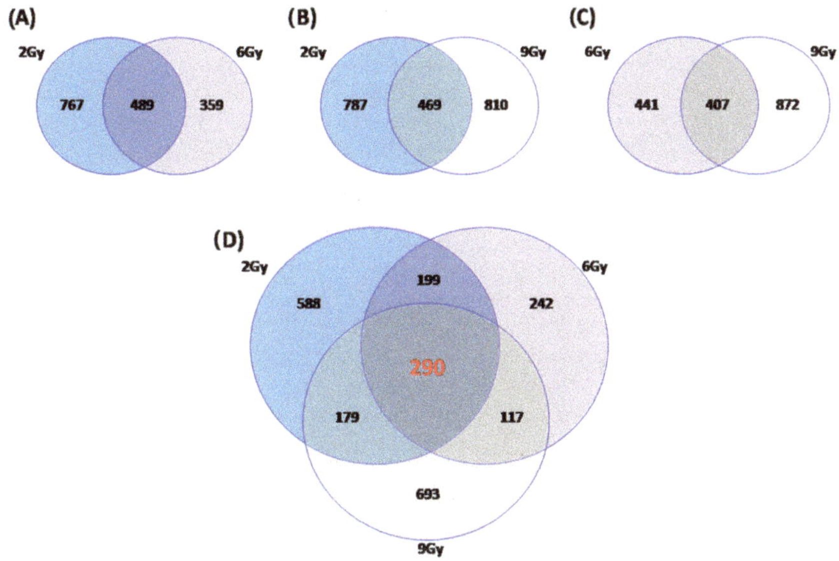

Figure 3. Venn diagram of the number of unique and shared differentially expressed genes after proton irradiation in MDA-MB-231 xenograft mice using 2, 6 and 9 Gy of radiation doses as follow: (**A**) unique and shared differentially expressed genes in samples treated with 2 and 6 Gy; (**B**) unique and shared differentially expressed genes in samples treated with 2 and 9 Gy; (**C**) unique and shared differentially expressed genes in samples treated with 6 and 9 Gy; (**D**) unique and shared differentially expressed genes in samples treated with 2, 6 and 9 Gy.

Table 6. Top five molecular pathways of differentially expressed genes shared between MDA-MB-231 xenograft mice exposed to 2, 6 and 9 Gy of proton irradiation.

Top 5 Molecular Pathways of Differentially Expressed Genes Shared Between MDA-MB-231 Xenograft Mice Exposed to 2, 6 and 9 Gy of Proton Irradiations				
	Term	**Count**	***p* Value**	**Genes**
1	Antigen processing and presentation	6	0.003	*HLA-DQB1, HLA-DRB1, KIR3DL3, HLA-DPA1, HLA-DPB1, CD74*
2	Graft-versus-host disease	4	0.009	*HLA-DQB1, HLA-DRB1, HLA-DPA1, HLA-DPB1*
3	Allograft rejection	4	0.01	*HLA-DQB1, HLA-DRB1, HLA-DPA1, HLA-DPB1*
4	Phagosome	6	0.04	*HLA-DQB1, HLA-DRB1, TUBAL3, HLA-DPA1, HLA-DPB1, ITGB3*
5	Complement and coagulation cascades	4	0.05	*F5, CD46, F13A1, C1S*

3. Discussion

One of the main aims of our research group is to fill the gap regarding the radiobiological knowledge of the molecular responses induced by PT and its potential advantages for treating radioresistant tumor subtypes. In addition, several authors have reported controversies with respect to the use of conventional RT for patients with TNBC (ER–/PR–/HER2–) [15,16]. In this sense, the absence of hormonal or targeted therapy against TNBC makes it a clinical challenge for oncologists in terms of patient management.

According to these assumptions, we decided to investigate the molecular responses induced by PT in MDA-MB-231 BC cells, chosen as the best model of a TNBC disease, by using in vitro and in vivo approaches, in order to clarify mechanisms involved in the treatment success. This knowledge could help us to identify possible targets to induce radio-sensibilization and shift the balance of cell fate towards death. Furthermore, we aimed to use protons because they offer the physical advantage of reaching the target site while saving the organs at risk, such as heart and lungs. In addition, as suggested by some authors, a variable instead of fixed RBE (Relative Biological Effectiveness) (1.1) should be considered along the proton energy deposition curve (Bragg curve), taking into account the RBE dependency on linear energy transfer (LET) and tissue properties (α/β values) [20,21]. Particularly, there is an inverted relationship between proton energy and LET, corresponding to an RBE increase especially in the distal SOBP (Spread-out Bragg peak). These properties make them more suitable for those radioresistant tumors.

The dose range of 2–9 Gy has been chosen with the aim of understanding the different molecular response observable with the dose escalation. In particular, the doses of 2 and 9 Gy were previously used in our in vitro experiments, both using photons/electrons or protons, allowing us to use the possibility to make a GEP analysis comparison. It has to be remembered that 2 Gy is the daily dose of a conventional RT fractionated treatment, whereas 9 Gy is used as a boost dose in the IOERT (IntraOperative Electron Radiation Therapy) technique.

In particular, we highlighted a strong activation of the immune response in MDA-MB-231 cells subjected to PT irradiation in a time-dependent manner [7]. Indeed, MDA-MB-231 cells showed the strongest potentially pro-inflammatory profile compared with other BC cell lines analyzed. The activation of intracellular inflammatory-related pathways justifies the MDA-MB-231 cells' capacity to release inflammatory molecules in tumor microenvironment, then driving cell fate balance, senescence mechanism, angiogenesis and cell migration. Taken together, these data strongly suggest that the TNBC cells' aggressive phenotype and their resistance to treatments are sustained by the activation of intracellular pro-inflammatory mechanisms, which are known to be a typical response to stress stimuli such as irradiation. Notably, this particular behavior is peculiar to MDA-MB-231 and it is not shared by other cell lines studied previously by our group, as for example in MCF7 cells, which had a very low activation of the immune response, both for quantity and types of released cytokines, and, instead, showed a molecular response to irradiation driven by the p53 pathway [7,8,22].

Then, here we used an immunohistochemical approach and a whole-genome cDNA microarray to better understand the real in vivo response to PT irradiation, in respect to what we previously observed in vitro, applying the same doses.

Table 1 displays the quantification of some key biomarkers involved in cell fate balance after irradiation, by IHC approach. These biomarkers are crucial actors in cell cycle regulation, stem cell regenerative capability or death processes, such as apoptosis or necrosis; for the latter the percentage extension was also determined.

Tables 2–4 and 6 display the top 5–15 significant pathways activated in the tumor tissues of our xenograft models in response to 2, 6 and 9 Gy, respectively, using untreated tumor tissues as control, resulting in a relative GEP quantification. It can be observed that the three dose configuration used (2, 6 and 9 Gy) share: (1) a strong cancer cell communication, driven by CAM molecules; (2) graft-versus-host disease due to the xenografts model creation; (3) the activation of stem cells pathway known to be involved in radioresistant phenotype; (4) an overall up-regulation of inflammation biomarkers confirming, once again, the MDA-MB-231 great inflammatory potential.

In comparison with our previous data on the MDA-MB-231 in vitro model, the first two above mentioned pathways are peculiar of the PT response by the animal model, due to the presence of a tridimensional tissue in this case, which amplify the cell–cell communication, and due to the insertion of a human tumor cell line into the mouse model, which activates the well-known host versus graft response, here observable despite the fact that we used nude immunosuppressed mice.

On the other hand, the other two pathways (3 and 4) are common in the in vitro and in vivo models, confirming the PT capacity, even at low doses, to activate a certain tumor counterattack strategy for radioresistance, using inflammation and stem cell activation, which try to direct cell fate towards survival and proliferation.

In this regard, even the literature describes the role of remaining tumor cancer stem cells before the RT starting and their ability to repopulate over the course of treatment plan in several conditions including hypoxia, stroma interaction among cells and variations in the intrinsic cells' sensitivity to radiation, as well as in the modulation of DNA repair or other cell survival pathways [23–27]. Thus, we better explored the stem cell marker modulation in our BC xenograft model during PRT plans, considering our GEP and IHC data (Tables 1 and 5). In particular, an up-regulation of CD24 and CD44 gene expression was assessed both by microarray technology and by qRT-PCR in all the three dose configurations assayed at 72 h post-RT. Similarly, using IHC, we reported the CD133 up-regulation after PT using all the doses 2, 6 and 9 Gy in the early time-window of 72 h. Instead, after 10 days post irradiation, a CD133 progressive reduction is observed, in respect to untreated mice, making us hypothesize that the tumor stem potential downregulation is a late effect of PRT. This opposite trend, between the two-time windows analyzed, is probably due to the scavenger process driven by macrophages, which is more evident after more days. Indeed, the macrophage CD68 marker was positive only in the 10 days post-RT samples, by IHC approach (Table 1), confirming their role in tissue remodeling during the days after PT.

We also investigated the cell cycle modulation after PT, observing the CDC20, CDC25 and CCNA decreased gene expression after the administration of the dose of 2 Gy (Table 5), and their upregulation 72 h post-PT with 9 Gy. However, a mild increase of Cyclin D1 is observed, by IHC approach, in samples treated with all three doses, both after 72 h or 10 days from PT exposure, with the only exception of the 9 Gy sample analyzed after 10 days, most likely due to the above-mentioned scavenger effect by macrophages.

These results make evident the role of irradiation in stimulating cell cycle progression and stem cell proliferation in MDA-MB-231 cells, thus activating radioresistance, especially observable at low doses in early post-treatment time windows.

Besides, we assayed the Ki67 immunomarker, a nuclear protein that is expressed exclusively during the active cell cycle phases, but not in resting cells. As known, Ki67 is widely used in pathology to assess cell proliferation within multiple different neoplasms including BC [28]. In BC, Ki67 has shown a promising role as an independent prognostic marker and as a predictive marker of responsiveness or resistance to therapies, with a consolidated prognostic utility [29]. In our experiment, no statistical variation among the PT configurations assayed was reported, as displayed in Table 1.

On the other hand, in order to analyze the activation of the cell death process after PT exposure, we assayed the amount of the main apoptosis key regulator by IHC: the CASP3 cleaved protein. As reported in Table 1 a strong increase of CASP3 was highlighted in a dose- and time-dependent manner, except for samples treated with 9 Gy at the time window of 10 days post-PT. Similarly to what we observed for the CD133 and Cyclin D1 biomarkers, this observation could be the result of scavenger action, which removed the dead cells. In line with the above-reported cell cycle and apoptosis trends, a BAX down-regulation was reported after 2 Gy, and, conversely, an up-regulation after 9 Gy of IR dose at 72 h post-PT (Table 5). In addition, considering the apoptotic count, assessed by morphology, a slight increase is observed with increasing dose in both samples analyzed at 72 h and 10 days post-treatment.

As well described in the literature, IR activates complex cross-linked intracellular networks, able to define cell fate in the choice between survival and death. It has become evident that, in particular for solid tumors, the inhibition of neoplastic cell proliferative capacity following irradiation can occur through different types of cell death (i.e., apoptosis, necrosis, mitotic catastrophe, autophagy and senescence) [30]. However, necrosis has generally been considered as a predominant cell death process after the administration of treatment with high IR doses, while at a lower doses, it has been recognized as a passive and unregulated event [31]. Then, we also evaluated the number of necrotic foci and the percentage of necrosis extension by morphology, as displayed in Table 1.

In the early time window of 72 h post-PT, a strong increase of necrotic extension is recognizable with the dose increase, whereas 10 days after PT, we can observe a strong increase with the dose of 2 Gy in respect to untreated sample, and a reduced necrotic extension using 6 and 9 Gy in respect to the 2 Gy sample. Again, this negative trend reported in the late time window of 10 days post-PT makes evident the scavenger activity to clear cellular debris, promoting phagocytosis, and mediating the recruitment and activation of other macrophages. Indeed, the CD68 positivity correlates with the dose delivered: 5–20% after 2 and 6 Gy; 20–50% after 9 Gy of IR doses, respectively.

A tumor size reduction post-PT was not revealed by digital caliper, since it would have been necessary to wait for longer than 10 days, during which tumors treated with low doses would have grown beyond the maximum limit imposed for the animal sacrifice, in order to avoid suffering. Therefore, the in-depth study of the above described biological processes has become necessary to clarify the PT effect in our xenograft TNBC models.

Among the early activated GEP within 72 h post-PT, we confirm the activation of a key transcription factor (TF) as FOS, known to be related to the response to radiation, but poorly described following unconventional treatment modalities, as after PT. TFs regulate a wide spectrum of genes involved in inflammation, apoptosis, invasion and angiogenesis processes, contributing to confer tumor cell radioresistance [30,32,33]. In particular, AP-1 proteins (assembled from JUN and FOS proteins), and, above all, c-Fos, play an important role in the induction and development of radiation late effects in normal tissues. The JunB gene is responsive to IR and is immediately induced early after stimulation [30].

Finally, the DAVID analysis of commonly deregulated genes and pathways, among the PT schedules, confirms the involvement of some immunological processes, described above as strictly linked to proton cell response in MDA-MB-231 xenograft model: antigen processing and presentation;

graft-versus-host disease; allograft rejection; phagosome and complement and coagulation cascades signaling (Table 6).

Moreover, we also confirm the activation of pathways poorly described in the literature, but previously observed by our group, such as the Rap1 signaling pathway and the Phagosome activation [34]. Rap1 gene encodes a protein that is part of a complex involved in the regulation of telomere length, possibly involved in the senescence process activation [35]. The regulation of phagosomes could be involved in radiation-induced autophagy, known to be able to enhance radioresistance and leading to activation of the survival pathway, as recently observed [36].

4. Material and Methods

4.1. Dose Evaluation and Distribution by Monte Carlo GEANT4 Toolkit

In order to optimize the irradiation procedures and dosimetry for in vivo experiments we decided to use a Monte Carlo approach using GEANT4 toolkit, as previously described by our group [8,21]. All proton treatments were performed at the INFN-LNS CATANA proton therapy facility in Catania (Italy) using a passive fixed horizontal proton beam line with an energy of 62 MeV/A. A beam shaping system was used to obtain a uniform dose distribution at the isocenter [21,37,38]. To optimize animal irradiation schedules and to obtain a precise and reproducible irradiation setting, we developed a dedicated positioning animal holder system [21,37,39]. All animals were irradiated using a degraded spread-out Bragg peak (SOBP), thanks to the use of a poly (methyl methacrylate) (PMMA) modulator wheel. The dosimetry was performed using a Markus ionization chamber (PTW Freiburg GmbH, Germany) and gafchromic EBT3 films (ISP Corp., New York, NY, USA) with an accuracy better than 3%. According to irradiation procedures conventionally used during PT in clinical practice, the dose delivery was monitored by a transmission ionization chamber placed along the beam line. All animal irradiations were performed at the same time interval of the day to guarantee no difference related to mice circadian rhythm, with a constant dose rate of 5 Gy/min.

Secondly, in order to perform an accurate and efficient prediction of the dose distribution inside tumors, a Monte Carlo simulation was applied. Moreover, the GEANT4 Monte Carlo toolkit was used to reduce irradiation of organs at risk (OAR). In turn, a previously developed and validated GEANT4-based application was used to perform a preliminary assessment of dose map, using the real mice anatomical structures. Thus, we evaluated linear energy transfer (LET) distribution [21] defining, voxel-by-voxel, the real target composition thanks to the use of a preclinical micro-PET/CT and DICOM micro-CT images (Albira Si, Bruker, Belgium), available at CAPiR (Centre for Advanced Preclinical in vivo Research), University of Catania, Italy [21]. The datasets were acquired using 600 views in high-resolution configuration, X-Ray energy of 45 kVp, a current of 400 μA and the dimension of each CT-voxel was equal to $125 \times 125 \times 125\ \mu m^3$.

4.2. Animal Model

Ethics Statement and Animal Model

The experiments were performed in accordance with the European Communities Council directive and Italian regulations (EEC Council 2010/63/EU and Italian D.Lgs. 26/2014). The project was approved by the Italian Ministry of Health (authorization number n. 527/2016-PR, approved on 26 May 2016). Efforts were employed to replace, reduce and refine the use of laboratory animals. To avoid irrelevant suffering to treated mice, euthanasia was performed as soon as the final score was reached. The endpoint used to determine if animals should undergo euthanasia was reached when tumor lesions showed a dimension higher than 1.2 cm and/or weight loss more than 20%.

All reasonable efforts were made to ameliorate suffering, avoiding the most painful procedures. To minimize suffering and mice distress, standard environmental enrichment of two nestles, a cardboard Fun Tunnel and one wooden chew block were provided.

Experiments were performed on 8 weeks old BALB/c Nude female mice (Charles River Laboratory), weighing 24 ± 3 g. Animals were housed in IVC cages at constant temperature (23–25 °C) under a 12/12 h light/dark cycle with ad libitum access to food and water. Mice were housed using a stocking density of 3 mice per cage in individual IVC cages.

A total of 4×10^6 MDA-MB-231 BC cells were inoculated in a total number of 24 mice (18 animals to treat with PT and 6 controls) [7,22,34]. Animal health and behavior were monitored twice a week together with body weight and clinical specific signs up to the sacrifice. After two weeks of growth the tumors reached the size of 8 +/− 2 mm, monitored by digital caliper, and irradiation treatments were performed. Proton irradiations were executed with doses of 2 Gy (6 mice), 6 Gy (6 mice) and 9 Gy (6 mice). So, 72 h and 10 days post-PT treatments, tumors were measured again using a digital caliper, then mice were sacrificed and treated tumors (as well as untreated ones used as control) were collected and stored at −80 °C until molecular analyses.

4.3. Immunohistochemistry (IHC)

Tumor sections from MDA-MB-231 xenografts in Balb/c nude mice were collected at specific time points (72 h and 10 days) after PT treatments using 2, 6 and 9 Gy of IR doses. Sections were processed as previously described [40]. Briefly, the slides were dewaxed in xylene, hydrated using graded ethanol and incubated for 30 min in 0.3% H_2O_2/methanol to quench endogenous peroxidase activity, then rinsed for 20 min with phosphate-buffered saline (PBS; Bio-Optica, Milan, Italy). The sections were heated (5 min × 3) in capped polypropylene slide-holders with citrate buffer (10 mM citric acid, 0.05% Tween 20, pH 6.0; Bio-Optica, Milan, Italy), using a microwave oven (750 W) to unmask antigenic sites. To reduce the commonly seen non-specific immunoreactivity due to endogenous biotin, sections were pretreated with 10 mg/mL of ovalbumin in PBS followed by 0.2% biotin in PBS, each for 15 min at room temperature. Then, the sections were incubated with rabbit monoclonal anti-Cyclin D1 antibody (sp4; Diagnostics Biosystems, Pleasanton, CA, USA), diluted 1:50 in PBS (Sigma, Milan, Italy); rabbit monoclonal anti-estrogen receptors (ERs) antibody (ep1; DAKO, Glostrup, Germany), diluted 1:50 in PBS (Sigma, Milan, Italy); rabbit monoclonal anti-progesterone receptors (PgRs) (PgR636; DAKO, Glostrup, Germany) diluted 1:50 in PBS (Sigma, Milan, Italy); mouse monoclonal anti-Ki-67 antibody (MIB-1; DAKO, Glostrup, Germany), diluted 1:100 in PBS (Sigma, Milan, Italy); rabbit polyclonal anti-c-erb-B2 oncoprotein antibody (DAKO, Glostrup, Germany), diluted 1:500 in PBS (Sigma, Milan, Italy); mouse monoclonal anti-CD68 antibody (pg-m1; DAKO, Glostrup, Germany), diluted 1:100 in PBS (Sigma, Milan, Italy); rabbit polyclonal anti-CD133 antibody (Abcam, Cambridge, UK), diluted 1:200 in PBS (Sigma, Milan, Italy); rabbit polyclonal anti-cleaved caspase-3 antibody (Abcam, Cambridge, UK), diluted 1:50 in PBS (Sigma, Milan, Italy). The secondary biotinylated anti-mouse antibody was applied for 30 min at room temperature, followed by the avidin-biotin–peroxidase complex (Vector Laboratories, Burlingame, CA, USA) for a further 30 min at room temperature. The immunoreaction was visualized by incubating the sections for 4 min in a 0.1% 3,3′-diaminobenzidine (DAB) and 0.02% hydrogen peroxide solution (DAB substrate kit, Vector Laboratories, CA, USA). The sections were lightly counterstained with Mayer's hematoxylin (Histolab Products AB, Göteborg, Sweden) mounted in GVA mountant (Zymed Laboratories, San Francisco, CA, USA) and observed with a Zeiss Axioplan light microscope (Carl Zeiss, Oberkochen, Germany).

Staining intensity score (IS) was obtained by two independent pathologists (RC, GB) and graded on a 0–3 scale, according to the following assessments: no detectable staining = 0, weak staining = 1, moderate staining = 2, strong staining = 3. Moreover, the percentage of immunopositive cells (Extent Score, ES) was scored in five categories: <5% (0); 5–30% (+); 31–50% (++); 51–75% (+++), and >75% (++++). Counting was performed at 200× magnification. In addition, staining intensity was multiplied by the percentage of positive cells to obtain the intensity reactivity score (IRS); IRS < 6 was considered as low expression (L-IRS), IRS > 6 was considered as high expression (H-IRS). The percentage of intratumoral CD68+ histiocytes was assessed according to the guidelines for evaluating the tumor-infiltrating lymphocytes (TILs) in breast cancer [41].

Only CD68 + histiocytes within the tumor's borders were evaluated. CD68 + cells outside of the tumor borders were not included. Necrotic areas were excluded from the evaluation. CD68 percentage was reported as the average of histiocytes in the stromal component of the tumor. Thus, 3 groups were identified: 0–10% stromal CD68 + cells (group 1); 20–40% (group 2); 50–90% (group 3).

In addition, as shown, the dosage and time of death, the number of necrotic foci, the necrosis extension (in terms of percentage of the total tumor tissue, %), were also reported in Table 1 for all the configurations assayed in this work.

4.4. Apoptotic Count

The apoptotic count was performed by two pathologists (RC; GB) on hematoxylin and eosin-stained sections. The apoptotic count was assessed as the total number of apoptotic cells per 10 high power fields (HPFs) at 40× magnification using a Zeiss Axioplan light microscope (Carl Zeiss, Oberkochen, Germany).

4.5. Whole Genome cDNA Microarray Expression Analysis

In this work, we performed whole-genome cDNA microarray gene expression analyses, to study the biological processes activated in MDA-MB-231 xenograft mice models following 72 h post-PT, highlighting specific key genes involved in cell response to radiation and to select potential new biomarkers of radiosensitivity and radioresistance as previously described [7,42,43]. Then, 72 h post-PT treatments, mice were sacrificed; treated and untreated (used as a control) tumor tissues, were collected and stored until processing of molecular analyses. cDNA Microsoft was performed as previously described. Total RNA was extracted from cells using Trizol and the RNeasy mini kit (Invitrogen) and RNA concentration and purity were determined using a Nanodrop ND-1000 (Thermo Scientific Open Biosystems, Lafayette, CO, USA). Samples with an RNA integrity number (RIN) value of 10, assessed by using a Bioanalyzer 2100 (Agilent Technologies, Santa Clara, CA, USA), were used for further microarray analyses. Thus, according to the Agilent Two-Color Microarray-Based Gene Expression Analysis protocol, we studied the GEPs induced by 2, 6 and 9 Gy of PRT in MDA-MB-231 xenograft mice models. So, cRNA synthesis, labelling with Cy dyes, hybridization onto Whole Human Genome 4 × 44 K microarray GeneChips (Agilent Technologies) and microarray image detections, were conducted as reported [44]. Finally, statistical data analysis, background correction and normalization of the GEPs were performed using Feature Extraction and GeneSpring GX 13.0 softwares (Agilent Technologies). Genes were identified as being differentially expressed if they showed a fold change (FC) of at least 2 with a p-value < 0.05 compared with radiation untreated tumor, used as reference. GEP data included in this paper are available in compliance with Minimum Information About a Microarray Experiment (MIAME) standards, using the following Gene Expression Omnibus (GEO) accession number: GSE149023.

Finally, we studied biological pathways regulated by the genes belonging to the differentially expressed gene lists obtained by GEP analyses, firstly using the Database for Annotation, Visualization and Integrated Discovery (DAVID) network building tool (https://david.ncifcrf.gov/tools.jsp), that provides a comprehensive set of functional annotation for investigators to study the biological content captured by high throughput technologies such as microarray analyses and secondly by using PubMatrix tool in order to confirm our assumptions [45].

4.6. qRT-PCR Analysis

Candidate genes for qRTPCR analysis were chosen based on the microarray results. Total RNA was reverse-transcribed into cDNA and then analyzed by real-time PCR in triplicate using a Fast 7500 Real-Time PCR System (Applied Biosystems, Carlsbad, CA, USA), as previously described [46]. The oligonucleotide primers chosen for qRT-PCR were selected with Primer3 software and tested as previously described [46]. Quantitative data, normalized versus that for the rRNA *18S* gene, were generated from three independent experiments and analyzed by the average of triplicate cycle threshold (Ct) according to the $2^{-\Delta\Delta ct}$ method using SDS software (version 1.4, Applied Biosystems,

Carlsbad, CA, USA). The data shown and the values are expressed as the mean ± SD relative to mRNA levels in the untreated MDA-MB-231 xenograft mice not exposed to PRT, used as the control sample.

5. Conclusions

As described above, the main aim of this work was to analyze the in vivo molecular response to PT and its efficacy in a MDA-MB-231 TNBC xenograft model, also describing for the first time to our knowledge specific PT molecular signatures linked to the dose delivered.

Summarizing, our study reveals the effect of single shots of PT doses on a TNBC xenograft model, in the two-time window of 72 h and 10 days post-PT, showing a detailed molecular response for each configuration studied. In particular, the GEP analysis reveals a great immune response activation by the tumor tissue itself, due to irradiation, as well as a tentative tumor radioresistance through the positive modulation of cell cycle and stem cell process.

Only the dose of 9 Gy showed evident effects of moving balance toward tumor cell death. Further evidence is the role of both innate immunity and macrophage cells which were actively attracted within the tumor to remove tumor dead cells and debris, a process evident even in our cell models of nude mice. Figure 4 displays in a synoptic way the conclusive remarks of this study, describing the trend of main cellular processes involved in the balance between death and survival, with respect to the increase of delivered dose (an activation of stem cell pathway; a modulation of cell cycle process; the apoptosis regulation and a strong increase of necrotic extension recognizable with the dose increase).

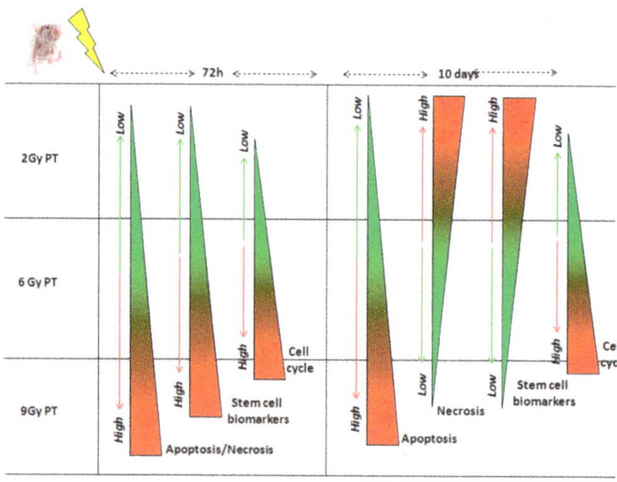

Figure 4. The figure displays the main remarks of this work. Green and red color represent the down- and up-regulation of the selected process after PT exposures, respectively. Stem cell activation: after 72 h post-PT with 2, 6 and 9 Gy, we assessed the up-regulation of CD24 and CD44 genes and by CD133 up-regulation. After 10 days post-irradiation, a CD133 progressive reduction was observed. Cell cycle modulation: a decreased gene expression followed by an up-regulation of CDC20, CDC25 and CCNA gene expression was measured according to the increasing dose delivered after 72 h post-PT. In addition, a mild increase of Cyclin D1 is observed, both after 72 h or 10 days from PT exposures. Apoptosis activation: a strong increase of CASP3 was highlighted in a dose- and time-dependent manners, except for samples treated with 9 Gy at the time window of 10 days post-PT. Moreover, the apoptotic count reported a slight increase according to the increasing dose, in both samples analyzed at 72 h and 10 days post-PT. Necrosis: after 72 h post-PT a strong increase of necrotic extension is recognizable with the dose increase, whereas 10 days after PT, a strong increase with the dose of 2 Gy in respect to untreated sample, and a reduced necrotic extension using 6 and 9 Gy in respect to the 2 Gy sample, were reported.

As suggested by some authors, the proton energy deposition along the Bragg curve permits evaluation of a variable instead of fixed RBE, especially higher in the distal SOBP. These properties make protons more suitable to deliver higher total doses and dose per fractions, with dedicated treatment plans for more radioresistant tumors, targeted with much higher precision while saving surrounding healthy tissues and organs at risk [20,21].

Author Contributions: All authors participated in the conception and elaboration of the study. In particular, G.R., P.P. performed the irradiation setups and together with F.P.C. and R.T. performed the PT mice treatments. R.C. and G.B. carried out the histopathological and immunohistochemical characterizations. V.B. performed the microarray experiments and the gene expression analyses. G.I.F., V.B., F.P.C., R.C., G.B., L.M., P.P.; M.C., P.P., G.R., R.T.; B.T., F.T., R.A., G.P., G.A.P.C. and G.C. were involved in the interpretation of the findings and approved the final content of the manuscript. All authors have read and agreed to the published version of the manuscript.

Funding: This work was supported by National Institute for Nuclear Physics (INFN)-LNS-funded ETHICS project and by GeSeTon project (funded by Italian MISE Grant No. 489 of 21/02/2018).

Conflicts of Interest: The authors declare that they have no competing interests.

References

1. Bravatà, V.; Cava, C.; Minafra, L.; Cammarata, F.P.; Russo, G.; Gilardi, M.C.; Castiglioni, I.; Forte, G.I. Radiation-Induced Gene Expression Changes in High and Low Grade Breast Cancer Cell Types. *Int. J. Mol. Sci.* **2018**, *19*, 1084. [CrossRef]
2. Nicolini, A.; Ferrari, P.; Duffy, M.J. Prognostic and predictive biomarkers in breast cancer: Past, present and future. *Semin. Cancer Biol.* **2018**, *52*, 56–73. [CrossRef] [PubMed]
3. Choi, C.; Park, S.; Cho, W.K.; Choi, D.H. Cyclin D1 is Associated with Radiosensitivity of Triple-Negative Breast Cancer Cells to Proton Beam Irradiation. *Int. J. Mol. Sci.* **2019**, *20*, 4943. [CrossRef] [PubMed]
4. Kim, K.; Park, H.J.; Shin, K.H.; Kim, J.H.; Choi, O.H.; Park, W.; Ahn, S.D.; Kim, S.S.; Kim, D.Y.; Kim, T.H.; et al. Breast Conservation Therapy Versus Mastectomy in Patients with T1-2N1 Triple-Negative Breast Cancer: Pooled Analysis of KROG 14-18 and 14-23. *Cancer Res. Treat.* **2018**, *50*, 1316–1323. [CrossRef]
5. Wang, J.; Xie, X.; Wang, X.; Tang, J.; Pan, Q.; Zhang, Y.; Di, M. Locoregional and distant recurrences after breast conserving therapy in patients with triple-negative breast cancer: A meta-analysis. *Surg. Oncol.* **2013**, *22*, 247–255. [CrossRef] [PubMed]
6. Abdulkarim, B.; Cuartero, J.; Hanson, J.; Deschênes, J.; Lesniak, D.; Sabri, S. Increased Risk of Locoregional Recurrence for Women with T1-2N0 Triple-Negative Breast Cancer Treated With Modified Radical Mastectomy Without Adjuvant Radiation Therapy Compared With Breast-Conserving Therapy. *J. Clin. Oncol.* **2011**, *29*, 2052–2050. [CrossRef]
7. Bravatà, V.; Cammarata, F.P.; Minafra, L.; Pisciotta, P.; Scazzone, C.; Manti, L.; Savoca, G.; Petringa, G.; Cirrone, G.A.P.; Cuttone, G.; et al. Proton-irradiated breast cells: Molecular points of view. *J. Radiat. Res.* **2019**, *60*, 451–465. [CrossRef]
8. Bravatà, V.; Minafra, L.; Cammarata, F.P.; Pisciotta, P.; Lamia, D.; Marchese, V.; Petringa, G.; Manti, L.; Ap Cirrone, G.; Gilardi, M.C.; et al. Gene expression profiling of breast cancer cell lines treated with proton and electron radiations. *Br. J. Radiol.* **2018**, *91*, 1089. [CrossRef]
9. Tommasino, F.; Durante, M.; D'Avino, V.; Liuzzi, R.; Conson, M.; Farace, P.; Palma, G.; Schwarz, M.; Cella, L.; Pacelli, R. Model-based approach for quantitative estimates of skin, heart, and lung toxicity risk for left-side photon and proton irradiation after breast-conserving surgery. *Acta Oncol.* **2017**, *56*, 730–736. [CrossRef]
10. Spetz, J.; Moslehi, J.; Sarosiek, K. Radiation-Induced Cardiovascular Toxicity: Mechanisms, Prevention, and Treatment. *Curr. Treat. Options Cardiovasc. Med.* **2018**, *20*, 31. [CrossRef]
11. Chowdhary, M.; Lee, A.; Gao, S.; Wang, D.; Barry, P.N.; Diaz, R.; Bagadiya, N.R.; Park, H.S.; Yu, J.B.; Wilson, L.D.; et al. Is Proton Therapy a "Pro" for Breast Cancer? A Comparison of Proton vs. Non-proton Radiotherapy Using the National Cancer Database. *Front. Oncol.* **2019**, *8*, 678. [CrossRef]
12. Cuaron, J.; Macdonald, S.M.; Cahlon, O. Novel applications of proton therapy in breast carcinoma. *Chin. Clin. Oncol.* **2016**, *5*, 52. [CrossRef] [PubMed]
13. Corbin, K.S.; Mutter, R.W. Proton therapy for breast cancer: Progress & pitfalls. *Breast Cancer Manag.* **2018**, *7*, BMT06. [CrossRef]

14. Orecchia, R.; Fossati, P.; Zurrida, S.; Krengli, M. New frontiers in proton therapy. *Curr. Opin. Oncol.* **2015**, *27*, 427–432. [CrossRef] [PubMed]
15. Parshad, R.; Suhani; Kazi, M.; Seenu, V.; Mathur, S.; Dattagupta, S.; Haresh, K.P. Triple-negative breast cancers: Are they always different from nontriple-negative breast cancers? An experience from a tertiary center in India. *Indian J. Cancer* **2018**, *54*, 658–663. [CrossRef]
16. Gong, C.; Ma, G.; Hu, X.; Zhang, Y.; Wang, Z.-H.; Zhang, J.; Zhao, Y.; Li, Y.; Xie, Y.; Yang, Z.; et al. Pretreatment 18 F-FDG Uptake Heterogeneity Predicts Treatment Outcome of First-Line Chemotherapy in Patients with Metastatic Triple-Negative Breast Cancer. *Oncologist* **2018**, *23*, 1144–1152. [CrossRef]
17. Russo, G.; Pisciotta, P.; Cirrone, G.; Romano, F.; Cammarata, F.; Marchese, V.; Forte, G.; Lamia, D.; Minafra, L.; Bravatá, V.; et al. Preliminary study for small animal preclinical hadrontherapy facility. *Nucl. Instrum. Methods Phys. Res. Sect.* **2017**, *846*, 126–134. [CrossRef]
18. Pauwels, A.-M.; Trost, M.; Beyaert, R.; Hoffmann, E. Patterns, Receptors, and Signals: Regulation of Phagosome Maturation. *Trends Immunol.* **2017**, *38*, 407–422. [CrossRef]
19. Iozzo, R.V.; Sanderson, R.D. Proteoglycans in cancer biology, tumour microenvironment and angiogenesis. *J. Cell. Mol. Med.* **2011**, *15*, 1013–1031. [CrossRef]
20. Tommasino, F.; Durante, M. Proton Radiobiology. *Cancers* **2015**, *7*, 353–381. [CrossRef]
21. Pisciotta, P.; Costantino, A.; Cammarata, F.P.; Torrisi, F.; Marchese, V.; Cirrone, G.A.P.; Petringa, G.; Forte, G.I.; Minafra, L.; Bravatà, V.; et al. Evaluation of Proton Beam Radiation-induced Skin injury in a murine model by end-part of a SOBP. *PLoS ONE* **2020**, *22*, e0233258.
22. Workman, P.; An ad hoc committee of the National Cancer Research Institute; Aboagye, E.O.; Balkwill, F.; Balmain, A.; Bruder, G.; Chaplin, D.J.; Double, J.A.; Everitt, J.; Farningham, D.A.H.; et al. Guidelines for the welfare and use of animals in cancer research. *Br. J. Cancer* **2010**, *102*, 1555–1577. [CrossRef] [PubMed]
23. Liu, L.; Zhang, Z.; Zhou, L.; Hu, L.; Yin, C.; Qing, D.; Huang, S.; Cai, X.; Chen, Y.; Qin, J. Cancer associated fibroblasts-derived exosomes contribute to radioresistance through promoting colorectal cancer stem cells phenotype. *Exp. Cell Res.* **2020**, *391*, 111956. [CrossRef] [PubMed]
24. Yadav, P.; Shankar, B.S. Radio resistance in breast cancer cells is mediated through TGF-β signalling, hybrid epithelial-mesenchymal phenotype and cancer stem cells. *Biomed. Pharmacother.* **2019**, *111*, 119–130. [CrossRef]
25. Langlands, F.E.; Horgan, K.; Dodwell, D.D.; Smith, L. Breast cancer subtypes: Response to radiotherapy and potential radiosensitisation. *Br. J. Radiol.* **2013**, *86*, 20120601. [CrossRef]
26. Willers, H.; Azzoli, C.G.; Santivasi, W.L.; Xia, F. Basic Mechanisms of Therapeutic Resistance to Radiation and Chemotherapy in Lung Cancer. *Cancer J.* **2013**, *19*, 200–207. [CrossRef]
27. Césaire, M.; Ghosh, U.; Austry, J.-B.; Muller, E.; Cammarata, F.P.; Guillamin, M.; Caruso, M.; Castéra, L.; Petringa, G.; Cirrone, G.A.P.; et al. Sensitization of chondrosarcoma cells with PARP inhibitor and high-LET radiation. *J. Bone Oncol.* **2019**, *17*, 100246. [CrossRef]
28. Niazi, M.K.K.; Senaras, C.; Pennell, M.L.; Arole, V.; Tozbikian, G.; Gurcan, M.N. Relationship between the Ki67 index and its area based approximation in breast cancer. *BMC Cancer* **2018**, *18*, 867. [CrossRef]
29. Niazi, M.K.K.; Downs-Kelly, E.; Gurcan, M.N. Hot spot detection for breast cancer in Ki-67 stained slides: Image dependent filtering approach. In *Medical Imaging 2014: Digital Pathology*; SPIE: Bellingham, WA, USA, 2014.
30. Di Maggio, F.M.; Minafra, L.; Forte, G.I.; Cammarata, F.P.; Lio, D.; Messa, C.; Gilardi, M.C.; Bravatà, V. Portrait of inflammatory response to ionizing radiation treatment. *J. Inflamm.* **2015**, *12*, 14. [CrossRef]
31. Minafra, L.; Bravatà, V. Cell and molecular response to IORT treatment. *Transl. Cancer Res.* **2014**, *3*, 32–47.
32. Aggarwal, B.B.; Sethi, G.; Ahn, K.S.; Sandur, S.K.; Pandey, M.K.; Kunnumakkara, A.B.; Sung, B.; Ichikawa, H. Targeting signal-transducer-and-activator-of-transcription-3 for prevention and therapy of cancer: Modern target but ancient solution. *Ann. N. Y. Acad. Sci.* **2006**, *1091*, 151–169. [CrossRef] [PubMed]
33. Shimamura, H.; Terada, Y.; Okado, T.; Tanaka, H.; Inoshita, S.; Sasaki, S. The PI3-kinase-Akt pathway promotes mesangial cell survival and inhibits apoptosis in vitro via NF-kappa B and Bad. *J. Am. Soc. Nephrol.* **2003**, *14*, 1427–1434. [CrossRef]
34. Bravatà, V.; Cammarata, F.P.; Minafra, L.; Musso, R.; Pucci, G.; Spada, M.; Fazio, I.; Russo, G.; Forte, G.I.; Cammarata, P.F. Gene Expression Profiles Induced by High-dose Ionizing Radiation in MDA-MB-231 Triple-negative Breast Cancer Cell Line. *Cancer Genom. Proteom.* **2019**, *16*, 257–266. [CrossRef]

35. Martínez, P.; Gómez-López, G.; Pisano, D.G.; Flores, J.M.; Blasco, M.A. A genetic interaction between RAP1 and telomerase reveals an unanticipated role for RAP1 in telomere maintenance. *Aging Cell* **2016**, *15*, 1113–1125. [CrossRef]
36. Lin, H.-J.; Liu, H.-H.; Lin, C.-D.; Kao, M.-C.; Chen, Y.-A.; Chiang-Ni, C.; Jiang, Z.-P.; Huang, M.-Z.; Lin, C.-J.; Lo, U.-G.; et al. Cytolethal Distending Toxin Enhances Radiosensitivity in Prostate Cancer Cells by Regulating Autophagy. *Front. Microbiol.* **2017**, *7*, 223. [CrossRef] [PubMed]
37. Gueulette, J.; Blattmann, H.; Pedroni, E.; Coray, A.; De Coster, B.M.; Mahy, P.; Wambersie, A.; Goitein, G. Relative biologic effectiveness determination in mouse intestine for scanning proton beam at Paul Scherrer Institute, Switzerland. Influence of motion. *Int. J. Radiat. Oncol.* **2005**, *62*, 838–845. [CrossRef] [PubMed]
38. Pisciotta, P.; Russo, G.; Marchese, V.; Cirrone, G.; Romano, F.; Lamia, D.; Cuttone, G.; Gilardi, M. Preliminary dosimetric study for preclinical small animal hadrontherapy. *Phys. Medica* **2016**, *32*, 53. [CrossRef]
39. Broggi, G.; Musumeci, G.; Puzzo, L.; Russo, A.; Reibaldi, M.; Ragusa, M.; Longo, A.; Caltabiano, R. Immunohistochemical Expression of ABCB5 as a Potential Prognostic Factor in Uveal Melanoma. *Appl. Sci.* **2019**, *9*, 1316. [CrossRef]
40. Salgado, R.; Denkert, C.; DeMaria, S.; Sirtaine, N.; Klauschen, F.; Pruneri, G.; Wienert, S.; Eynden, G.V.D.; Baehner, F.L.; Penault-Llorca, F.; et al. The evaluation of tumor-infiltrating lymphocytes (TILs) in breast cancer: Recommendations by an International TILs Working Group 2014. *Ann. Oncol.* **2015**, *26*, 259–271. [CrossRef]
41. Bravatà, V.; Minafra, L.; Russo, G.; Forte, G.I.; Cammarata, F.P.; Ripamonti, M.; Casarino, C.; Augello, G.; Costantini, F.; Barbieri, G.; et al. High-dose Ionizing Radiation Regulates Gene Expression Changes in the MCF7 Breast Cancer Cell Line. *Anticancer. Res.* **2015**, *35*, 2577–2591.
42. Minafra, L.; Bravatà, V.; Russo, G.; Forte, G.I.; Cammarata, F.P.; Ripamonti, M.; Candiano, G.; Cervello, M.; Giallongo, A.; Perconti, G.; et al. Gene Expression Profiling of MCF10A Breast Epithelial Cells Exposed to IOERT. *Anticancer. Res.* **2015**, *35*, 3223–3234. [PubMed]
43. Minafra, L.; Porcino, N.; Bravatà, V.; Gaglio, D.; Bonanomi, M.; Amore, E.; Cammarata, F.P.; Russo, G.; Militello, C.; Savoca, G.; et al. Radiosensitizing effect of curcumin-loaded lipid nanoparticles in breast cancer cells. *Sci. Rep.* **2019**, *9*, 1–16. [CrossRef] [PubMed]
44. Becker, K.; Hosack, D.A.; Dennis, G.; Lempicki, R.A.; Bright, T.J.; Cheadle, C.; Engel, J. PubMatrix: A tool for multiplex literature mining. *BMC Bioinform.* **2003**, *4*, 61. [CrossRef] [PubMed]
45. Minafra, L.; Bravatà, V.; Cammarata, F.P.; Russo, G.; Gilardi, M.C.; Forte, G.I. Radiation Gene-expression Signatures in Primary Breast Cancer Cells. *Anticancer. Res.* **2018**, *38*, 2707–2715. [CrossRef] [PubMed]
46. Cirrone, G.; Margarone, D.; Maggiore, M.; Anzalone, A.; Borghesi, M.; Jia, S.B.; Bulanov, S.S.; Bulanov, S.; Carpinelli, M.; Cavallaro, S.; et al. ELIMED: A new hadron therapy concept based on laser driven ion beams. In *Laser Acceleration of Electrons, Protons, and Ions II; and Medical Applications of Laser-Generated Beams of Particles II; and Harnessing Relativistic Plasma Waves III*; International Society for Optics and Photonics: Bellingham, WA, USA, 2013.

International Journal of
Molecular Sciences

Article

Checkpoint Kinase 1 (CHK1) Inhibition Enhances the Sensitivity of Triple-Negative Breast Cancer Cells to Proton Irradiation via Rad51 Downregulation

Changhoon Choi [1,†], Won Kyung Cho [1,†], Sohee Park [1], Sung-Won Shin [1,2], Won Park [1,2], Haeyoung Kim [1,2] and Doo Ho Choi [1,2,*]

1 Department of Radiation Oncology, Samsung Medical Center, Seoul 06351, Korea; chchoi93@gmail.com (C.C.); wklove.cho@samsung.com (W.K.C.); psh3842@nate.com (S.P.); camuserik@gmail.com (S.-W.S.); wonro.park@samsung.com (W.P.); haeyoung0131.kim@samsung.com (H.K.)
2 Department of Radiation Oncology, Sungkyunkwan University School of Medicine, Seoul 06351, Korea
* Correspondence: doho.choi@samsung.com; Tel.: +82-2-3410-2436
† These authors contributed equally to this study.

Received: 27 March 2020; Accepted: 12 April 2020; Published: 13 April 2020

Abstract: Due to a superior dose conformity to the target, proton beam therapy (PBT) continues to rise in popularity. Recently, considerable efforts have been directed toward discovering treatment options for use in combination with PBT. This study aimed to investigate the targeting of checkpoint kinase 1 (CHK1), a critical player regulating the G2/M checkpoint, as a promising strategy to potentiate PBT in human triple-negative breast cancer (TNBC) cells. Protons induced cell-cycle arrest at the G2/M checkpoint more readily in response to increased CHK1 activation than X-rays. A clonogenic survival assay revealed that CHK1 inhibition using PF-477736 or small interfering RNA (siRNA) enhanced the sensitivity toward protons to a greater extent than toward X-rays. Western blotting demonstrated that PF-477736 treatment in the background of proton irradiation increased the pro-apoptotic signaling, which was further supported by flow cytometry using annexin V. Immunofluorescence revealed that proton-induced DNA double-strand breaks (DSBs) were further enhanced by PF-477736, which was linked to the downregulation of Rad51, essential for the homologous recombination repair of DSBs. Direct inactivation of Rad51 resulted in enhanced proton sensitization. Collectively, these data suggest that targeting CHK1 may be a promising approach for improving PBT efficacy in the treatment of TNBC.

Keywords: proton therapy; triple-negative breast cancer; CHK1; Rad51

1. Introduction

Triple-negative breast cancer (TNBC) is a breast cancer subtype characterized by the lack of estrogen receptor (ER) and progesterone receptor (PR) expression, and absence of human epidermal growth factor receptor 2 (HER2) amplification. Moreover, TNBC constitutes approximately 10-20% of all breast cancers [1,2]. Subsets of TNBC have demonstrated a remarkable response to chemotherapy, resulting in a better prognosis. In contrast, other subsets of TNBC have shown a poor response to treatment, consequently resulting in a poor prognosis [1]. Thus, several therapeutic agents such as poly ADP-ribose polymerase (PARP) inhibitors and PI3K/mTOR/AKT inhibitors are being tested as potential treatments for patients with TNBC [3–5]; however, chemotherapy remains the most beneficial form of treatment owing to the lack of approved targeted drugs.

Radiation therapy (RT) is a treatment method that locally destroys cancer cells using high energy photons or particles. Many preclinical and clinical studies have been performed to determine RT response in different molecular subtypes of breast cancer, and sensitivity to radiation seems to be higher in hormone-positive tumors as compared to that of HER2-type or TNBC [6]. To enhance the effects of RT in TNBC, several agents have been investigated in preclinical studies [5,7]. Cell cycle checkpoints are often deregulated in cancer cells, and thus, have been used as therapeutic targets. Mutations in TP53, a key protein regulating the G1 checkpoint, are more frequently seen in TNBCs than in ER-positive breast cancers. Thus, RT in combination with targeting of the G2/M checkpoint is an advantageous strategy for the treatment of TNBC. Furthermore, G2/M checkpoint abrogation using checkpoint kinase 1 (CHK1) inhibitors is known to increase the RT efficacy in preclinical models of pancreatic cancer [8], colorectal cancer [9], bladder cancer [10], and TNBC [11].

Proton beam therapy (PBT) is gaining popularity mainly due to the physical superiority of its Bragg peak, which leaves the healthy organs unaffected [12]. In breast cancer, this unique energy deposit profile can reduce heart and lung exposure to radiation, thereby potentially decreasing risk of cardiac toxicity and pneumonitis [12]. In radiobiology, the understanding of the advantages of proton beams over photon beams is still limited, although genetic alterations could modify the biological effectiveness of protons relative to that of photons in the lung, head and neck, liver, and breast cancer cells [13–16]. Proton beams induce more complex DNA damage, resulting in more prolonged DNA-damage repair and cell-cycle arrest than that observed in response to X-ray irradiation [17]. Furthermore, proton beams are capable of eliminating homologous recombination repair (HRR)-deficient tumor cells more effectively than photon beams [18]. Our recent study demonstrated that the sensitivity to protons varies across a panel of breast cancer cell lines, and that a cyclin D1/CDK4/RB1 axis may be implicated in determining proton sensitivity in TNBC cells. Considerable efforts are being directed towards discovering, or screening targeted drugs in combination with charged particles, including PBT [19]. This study aimed to evaluate the targeting of CHK1, a critical factor regulating the G2/M checkpoint, as a promising strategy to potentiate PBT in human TNBC cells.

2. Results

2.1. X-Rays and Protons Differently Affect ATR/CHK1 Activation in Human TNBC Cell Lines

To further understand the biological impact of proton irradiation in TNBC cells, we investigated the differential cell cycle checkpoint activation by X-rays and proton irradiation in two different TNBC cell lines, MDA-MB-231 and Hs578T. As both cell lines have a mutation in the TP53 gene, an essential gene for the G1/S checkpoint, we focused on the ataxia telangiectasia and rad3-related (ATR)/CHK1 signaling and the G2/M checkpoint. The samples were prepared 0.5 h, 2 h, and 24 h after exposure to 4 Gy of X-rays or protons. Western blotting revealed that in MDA-MB-231 cells, phosphorylation of ATR at Ser428 and BRCA1 at Ser1524 was strongly induced by protons at 0.5 h, with a concomitant increase in γH2AX expression, a surrogate marker for DNA double-strand breaks (DSBs) (Figure 1A). In addition, protons increased the phosphorylation of CHK1 at Ser345 to a greater extent than X-rays (Figure 1A). In Hs578T cells, there was a substantial increase in the phosphorylation of CHK1 and H2AX in response to both radiations. However, the difference between the two radiations was less pronounced in Hs578T cells than that in MDA-MB-231 cells (Figure 1B). Cell-cycle analysis showed that both radiations increased the cell population at the G2/M phase with a decrease in the G1 phase population in TNBC cells (Figure 1C,D). The effect of protons on the G2/M arrest was more prominent in MDA-MB-231 cells (Figure 1C) than that in Hs578T cells (Figure 1D). These data indicate that proton irradiation leads to the activation of CHK1 to a greater extent than that induced by X-rays, possibly due to more complex DNA damage, thereby resulting in a more pronounced G2/M arrest.

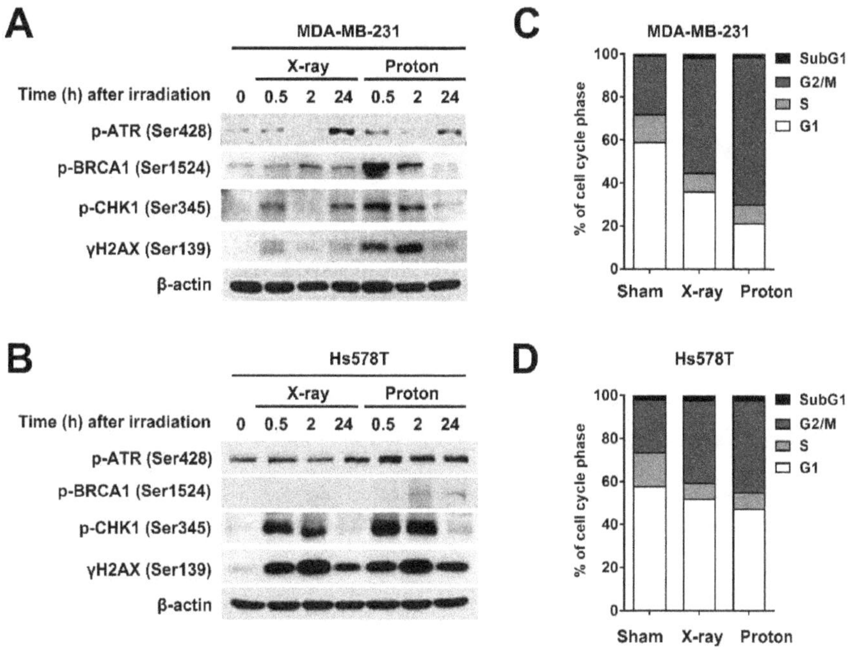

Figure 1. Differential checkpoint kinase 1 (CHK1) activation and cell cycle redistribution in response to irradiation with X-rays and protons in human triple-negative breast cancer (TNBC) cells. Western blotting showed differential activation of ataxia telangiectasia and rad3-related (ATR)/CHK1 signaling in response to X-ray and proton irradiation in MDA-MB-231 cells (**A**) and Hs578T cells (**B**). Cells were harvested at the indicated times after irradiation with 4 Gy of X-rays or protons. β-actin was used as a loading control. Cell-cycle distribution in MDA-MB-231 cells (**C**) and Hs578T (**D**) after irradiation with 4 Gy of X-rays or protons. Cells were harvested 24 h post-irradiation and subjected to flow cytometry.

2.2. *Pharmacological Inhibition of CHK1 Induces DNA Damage and Apoptosis in TNBC Cells*

Based on our findings that protons activated CHK1 and the G2/M checkpoint more readily than X-rays (Figure 1A), we targeted CHK1 using a pharmacological inhibitor and evaluated its effect on proton sensitization. Before investigating the efficacy of proton irradiation with respect to CHK1 inhibition in human breast cancer cells, we tested the cytotoxicity of a selective pharmacological CHK1 inhibitor, PF-477736 in TNBC cells. The results of the Cell Counting Kit-8 (CCK-8) assay showed a dose-dependent inhibition of cell viability in response to PF-477736 treatment, with the half maximal-inhibitory concentration (IC_{50}) being 0.84 μM and 0.33 μM for MDA-MB-231 and Hs578T cells, respectively (Figure 2A,B). The inhibitory effect of PF-477736 on CHK1 activity was evident from the decreased levels of phospho-Ser296 CHK1 and an increase in phospho-Ser345 CHK1 levels in MDA-MB-231 cells (Figure 2C). Inhibition of CHK1 activity impairs CHK1 autophosphorylation at Ser296, stabilizes cdc25a, and potentiates ATR-mediated CHK1 phosphorylation at Ser345, which provides a docking site for E3 ligase, leading to ubiquitin-proteasome-dependent degradation of CHK1 (Figure 2C,D). Additionally, PF-477736 also decreased the expression of phospho-Tyr15-cdc2 (a target of CHK1) and Rad51. These data are consistent with previous results on another CHK1 inhibitor, AZD7762 [8]. PF-477736 increased DNA damage and apoptosis in a dose-dependent manner, which was assessed based on the increase in the levels of γH2AX and cleaved PARP (Figure 2E,F). Moreover, Hs578T cells demonstrated a DNA damage response and apoptotic cell death at 100 nM of PF-477736, whereas MDA-MB-231 cells did not. Another TNBC cell line, MDA-MB-453, also showed sensitivity to PF-477736 in a manner similar to Hs578T (Figure S1A,B), suggesting that MDA-MB-231 cells are relatively resistant to the CHK1 inhibitor.

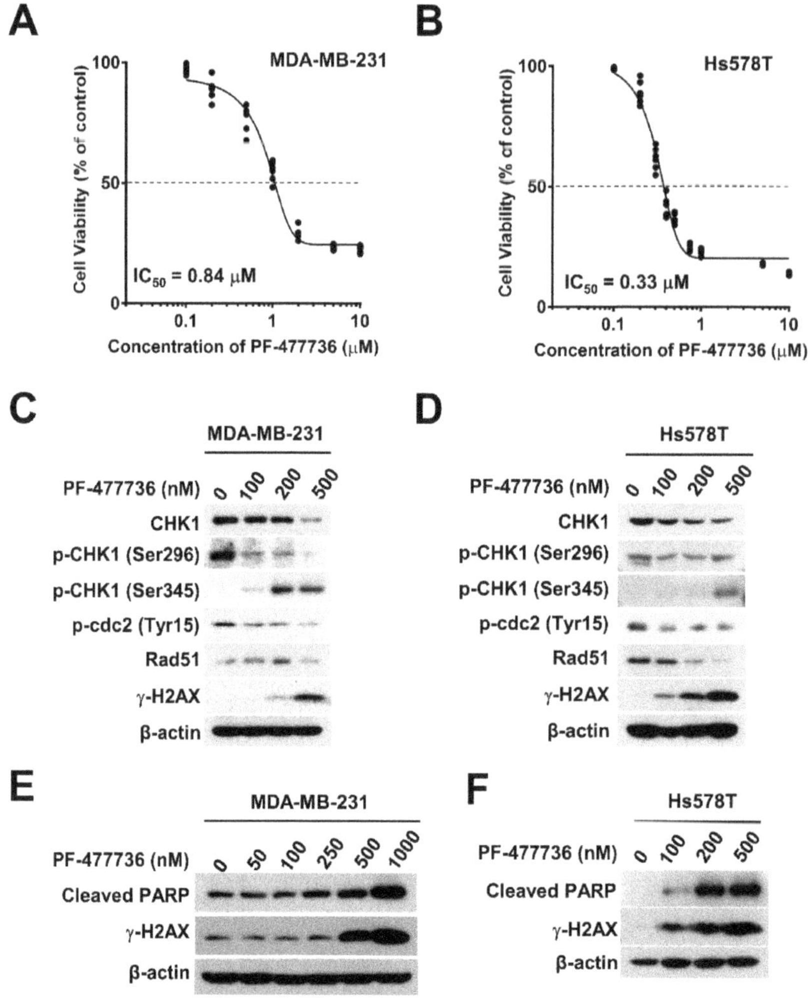

Figure 2. The cytotoxicity of the CHK1 inhibitor, PF-477736 is different in human TNBC cells. Cytotoxic effects of PF-477736 on MDA-MB-231 cells (**A**) and Hs578T cells (**B**). Cell proliferation was measured using the Cell Counting Kit-8 (CCK-8) assay kit after 72 h of PF-477736 treatment. Concentration-dependent inhibition of CHK1 activity in response to PF-477736 treatment in MDA-MB-231 cells (**C**) and Hs578T cells (**D**). Cells were incubated with the indicated concentrations of PF-477736 for 48 h. β-actin was used as a loading control. Concentration-dependent cellular damage in response to PF-477736 treatment in MDA-MB-231 cells (**E**) and Hs578T cells (**F**). Cleaved polymerase (PARP) and γ-H2AX were used as markers of apoptosis and DNA damage. β-actin was used as a loading control.

2.3. CHK1 Inhibition in Response to PF-477736 Treatment Sensitizes TNBC Cells to Proton Irradiation

In order to determine the effect of CHK1 inhibition on the sensitivity to radiation, we performed a clonogenic survival assay. MDA-MB-231 cells were pre-treated with 100 nM PF-477736 for 3 h and then exposed to 4 Gy of X-rays or protons. The surviving colonies were stained and counted after 14 d of irradiation. Both X-rays and protons effectively inhibited the clonogenic survival of MDA-MB-231 cells; protons significantly affected the survival as compared to X-rays ($p < 0.01$; Figure 3A,B). The effect of the combination of PF-477736 and irradiation was significant in proton-irradiated cells ($p < 0.01$) than that in X-ray-irradiated cells ($p < 0.05$; Figure 3B). Cell counts revealed significantly lower clonogenic survival of MDA-MB-231 cells in response to combinatorial treatment with 100 nM PF-477736 and proton irradiation than with X-ray irradiation ($p < 0.01$; Figure 3B). Next, the effect of PF-477736 on radiation-induced cell-cycle redistribution was determined (Figure 3C). As shown in Figure 1C, proton irradiation led to a marked arrest at the G2/M phase. PF-477736 dramatically increased the cell population in the S-phase, indicating the abrogation of radiation-induced G2/M arrest (Figure 3C).

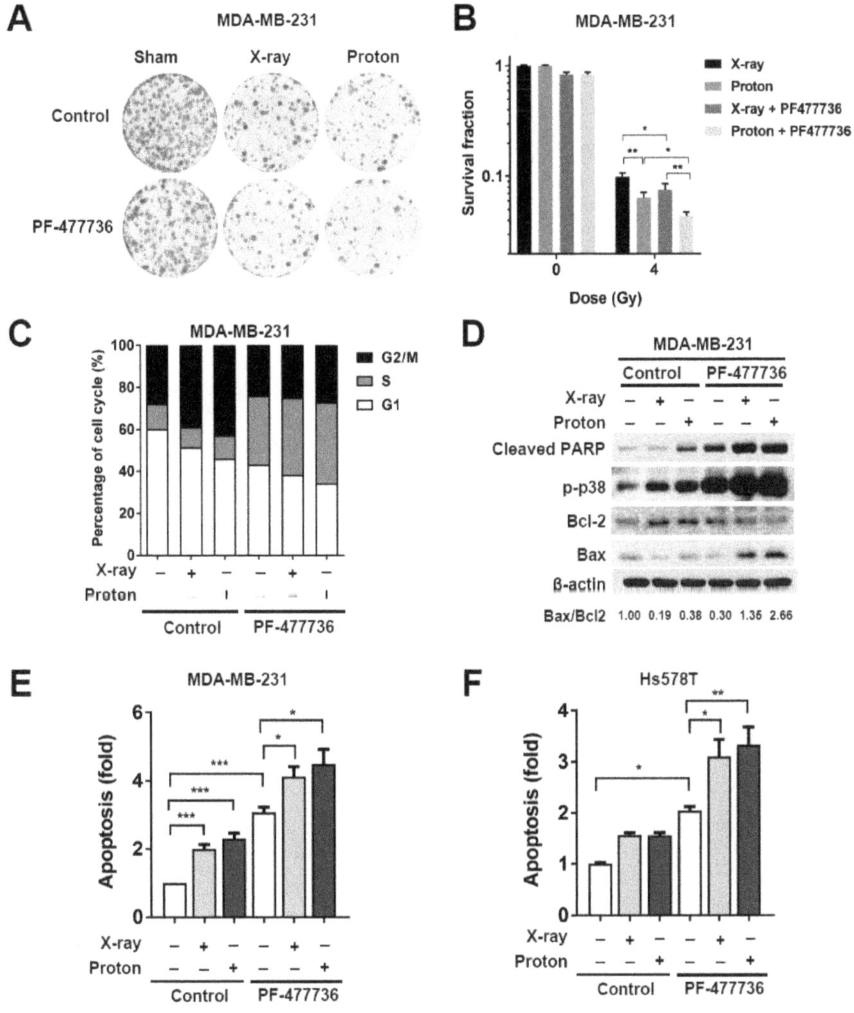

Figure 3. CHK1 inhibition in response to PF-477736 treatment sensitizes TNBC cells to proton irradiation. (**A**) Effect of co-treatment with PF-477736 and 4 Gy of X-rays or protons on clonogenic survival. MDA-MB-231 cells were pre-treated with 100 nM PF-477736 for 3 h, followed by irradiation with 4 Gy of X-rays or protons. Colonies were stained with crystal violet after 14 days; (**B**) Quantification

of survived colonies. Data are shown as mean ± S.D. from two independent experiments. * $p < 0.05$; ** $p < 0.01$; (**C**) Cell-cycle distribution after combinatorial treatment with PF-477736 and X-rays or protons. MDA-MB-231 cells pre-treated with 500 nM PF-477746 for 3 h were irradiated with 4 Gy of X-rays or protons and were harvested 24 h after irradiation for flow cytometric analysis. X-ray and proton irradiations led to a significant increase in the cell population at the G2/M phase, and their combination with PF-477736 increased the cell population at the S phase; (**D**) Western blotting showed that pre-treatment of 500 nM PF-477736 for 3 h, followed by 4 Gy radiation increased apoptotic signaling, as compared to that seen upon irradiation alone. Cells were harvested 72 h post-irradiation. β-actin was used as a loading control. Densitometric analysis showed increased Bcl-2-associated X (Bax)/B-cell lymphoma 2 (Bcl-2) ratio after the combined treatment. Enhanced apoptosis in response to combinatorial treatment with 500 nM PF-477736 and 4 Gy radiation in MDA-MB-231 cells (**E**) and Hs578T cells (**F**). MDA-MB-231 cells and Hs578T cells were pre-treated for 3 h with 500 nM and 100 nM of PF477736, respectively, followed by irradiation with 4 Gy of X-rays or protons. Cells were harvested 72 h post-irradiation and apoptotic population was determined as described in Materials and Methods. Quantification data were shown. Data are shown as mean ± S.D. from three independent experiments * $p < 0.05$; ** $p < 0.01$; *** $p < 0.001$.

Furthermore, we found that treatment with 500 nM PF-477736 increased the levels of pro-apoptotic proteins such as Bcl-2-associated X (Bax), phospho-p38, and cleaved PARP in MDA-MB-231 cells, which were further enhanced in response to combinatorial treatment with PF-477736 and 6 Gy of either X-rays or protons (Figure 3D). The densitometric analysis showed that the Bax/B-cell lymphoma 2 (Bcl-2) ratio was the highest upon co-treatment with protons and PF-477736. Analysis of apoptosis using annexin V/propidium iodide double-staining confirmed that apoptosis of MDA-MB-231 cells was significantly increased in response to irradiation with either X-rays ($p < 0.001$) or protons ($p < 0.001$; Figure 3E). Further, treatment with 500 nM PF-477736 increased the number of apoptotic cells ($p < 0.001$), which was further increased upon combination with the two radiations ($p < 0.05$). However, no significant difference between the combinatorial treatments using either proton or X-ray irradiation was observed with respect to the number of apoptotic cells (Figure 3E). In Hs578T cells, treatment with 100 nM PF-477736 induced apoptosis ($p < 0.05$) and both radiations enhanced apoptosis when combined with 100 nM PF-477736 ($p < 0.05$ and $p < 0.01$ for X-rays and protons, respectively; Figure 3F). MDA-MB-453 cells also showed enhanced apoptotic cell death in response to combinatorial treatment with 100 nM PF-477736 and protons (Figure S1C).

2.4. CHK1 Knockdown via Small Interfering RNA (siRNA) Treatment Sensitizes MDA-MB-231 Cells to Proton Irradiation

Increased S-phase cells upon treatment with PF-477736 alone was not different from those upon its combination treatment with radiations (Figure 3C), suggesting that PF-477736 might have non-specific toxicity. To determine whether proton radiosensitization by PF-477736 may be off-target effects, we performed gene silencing experiments for CHK1. Western blotting confirmed siRNA-mediated knockdown of CHK1 in MDA-MB-231 cells (Figure 4A). Radiosensitization in response to CHK1 knockdown was determined using the clonogenic assay (Figure 4B). As compared to the control siRNA, CHK1 siRNA lowered the fraction of survived cells in irradiated MDA-MB-231 cells, which was further evident upon proton irradiation (Figure 4B). With respect to pharmacological inhibition, siRNA-mediated CHK1 knockdown enhanced apoptosis, as evidenced by the increased levels of cleaved caspase-3. As compared to the control siRNA, CHK1 siRNA increased the expression of the pro-apoptotic protein Bak with a concomitant decrease in the expression of the anti-apoptotic protein Bcl-XL. In addition, the ratio of Bak/Bcl-XL was the highest in the cells co-treated with siCHK1 and protons (Figure 4C). The effects of CHK1 knockdown on apoptotic cell death were similar to those of PF-477736; CHK1 knockdown enhanced proton-induced apoptosis, as evidenced by annexin V staining in flow cytometry (Figure 4D). Together, these data suggest that CHK1 inhibition by either a small molecule inhibitor or siRNA increased proton radiosensitivity of TNBC cells via induction of apoptosis.

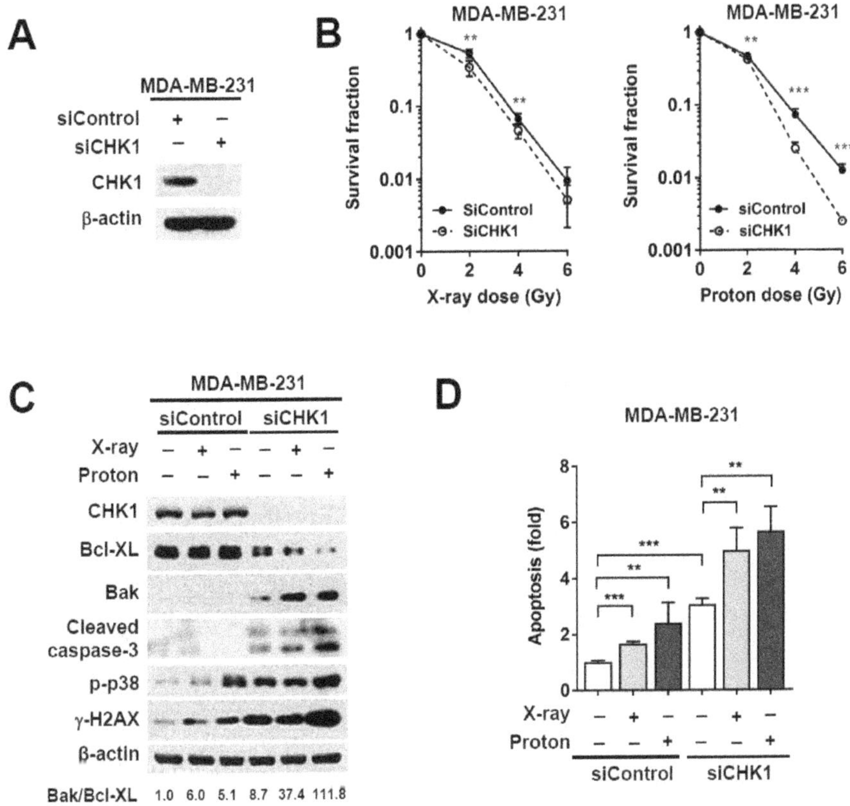

Figure 4. CHK1 silencing using small interfering RNA (siRNA) sensitizes MDA-MB-231 cells to proton irradiation. (**A**) siRNA-mediated CHK1 knockdown was confirmed using western blotting. MDA-MB-231 cells were treated with 10 nM siRNA, and CHK1 levels were determined after 24 h. β-actin was used as loading control; (**B**) The clonogenic survival assay showed that siRNA-mediated CHK1 knockdown resulted in enhanced proton radiosensitivity of MDA-MB-231 cells. Data are shown as mean ± S.D. from two independent experiments. ** $p < 0.01$; *** $p < 0.001$; (**C**) Western blot showed that CHK1 knockdown using siRNA further increased the radiation-induced pro-apoptotic signaling. Cells were treated with 10 nM siRNA for 24 h, followed by irradiation with 4 Gy of X-rays or protons. Cells were harvested 72 h post-irradiation. β-actin was used as a loading control. Densitometric analysis showed increased Bak/Bcl-XL ratio after co-treatment with CHK1 siRNA and protons; (**D**) Flow cytometric analysis revealed that apoptosis was enhanced in response to combinatorial treatment with CHK1 siRNA and 4 Gy of proton irradiation. Data are shown as mean ± S.D. from two independent experiments. ** $p < 0.01$; *** $p < 0.001$.

CHK1 plays a critical role in the DNA damage response as well as checkpoint activation. In order to understand the effect of CHK1 inhibition on proton-induced DNA damage, we evaluated the number of γH2AX foci in irradiated MDA-MB-231 cells in the presence or absence of PF-477736. Cells were fixed and stained with anti-γ-H2AX antibodies 24 h post-irradiation. Immunofluorescence showed that independent treatment with PF-477736 increased the number of γ-H2AX foci, as compared to those seen in the untreated control cells ($p < 0.01$; Figure 5A,B). Both radiations dramatically increased γH2AX foci formation ($p < 0.001$); proton irradiation induced γ-H2AX foci formation to a greater extent than X-ray irradiation ($p < 0.01$). Combinatorial treatment using PF-477736 and radiation further augmented the number of γ-H2AX foci, as compared to either treatment alone ($p < 0.001$); a greater number of γ-H2AX foci were seen in proton-irradiated cells than in X-ray-irradiated cells ($p < 0.01$; Figure 5A,B). Immunofluorescence of γ-H2AX at 30 min post-irradiation showed that both radiations increased γH2AX foci formation, but PF-477736 did not augment the radiation-induced γH2AX foci formation at such an early time point (Figure S2).

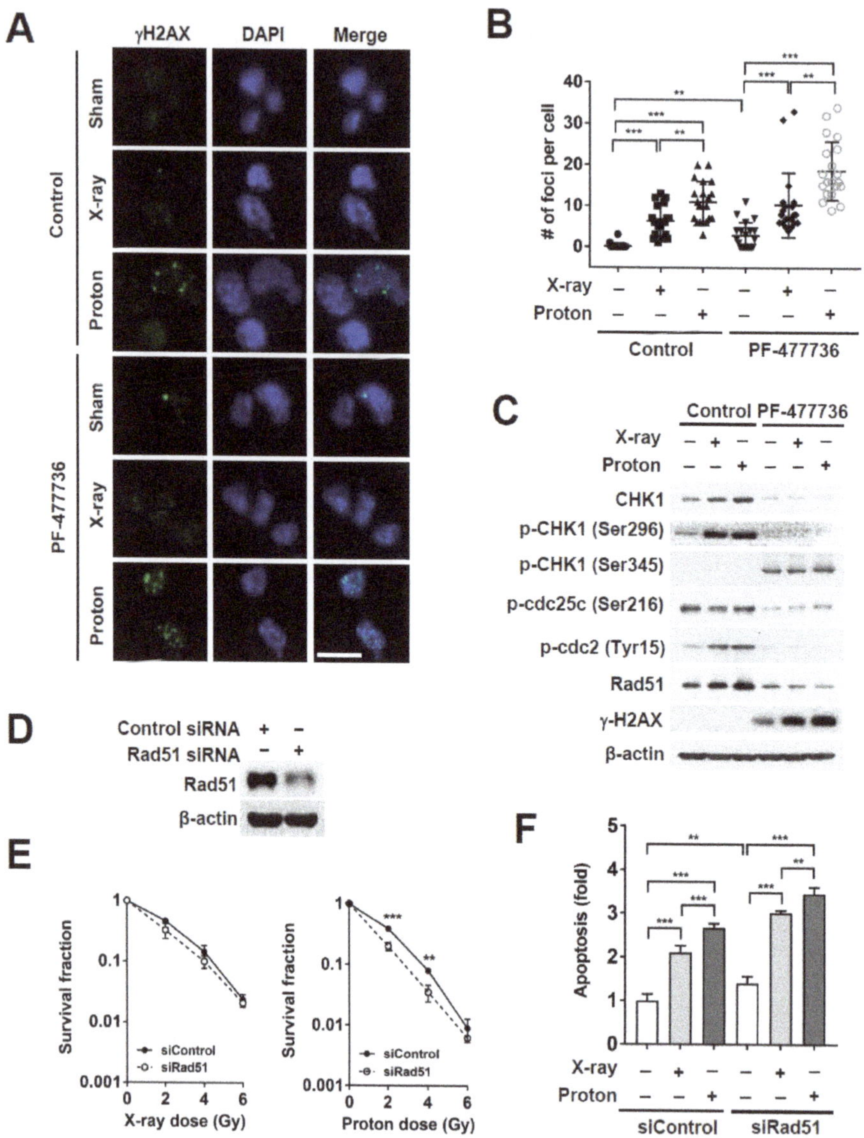

Figure 5. CHK1 inhibition in response to PF-477736 treatment augments proton irradiation-induced DNA damage in MDA-MB-231 cells via Rad51 downregulation. (**A**) Representative immunofluorescence

images showing enhanced proton-induced DNA damage upon CHK1 inhibition. MDA-MB-231 cells were pre-treated with 500 nM PF-477736 for 3 h, followed by irradiation with 4 Gy of X-rays or protons. After 24 h of irradiation, cells were fixed and probed using the γ-H2AX antibody (green) and stained with 4′, 6-diamidino-2-phenylindole (DAPI) (blue). The scale bar indicates 10 μm; (**B**) Quantification data showing a significant increase in the number of γ-H2AX foci in the nucleus upon pre-treatment with 500 nM PF-477736 for 3 h, followed by irradiation with 4 Gy of protons. Data are shown as mean ± S.D (n = 55) from two independent experiments. ** $p < 0.01$; *** $p < 0.001$; (**C**) Induction of Rad51 expression in response to proton irradiation was suppressed upon pre-treatment with PF-477736; (**D**) Western blotting verified siRNA-mediated knockdown of Rad51 in MDA-MB-231 cells. β-actin was used as a loading control; (**E**) Knockdown of Rad51 via siRNA increased the radiosensitization of MDA-MB-231 cells to protons as compared to that observed upon using control siRNA; however, it did not affect the radiosensitization to X-rays. Data are shown as mean ± S.D. from two independent experiments. ** $p < 0.01$; *** $p < 0.001$; (**F**) Knockdown of Rad51 via siRNA enhanced proton-induced apoptosis as compared to that observed upon using control siRNA. Data are shown as mean ± S.D. from two independent experiments. ** $p < 0.01$; *** $p < 0.001$.

Both X-ray and proton irradiation increased the levels of CHK1 and phospho-cdc2 (Tyr15). However, PF-477736 treatment abolished the increase in levels of these proteins (Figure 5C). The expression of Rad51, a DNA damage repair protein, was markedly reduced in PF-477736-treated cells, regardless of the radiation treatment (Figure 5C). In order to determine whether Rad51 might be implicated in proton sensitization upon CHK1 inhibition, we knocked down Rad51 using RNA interference (Figure 5D). The clonogenic survival assay showed that Rad51 siRNA enhanced radiosensitization of MDA-MB-231 cells to proton irradiation to a greater extent than X-ray irradiation, relative to that observed upon treatment with control siRNA (Figure 5E). Apoptosis assay confirmed that Rad51 knockdown further augmented radiation-induced apoptosis ($p < 0.001$); in the background of Rad51 knockdown, proton irradiation resulted in a higher number of apoptotic populations than X-ray irradiation ($p < 0.01$; Figure 5F). Furthermore, treatment with B02, a selective Rad51 inhibitor, increased the level of DNA damage and apoptosis (Figure S3A), leading to proton sensitization, as evidenced by the results of the clonogenic survival assay (Figure S3B,C). These data suggest that the downregulation of Rad51 may be associated with a causal link between CHK1 inhibition and proton sensitization.

3. Discussion

CHK1 is a serine/threonine-protein kinase that plays an important role in maintaining genomic stability. It is phosphorylated in response to DNA damage and facilitates two key functions, cell cycle arrest and DNA repair through the homologous recombinational repair (HRR) pathway [4,20,21]. CHK1 is frequently deregulated in various types of cancers, and its inhibition in combination with antimetabolites or DNA-damaging agents enhances cytotoxicity [22]. CHK1 inhibitors potentiate RT in various preclinical models [8–10,23], especially in being effective radiosensitizers in p53-deficient tumor cells by targeting the G2/M checkpoint [8,22]. In head and neck squamous cell carcinoma (HNSCC) cells, PF-477736 radiosensitized human papillomavirus (HPV)-positive cancer cells lacking TP53 by abrogating radiation-induced G2/M arrest. However, it did not show any such effect in normal fibroblasts [24]. Furthermore, CHK1 inhibitors have been shown to sensitize p53-expressing colorectal cancer HCT116 cells to radiation [9]. Although enhanced efficacy of conventional photon RT by CHK1 inhibitors has been well described, thus far, no data are available regarding the effect of proton beams. In this study, we directly compared the effectiveness protons and X-rays with respect to CHK1 inhibition in various TNBC cells.

TNBC is a heterogeneous group of tumors, which have the worst prognosis among breast cancer subtypes [1]. There are several hypotheses regarding the resistance of TNBC to treatments, including radiotherapy. Moreover, TP53 mutation is more frequently observed in TNBC than in other breast cancer subtypes and is associated with radioresistance. For example, TP53 mutation is seen in 30% of all breast cancers, and is most frequently observed in the basal-like type of TNBC (88%), whereas it is less common in the luminal A subtype (17%) [25]. TP53 mutations are associated with radioresistance; this is mediated via activation of DNA repair genes including Rad51, or by inhibition of p53-induced apoptosis [26]. Further, TP53 mutations in TNBC also result in the loss of the G1 checkpoint, as well as the dependence on CHK1 for responding to DNA damage. Moreover, CHK1 is significantly overexpressed in TNBC tissues as compared to that in non-TNBC tissues [27]. Thus, targeting CHK1 in TNBC might be a plausible approach to enhance the efficacy of RT.

Proton therapy biologically shows a 10% higher effectiveness than conventional photon therapy, and the current use of proton therapy largely relies on its physical superiority, such as the Bragg peak [28]. Accumulating evidence suggests that there is a difference in the biological efficacy between proton and photon therapies [13,15,16,18,29,30]. In the form of a particle beam, protons cause more DNA damage than photons, as photons lack mass. Recent preclinical trial data suggest that HRR appears to be crucial for proton-induced DNA damage repair, whereas, non-homologous end joining (NHEJ) is important for photon-induced DNA damage repair [18]. Impairment of HRR is one of the mechanisms by which CHK1 inhibitors function in cancer cells. Thus, we hypothesized that CHK1 inhibition might be more effective in combination with proton therapy than with photon therapy. The clonogenic survival assay showed that the surviving colony numbers of MDA-MB-231 cells were significantly reduced in response to combination treatment (PF-477736 and radiation) as compared to PF-477736 treatment or radiation alone. As expected, the combination of PF-477736 treatment along with proton therapy showed a pronounced effect on clonogenic survival than X-rays. Considering that PF-477736 has a strong affinity for CHK2, in a manner similar to other CHK1 inhibitors [22], we knocked down CHK1 using siRNA and confirmed that the proton-radiosensitizing effect of PF-477736 is mainly due to the selective inhibition of CHK1 activity.

Morgan et al. [8] reported that CHK1 inhibition using AZD7762 not only abrogates the G2 checkpoint but also impairs HRR, as indicated by the reduced formation of Rad51 foci and persistent γH2AX expression. Consistently, our data showed that PF-477736 abrogated the G2/M checkpoint and accumulated DNA damage, as evidenced by the reduced Rad51 expression and increased γH2AX levels. Immunofluorescence revealed that combinatorial treatment with PF-477736 and proton therapy generated more γH2AX foci than combinatorial treatment with PF-477736 and X-rays, suggesting that PF-477736 might be involved in enhancing proton-induced DNA DSBs. CHK1 phosphorylates DNA repair proteins, including Rad51, FANCD2, and FANCE, and activates HRR [22]. Amongst these, Rad51 is an essential protein for HRR, and its recruitment to the sites of DNA damage triggers the initiation of HRR. Knockdown of Rad51 or Rad51 inhibitor treatment enhanced proton-mediated clonogenic cell death and apoptosis of TNBC cells, supporting the hypothesis that the downregulation of Rad51 in response to PF-477736 treatment might result in enhanced sensitivity to proton therapy. The radiosensitizing effect of PF-477736 with respect to X-rays was relatively weak, which may be attributed to residual Rad51 expression.

Our data support previous findings that HRR-deficient tumor cells are more susceptible to proton irradiation than to photon irradiation [18]. TNBC is frequently associated with mutations in DNA damage repair-associated genes, which are being utilized as therapeutic targets [31]. For example, BRCA mutations have been reported in 20% of the TNBC cases, and the BRCA1 promoter CpG island methylation is highly prevalent in patients with TNBC [32]. Synthetic lethality in response to PARP inhibitors, such as olaparib, has been tested in BRCA-deficient breast cancer patients [33]. Our findings suggest that proton therapy could be considered a superior alternative to conventional X-ray therapy for the elimination of residual TNBC cells, especially those with an impaired HRR pathway. CHK1 is one of the therapeutic target candidates in TNBC, which has no approved targeted therapy. Our data suggest that CHK1 inhibition sensitizes TNBC cells to proton therapy both by abrogating the radiation-induced G2/M arrest and by impairing HRR through Rad51 downregulation. To our knowledge, this study is the first of its kind to provide evidence for increased efficacy of proton therapy when used in combination with a CHK1 inhibitor in TNBC cells. Further investigation should be made regarding the use of the combinatorial treatment using proton therapy and CHK1 inhibition as an alternative strategy for the treatment of refractory TNBC.

4. Materials and Methods

4.1. Cell Culture

Three human TNBC cell lines, MDA-MB-231, Hs578T, and MDA-MB-453, were used in this study. MDA-MB-231 cells were purchased from the American Type Culture Collection (ATCC, Manassas, VA, USA), and Hs578T cells and MDA-MB-453 cells were purchased from the Korean Cell Line Bank (Seoul National University, Seoul, Korea). MDA-MB-231 and MDA-MB-453 cells were cultured in Dulbecco's Modified Eagle's Medium (DMEM) supplemented with 10% fetal bovine serum (FBS), and Hs578T cells were cultured in DMEM supplemented with 10% FBS and 25 mM 4-(2-hydroxyethyl)-1-piperazineethanesulfonic acid (HEPES). Cultures were maintained in a humidified atmosphere comprising 95% air and 5% CO2 at 37 °C.

4.2. Reagents and Antibodies

Anti-phospho-H2AX (Ser139) antibodies were purchased from Millipore (Burlington, MA, USA). Anti-cleaved PARP, anti-cleaved caspase-3, anti-phospho-p38 MAPK (Thr180/Tyr182), anti-CHK1, anti-phospho-CHK1 (Ser345 and Ser296), anti-phospho-cdc2 (Tyr15), anti-phospho-BRCA1 (Ser1524), anti-Bcl2, anti-Bax, anti-Bak, anti-Bcl-XL, and anti-Rad51 antibodies were purchased from Cell Signaling Technology (Danvers, MA, USA). Anti-β-actin antibodies were purchased from Sigma Aldrich (St. Louis, MO, USA). Anti-Rad51 antibodies were purchased from Santa Cruz Biotechnology (Santa Cruz, CA, USA). For gene knockdown experiments, control siRNA (sc-37007), CHK1 siRNA (sc-29269), and Rad51 siRNA (sc-36361) were purchased from Santa Cruz Biotechnology. RNAiMax was purchased from Invitrogen (Carlsbad, CA, USA). PF-477736, a selective CHK1 inhibitor, and B02, a Rad51 inhibitor, were purchased from Sigma Aldrich.

4.3. Cell Viability Assay

The Cell Counting Kit-8 (CCK-8, Dojindo Laboratories, Kumamoto, Japan) was used to determine cell viability according to the manufacturer's instructions. Briefly, MDA-MB-231 cells and Hs578T cells were seeded at a density of 1×10^3 cells/well into a 96-well plate, and on the next day were treated with various concentrations of PF-477736 dissolved in dimethyl sulfoxide (DMSO). Cells were incubated with the CCK-8 solution 72 h post-treatment with PF-477736, and cell viability was determined by measuring the absorbance at 450 nm using the SpectraMax i3 microplate reader (Molecular Devices, Sunnyvale, CA, USA).

4.4. Irradiation Experiments

X-ray beams and proton beams were made incident using a linear accelerator Varian Clinac 6EX (Varian Medical Systems, Palo Alto, CA, USA), and a proton therapy system (Sumitomo Heavy Industries, Tokyo, Japan) respectively, at the Samsung Medical Center in Seoul, Korea [34]. For proton irradiation, cell plates were positioned in the center of the spread-out Bragg peak (SOBP) measuring 11.2 cm within a field collimated by 18 × 12 cm Brass blocks. Cells were irradiated with graded doses of 230 MeV protons, for which the beam range was 22.8 cm (distal 90%) [35]. The absolute proton beam dose was verified to an accuracy of 1% for proton therapy, in accordance with TRS-398. For X-ray irradiation, cells were placed under a 2 cm thick solid water phantom at a source surface distance of 100 cm and a field size of 30 × 30 cm. The absolute X-ray dose was verified in accordance with TG-51 using the Gafchromic film to an accuracy of 1%.

4.5. Clonogenic Survival Assay

Cells were seeded at a low density (800–3200 cells/60 mm dish) 24 h before irradiation and were irradiated with various doses of X-rays or protons. After incubation for 14 d, surviving colonies consisting of more than 50 cells were visualized by staining with 0.5% crystal violet (Sigma-Aldrich) and were then counted. The survival fraction (SF) was calculated as the ratio of the plating efficiency (ratio of the number of surviving colonies to the number of seeded cells) of the irradiated cells to the plating efficiency of un-irradiated cells in order to quantify the cell survival [36].

4.6. Immunoblotting

Cells were lysed using modified RIPA buffer (20 mM Tris (pH 8.0), 137 mM NaCl, 10% glycerol, 1% Nonidet *p*-40, 10 mM EDTA, 100 mM NaF, 1 mM phenylmethylsulfonyl fluoride, and 10 mg/mL leupeptin), and the supernatant was collected after centrifugation at 15,871 *g* for 15 min. Protein concentration was measured using the Bio-Rad protein assay reagent (Bio-Rad, Richmond, CA, USA), according to the manufacturer's instructions. Equal amounts of protein were subjected to sodium dodecyl sulfate-polyacrylamide gel electrophoresis (SDS-PAGE). Separated proteins were transferred onto nitrocellulose membranes (Bio-Rad); blots were blocked overnight with 5% skimmed milk prepared in PBS, pH 7.4 at 4 °C and probed with primary antibodies overnight. Protein bands were detected using the Amersham enhanced chemiluminescence detection reagents (GE Healthcare, Piscataway, NJ, USA).

4.7. Flow Cytometry

Flow cytometry was performed in order to measure the number of apoptotic cells. Harvested cells were stained with annexin V-fluorescein isothiocyanate (FITC) (BD Pharmingen, San Diego, CA, USA) and 2 μg/mL PI in the annexin V binding buffer (10 mM HEPES, pH 7.4, 140 mM NaCl, 2.5 mM $CaCl_2$) for 15 min at 37 °C in the dark. The number of apoptotic cells was quantified using BD FACSVerse and the BD FACSuite software.

4.8. Immunofluorescence

Image acquisition and analyses were conducted as described previously [37]. Briefly, 2 × 104 MDA-MB-231 cells were seeded on a cover slip (Marinfild Inc., Rochester, NY, USA) a day before irradiation. Cells were pre-treated with 500 nM PF-477736 for 3 h, exposed to photon or proton beams, and then fixed with 4% formaldehyde at 24 h. After permeabilization using 0.01% Triton X-100, and blocking with 2% bovine serum albumin (BSA) for 30 min, cells were incubated with the phospho-S193 H2AX antibody for 2 h, followed by incubation with the Alexa Fluor 488-conjugated secondary antibody (Life Technologies, Paisley, UK) and 4′,6-diamidino-2-phenylindole (DAPI) for 30 min at room temperature. After mounting the cells using the SlowFade anti-fade reagent (Molecular Probes, Eugene, OR, USA), images were acquired using a fluorescence microscope (Zeiss Observer D1, Carl Zeiss Co., Ltd., Jena, Germany).

4.9. Gene Knockdown Experiments

Small interfering RNAs (siRNAs) were used to knockdown CHK1 and Rad51 proteins. For transfection, 1 × 105 MDA-MB-231 cells were seeded in 6-well plates and incubated with a mixture of 10 nM siRNA and Lipofectamine RNAiMax (1:1 ratio) for 4 h, and then the culture medium was replaced with fresh medium. Reduction in the expression of the proteins of interest was confirmed using immunoblotting.

4.10. Statistical Analysis

Data are presented as the mean ± standard deviation (S.D.). Statistical analysis was performed using GraphPad Prism 7.02. Significance of differences between two groups was determined using the unpaired, two-tailed Student's t-test. For a comparison of more than two groups, one-way ANOVA with Tukey's multiple comparison test was used. $p < 0.05$ was considered to be statistically significant.

Supplementary Materials: The following are available online at http://www.mdpi.com/1422-0067/21/8/2691/s1.

Author Contributions: Conceptualization, C.C., W.K.C., and D.H.C.; methodology, C.C. and S.P.; validation, C.C.; formal analysis, C.C. and S.P.; investigation, C.C., S.P., and S.-W.S.; data curation, C.C., S.P., and W.K.C.; writing original draft preparation, C.C. and W.K.C.; writing—review and editing, C.C., W.K.C., H.K., and D.H.C.; visualization, C.C., S.P., S.-W.S.; supervision, W.P. and D.H.C.; project administration, W.P. and D.H.C.; funding acquisition, C.C. and D.H.C. All authors have read and agreed to the published version of the manuscript.

Funding: This study was supported by a Samsung Medical Center grant (SMX1170211) and the Basic Science Research Program through the National Research Foundation of Korea funded by the Ministry of Education (NRF-2016R1D1A1B03935165 and NRF-2018R1D1A1B07042738).

Acknowledgments: We would like to thank Editage (www.editage.co.kr) for English language editing.

Conflicts of Interest: The authors declare no conflict of interest.

Abbreviations

ANOVA	Analysis of variance
ATR	Ataxia telangiectasia and rad3-related
Bax	Bcl-2-associated X
Bcl-2	B-cell lymphoma 2
BSA	Bovine serum albumin
CHK1	Checkpoint kinase 1
DAPI	4′,6-diamino-2-phenylindole
FITC	Fluorescein isothiocyanate
PBT	Proton beam therapy
DSB	Double strand break
HRR	Homologous recombination repair
NHEJ	Non-homologous end joining
PARP	Poly(ADP-ribose)polymerase
RT	Radiation therapy
siRNA	Small interfering ribonucleic acid
TNBC	Triple negative breast cancer

References

1. Foulkes, W.D.; Smith, I.E.; Reis-Filho, J.S. Triple-negative breast cancer. *New Engl. J. Med.* **2010**, *363*, 1938–1948. [CrossRef]
2. Cancer Genome Atlas, N. Comprehensive molecular portraits of human breast tumours. *Nature* **2012**, *490*, 61–70. [CrossRef]
3. Robson, M.; Im, S.A.; Senkus, E.; Xu, B.; Domchek, S.M.; Masuda, N.; Delaloge, S.; Li, W.; Tung, N.; Armstrong, A.; et al. Olaparib for Metastatic Breast Cancer in Patients with a Germline BRCA Mutation. *New Engl. J. Med.* **2017**, *377*, 523–533. [CrossRef]
4. Turk, A.A.; Wisinski, K.B. PARP inhibitors in breast cancer: Bringing synthetic lethality to the bedside. *Cancer* **2018**, *124*, 2498–2506. [CrossRef]
5. Fatehi, D.; Soltani, A.; Ghatrehsamani, M. SRT1720, a potential sensitizer for radiotherapy and cytotoxicity effects of NVB-BEZ235 in metastatic breast cancer cells. *Pathol. Res. Pract.* **2018**, *214*, 889–895. [CrossRef] [PubMed]
6. Horton, J.K.; Jagsi, R.; Woodward, W.A.; Ho, A. Breast Cancer Biology: Clinical Implications for Breast Radiation Therapy. *Int. J. Radiat. Oncol. Biol. Phys.* **2018**, *100*, 23–37. [CrossRef] [PubMed]
7. Jang, N.Y.; Kim, D.H.; Cho, B.J.; Choi, E.J.; Lee, J.S.; Wu, H.G.; Chie, E.K.; Kim, I.A. Radiosensitization with combined use of olaparib and PI-103 in triple-negative breast cancer. *BMC Cancer* **2015**, *15*, 89. [CrossRef] [PubMed]

8. Morgan, M.A.; Parsels, L.A.; Zhao, L.; Parsels, J.D.; Davis, M.A.; Hassan, M.C.; Arumugarajah, S.; Hylander-Gans, L.; Morosini, D.; Simeone, D.M.; et al. Mechanism of radiosensitization by the Chk1/2 inhibitor AZD7762 involves abrogation of the G2 checkpoint and inhibition of homologous recombinational DNA repair. *Cancer Res.* **2010**, *70*, 4972–4981. [CrossRef]
9. Tao, Y.; Leteur, C.; Yang, C.; Zhang, P.; Castedo, M.; Pierre, A.; Golsteyn, R.M.; Bourhis, J.; Kroemer, G.; Deutsch, E. Radiosensitization by Chir-124, a selective CHK1 inhibitor: Effects of p53 and cell cycle checkpoints. *Cell Cycle* **2009**, *8*, 1196–1205. [CrossRef]
10. Patel, R.; Barker, H.E.; Kyula, J.; McLaughlin, M.; Dillon, M.T.; Schick, U.; Hafsi, H.; Thompson, A.; Khoo, V.; Harrington, K.; et al. An orally bioavailable Chk1 inhibitor, CCT244747, sensitizes bladder and head and neck cancer cell lines to radiation. *Radiother. Oncol.* **2017**, *122*, 470–475. [CrossRef]
11. Zhou, Z.R.; Yang, Z.Z.; Wang, S.J.; Zhang, L.; Luo, J.R.; Feng, Y.; Yu, X.L.; Chen, X.X.; Guo, X.M. The Chk1 inhibitor MK-8776 increases the radiosensitivity of human triple-negative breast cancer by inhibiting autophagy. *Acta Pharm. Sin.* **2017**, *38*, 513–523. [CrossRef] [PubMed]
12. Braunstein, L.Z.; Cahlon, O. Potential Morbidity Reduction with Proton Radiation Therapy for Breast Cancer. *Semin. Radiat. Oncol.* **2018**, *28*, 138–149. [CrossRef] [PubMed]
13. Liu, Q.; Ghosh, P.; Magpayo, N.; Testa, M.; Tang, S.; Gheorghiu, L.; Biggs, P.; Paganetti, H.; Efstathiou, J.A.; Lu, H.M.; et al. Lung cancer cell line screen links fanconi anemia/BRCA pathway defects to increased relative biological effectiveness of proton radiation. *Int. J. Radiat. Oncol. Biol. Phys.* **2015**, *91*, 1081–1089. [CrossRef]
14. Liu, Q.; Underwood, T.S.; Kung, J.; Wang, M.; Lu, H.M.; Paganetti, H.; Held, K.D.; Hong, T.S.; Efstathiou, J.A.; Willers, H. Disruption of SLX4-MUS81 Function Increases the Relative Biological Effectiveness of Proton Radiation. *Int. J. Radiat. Oncol. Biol. Phys.* **2016**, *95*, 78–85. [CrossRef] [PubMed]
15. Choi, C.; Son, A.; Lee, G.H.; Shin, S.W.; Park, S.; Ahn, S.H.; Chung, Y.; Yu, J.I.; Park, H.C. Targeting DNA-dependent protein kinase sensitizes hepatocellular carcinoma cells to proton beam irradiation through apoptosis induction. *PLoS ONE* **2019**, *14*, e0218049. [CrossRef] [PubMed]
16. Choi, C.; Park, S.; Cho, W.K.; Choi, D.H. Cyclin D1 is Associated with Radiosensitivity of Triple-Negative Breast Cancer Cells to Proton Beam Irradiation. *Int. J. Mol. Sci.* **2019**, *20*, 4943. [CrossRef] [PubMed]
17. Carter, R.J.; Nickson, C.M.; Thompson, J.M.; Kacperek, A.; Hill, M.A.; Parsons, J.L. Complex DNA Damage Induced by High Linear Energy Transfer Alpha-Particles and Protons Triggers a Specific Cellular DNA Damage Response. *Int. J. Radiat. Oncol. Biol. Phys.* **2018**, *100*, 776–784. [CrossRef]
18. Fontana, A.O.; Augsburger, M.A.; Grosse, N.; Guckenberger, M.; Lomax, A.J.; Sartori, A.A.; Pruschy, M.N. Differential DNA repair pathway choice in cancer cells after proton- and photon-irradiation. *Radiother. Oncol.* **2015**, *116*, 374–380. [CrossRef]
19. Konings, K.; Vandevoorde, C.; Baselet, B.; Baatout, S.; Moreels, M. Combination Therapy With Charged Particles and Molecular Targeting: A Promising Avenue to Overcome Radioresistance. *Front. Oncol.* **2020**, *10*, 128. [CrossRef]
20. Zhang, Y.; Hunter, T. Roles of Chk1 in cell biology and cancer therapy. *Int. J. Cancer* **2014**, *134*, 1013–1023. [CrossRef]
21. Qiu, Z.; Oleinick, N.L.; Zhang, J. ATR/CHK1 inhibitors and cancer therapy. *Radiother. Oncol.* **2018**, *126*, 450–464. [CrossRef] [PubMed]
22. Ma, C.X.; Janetka, J.W.; Piwnica-Worms, H. Death by releasing the breaks: CHK1 inhibitors as cancer therapeutics. *Trends Mol. Med.* **2011**, *17*, 88–96. [CrossRef] [PubMed]
23. Wang, X.; Ma, Z.; Xiao, Z.; Liu, H.; Dou, Z.; Feng, X.; Shi, H. Chk1 knockdown confers radiosensitization in prostate cancer stem cells. *Oncol. Rep.* **2012**, *28*, 2247–2254. [CrossRef] [PubMed]
24. Busch, C.J.; Kriegs, M.; Laban, S.; Tribius, S.; Knecht, R.; Petersen, C.; Dikomey, E.; Rieckmann, T. HPV-positive HNSCC cell lines but not primary human fibroblasts are radiosensitized by the inhibition of Chk1. *Radiother. Oncol.* **2013**, *108*, 495–499. [CrossRef] [PubMed]
25. Bertheau, P.; Lehmann-Che, J.; Varna, M.; Dumay, A.; Poirot, B.; Porcher, R.; Turpin, E.; Plassa, L.F.; de Roquancourt, A.; Bourstyn, E.; et al. p53 in breast cancer subtypes and new insights into response to chemotherapy. *Breast (Edinb. Scotl.)* **2013**, *22* (Suppl. 2), S27–S29. [CrossRef] [PubMed]
26. Cuddihy, A.R.; Bristow, R.G. The p53 protein family and radiation sensitivity: Yes or no? *Cancer Metastasis Rev.* **2004**, *23*, 237–257. [CrossRef]

27. Albiges, L.; Goubar, A.; Scott, V.; Vicier, C.; Lefebvre, C.; Alsafadi, S.; Commo, F.; Saghatchian, M.; Lazar, V.; Dessen, P.; et al. Chk1 as a new therapeutic target in triple-negative breast cancer. *Breast (Edinb. Scotl.)* **2014**, *23*, 250–258. [CrossRef]
28. Paganetti, H. Relating proton treatments to photon treatments via the relative biological effectiveness-should we revise current clinical practice? *Int. J. Radiat. Oncol. Biol. Phys.* **2015**, *91*, 892–894. [CrossRef]
29. Wang, L.; Wang, X.; Li, Y.; Han, S.; Zhu, J.; Wang, X.; Molkentine, D.P.; Blanchard, P.; Yang, Y.; Zhang, R.; et al. Human papillomavirus status and the relative biological effectiveness of proton radiotherapy in head and neck cancer cells. *Head Neck* **2017**, *39*, 708–715. [CrossRef]
30. Choi, C.; Lee, C.; Shin, S.W.; Kim, S.Y.; Hong, S.N.; Park, H.C. Comparison of Proton and Photon Beam Irradiation in Radiation-Induced Intestinal Injury Using a Mouse Model. *Int. J. Mol. Sci.* **2019**, *20*, 1894. [CrossRef]
31. De Summa, S.; Pinto, R.; Sambiasi, D.; Petriella, D.; Paradiso, V.; Paradiso, A.; Tommasi, S. BRCAness: A deeper insight into basal-like breast tumors. *Ann. Oncol.* **2013**, *24* (Suppl. 8), viii13–viii21. [CrossRef]
32. Xu, Y.; Diao, L.; Chen, Y.; Liu, Y.; Wang, C.; Ouyang, T.; Li, J.; Wang, T.; Fan, Z.; Fan, T.; et al. Promoter methylation of BRCA1 in triple-negative breast cancer predicts sensitivity to adjuvant chemotherapy. *Ann. Oncol.* **2013**, *24*, 1498–1505. [CrossRef]
33. Lord, C.J.; Tutt, A.N.; Ashworth, A. Synthetic lethality and cancer therapy: Lessons learned from the development of PARP inhibitors. *Annu. Rev. Med.* **2015**, *66*, 455–470. [CrossRef]
34. Chung, K.; Han, Y.; Kim, J.; Ahn, S.H.; Ju, S.G.; Jung, S.H.; Chung, Y.; Cho, S.; Jo, K.; Shin, E.H.; et al. The first private-hospital based proton therapy center in Korea; status of the Proton Therapy Center at Samsung Medical Center. *Radiat. Oncol. J.* **2015**, *33*, 337–343. [CrossRef]
35. Yu, J.I.; Choi, C.; Shin, S.W.; Son, A.; Lee, G.H.; Kim, S.Y.; Park, H.C. Valproic Acid Sensitizes Hepatocellular Carcinoma Cells to Proton Therapy by Suppressing NRF2 Activation. *Sci. Rep.* **2017**, *7*, 14986. [CrossRef] [PubMed]
36. Choi, C.; Son, A.; Lee, H.S.; Lee, Y.J.; Park, H.C. Radiosensitization by Marine Sponge Agelas sp. Extracts in Hepatocellular Carcinoma Cells with Autophagy Induction. *Sci. Rep.* **2018**, *8*, 6317. [CrossRef] [PubMed]
37. Shin, S.W.; Choi, C.; Lee, G.H.; Son, A.; Kim, S.H.; Park, H.C.; Batinic-Haberle, I.; Park, W. Mechanism of the Antitumor and Radiosensitizing Effects of a Manganese Porphyrin, MnHex-2-PyP. *Antioxid. Redox. Signal.* **2017**, *27*, 1067–1082. [CrossRef] [PubMed]

International Journal of *Molecular Sciences*

MDPI

Article

Valine Radiolysis by H^+, He^+, N^+, and S^{15+} MeV Ions

Cíntia A. P. da Costa [1], Gabriel S. Vignoli Muniz [2], Philippe Boduch [3], Hermann Rothard [3,*] and Enio F. da Silveira [1]

1 Physics Department, Pontifícia Universidade Católica do Rio de Janeiro, Rua Marquês de São Vicente 225, Rio de Janeiro 22451-900, Brazil; cintia-apc@hotmail.com (C.A.P.d.C.); enio@vdg.fis.puc-rio.br (E.F.d.S.)

2 Physics Institute, Universidade de São Paulo, Rua do Matão, Sao Paulo 1371–05508-090, Brazil; gabrivignoli@gmail.com

3 Centre de Recherche sur les Ions; les Matériaux et la Photonique. Normandie Université, ENSICAEN, UNICAEN, CEA, CNRS, CIMAP, 14000 Caen, France; boduch@ganil.fr

* Correspondence: rothard@ganil.fr

Received: 27 January 2020; Accepted: 3 March 2020; Published: 10 March 2020

Abstract: Radiolysis of biomolecules by fast ions has interest in medical applications and astrobiology. The radiolysis of solid D-valine (0.2–2 μm thick) was performed at room temperature by 1.5 MeV H^+, He^+, N^+, and 230 MeV S^{15+} ion beams. The samples were prepared by spraying/dropping valine-water-ethanol solution on ZnSe substrate. Radiolysis was monitored by infrared spectroscopy (FTIR) through the evolution of the intensity of the valine infrared 2900, 1329, 1271, 948, and 716 cm^{-1} bands as a function of projectile fluence. At the end of sample irradiation, residues (tholins) presenting a brownish color are observed. The dependence of the apparent (sputtering + radiolysis) destruction cross section, σ_d, on the beam stopping power in valine is found to follow the power law $\sigma_d = aS_e^n$, with n close to 1. Thus, σ_d is approximately proportional to the absorbed dose. Destruction rates due to the main galactic cosmic ray species are calculated, yielding a million year half-life for solid valine in space. Data obtained in this work aim a better understanding on the radioresistance of complex organic molecules and formation of radioproducts.

Keywords: amino acid; valine; MeV ion irradiation; radiolysis; infrared absorption spectroscopy; destruction cross section; stopping power dependence

1. Introduction

Amino acids are building-blocks of proteins, essential in the dynamics of all living organisms. Moreover, they also play an important role in cell signalization, and in the innate immune system. These stable organic molecules are constituted by carbon, hydrogen, oxygen, and nitrogen. Sulfur is also present in some species. The amino acids molecular structure is constructed by one central carbon atom, called α-carbon, linked to four groups: the carboxyl radical, –COOH, the amino radical, $–NH_2$, a hydrogen atom, –H, and a side chain, –R. Two important chemical properties are caused by the interaction between the carboxyl and amino radicals. The first one is related to the dipole moment of the amino acid, which is different in solid and gas phases. Indeed, in solid phase, the mentioned structure is preserved (particularly at high temperatures). For both polar and apolar side chains, the molecule has a significant dipole moment and is called zwitterion. In gas phase, the dipole moment is weak and the molecule is called "neutral". In condensed phase, the proton of the –COOH group migrates to the amino one forming the zwitterion. The second chemical property concerns to the peptide bond which occurs when the hydroxyl radical of the COOH group is induced to react with the NH_2 radical attached to another amino acid molecule, delivering a H_2O molecule and synthesizing a peptide through a bond between the CO and NH radicals. Endoergic reactions may occur several times with nearby amino acids producing larger peptides [1].

The discovery of amino acids in meteorites [2,3], in comets [4], and possibly in the interstellar medium [5–8] inscribed them into the portfolio of astrophysical materials. Nevertheless, finding the origin of amino acids (as well as any prebiotic material, in general) is a central question in Astrobiology. The asteroids and comets that heavily bombarded the Earth circa 4 billion years ago [9,10] probably transported prebiotics. How relevant were these molecular species for triggering life on Earth? Furthermore, given the omnipresent radiation in outer space and around the giant planets [11], once formed in astrophysical environment, can amino acids survive the interplanetary journey? How could they survive the harsh conditions of the early Earth?

Ionizing radiation processes are largely used in industry, in the production and/or to increase the lifetime of biological materials, e.g., ionizing radiation induces genetic modifications in plant seeds. Therefore, knowledge about the radioresistance of complex organic molecules and the subsequent formation of radioproducts may also have importance in medical science and in the industry, as for instance in food production [12], pharmaceutics [13], and particularly in cancer treatment, shielding astronauts in space, and treating victims of irradiation by nuclear materials. Except for glycine, amino acids can exist in two optical isomeric forms, known as L and D; only L-amino acids form proteins; D-amino acids can be found in antibiotics [14] and in some bacteria cell walls [1,15]. Studies on amino acid radiolysis have already been performed. In particular, for glycine, the simplest amino acid, Gerakines et al. studied thermal effects on its dissociation by energetic protons [16]. Pilling et al. irradiated glycine with 1 MeV H^+ in order to study the stability of different polymorph configurations [17]. Pilling et al. [18], Maté et al. [19], and Souza-Corrêa et al. [20] studied its dissociation by keV electrons. UV radiolysis was performed by Peeters et al. [21] and by Ferreira-Rodrigues et al. [22]. For valine, electronic sputtering by MeV heavy ions was discussed by Salehpour et al. [23], Becker et al. [24], and Sundqvist et al. [25].

The current work aims at investigating some aspects of amino acid radiolysis. Specifically, the goal is to determine amino acid destruction cross sections by light and heavy energetic projectiles over a large projectile energy range. This huge experimental task is strongly reduced if amino acid radiolysis proceeds similarly as for condensed gases: their destruction cross sections depend only on the electronic stopping power (S_e) of the ion-target system and not on specific quantities such as kinetic energy, mass, and atomic number of the projectile, as well as molecular and crystalline structures of the target [26–30]. Based on these premises, we have selected the amino acid D-valine (Val) to be irradiated with MeV H^+, He^+, N^+ and 230 MeV S^{15+} ions, the latter for obtaining a much higher S_e. This compound is hydrophilic, has molar mass 117.15 g·mol^{-1}, density 1.32 g·cm^{-3}, and chirality properties [31]. Although observed in meteorites [1,2], D-valine does not participate in the known biological processes; L-valine, the chiral form of the D-valine, is the isotopomer existent in living organisms, also occurs in astrophysical environment and – except for the optical activity—has the same characteristics. Fourier transform infrared absorption spectroscopy (FTIR) at room temperature was used to monitor the valine radiolysis.

Data obtained in this work are a contribution for a better understanding on the radioresistance of complex organic molecules, as well as on the formation of radioproducts; our findings may have consequences to Medical Science [32] and to Astrobiology [33]. The work also shows that, although the sample degradation could be adequately monitored by any vibrational band, the absorbance evolutions as a function of fluence exhibit variations for different bands. This finding may be particularly relevant for astrophysical observations.

Irradiations with low electronic stopping power projectiles were performed using 1.5 MeV H^+, He^+, and N^+ ion beams provided by the Van de Graaff Laboratory at the Pontifícia Universidade Católica do Rio de Janeiro (PUC-Rio). Irradiations with 230 MeV S^{15+} projectiles were carried out at GANIL (Grand Accélérateur National d'Ions Lourds), Caen, France.

2. Results

When a solid is bombarded by ion beams, at least three relevant material processes occur: (a) structural rearrangement of molecules, (b) induced chemical reactions (rearrangement of atoms) and (c) sputtering (emission of atoms and molecules). All the three processes can be monitored by infrared spectroscopy. Successive acquisition of valine infrared spectra during bombardment with different ion beams show that they are very similar: for this reason, only the FTIR spectra obtained with the H^+ and S^{15+} beams are displayed. In fact, the major difference observed in the experiments is the absorbance decrease rate: for the same beam flux, the higher the stopping power of the projectile in the sample, the greater is the disappearance rate of valine molecules from the sample. The crystalline rearrangement (crystallization, amorphization, pore collapsing/compaction) is a process observable by FTIR in the very beginning of irradiation and does not involve the dissociation or removal of valine molecules.

2.1. *Results from the Van de Graaff Laboratory: H^+, He^+ and N^+ Ion Beams*

2.1.1. 1.5 MeV H^+ Beam Irradiation

Figure 1 shows the FTIR spectrum evolution for three fluences of the 1.5 MeV H^+ beam: (i) zero fluence or virgin sample (black line, with the highest absorbances); (ii) F = 1.4×10^{15} ions cm^{-2}, an intermediate fluence (red line, spectrum in the middle); and (iii) F = 9.5×10^{15} ions cm^{-2}, the highest fluence (blue line, the lowest absorbances). For the highest fluence, all valine molecules of the sample have been degraded: the still existing bands are attributed to daughter molecules (products). The only exception is the double peak at ~2350 cm^{-1} which is due to atmospheric CO_2 existing outside the chamber but inside the sight of the spectrometer IR beam (its absorbance varies with time, not with fluence).

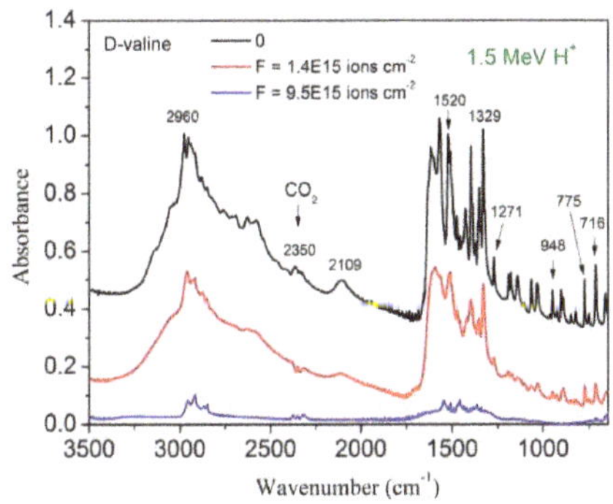

Figure 1. Evolution of the valine infrared spectrum as the projectile fluence increases. The upper spectrum corresponds to the virgin sample; the two others were acquired at fluences 1.4×10^{15} and 9.5×10^{15} ions cm^{-2}, respectively. The feature at ~2350 cm^{-1} is due to atmospheric CO_2 outside the chamber but inside the FTIR spectrometer; its absorbance varies up and down with time.

In Figure 2, the spectrum evolution with fluence is presented for some selected bands. Data are the same as Figure 1 but, in this case, spectra for nine fluences are shown for: (a) 3400–2400 cm^{-1} and (b) 1650–1300 cm^{-1} ranges, and (c) 1271, (d) 948, and (e) 716 cm^{-1} bands. At first sight, all bands seem to decrease proportionally with fluence; but, after detailed inspection, different behaviors can be remarked specially at the beginning (due to compaction) and at the end (due to the appearance of product bands, Figure 2f) of irradiation. At the end of irradiation the samples (residues) present a brownish color. In reference to the pioneer work of Sagan and Khare [34], these radioproducts (daughter molecules) are referred to as tholins.

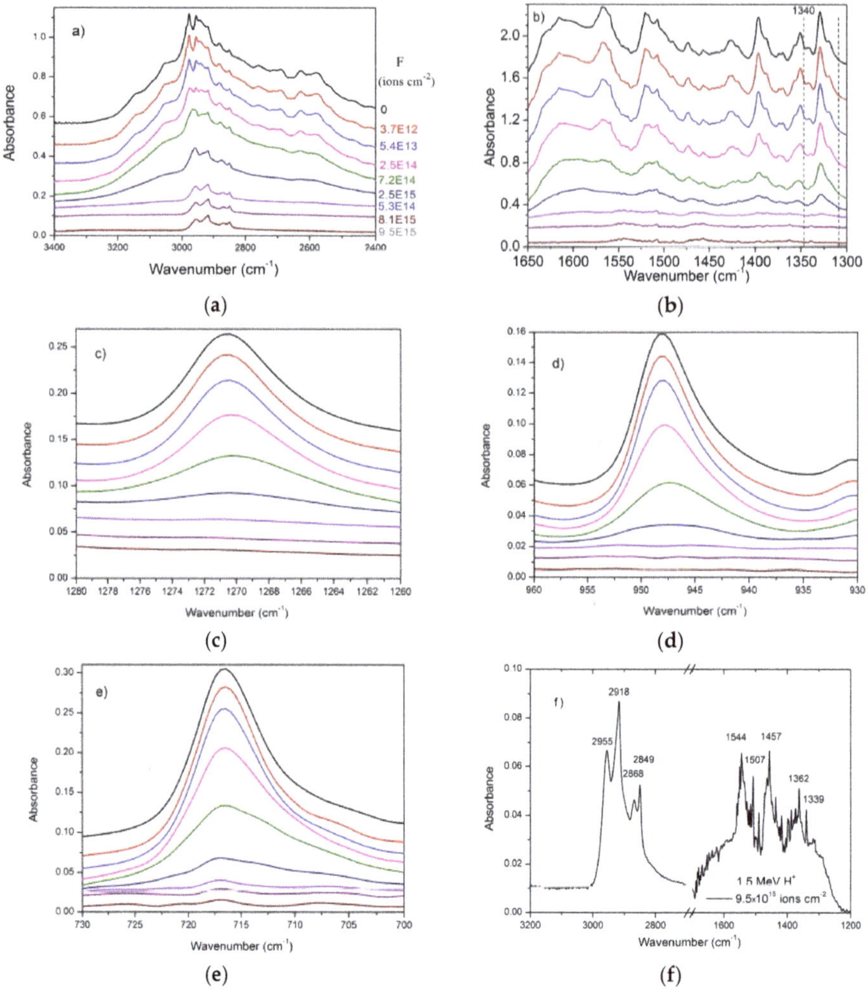

Figure 2. Decrease of valine absorbance as the 1.5 MeV H$^+$ beam fluence increases for five selected IR regions (**a**–**e**). Note in particular that absorbances of 1329, 1271 and 948 cm^{-1} bands disappear completely, while those of 2955, 1507 and 716 cm^{-1} bands do not. The fluence legend for Figure 2a is the same for all figures. Panel (**f**) is a zoom of (**a**) and (**b**), for the highest fluence.

The fact that the 1544 and 1457 cm^{-1} features increase along the irradiation (dominating the IR spectrum at the end) indicates that they are not valine bands; instead, they belong to valine daughter molecules. Furthermore, the 3300–2400 and the 1500–1300 cm^{-1} spectral regions, as well as the 716 cm^{-1} band, present a slower decrease rate and do not disappear at the final fluence; they should be formed by two overlapping components, one due to precursor molecule bands, the other due to product bands. Table 1 presents the wavenumber of the main observed valine bands. Here, the third column displays the wavenumber of products after H$^+$ irradiation.

Table 1. Evolution of valine bands as the 1.5 MeV H^+ and 230 MeV S^{15+} projectile fluences increase. Band identification and their band wavenumbers for valine before irradiation; wavenumber of products after H^+ and S^{15+} irradiation. (a) split band at low temperature: da Costa and da Silveira [33]; (b) Kumar [35]; (c) Façanha Filho et al. (2008) [36]. Attributions are tentative. (*) strong absorbance; (**) low absorbance.

Vibration Mode	Band (cm^{-1})	Band Collapses into (Possible Species)	
		H^+	S^{15+}
NH_3^+ asy str	3150, 3050	-	-
CH_3^+ asy str	2960, 2940	2955, 2918 (C_2H_6 *)	2961, 2933 (C_2H_6 *)
CH_3^+ asy str	2880, 2850	2868, 2849 (C_2H_6 *)	2874 (C_2H_6 *)
N-H ... O	2690, 2630, 2580	-	-
CH_3 bend + NH_2 rocking [b]	~2109 (2153 + 2013) [a]	-	-
CO_2^- stretch [c]	1640	-	-
NH_3^+ asy def.	1600, 1555	-	1595 (Amine)
-	-	1544, 1507	1516 (Nitro)
COO^+ sy str	1520, 1430, 1390	1457	1469 (C_2H_6), 1394
-	-	-	1355 (Nitro)
COO^+/CO	1340, 1320	1339	1329
CH_3^+ def.	1271, 948	-	1271
			885 (C_2H_6 **)
CO_2 Bend [b]	775	-	
C-H out bend	716	716	

The integrated absorbance evolution with H^+ fluence is presented in Figure 3 for several valine bands. The absorbance is normalized to 1 at F = 0, in order to compare compaction effects on different bands. Two groups can be recognized: the first one, the "tholin" group [34], is constituted by the 2900 (integrated from 3300 to 2400 cm^{-1}) and 716 cm^{-1} bands which present a slow decrease with fluence; another, the "pure valine group", is constituted by the 1329, 1271 and 948 cm^{-1} bands. The tholin group includes bands of daughter molecules, which absorb IR wavelengths close to the valine bands. Consequently, the bands assigned to this group do not vanish completely, in contrast to the behavior of pure valine group. The absorbance decrease at low fluences is emphasized in Figure 3b. At the very beginning of irradiation, absorbances of the pure valine group exhibit an exponential behavior, while absorbances of the 716 cm^{-1} band and in particular of the 2900 cm^{-1} band are approximately constant with increasing fluence. Our interpretation is that these two bands are sensitive to compaction, while the other three are not. This means that valine porous samples are more transparent than compacted ones in the 3300–2400 cm^{-1} IR region.

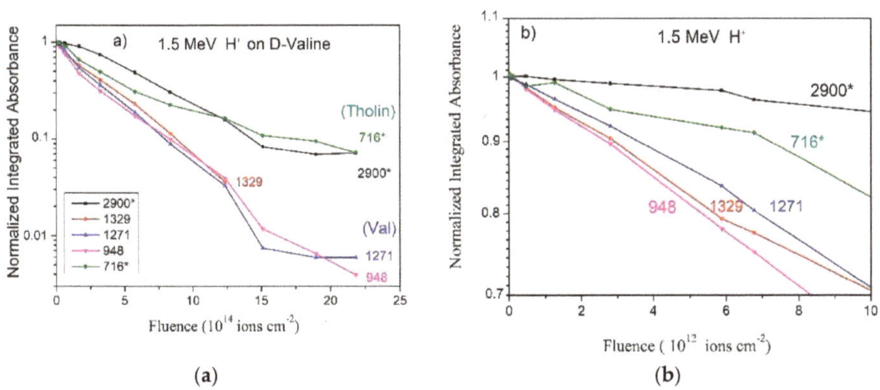

(a) (b)

Figure 3. Dependence on fluence of the integrated absorbance for several bands observed in the H^+ beam irradiation:(**a**) up to the fluence 25 × 10^{14} ions cm^{-2}; (**b**) up to the fluence 1 × 10^{13} ions cm^{-2}.

Data are normalized to 1 at F = 0. The smaller slopes at high fluences exhibited by the 2900* (3300–2400, shown in Figure 2a) and 716* (Figure 2e) cm^{-1} bands are attributed to the rising of product's bands at very close wavenumbers. (*) means overlapping with product's band. Colored lines are guides for eyes.

2.1.2. 1.5 MeV He^+ and N^+ Beam Irradiations

Absorbance evolution obtained for the 1.5 MeV He^+ and N^+ irradiations are presented in Figures 4 and 5, respectively. Data for the same five bands analyzed for the H^+ irradiation are again depicted. For the He^+ beam data, the integrated absorbances are normalized at F = 0, to facilitate a comparison of their slopes. Compaction and product formation (for the 2900 and 716 cm^{-1} bands) are observed for low and high fluences, respectively.

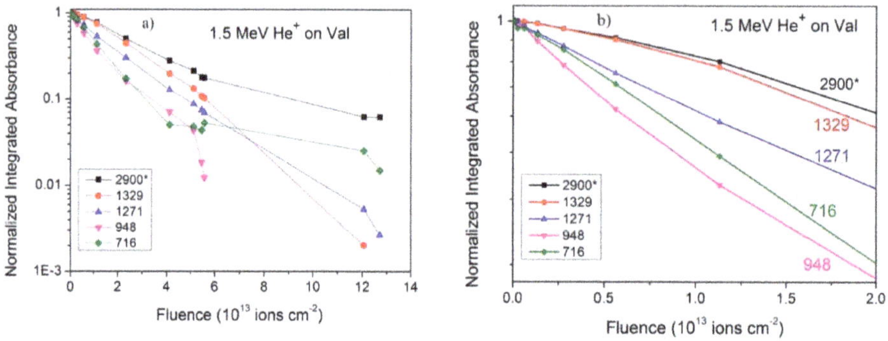

(a) (b)

Figure 4. (**a**) Dependence on fluence of the integrated absorbance for several bands observed in the He^+ beam irradiation. Data are normalized to 1 at F = 0. The asterisk indicates overlapping with product's bands. (**b**) Zoom into the low fluence region. Colored lines are guides for eyes.

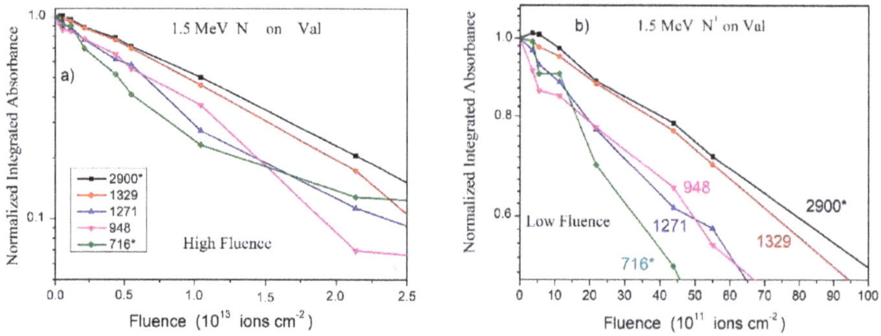

(a) (b)

Figure 5. (**a**) Dependence on fluence of the integrated absorbance for several bands observed in the N^+ beam irradiation. (**b**) Zoom into the low fluence region. Data are normalized to 1 at F = 0. Colored lines are guides for eyes.

Concerning the N^+ beam data, Figure 5a shows the absorbance evolution of five valine bands, normalized at F = 0. In agreement with observations for irradiations with other ion beams, the slopes of the 2900 and 716 cm^{-1} bands are lower, especially at high fluences, due to the contribution of product absorbances.

Compaction effects are clearly seen for H^+ (Figure 3) at the beginning of irradiation; for He^+ (Figure 4) and N^+ beams (Figure 5a and its zoom in Figure 5b), this effect is particularly noticeable for the 2900* cm^{-1} band.

2.2. Results from GANIL (CIMAP Laboratory)

The overall IR spectrum evolution of valine bombardment by 230 MeV S^{15+} projectiles is presented in Figure 6a. As irradiation proceeds, absorbance decreases equally for all bands. This is a hint that their correspondent A-values do not depend strongly on fluence. In other words, a given band does not disappear faster than another, so that the deduced valine destruction cross section is about the same regardless of the band considered. This is true for both narrow and broad bands. The latter include large and complex structures, such as the 3200–2400 cm^{-1} and 1700–1200 cm^{-1} regions: the whole feature belongs to valine and collapses into product bands as fluence increases (Figure 6b, acquired for the highest fluence). As an illustration, Figure 7 compares the decrease of three integrated absorbances with fluence: the (I) region corresponding to the well-defined 2109 cm^{-1} band, the (II) region of the corresponding baseline structure, and the (III) region next to (II) with the same wavenumber width (Figure 7a). At high fluences, all of the three integrated absorbances decrease exponentially and proportionally to each other (Figure 7b). At low fluences, compaction effects are not visible for absorbance of (I) region but are clearly seen for (II) and (III) regions. The sample was irradiated up to 0.92×10^{13} ions cm^{-2}, corresponding to a dose of 23 eV $molecule^{-1}$. The 1329, 1271 and 948 cm^{-1} bands have disappeared at the end of irradiation, evidence that valine was totally degraded. On the other hand, as presented in Figure 6b for the 3200–2200 and 1700–1200 cm^{-1} regions, several bands are still seen at the end of irradiation, reason why they are attributed to daughter molecules. In Table 1, the fourth column lists the main product's bands observed at the highest fluence with possible attributions.

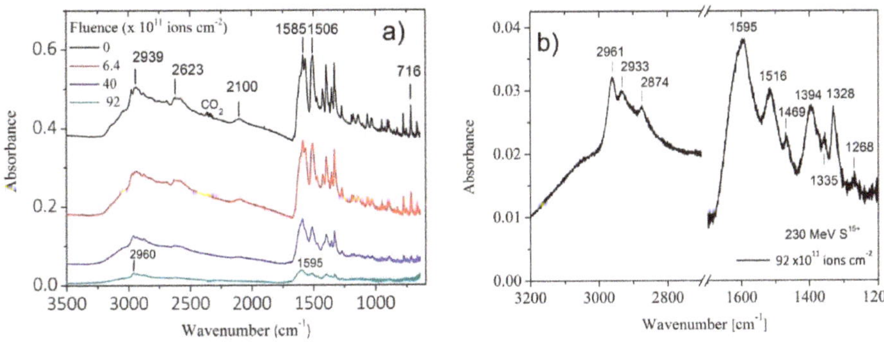

(a) (b)

Figure 6. (**a**) Valine IR spectra as a function of projectile fluence. The first spectrum at the top corresponds to the non-irradiated sample and the 4th one was acquired at the end of the irradiation. (**b**) Zoom of the 3200–2700 and 1700–1200 cm^{-1} regions for the highest fluence measurement.

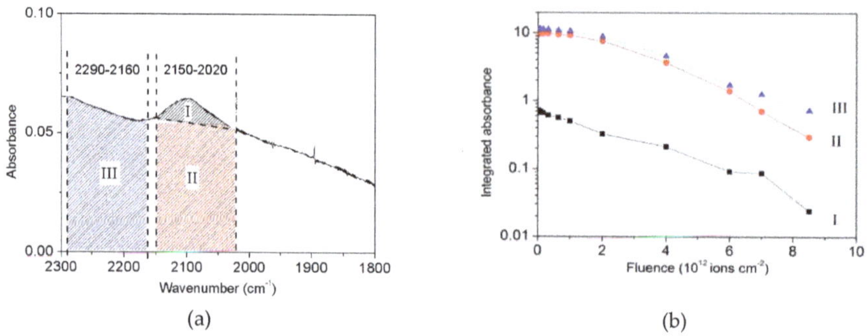

Figure 7. (**a**) Valine IR spectrum around the 2109 cm^{-1} band. The three regions selected correspond to: (I) the 2109 cm^{-1} band, (II) background below this band, and (III) background next to it. (**b**) The integrated absorbance evolution with projectile fluence for these three spectrum regions.

Figure 8 depicts the same data as in Figure 6 but zooming into the 3200–2400 and 1340–1315 cm^{-1} regions; intermediate fluences are now included. Note that: (a) no significant peak shifts are observed; (b) peaks become moderately broader at high fluences (the thicker line for fluence ~2×10^{12} ions cm^{-2} may be taken as a reference); (c) some peaks (e.g., at 1329 cm^{-1}) disappear at high fluences but others (e.g., at 2961 cm^{-1}) do not, which provides evidence that these are residues. The peak broadening causes the disappearance of minima between some bands.

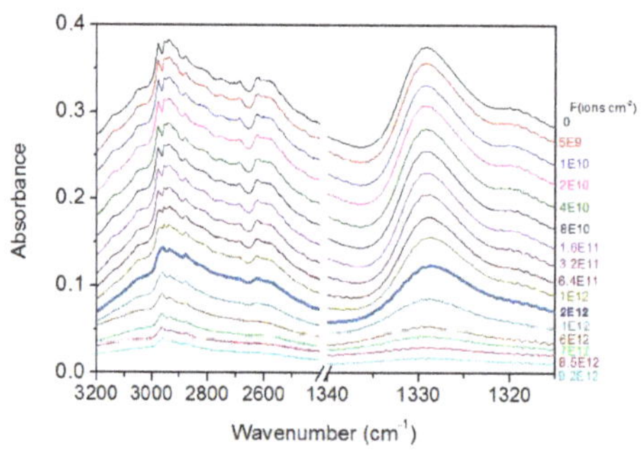

Figure 8. Absorbance decrease for the 3200–2400 and 1340–1315 cm^{-1} regions, as a function of projectile fluence. The first spectrum (top) and the one marked by the thicker line correspond to the non-irradiated sample and to a fluence of 2×10^{12} ions cm^{-1}, respectively. Note the disappearance of minima at high fluences. The legend at the right side expresses fluence in ions cm^{-2}.

The dependences of the integrated absorbances on fluence for the 3300–2400, 1335–1304, 1279–1261, 957–937, and 726–705 cm^{-1} bands are depicted in Figure 9 for (a) high fluences and (b) low fluences. Clearly, the slopes are close but not identical. Four possible reasons are: (i) compaction, (ii) growth of products, (iii) uncertainty of the baseline selected for the absorbance calculation, and (iv) dependence of A-values on fluence. In detail:

1. Compaction effects have been observed to occur at low fluences and differently for distinct bands. However, since this is a property of the material, the different absorbance evolutions are expected to be similar for any ion beam. Indeed, comparing Figure 3b with Figure 9b, it can be seen that—for both irradiations—the 948 cm^{-1} band absorbance is the one that decreases the most, while those of the 2900 and 1329 cm^{-1} bands decrease much less. The compaction effect modifies absorbances up to 20%.

2. Product bands: For low fluences, the concentration of daughter molecules is still too low for disturbing the chemical environment in such a way to modify precursor's absorbances (this effect is nevertheless expected to be seen at high fluences). Furthermore, the absorbance slopes do not seem to be correlated with product formation. Figures 8 and 6b show that the 1271, 948, and 716 cm^{-1} bands are due only to valine and, consequently, their absorbances decrease to zero at the end of irradiation; the 3300–2400 and 1700–1300 cm^{-1} (large) regions contain contributions of daughter molecules and their absorbances decrease more slower. The 716 cm^{-1} band is well seen in the non-irradiated valine spectrum, but contrarily to what happens with H^+ irradiation, this band is not observed at the last fluence (Figure 6a). This may be explained by the fact that thickness of the sample irradiated by the S^{15+} beam was half of that irradiated by the H^+ beam, preventing small (valine or product) peaks to be observed in the highest fluence spectrum. The large background in the 1700–1200 cm^{-1} region is probably due to amide compounds. Consistently, Figure 9a shows that the 2900 and 716 cm^{-1} bands have a lower slope at high fluences, indicating overlapping with product's bands.
3. Baseline selection: Figure 7a,b show that the baseline is not critical, since the baseline and peak evolutions are similar. However, once baseline is generated by very large vibrational bands, these may have distinct sensitivities to compaction.
4. Chemical environment changes: Irradiation modifies the crystalline structure (compaction, amorphization and crystallization). Therefore, a moderate dependence of A-value on constituent concentrations is expected. An overview of the A-value variation, before and after compaction, is presented in Table 2 for the four different ion beams.

Inspection of the data presented in Table 2 reveals that:

1. For non-irradiated samples, the dispersion of the relative A-values are about 20%. This is probably due to non-homogeneous samples. Porosities may be different. No particular correlation between the relative A-values with sample thickness is observed.
2. For non-irradiated samples, the relative A-value variations are band-dependent. For the H^+ beam, the relative A-values for the 2900 and 716 cm^{-1} bands have increased by a factor ~2 from $F = 0$ to $F = F_{ref}$; these two bands are compaction sensitive (Figure 3b); for the other bands, the relative A-values are close.
3. Relative A-value variations are lower for the N^+ and S^{15+} beams. A possible explanation is that these beams have larger stopping power than the two others.

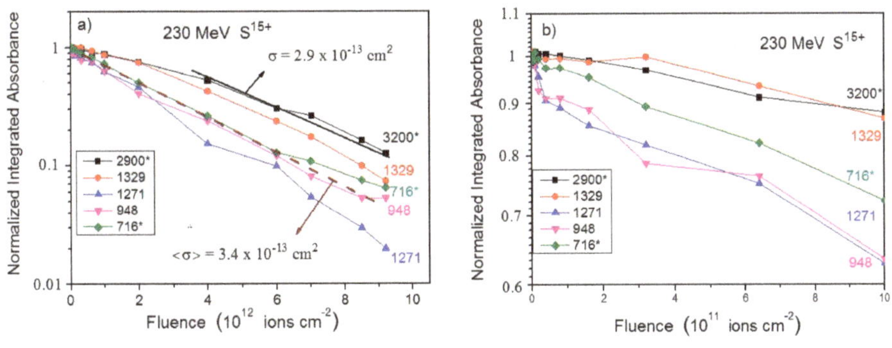

(a) (b)

Figure 9. (**a**) Integrated absorbance decrease of five valine bands as a function of 230 MeV S^{15+} projectile fluence. Data have been normalized to unity for the virgin sample. The dashed line represents degradation with the average cross section <σ>. (**b**) Zoom of the same data for the low fluence region. (*) means overlapping with product's band.

Table 2. Relative A-values (integrated absorbance ratios), before and during irradiation. Ratios refer to the 948 cm^{-1} band, taken as reference. F = 0 means non irradiated sample and F_{ref} is the fluence for which the absorbance ratios were measured (4th line). Sample thickness values are also shown.

	Ion Beam:	**A-Value/A-Value (F_{ref})**			
		H^+ 1.5 MeV	**He^+ 1.5 MeV**	**N^+ 1.5 MeV**	**S^{15+} 230 MeV**
	Sample Thickness	**2.4 μm**	**0.96 μm**	**0.23 μm**	**0.58 μm**
Band (cm^{-1}) (interval)	**F_{ref} (ions cm^{-2})**	**5.8×10^{14}**	**1.1×10^{13}**	**4.39×10^{12}**	**3.2×10^{11}**
2900* (3300–2400)	0	273	282	235	246
	F_{ref}	768	602	270	303
1329 (1335–1301)	0	3.94	3.07	9.51	10.8
	F_{ref}	5.23	6.34	10.5	13.8
1271 (1279–1261)	0	0.876	1.16	0.981	0.984
	F_{ref}	0.955	1.70	0.997	1.22
948 (957–937)	reference	1	1	1	1
	reference	1	1	1	1
716* (726–705)	0	2.80	2.70	2.35	3.39
	F_{ref}	5.05	3.31	2.54	3.85

This analysis confirms that a procedure to separate compaction effects from radiolysis/sputtering degradation is to analyze absorbance at different fluences. At the beginning of irradiation, effects due to all three processes are observable; after a certain fluence, which depends on the beam stopping power, the compaction process is over. Annealing would be another procedure to promote compaction and inducing crystallization.

2.3. Cross Section Measurements

Figure 10 illustrates the concepts presented in Section 4.1 for the S^{15+} and He^+ ion beam irradiations, using absorbances of the 775 cm^{-1} band. The measured apparent destruction cross sections are quite distinct, $\sigma_d{}^{ap}(S)/\sigma_d{}^{ap}(He) = 8.5$, while the compaction ones are close to each other, $\sigma_c(S)/\sigma_c(He) = 0.60$. The relative porosities of the two samples used for irradiation with the S^{15+} and He^+ ion beams are $\zeta = 0.32$ and $\zeta = 0.12$, respectively. Note that, for both irradiations, the sample porosity has the effect to decrease the absorbances at wavenumbers around 775 cm^{-1}. Once compaction starts to remove pores at the beginning of irradiation, the absorbance increases even if the real column density decreases due to radiolysis and sputtering. For low fluence He^+ beam, the compaction effect dominates over those of the other processes, so that $S(F)/S_p$ remains greater than 1 up to $F \sim 2/\sigma_c = 2.4 \times 10^{12}$ ions cm^{-2}. For much higher fluences, the S(F) evolution is dominated by the radiolysis and sputtering.

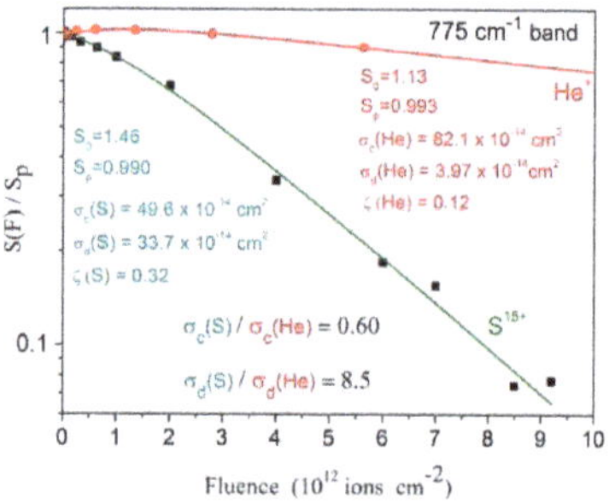

Figure 10. Integrated absorbance decrease for the 775 cm^{-1} band as a function of fluence for the 1.5 MeV He^+ and the 230 MeV S^{15+} beams. Fittings are performed with Equations (7) and (8) (Section 4.2).

Table 3 offers an overview of the measured apparent destruction cross sections for the all the ion beams, based on five selected bands; as illustration, one extra band (782–763 cm^{-1}) was analyzed for the He^+ and S^{15+} ion beams to extract σ_c. The conclusion, already anticipated by the results exhibited in Figures 3–5, is that cross sections vary according the selected band. These variations should not be attributed to statistical fluctuations since distinct irradiations yield similar discrepancies. For describing this systematic variation supported by experimental data, the parameter $\Delta\sigma_j$, defined in Equation (9) (Section 4.2), is introduced.

Table 3. Band interval used for the integrated absorbance calculation. Apparent destruction cross sections obtained from individual fittings for 1.5 MeV, H^+, He^+ and N^+, and 230 MeV S^{15+} ion irradiations. For S^{15+}, $\Delta\sigma_j$ is the discrepancy between the mean value and each individual $\sigma_d{}^{ap}$. (*) Overlapping with product's band.

Band (cm^{-1})		Ion Beam					
		1.5 MeV			230 MeV		
		H^+	He^+	N^+	S^{15+}		
Band Interval	Band Maximum	$\sigma_d{}^{ap}$ (10^{-14} cm^2)				$\Delta\sigma_j = (\sigma_d{}^{ap})_{mean} - (\sigma_d{}^{ap})_j$ (10^{-14} cm^2)	Observations
3300–2400	2900*	14	3.3	7.2	29	4.7	no compaction, tholins at the end
1335–1304	1329	16	3.8	7.6	30	3.7	no compaction
1279–1261	1271	33	4.9	11	39	−5.3	disappears at the end
957–937	948	27	7.8	13	37	−3.3	
782–763	775	-	4.0	-	34	-0.3	$\sigma_c(S) = 50 \times 10^{-14}$ $cm^2 \sigma_c(He) = 82 \times 10^{-14}$ cm^2
726–705	716 *	42	7.2	15	33	0.7	no compaction, tholins peaks too small
Mean value	$\sigma_{d,j}{}^{ap} \pm \Delta\sigma_j$	26 ± 16	5.4 ± 2.1	11 ± 4	33.7 ± 4	0	12% rms error

The 230 MeV S^{15+} beam data were analyzed in the same way. In order to evaluate the influence of the chemical environment on absorbance for each vibration mode, $\sigma_d{}^{ap}$ was determined for the S^{15+} irradiation, based on data plotted in Figure 9a. Results are presented in Table 3. The mean value is $\sigma_d{}^{ap} = (33.7 \pm 4) \times 10^{-14}$ cm^2. Data presented in Figure 9a are evidence that absorbance evolutions of bands are not equal to each other but similar, so that the $\sigma_d{}^{ap}$ extracted from them are distinct. Futhermore, since compaction and product formation are not responsible for such discrepancies, the conclusion is that A-values of band j should depend on constituent concentrations. For the S^{15+} irradiation, $\Delta\sigma_j$ was determined by fitting data with Equation (9) and the obtained values are also shown in Table 3. According to this model, $\Delta\sigma_j$ should be a characteristic of the band and, therefore, the evolution pattern of the bands should be the same for all ion beams. A comparison of Figure 9a with Figure 3a, Figure 4, and Figure 5a shows that this is indeed the case.

2.4. Summary of the Experimental Results

The methodology used to obtain the results of Figure 10 and Table 3 (see Section 4.2) was also applied to the other measurements. The corresponding cross sections and relevant experimental characteristics of the ion beams interacting with valine are summarized in Table 4. Average values of the initial column density, N_0, and of the thickness, T_k, are obtained from analysis of the 716 cm^{-1} ($A_v = 3.37 \times 10^{-18}$ cm/molec) and the 948 cm^{-1} ($A_v = 1.04 \times 10^{-18}$ cm/molec) bands; $\sigma_d{}^{ap}$ is determined by the average behavior of several bands.

Table 4. Relevant characteristics of the four irradiations: beam energy, electronic and nuclear stopping powers for valine, initial absorbances for the 716 and 948 cm^{-1} bands, initial column density, initial sample thickness, maximum beam penetration (range) and measured apparent destruction cross section.

Ion Beam	H^+	He^+	N^+	S^{15+}
Energy (MeV)	1.5	1.5	1.5	230
S_e (keV μm^{-1})	26.7	252	998	1690
S_n (keV μm^{-1})	0.0187	0.256	7.84	1.03
S_p 716 cm^{-1} (cm^{-1})	1.82	0.903	0.20	0.688
S_p 948 cm^{-1} (cm^{-1})	0.173	0.345	0.0849	0.151
Valine sample				
N_0 (10^{17} molec/cm^2)	8.12	6.56	1.61	4.00
Tk (μm)	2.4	0.96	0.23	0.58
Range (μm)	35	5.8	2.4	97
$\sigma_d{}^{ap}$ (10^{-14} cm^2)	26 ± 16	5.4 ± 2.1	11 ± 4	34 ± 4

3. Discussion

As fluence increases, three evident characteristics can be seen in the spectra presented in Figures 1 and 2: (i) the overall spectrum shape remains approximately the same, (ii) the spectral resolution deteriorates, and (iii) there is an exponential decrease of absorbance.

In general spectrum shapes are preserved during irradiation, evidence that the absorbance decrease on fluence is quite uniform over the analyzed spectrum range. Comparing the spectrum of the virgin sample with the one obtained at F = 1.4×10^{15} ions/cm^2 for H^+ ions, small differences can be recognized: in the irradiated sample spectrum, peaks are larger and peak/baseline ratio is lower. The shoulder at ~3200 cm^{-1} visible in the virgin sample spectrum disappears at F = 1.4×10^{15} ions/cm^2. The same occurs for the 2109 cm^{-1} band, which is clearly seen in the former but is well reduced in the latter. Features visible around 2300 cm^{-1} are due to CO_2 contamination outside the chamber, but inside the spectrometer. Figures 3–5 and 9, drawn in a semi-log plot, show that the trend of the integrated absorbance decrease is exponential. Discrepancies from this function are due to compaction (at low doses) and overlapping with daughter molecule bands (mainly at high fluences). With different ion beams, for already compacted samples and for pure precursor bands, slopes for distinct bands are systematically different from each other. Since this fact happens for precursor bands that do not overlap with product ones, it strongly suggests that A-values are not constant during irradiation. We propose that A-values have a weak exponential dependence on fluence, which has to be taken into account due to the exponential decrease of the precursor concentration and, consequently, due to the exponential increase of the concentration of products.

Literature data on amino acid radiolysis by MeV ions are poor. Two similar works on glycine bombarded by MeV H^+ are available: Gerakines et al., 2012 and Pilling et al., 2013 [16,17]. In the former, for glycine at 140 K irradiated by 0.8 MeV H^+, destruction cross section was determined to be 0.12×10^{-14} cm^2; it is expected that at 300 K this value be slightly higher. In the latter, for glycine at 300 K irradiated by 1.0 MeV H^+, the cross sections 0.1 to 0.5×10^{-14} cm^2 were determined for β-Gly and 2 to 3×10^{-14} cm^2 for α-Gly, respectively; cross sections for α-Gly are higher, but they may be affected by compaction, a process better understood nowadays. These results are in good agreement with the values 0.14 to 0.42×10^{-14} cm^2 obtained for valine in the current work for 1.5 MeV H^+.

3.1. *Dependence of Cross Sections on the Electronic Stopping Power*

The dissociation of a molecule in a highly excited electronic state is very likely and therefore the dependence of its destruction cross section on the electronic stopping power S_e is expected. Figure 11 depicts the evolution of S_e with the projectile energy for the four beams used. A pertinent question is whether this dependence is linear or not, revealing if dose is the relevant quantity in degradation by radiolysis. This analysis may be performed dividing absorbance values by S_e and comparing the normalized integrated absorbance evolution with fluence for the different ion beams.

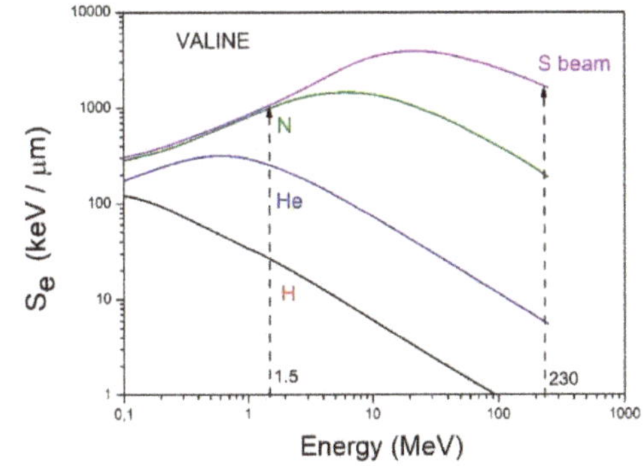

Figure 11. Dependence of electronic stopping power on projectile (H, He, N and S) energy. Dash lines indicate the two ion beam energies used in the current work.

Figure 12a shows how the integrated absorbance of the 948 cm^{-1} band varies with fluence for each irradiation; all data are divided by the initial integrated absorbance, S_0, which means that they are normalized to unity for the virgin sample. The exponential decrease is evident for the four measurements. It is also clear that the higher the stopping power the steeper the absorbance decrease. In order to compensate this effect, absorbance is plotted as a function of absorbed dose, D, defined as the projectile energy, deposited in a given volume of the sample, divided by the mass in that volume: $D = S_e F/\rho$, where ρ is mass density. Figure 12b presents the same data exhibited in Figure 12a but now as a function of D. The span in the evolution due to the different beams is strongly reduced: except for the He beam, data fluctuate around the dashed line, corresponding to the average function $S/S_0 = \exp(-\sigma_d^{ap} F) = \exp(-D/D_{mean})$, where $D_{mean} = S_e/(\rho\ \sigma_d^{ap})$ is the mean dose to destroy valine. For instance, for the 1.5 MeV N^+ beam, $<\sigma_d^{ap}> = (11 \pm 4)\ 10^{-14}\ cm^2$, $D_{mean} = 6.9 \times 10^{19}$ keV/g = 1.1×10^7 Gy.

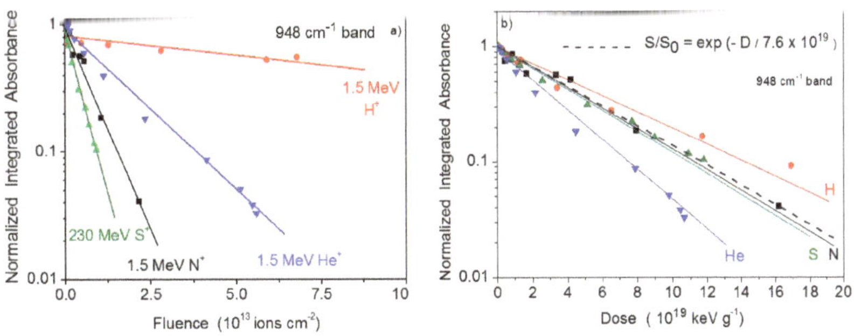

(a) (b)

Figure 12. (**a**) Evolution of the 948 cm^{-1} band integrated absorbance on fluence for H, He, N and S ion beams. In semi-log plot, angular coefficients represent cross sections. Integrated absorbances have been normalized to 1 at F = 0. (**b**) Same data, but plotted as a function of the absorbed dose.

As discussed before, radiolysis and sputtering are supposed to have the same dependence on fluence and cannot be (easily) determined individually: what is experimentally determined by this methodology is the sum of the two effects, namely the apparent destruction cross section of valine. This quantity, $\sigma_d{}^{ap}$, is displayed in Figure 13. The current measurements indicate that $\sigma_d{}^{ap}$ is approximately proportional to the electronic stopping power:

$$\sigma_d{}^{ap} = a\,(S_e)^n \tag{1}$$

where the best fitting yields $a = 1.8 \times 10^{-20}$ $cm^{2.936}/keV^{0.936}$ and $n = 0.936$; $\sigma_d{}^{ap}$ and S_e are expressed in cm^2 and keV/μm, respectively. This result is equivalent to the statement that the destruction cross section and the sputtering yield of valine are approximately proportional to the deposited dose in the material.

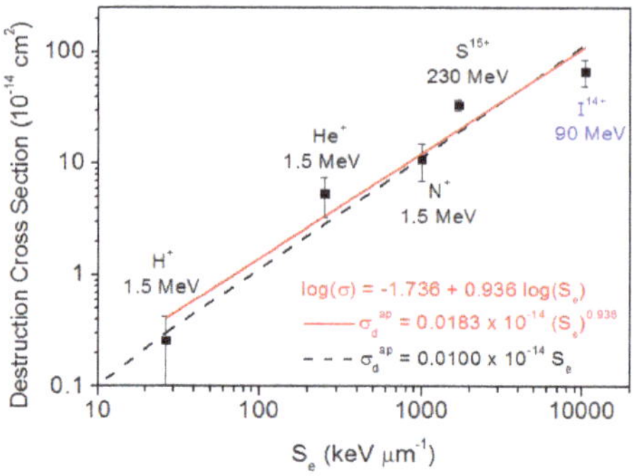

Figure 13. Dependence of the apparent destruction cross section on electronic stopping power. Dash line corresponds to $\sigma_d{}^{ap}$ proportional to S_e. For the fitting parameters, $\sigma_d{}^{ap}$ and S_e are expressed in cm^2 and keV/μm, respectively. 90 MeV ^{127}I projectile data were obtained with secondary ion mass spectrometry.

In the data shown in Figure 13, one extra cross section is the value corresponding to valine irradiated by 90 MeV $^{127}I^{14+}$ projectiles. This result was obtained by Salehpour et al. [23], using low flux projectiles (1000 ions/s) and acquiring the time-of-flight spectrum of the desorbed ions. The electronic stopping power of 90 MeV $^{127}I^{+14}$ ions in valine is 1.04×10^4 keV/μm. Besides the desorption valine products, the ejected ions were composed by protonated valine $(M + H)^+$ and by the cluster ion series $(M_n + H)^+$, where n runs up to ~20. During the irradiation with up to 2×10^{11} projectiles, the yield of emitted cluster ions is exponentially quenched. The damage (destruction) cross sections of the protonated $(M + H)^+$ and deprotonated $(M\text{-}H)^-$ species were determined to be 68 (±18) × 10^{-14} cm^2 and 44 (±15) × 10^{-14} cm^2, respectively. We note that this is a direct sputtering measurement of charged particles. Although neutral particles are dominant in the induced desorption, the measured cross section corresponds to *radiolysis*, that is, only to the destruction cross section. Indeed, the sputtering yield (of ion species) reflects the evolution of the valine concentration on the sample surface which is ruled by its radiolysis. Therefore, the radiolysis cross section for the 90 MeV $^{127}I^{14+}$ beam shown in Figure 13 appears below the line which represent values produced by both sputtering and radiolysis.

3.2. Astrophysical Implications

The assumption that Equation (1) holds for the destruction of solid valine by all cosmic rays regardless of their energy allows evaluating its absolute destruction rate. For each cosmic ray constituent, j, the corresponding partial destruction rate is:

$$R_j = \int_0^{\infty} \frac{d\Phi_j}{dE}(E)\,\sigma_{d,j}^{ap}(E)\,dE \tag{2}$$

where E is the kinetic energy of species j, $\Phi_j(E)$ is its flux density and $\sigma_{d,j}(E)$ its apparent destruction cross section given by Equation (1); for the current calculations, we have considered a = 1.0×10^{-20} cm^3/keV and n = 1.0. Shen et al. [37] proposed a model in which the galactic cosmic ray (GCR) flux densities present the analytical expression:

$$\frac{d\Phi_j}{dE}(E) = C_j \frac{E^{0.3}}{(E+E_0)^3} \tag{3}$$

where E is the GCR energy per nucleon, E_0 is a parameter considered here as equal to 400 MeV, and C_j is determined from the GCR abundances. For the current calculations, we considered abundances and fluxes presented in Table 1 and Figure 2 of ref. Shen et al. [37]; the fluxes are multiplied by 2π sr, for taking into account the isotropic GCR incidence on a spot located in a flat surface. The flux density dependence on E/m (Equation (3)) is presented in Figure 14a and the destruction rate dependence on E/m (Equation (2)) is shown in Figure 14b. The dominant effect of H and He ions at low GCR energy (and of Fe ions at high energy) is evident. The integral in Equation (2) was performed from 10 keV up to 10 GeV/u. Table 5 presents the C_j values (integrated over 2π sr) used, the R_j results for j = H. He, C, O, Ne and Fe, as well as the respective partial half-lives, defined as ln(2)/R_j, the total destruction rate and total half-life.

The valine radiolysis and sputtering by GCR are ruled by their iron constituents. The half-life of valine in the ISM (InterStellar Medium) is estimated to be about one million years. Therefore, since no IR (infrared) signal of valine has been detected yet, we would like to encourage observations of valine rotational features by radio spectroscopy.

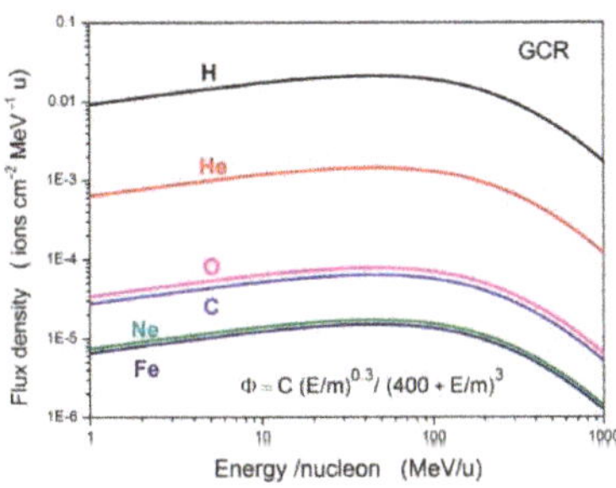

(a)

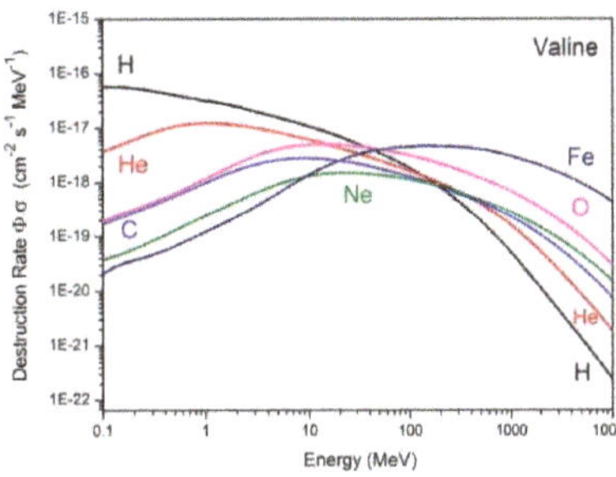

(b)

Figure 14. (**a**) Galactic cosmic ray (GCR) flux densities as predicted by Shen et al. [37]. (**b**) Valine destruction rate dependence on cosmic ray energy.

Table 5. The C_j flux parameter, the destruction rate R_j (in second^{-1} and in million-year^{-1}), and the half-live (in million-year) for each GCR species j.

j	C_j Ions cm^{-2} s^{-1} MeV$^{-1.7}$	R_j (10^{-16} s^{-1})	R_j (Ma^{-1})	$\tau_{1/2}$ = ln(2)/R_j (Ma)
H	5.96×10^5	7.1	0.022	31
He	4.11×10^4	7.5	0.024	29
C	1.79×10^3	8.8	0.028	25
O	2.22×10^3	28	0.088	7.9
Ne	438	12	0.038	18
Fe	425	170	0.54	1.2
Total	6.42×10^5	230	0.74	0.94

4. Materials and Methods

4.1. Ion Beam Irradiation of Valine

Irradiations using 1.5 MeV H^+, He^+, and N^+ ion beams were performed at PUC-Rio and those with 230 MeV S^{15+} projectiles were carried out at GANIL. Pilling et al. [17] and Vignoli Muniz et al. [30] published detailed descriptions of the experimental set-ups, respectively. Chamber pressure was lower than 10^{-6} mbar. Samples were irradiated at room temperature (~300 K) and effects for lower temperature (40–300 K) on shape and integrated absorbance of valine vibrational bands were reported recently by da Costa and da Silveira [33].

Samples were prepared as follows: D-Valine (99.5% of purity, delivered by Sigma Aldrich, São Paulo, Brazil) was dissolved in a solution of water (40% *v*/*v*) and ethanol (60% *v*/*v*). The mixture was shaken in ultrasound machine up to a complete and homogeneous solution has been obtained. For the experiments at the Van de Graaff Laboratory, this solution was sprayed onto a ZnSe window, a 2 mm thick and 13 mm diameter disk. For the GANIL experiment, drops of this solution were deposited onto the ZnSe window. Substrate and solution were heated at 100 °C for enhancing the solvent evaporation. Sample thicknesses, T_k, were determined from $T_k = S \ln(10)\, M/\rho\, N_A\, A_v$, where S is the integrated absorbance of a given band, A_v is its A-value (band strength), ln(10) = 2.303, M = 117.15 g/mol the valine molecular weight, $\rho = 1.32$ g/cm^3 the valine mass density and N_A the Avogadro number. Two bands have been used for this calculation: the one at 948 cm^{-1} with $A_v = 1.04 \times 10^{-18}$ cm/molec, and the one at 716 cm^{-1} with $A_v = 3.37 \times 10^{-18}$ cm/molec. The results agree within 50% for several samples.

At the Van de Graaff Laboratory, the samples were kept inside the analytical chamber on a stainless steel sample-holder at room temperature (pressure below 10^{-6} mbar). Ion beam fluence was determined from current measurements via two nanoampere meters connected to a beam collimator and to a Faraday cup, respectively; both electrodes were biased to +90 V, for minimizing secondary electron emission. Infrared spectra were acquired in transmission mode by a JASCO FTIR-4100 spectrometer. The 4000–600 cm^{-1}spectral region was inspected by averaging 70 scans with 2.0 cm^{-1} resolution. Absorbances were measured by integrating areas after subtracting a linear background defined within the band limits. At GANIL, the Casimir chamber was installed at the SME irradiation beamline. This is a dedicated beamline for sample irradiation with homogeneous irradiation field (*xy*-scanning of the ion beam) and permanent control of the beam flux, which was about 1×10^9 cm^{-2}·s^{-1} for MeV S^{15+} projectiles· The residual pressure was about 2×10^{-6} mbar; FTIR acquisition was done by a Nicolet Magna 550 spectrometer over the 4000–600 cm^{-1} range with 1-cm^{-1} resolution.

The observed CO_2 bands are produced by molecules outside the chamber and do not interfere with the valine radiolysis. The ion beam electronic (S_e) and nuclear (S_n) stopping powers for valine were determined through the software SRIM [38]; $S_e >> S_n$ for all used ion beams, ion–atom collisions occurred in the electronic energy loss regime.

4.2. Cross Sections from Modelling of Fluence Dependent Absorption

Determination of molecular destruction and formation cross sections requires monitoring abundance rates of molecular species. This can be done by infrared spectroscopy, using the Beer–Lambert law, $N = \ln 10\, S/A_v$, which connects the integrated absorbance S with the molecular column density N for each substance in the sample—provided that the A-value parameter, A_v, is known. A major concern is that this parameter varies with the chemical environment of the molecule, which in turn depends on temperature, solid state phase (crystallization/amorphization) and relative abundance of molecular constituents of the sample, quantities that change with irradiation.

Common models describing the dependence of integrated absorbance on projectile fluence consider that S(F) variation is due to sputtering (ejection of valine or fragments), radiolysis (chemical reactions) and—sometimes—by beam induced phase transition (molecular rearrangement in the sample, such as crystallization, amorphization, and compaction). Since molecular rearrangement processes are accomplished faster than the sample destruction, data analysis may be separated in two parts: a first one, when absorption changes by the rearrangement and by other processes, and a second one, when only sputtering and radiolysis proceed. Since the compaction process does not alter the column density N of a molecular film (although its thickness may vary), absorbance changes due to A-value variations are considered in the current model to avoid misinterpretation of destruction cross sections. Inside the material (where there is no sputtering), the decrease rate of the precursor concentration is proportional to the concentration itself and, therefore, this concentration decreases as $\exp(-\sigma_d F)$ with fluence. Accordingly, at the material surface, the precursor sputtering yield, Y, also decreases exponentially for the same reason: the radiolysis dictates the reduction of the precursor concentration (Sundqvist et al. [25]).

Infrared spectroscopy in transmission mode measures the precursor column density as a function of fluence. Hereby, radiolysis and sputtering effects are simultaneously monitored:

$$\frac{dN}{dF} = -\sigma_d N - Y \tag{4}$$

Considering the precursor sputtering yield $Y = Y_0 (C/C_0) \approx Y_0 (N/N_0)$, where Y_0, C_0 and N_0 are the sputtering yield, the concentration at surface and the column density of the virgin sample. For fluences that are not too high, Equation (4) can be rewritten as:

$$\frac{dN}{dF} \approx -\sigma_d N - Y_0 \frac{N}{N_0} \equiv -\sigma_d^{ap} N \tag{5}$$

with the solution:

$$N(F) \approx N_0 \exp\left(-\sigma_d^{ap} F\right) \tag{6}$$

Equation (6) shows that the apparent destruction cross section $\sigma_d^{ap} = \sigma_d + Y_0/N_0$ is the quantity actually determined in FTIR measurements [29]. This experimental inability should not be regarded as a methodological deficiency. Rather, it reveals that the laboratorial simulation mimics correctly the combined effect of molecular dissociation and molecular sputtering that occurs when swift heavy ions impinge on a material surface in space.

In order to describe Equation (6) in terms of absorbance, the Beer–Lambert law is applied: $N(F) = \ln 10\, S(F)/A_v(F)$ and, $N_0 = \ln 10\, S_p/A_v(0)$:

$$\frac{S(F)}{S_p} = \frac{A_v(F)}{A_v(0)} \exp\left(-\sigma_d^{ap} F\right) \tag{7}$$

where S_p is the "porous" initial absorbance, that is, the absorbance for the fresh sample as prepared. The ratio $A_v(0)/A_v(F)$ describes compaction effects, and also explains why the different precursor bands do not evolve proportionally to each other as fluence increases. For compaction, it was found empirically that [39,40]:

$$\frac{A_v(F)}{A_v(0)} = \frac{1 - \zeta \exp(-\sigma_c F)}{1 - \zeta} \tag{8}$$

where $\zeta = (S_0 - S_p)/S_0$ is the relative porosity, understood as a (beam-dependent) parameter able to quantify the effect of porosity on the absorbance evolution of a given band. S_0 is interpreted as being the initial integrated absorbance of a compacted sample; for non-compac ted materials, this value is obtained by extrapolating the integrated absorbance decay to F = 0, but using absorbances for an already compacted sample. Equation (8) takes into account the infrared spectrum change of a porous sample when it suffers compaction either by irradiation or by annealing (before irradiation); each band shows this absorbance change differently, reason why each band has its own ζ for a specific ion beam. The parameter σ_c is called compaction cross section; for $\sigma_c F << 1$, Equation (8) writes $A_v(F)/A_v(0) \sim 1 + (\zeta/1-\zeta)\, \sigma_c F$, another way of seeing that absorbance may increase or decrease with fluence depending whether ζ is positive or negative, respectively.

Independently of compaction or other phase changes, $A_v(F)$ may also have a dependence on fluence due to the persistent increase of daughter molecules in the sample. Since concentrations vary exponentially, it is reasonable to assume in first order approximation that $A_v^j(F) = A_v^j(0) \exp(-\Delta\sigma_j F)$ for each vibrational band j. In this case, Equation (7) writes, excluding compaction:

$$\frac{S_j(F)}{S_{p,j}} = \frac{A_v{}^j(F)}{A_v{}^j(0)} \exp\left(-\sigma_{d,j}^{ap} F\right) = \exp\left(-\left(\sigma_{d,j}^{ap} + \Delta\sigma_j\right)F\right) \tag{9}$$

The prediction is that absorbances decrease exponentially but not with the same $\sigma_d{}^{ap}$ for all bands; the dispersion is nevertheless small, once $\Delta\sigma_j << \sigma_{d,j}{}^{ap}$. Again, comparison among different systems should be made with caution, comparing cross sections for the same band or mean values.

5. Conclusions

Thin films of D-valine were irradiated by four MeV ion beams. Compaction, radiolysis and sputtering processes were analyzed by FTIR spectroscopy at room temperature. The main conclusions are:

1. Compaction effects are seen via absorption modification, which means band strength modification. This process is due to the destruction of pores and to phase changes in the solid sample.
2. The chemical effects of the irradiation by distinct ion beams on valine are similar except for their molecular destruction rates (or destruction cross sections), which vary according to the deposited dose.
3. The elimination of valine molecules from the irradiated sample proceeds by radiolysis and by sputtering. Radiolysis imposes that the sample column density decreases exponentially with beam fluence. Another consequence is that the precursor concentration on the sample surface varies also exponentially and, in turn, the precursor sputtering yield decreases accordingly.
4. Strong evidence that the sputtering yield decays exponentially during irradiation is given by mass spectrometry measurements. Indeed, Salehpour et al. [23] found this behavior for valine bombarded by $^{127}I^{14+}$ ions. Furthermore, Ferreira-Rodrigues et al. reported similar results analyzing the sample surface by TOF-^{252}Cf–PDMS for glycine radiolysis [22]. FTIR spectroscopy determines the loss rate of precursors: the technique cannot distinguish those sputtered from those dissociated. Accordingly, the apparent destruction cross section, $\sigma_d{}^{ap}$, is measured: it quantifies the combined effect of both processes. It is found that $\sigma_d{}^{ap}$ is approximately proportional to the electronic stopping power and, therefore, to the absorbed dose. This is an unexpected finding, since for condensed gases the literature indicates $\sigma_d{}^{ap}$ proportional to $S_e{}^n$, where $n \sim 3/2$ [28].
5. At the end of irradiation, valine was destroyed and their bands vanished. The still visible bands are due to valine's daughter molecules. Structural attributions are attempted.
6. Band strengths evolve with fluence but not linearly; moreover, some A-values are more sensitive to fluence than others. This feature does not depend on the processing ion and is attributed to compaction and to the increase of product concentrations.
7. Concerning Astrophysics, using theoretical GCR flux distribution, the solid valine half-life is predicted to be about one million years. Recent work estimated solid adenine half-life to be 10 ± 8 million years (Vignoli Muniz et al. [30]). These results suggest that the search for valine by radio astronomy should be envisaged.
8. Concerning Astrobiology, 1.1×10^7 Gy is the mean dose to destroy solid valine. This is a huge value for Biology standards. For human beings, 50 Gy is a typical lethal dose.
9. Concerning radiotherapy, the current results, obtained for valine, are actually typical for amino acids in general and even for biological material. Stopping powers are calculated for ion-atom interactions and biological materials are mostly formed by carbon, oxygen, nitrogen and hydrogen—the same atoms of the molecular structure of valine; biological material mass densities are close to 1 g/cm^3, so that penetration depths are similar to valine. Therefore, the findings of

this work may be a useful contribution for the understanding of microscopic processes in the damage of biological targets, and in particular, radiation protection and ion beam therapy.

Author Contributions: Conceptualization, C.A.P.d.C., G.S.V.M., P.B. and H.R.; Data curation, C.A.P.d.C. and E.F.d.S.; Formal analysis, C.A.P.d.C. and G.S.V.M.; Funding acquisition, P.B. and E.F.d.S.; Investigation, C.A.P.d.C., G.S.V.M., P.B., H.R. and E.F.d.S.; Methodology, C.A.P.d.C., G.S.V.M., P.B., H.R. and E.F.d.S.; Project administration, E.F.d.S.; Resources, P.B.; Supervision, P.B. and H.R.; Visualization, C.A.P.d.C. and H.R.; Writing—original draft, C.A.P.d.C., H.R. and E.F.d.S.; Writing—review & editing, H.R. All authors have read and agreed to the published version of the manuscript.

Funding: The authors would like to acknowledge the partial support by the Brazilian agencies CNPq (INEspaço and Science without Borders) and FAPERJ (E-26/202.843/2018), as well as the CAPES-COFECUB French-Brazilian exchange program.

Acknowledgments: We thank the staff of Van de Graaff Laboratory at PUC-Rio, CIMAP and GANIL.

Conflicts of Interest: The authors declare no conflict of interest.

References

1. Marzzoco, A.; Torres, B.B. *Bioquímica Básica*, 4th ed.; Guanabara Koogan: Rio de Janeiro, Brazil, 2015.
2. Pizzarello, S.; Feng, X.; Epstein, S.; Cronin, J.R. Isotopic analyses of nitrogenous compounds from the Murchison meteorite: Ammonia, amines, amino acids, and polar hydrocarbons. *Geochim. Cosmochim. Acta* **1994**, *58*, 5579–5587. [CrossRef]
3. Botta, O.; Bada, J.L. Extraterrestrial Organic Compounds in Meteorites. *Surv. Geophys.* **2002**, *23*, 411–467. [CrossRef]
4. Elsila, J.E.; Glavin, D.P.; Dworkin, J.P. Cometary glycine detected in samples returned by Stardust. *Meteorit. Planet. Sci.* **2009**, *44*, 1323–1330. [CrossRef]
5. Kuan, Y.; Charnley, S.B.; Huang, H.; Tseng, W.; Kisiel, Z. Interstellar Glycine. *Astrophys. J.* **2003**, *593*, 848. [CrossRef]
6. Snyder, L.E.; Lovas, F.J.; Hollis, J.M.; Friedel, D.N.; Jewell, P.R.; Remijan, A.; Ilyushin, V.V.; Alekseev, E.A.; Dyubko, S.F. A Rigorous Attempt to Verify Interstellar Glycine. *Astrophys. J.* **2005**, *619*, 914. [CrossRef]
7. Cunningham, M.R.; Jones, P.A.; Godfrey, P.D.; Cragg, D.M.; Bains, I.; Burton, M.G.; Calisse, P.; Crighton, N.H.M.; Curran, S.J.; Davis, T.M.; et al. A search for propylene oxide and glycine in Sagittarius B2 (LMH) and Orion. *Mon. Not. R. Astron. Soc.* **2007**, *376*, 1201–1210. [CrossRef]
8. Jones, P.A.; Cunningham, M.R.; Godfrey, P.D.; Cragg, D.M. A Search for biomolecules in Sagittarius B2 (LMH) with the Australia Telescope Compact Array. *Mon. Not. R. Astron. Soc.* **2007**, *374*, 579–589. [CrossRef]
9. Abramov, O.; Mojzsis, S.J. Microbial habitability of the Hadean Earth during the late heavy bombardment. *Microb. Habitability Hadean Earth Late Heavy Bombard.* **2009**, *7245*, 419–422. [CrossRef]
10. Chyba, C.; Sagan, C. Endogenous production, exogenous delivery and impact-shock synthesis of organic molecules: An inventory for the origins of life. *Nature* **1992**, *355*, 125–132. [CrossRef]
11. Rothard, H.; Domaracka, A.; Boduch, P.; Palumbo, M.E.; Strazzulla, G.; da Silveira, E.F.; Dartois, E. Modification of ices by cosmic rays and solar wind. *J. Phys. B At. Mol. Opt. Phys.* **2017**, *50*, 062001. [CrossRef]
12. Diehl, J.F. Food irradiation—Past, present and future. *Radiat. Phys. Chem.* **2002**, *63*, 211–215. [CrossRef]
13. Silindir, M.; Özer, Y. The Effect of Radiation on a Variety of Pharmaceuticals and Materials Containing Polymers. *PDA J. Pharm. Sci. Technol.* **2012**, *66*, 184–199. [CrossRef] [PubMed]
14. Adaligil, E.; Patil, K.; Rodenstein, M.; Kumar, K. Discovery of Peptide Antibiotics Composed of D-Amino Acids. *ACS Chem. Biol.* **2019**, *14*, 1498–1506. [CrossRef]
15. Lam, H.; Oh, D.-C.; Cava, F.; Takacs, C.N.; Clardy, J.; de Pedro, M.A.; Waldor, M.K. D-Amino Acids Govern Stationary Phase Cell Wall Remodeling in Bacteria. *Science* **2009**, *325*, 1552–1555. [CrossRef] [PubMed]
16. Gerakines, P.A.; Hudson, R.L.; Moore, M.H.; Bell, J.-L. In situ measurements of the radiation stability of amino acids at 15–140 K. *Icarus* **2012**, *220*, 647–659. [CrossRef]
17. Pilling, S.; Mendes, L.A.V.; Bordalo, V.; Guaman, C.F.M.; Ponciano, C.R.; da Silveira, E.F. The Influence of Crystallinity Degree on the Glycine Decomposition Induced by 1 MeV Proton Bombardment in Space Analog Conditions. *Astrobiology* **2013**, *13*, 79–91. [CrossRef]

18. Pilling, S.; Nair, B.G.; Escobar, A.; Fraser, H.; Mason, N. The temperature effect on the glycine decomposition induced by 2 keV electron bombardment in space analog conditions. *Eur. Phys. J. D* **2014**, *68*, 58. [CrossRef]
19. Maté, B.; Tanarro, I.; Escribano, R.; Moreno, M.A.; Herrero, V.J. Stability of extraterrestrial glycine under energetic particle radiation estimated from 2 kev electron bombardment experiments. *Astrophys. J.* **2015**, *806*, 151. [CrossRef]
20. Souza-Corrêa, J.A.; da Costa, C.A.P.; da Silveira, E.F. Compaction and Destruction Cross-Sections for α-Glycine from Radiolysis Process via 1.0 keV Electron Beam as a Function of Temperature. *Astrobiology* **2019**, *19*, 1123–1138. [CrossRef] [PubMed]
21. Peeters, Z.; Botta, O.; Charnley, S.B.; Ruiterkamp, R.; Ehrenfreund, P. The Astrobiology of Nucleobases. *Astrophys. J.* **2003**, *593*, L129. [CrossRef]
22. Ferreira-Rodrigues, A.M.; Homem, M.G.P.; Naves de Brito, A.; Ponciano, C.R.; da Silveira, E.F. Photostability of amino acids to Lyman α radiation: Glycine. *Int. J. Mass Spectrom.* **2011**, *306*, 77–81. [CrossRef]
23. Salehpour, M.; Håkasson, P.; Sundqvist, B. Damage cross sections for fast heavy ion induced desorption of biomolecules. *Nucl. Instrum. Methods Phys. Res. Sect. B Beam Interact. Mater. At.* **1984**, *2*, 752–756. [CrossRef]
24. Becker, O.; Della-Negra, S.; Le Beyec, Y.; Wien, K. MeV Heavy Ion induced desorption from insulating films as function of projectile velocity. *Nucl. Instrum. Methods Phys. Res.* **1986**, *16*, 321–333. [CrossRef]
25. Sundqvist, B.; Hedin, A.; Håkasson, P.; Salehpour, M.; Säve, G. Sputtering of Biomolecules by fast heavy ions. *Nucl. Instrum. Methods Phys. Res. Sect. B Beam Interact. Mater. At.* **1986**, *B14*, 429–435. [CrossRef]
26. De Barros, A.L.F.; Domaracka, A.; Andrade, D.P.P.; Boduch, P.; Rothard, H.; da Silveira, E.F. Radiolysis of frozen methanol by heavy cosmic ray and energetic solar particle analogues: Radiolysis of frozen methanol by heavy cosmic ray and energetic solar particle analogues. *Mon. Not. R. Astron. Soc.* **2011**, *418*, 1363–1374. [CrossRef]
27. Godard, M.; Féraud, G.; Chabot, M.; Carpentier, Y.; Pino, T.; Brunetto, R.; Duprat, J.; Engrand, C.; Bréchignac, P.; d'Hendecourt, L.; et al. Ion irradiation of carbonaceous interstellar analogues: Effects of cosmic rays on the 3.4 μ m interstellar absorption band. *A&A* **2011**, *529*, A146.
28. Andrade, D.P.P.; de Barros, A.L.F.; Pilling, S.; Domaracka, A.; Rothard, H.; Boduch, P.; da Silveira, E.F. Chemical reactions induced in frozen formic acid by heavy ion cosmic rays. *Mon. Not. R. Astron. Soc.* **2013**, *430*, 787–796. [CrossRef]
29. Mejía, C.F.; de Barros, A.L.F.; Bordalo, V.; da Silveira, E.F.; Boduch, P.; Domaracka, A.; Rothard, H. Cosmic ray–ice interaction studied by radiolysis of 15 K methane ice with MeV O, Fe and Zn ions. *Mon. Not. R. Astron. Soc.* **2013**, *433*, 2368–2379. [CrossRef]
30. Vignoli Muniz, G.S.; Mejía, C.F.; Martinez, R.; Augé, B.; Rothard, H.; Domaracka, A.; Boduch, P. Radioresistance of Adenine to Cosmic Rays. *Astrobiology* **2017**, *17*, 298–308. [CrossRef]
31. Weast, R.C. *CRC Handbook of Chemistry and Physics*, 62nd ed.; CRC Press: Boca Raton, FL, USA, 1981; ISBN 0-8493-0462-8.
32. Ada Bibang, P.C.J.; Aditya, N.A.; Augé, B.; Boduch, P.; Desfrançois, C.; Domaracka, A.; Lecomte, F.; Manil, B.; Martinez, R.; Muniz, G.S.V.; et al. Ion radiation in icy space environments: Synthesis and Radioresistance of Complex Organic Molecules. *Low Temp. Phys.* **2019**, *45*, 590–597. [CrossRef]
33. Da Costa, C.A.P.; da Silveira, E.F. Valine infrared absorbance at cryogenic temperatures. *Low Temp. Phys.* **2019**, *45*, 649–655. [CrossRef]
34. Sagan, C.; Khare, B.N. Tholins: Organic chemistry of interstellar grains and gas. *Nature* **1979**, *277*, 102–107. [CrossRef]
35. Kumar, S. Spectroscopic studies of valine and leucine molecules a comparative study. *Vib. Spectrosc.* **2011**, *39*, 4996–4999.
36. Façanha Filho, P.F.; Freire, P.T.C.; Lima, K.C.V.; Mendes Filho, J.; Melo, F.E.A.; Pizani, P.S. High temperature Raman spectra of L-leucine crystals. *Braz. J. Phys.* **2008**, *38*, 131–137. [CrossRef]
37. Shen, C.J.; Greenberg, J.M.; Schutte, W.A.; van Dishoeck, E.F. Cosmic ray induced explosive chemical desorption in dense clouds. *A&A* **2004**, *415*, 203–215.
38. Ziegler, J.F.; Ziegler, M.D.; Biersack, J.P. Interactions of Ions with Matter. Available online: srim.org (accessed on 10 August 2019).

39. Mejía, C.; de Barros, A.L.F.; Seperuelo Duarte, E.; da Silveira, E.F.; Dartois, E.; Domaracka, A.; Rothard, H.; Boduch, P. Compaction of porous ices rich in water by swift heavy ions. *Icarus* **2015**, *250*, 222–229. [CrossRef]
40. De Barros, A.L.F.; da Silveira, E.F.; Fulvio, D.; Rothard, H.; Boduch, P. Ion irradiation of ethane and water mixture ice at 15 k: Implications for the solar system and the ism. *Astrophys. J.* **2016**, *824*, 81. [CrossRef]

International Journal of
Molecular Sciences

MDPI

Article

In Vitro Comparison of Passive and Active Clinical Proton Beams

Anna Michaelidesová [1,2,3,*], Jana Vachelová [1], Jana Klementová [1,4], Tomáš Urban [3], Kateřina Pachnerová Brabcová [1], Stanislav Kaczor [3], Martin Falk [5], Iva Falková [5], Daniel Depeš [5], Vladimír Vondráček [2] and Marie Davídková [1,*]

1 Nuclear Physics Institute of the Czech Academy of Sciences, Řež 130, 25068 Řež, Czech Republic; vachelova@ujf.cas.cz (J.V.); jana.klementova@img.cas.cz (J.K.); brabcova@ujf.cas.cz (K.P.B.)
2 Proton Therapy Center Czech, Budínova 2437/1a, 180 00 Prague, Czech Republic; vladimir.vondracek@ptc.cz
3 Faculty of Nuclear Sciences and Physical Engineering, Czech Technical University in Prague, Břehová 7, 11519 Prague, Czech Republic; tomas.urban@fjfi.cvut.cz (T.U.); stanislavkczr@gmail.com (S.K.)
4 Institute of Molecular Genetics of the Czech Academy of Sciences, Vídeňská 1083, 14220 Prague 4, Czech Republic
5 Institute of Biophysics of the Czech Academy of Sciences, Královopolská 135, 61265 Brno, Czech Republic; falk@ibp.cz (M.F.); ivafalk@ibp.cz (I.F.); depesd26@gmail.com (D.D.)
* Correspondence: michaelidesova@ujf.cas.cz (A.M.); davidkova@ujf.cas.cz (M.D.)

Received: 30 June 2020; Accepted: 4 August 2020; Published: 6 August 2020

Abstract: Nowadays, the irradiation methodology in proton therapy is switching from the use of passively scattered beams to active pencil beams due to the possibility of more conformal dose distributions. The dose rates of active pencil beams are much higher than those of passive beams. The purpose of this study was to investigate whether there is any difference in the biological effectiveness of these passive and active irradiation modes. The beam qualities of double scattering and pencil beam scanning were measured dosimetrically and simulated using the Monte Carlo code. Using the medulloblastoma cell line DAOY, we performed an in vitro comparison of the two modes in two positions along the dose–deposition curve plateau and inside the Bragg peak. We followed the clonogenic cell survival, apoptosis, micronuclei, and γH2AX assays as biological endpoints. The Monte Carlo simulations did not reveal any difference between the beam qualities of the two modes. Furthermore, we did not observe any statistically significant difference between the two modes in the in vitro comparison of any of the examined biological endpoints. Our results do not show any biologically relevant differences related to the different dose rates of passive and active proton beams.

Keywords: proton therapy; pencil beam scanning; double scattering; cell survival

1. Introduction

The clinical advantages of a proton beam were firstly suggested by Wilson in 1946 in his paper on the radiological use of high-energy protons [1]. Since the implementation of protons in cancer treatment in 1954 [2], an increased interest in proton therapy has been observed worldwide. The most important advantage of protons compared to high-energy photons, depositing their energy exponentially after a maximum located couple of centimeters inside the patient's body, is that protons deposit most of their energy in a precisely defined depth of the patient's body given by their initial kinetic energy. Their depth–dose distribution is called the Bragg curve, with a maximum known as the Bragg peak.

The first step in proton therapy is the generation of protons (hydrogen ionization) and their acceleration inside a particle accelerator, typically a cyclotron or a synchrotron. For treatment purposes, the beam needs to be spread longitudinally by superimposing beams of different energies and weights. To obtain a lower energy than the fixed energy coming out of the cyclotron, an adjustable amount of material must be placed into the beam path. This can be carried out by a degrader inserted right after the beam extraction or by placing a stack with a variable number of plates (range shifter), a plate with ripples (ridge filter), or a rotating wheel with an azimuthally changing thickness (range modulation wheel) inside the nozzle in the irradiation room. In the case of synchrotron, the energy is adjusted in the accelerator and therefore there is no need for any additional devices. This longitudinally spread proton beam is known as the Spread Out Bragg Peak (SOBP) [3].

Along with the longitudinal spread, the beam has to be spread also laterally, which can be achieved by passive or active modes. Examples of passive modes are Single or Double Scattering (SiS or DS), while an example of an active mode is Pencil Beam Scanning (PBS). For the passive modes, the beam passes through the scatters (one or two, SiS or DS, respectively) and is subsequently spread in the lateral direction. Active modes are based on the use of scanning magnets, which redirect the narrow proton beam to several positions according to the treatment plan. The dose is then delivered to each layer of the volume spot by spot [3].

Currently, there is an increased interest in the use of active beam modes for proton therapy due to the possibility of delivering more conformal dose distributions to patients. Another advantage of the active irradiation modes is the production of lower levels of secondary-induced radiation, mostly neutrons produced by the proton beam interacting with the components of the technological constructions or patient-specific devices, such as collimators or compensators, which are not used in the case of the active modes [4,5]. Thus, the growing interest in active beams is not surprising, and the issue of the comparison of passive and active modes is very topical.

The majority of biological studies on proton beams were performed using passive modes. The proton beam profiles can be adjusted to be almost identical, regardless of the mode used. However, Monte Carlo simulations on patients have shown considerably increased dose-averaged linear energy transfer (LET_d) values at the distal fall-off of the proton beams in the case of the active beams compared to the passive beams, which can affect the biological response of tissues situated in this position during proton therapy [6]. Moreover, the dose rates of the active proton beams are several orders of magnitude higher than a few Gy/min for the passively scattered beams. Generally, high and ultrahigh dose rates are associated with reduced biological damage, which can be explained by oxygen depletion and radical recombination [7–9].

On the other hand, in more recent studies dealing with various dose rates of photon radiation, a decreasing cell survival has been found in AG01522 cells, unlike in more radioresistant DU145 cells, which stayed unaffected by the increasing dose rate [10]. More efficient cell killing with higher dose per pulse using 10 MV X-rays has been reported in the case of glioblastoma cell lines (T98G, U87-MG) [11]. Furthermore, studies on FLASH radiotherapy using ultrahigh dose rates of above 40 Gy/s confirmed an increase in the differential response in normal and tumor tissues [12–14].

For protons, Iwata et al. have found almost identical values of relative biological effectiveness (RBE) in SOBPs formed by passive and active irradiation modes [15], and a comparison of a continuous beam with an ultra-high dose rated beam (10 MGy/s) produced by laser acceleration displayed only a statistically insignificant increase in the relative RBE in the latter case [16]. However, the irradiation of HLE cells by active proton beams caused a decrease in DNA synthesis, unlike the irradiation by passive beam [17]. It has also been observed that the delivery of higher doses per pulse can reduce cell survival due to an increase in the DNA damage, mostly in the yield of double-strand breaks (DSB) [18]. Other studies suggest that the higher dose rates can induce modifications of proteins, fatty acids, and cell organelles in the cytoplasm, which can lead to cell death [19–21].

The aim of the present study was to compare the influence on the biological response on different dose rates associated with the PBS (active) and DS (passive) modes at the Proton Therapy Center Czech (PTC). PTC is equipped with an isochronous cyclotron Proteus-235 (Ion Beam Applications, IBA, Belgium), which accelerates protons to an energy of 230 MeV. The comparison was held to investigate several biological endpoints after irradiation in two different positions in Bragg curves with similar dosimetric properties in both irradiation modes. The beam properties were simulated using the Monte Carlo code to reveal possible differences in the beam qualities.

2. Results

2.1. Monte Carlo Simulations

The LET_d values for protons, electrons, positrons, deuterons, tritons, helions, and alphas as well as the LET_d values of all these particles together for the two modes (PBS, DS) and the two positions (Position 1 and Position 2) are given in Table 1. Furthermore, in Tables 2 and 3, information about the neutron production for the two modes and the two positions is provided. In Figure 1, the calculated neutron spectra and the radiation weighting factors (w_R) according to the Publication 103 of the International Commission on Radiological Protection (ICRP) [22] are shown.

Table 1. The dose-averaged linear energy transfer (LETd) values for the two modes and the two positions.

	LET_d (keV/μm)			
	Position 1		Position 2	
	PBS	DS	PBS	DS
Protons	0.45	0.45	3.17	3.43
Electrons	0.28	0.28	0.19	0.19
Positrons	0.19	0.19	0.18	0.18
Deuterons	5.74	5.57	8.68	7.40
Tritons	7.54	8.27	8.08	11.16
Helions	32.00	32.80	59.73	42.79
Alphas	51.79	51.76	102.58	103.00
All particles	0.44	0.44	3.11	3.34

Table 2. Summarized neutron production at Position 1 and their categorization according to the ICRP 103 report [22].

E (MeV)	No. PBS (-)	No. DS (-)	No. PBS (%)	No. DS (%)
≤1	4.36×10^8	6.20×10^8	1.00×10^2	1.00×10^2
1–50	2.61×10^1	3.74×10^1	5.99×10^{-6}	6.03×10^{-6}
≥50	6.81×10^0	7.70×10^0	1.56×10^{-6}	1.24×10^{-6}
All energies	4.36×10^8	6.20×10^8	1.00×10^2	1.00×10^2

Table 3. Summarized neutron production at Position 2 and their categorization according to the ICRP 103 report [22].

E (MeV)	No. PBS (-)	No. DS (-)	No. PBS (%)	No. DS (%)
≤1	2.94×10^8	2.63×10^8	1.00×10^2	1.00×10^2
1–50	1.33×10^2	1.21×10^2	4.52×10^{-5}	4.61×10^{-5}
≥50	2.67×10^1	2.44×10^1	9.09×10^{-6}	9.27×10^{-6}
All energies	2.94×10^8	2.63×10^8	1.00×10^2	1.00×10^2

2.2. Cell Survival

The parameters of the Linear-Quadratic (LQ) model and the cell survival data from the proton irradiations and those from the irradiation by a ^{60}Co source are presented in Table 4 and Figure 2.

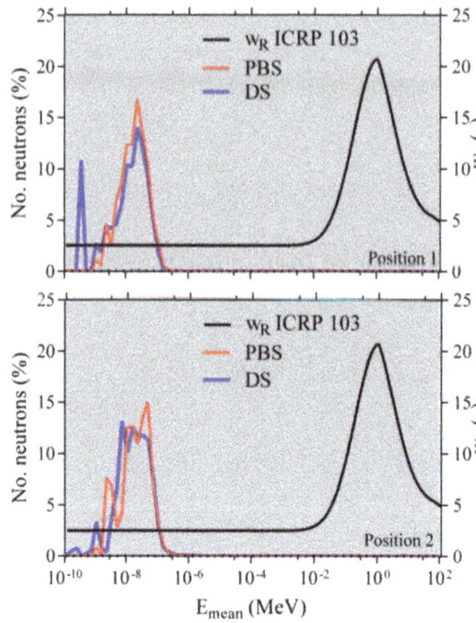

Figure 1. The normalized neutron spectra (the total number of neutrons is taken as 100%, as is shown in Tables 2 and 3) and the representation of the radiation weighting factors (w_R) according to the ICRP Publication 103 [22].

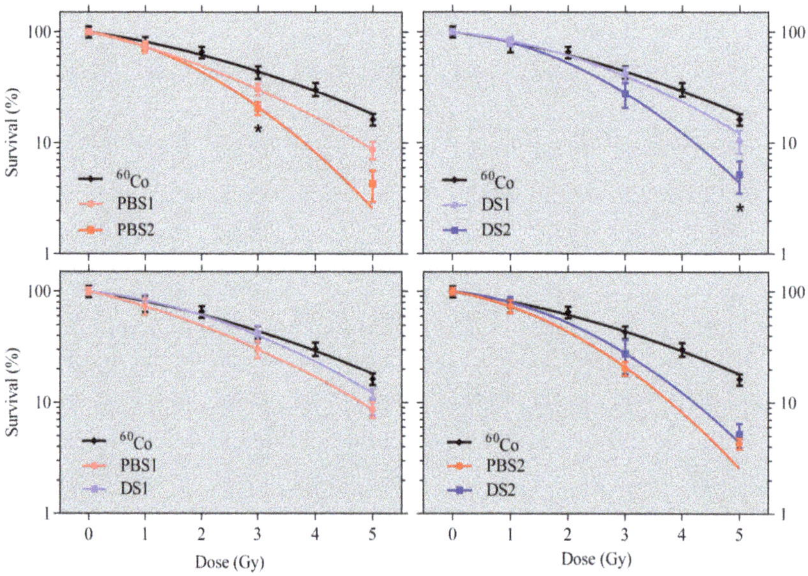

Figure 2. Clonogenic cell survival data of cells irradiated by Pencil Beam Scanning (PBS) and Double Scattering (DS) proton modes in the two irradiation positions (1—plateau, 2—Bragg peak) and by a ^{60}Co gamma source. Up: a comparison of the two positions for each of the modes. Down: a comparison of the two modes for each of the positions. Points correspond to the experimental averages, with their standard deviations and lines corresponding to the LQ model fit. The asterisk (*) corresponds to *t*-tests lower than 0.05.

Table 4. The parameters of the Linear-Quadratic (LQ) model (α, β), their standard errors (σ_α, σ_β), and the correlation parameter R^2.

Data Set	α (Gy^{-1})	σ_α (Gy^{-1})	β (Gy^{-2})	σ_β (Gy^{-2})	R^2
^{60}Co	0.17	3.25×10^{-2}	0.03	9.99×10^{-3}	0.99
PBS1	0.26	1.13×10^{-3}	0.05	2.77×10^{-2}	1.00
PBS2	0.20	3.95×10^{-4}	0.11	1.27×10^{-2}	1.00
DS1	0.13	2.87×10^{-2}	0.06	9.49×10^{-3}	1.00
DS2	0.12	1.21×10^{-2}	0.10	4.99×10^{-3}	1.00

2.3. Apoptosis

The percentages of total apoptotic cells for the three different periods of time post irradiation after deducting the percentages of apoptosis in the control samples are shown in Figure 3. The total apoptosis includes cells (Caspase 3/7-positive) undergoing both early and late apoptosis.

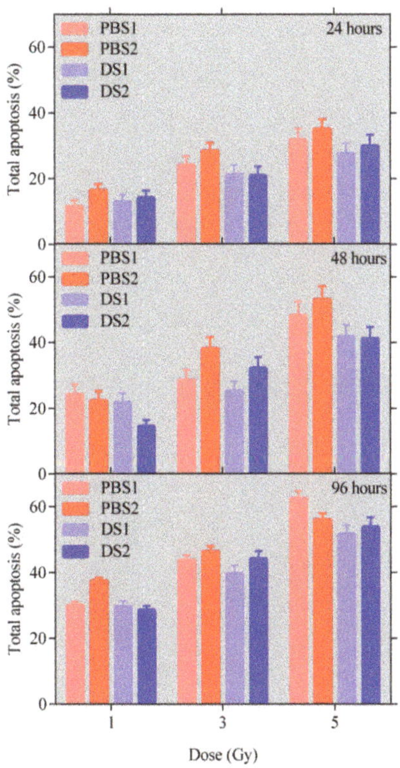

Figure 3. The percentage of Caspase 3/7-positive cells undergoing apoptosis (total apoptosis = early + late) determined for the two modes and two positions at three different periods of time post irradiation. No statistically significant differences have been found using the *t*-test.

2.4. Micronuclei Assay

The average percentages with their standard deviations of binuclear cells (BNC) containing micronuclei (MN) are shown on the upper side of Figure 4 for both modes and both positions in the Bragg peak. On the lower side of Figure 4, the micronuclei frequencies are given.

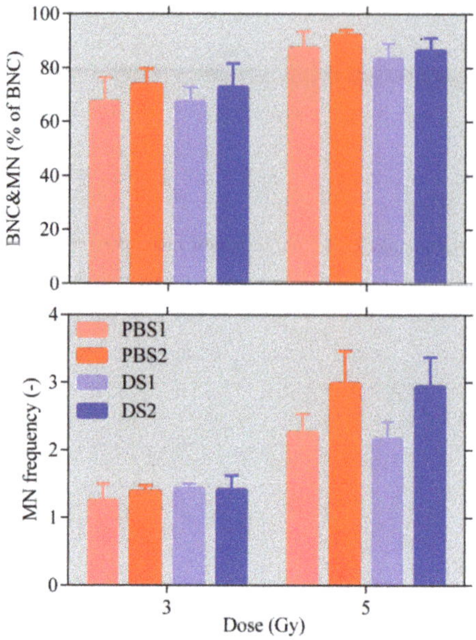

Figure 4. The percentage of binuclear cells (BNC) containing micronuclei (MN) from the total number of BNC and the micronuclei frequencies for the two modes and the two positions in the proton beam for doses of 3 and 5 Gy. No statistically significant differences have been found using the *t*-test.

2.5. *γ H2AX Assay*

The average number of γH2AX foci two hours after the irradiation is shown in Figure 5.

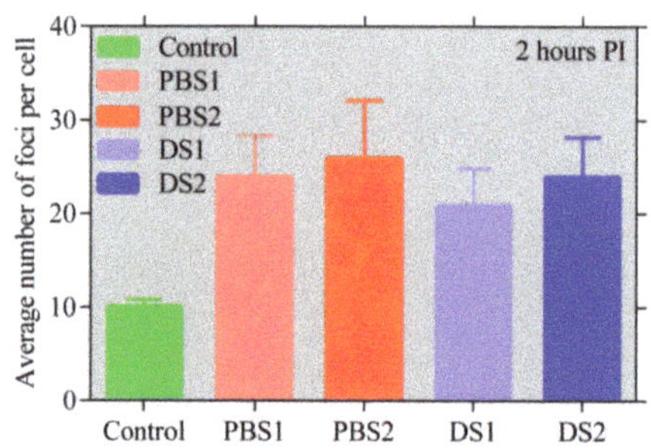

Figure 5. The average numbers of γH2AX foci per nucleus with their standard deviations compared for the two modes and the two positions in the proton beam path; measured two hours post irradiation by dose 1 Gy. No statistically significant differences have been found using the *t*-test.

3. Discussion

The comparison of the two modes of proton beam irradiation has been performed using the medulloblastoma cell line DAOY and several selected biology endpoints. The DAOY cell line has been selected as a representative brain/cerebellum tumor highly relevant from the clinical point of view. The inclusion of another cell line in the study would be important to obtain more general results, but on the other hand would further decrease the statistical relevance of our results, since we have used the maximum beam time available. All the irradiations and sample processing have been repeated in two independent experiments. The independent runs have been realized with a time separation of several months to exclude unpredictable effects of the cell culturing or accelerator performance. Cryopreserved cells have been defrosted and cultured to comparable low passages prior to experimental processing in each run, and several sample replicates for each dose/time point and the irradiation method have been used. Although we are not able to extend further the present study for more independent experimental repeats or different cell lines, we believe our manuscript presents relevant knowledge when it is interpreted in combination with other studies.

In Table 1, the simulated LET_d values for protons, electrons, positrons, deuterons, tritons, helions, and alphas as well as the LET_d values for all the particles together for the two modes (PBS, DS) and two positions (Position 1 and Position 2) are presented. The LET_d values do not manifest any significant differences between the two modes. The same can be stated for neutrons. A summary of the neutron production along the primary beam path is given in Tables 2 and 3. In the first position, a higher number of neutrons is produced. While looking at Figure 1, we can observe that almost all the produced neutrons are located in a region where the w_R, according to the ICRP 103, is equal to 2.5. This means that the produced neutrons from both modes in both positions are biologically almost identical, with not a very high biological relevance. We thus believe that we can consider the beam qualities of both modes to be the same, and possible biological differences between the two modes must be assigned to the different dose rates. From the information provided by IBA staff about the current from the cyclotron, the beam line efficiency, and other beam parameters, we could calculate that the dose rate in the PBS mode is about 4000 times higher than that in the DS mode.

The data fitted by the LQ model have revealed statistically significant differences in cell survival between the two positions for each mode (Figure 2 upper side, Table 4). However, no statistically significant differences have been found between the two modes, despite the fact that a slightly lower survival for the PBS mode is detectable in both irradiation positions (Figure 2 lower side, Table 4). Using the α and β parameters estimated from the LQ model fit, the RBE values in comparison to ^{60}Co have been estimated for both modes and both positions at the 50% cell survival level. The $RBE_{50\%}$ for PBS1 was found to be equal to 1.34 ± 0.16; for PBS2 it was 1.48 ± 0.20, for DS1 it was 1.05 ± 0.16, and for DS2 it was 1.26 ± 0.16.

The apoptosis level and the number of BNC containing MN have been found to be almost identical for the two modes. In terms of micronuclei frequencies and the 5-Gy-dose point, the frequencies are higher for the peak position in comparison to the proximal position for both modes, which is in agreement with the expected more complex DNA damage at the peak position (see [23]). The micronuclei frequencies appear to be almost the same for the 3-Gy-dose point and slightly higher for PBS for the 5-Gy-dose point, but no statistically significant differences have been found using the *t*-test in both cases.

Our results suggest that the in vitro biological effectiveness of PBS in comparison to DS is almost identical for doses above 1 Gy, even if the two modes have very different dose rates. This statement is based on the Monte Carlo simulations of the beams used for the biological study and on the results from all the studied biological endpoints (cell survival, apoptosis, γH2AX, and micronuclei assay). The combination of the MN test and γH2AX assay allowed us to sensitively monitor the effects of both lower and higher doses in parallel. While γH2AX is currently the most sensitive method to detect DSBs, the micronucleus test is much less sensitive and is inaccurate at low doses. As different doses and experimental approaches revealed the same trends, our results suggest that the radiation dose in the range studied does not influence the results, so we can jointly interpret our results irrespective of the dose.

Our results are in agreement with a previously published study examining the differences in the PBS and DS in SOBP [15], where not only the dose rate, but also other effects can play a role in the biological comparison of the two modes. Since we used almost monoenergetic beams, our results show that the dose rate does not play any significant role in the in vitro biological effectiveness of protons (including secondary particles) in a dose range relevant for the proton therapy of cancer.

4. Materials and Methods

4.1. Cell Culture

The used human medulloblastoma cells DAOY (ATCC® HTB-186™, Manassas, VA, USA) were originally derived from a tumor in the posterior fossa of a four-year old male. The adherent DAOY cell line exhibited a doubling time of 24 h and a plating efficiency of 30%.

The cells were grown in Improved Minimum Essential Medium (GIBCO, Gaithersburg, MD, USA) supplied with 10% fetal bovine serum (BIOSERA, Nuaille, France), 100 U/mL of penicillin, and 0.1 mg/mL of streptomycin (Sigma-Aldrich, St. Louis, MO, USA) in T25 culture flasks (TPP Techno Plastic Products AG, Trasadingen, Switzerland) placed in an incubator with a 5% CO_2 atmosphere and a temperature of 37 °C. Non-synchronized cell populations were used for the irradiations to mimic the situation in the patients' tissues better.

4.2. Proton Beam Description and Monte Carlo Simulations

The proton irradiations were performed using the PBS and DS modes at the PTC using depth–dose curves that were as similar as possible and had narrow energy spectra. We chose to use narrow energy spectra to achieve well-defined beams and maximal dose rates in both modes (in case of PBS, the goal was to irradiate the whole sample at once without the need to switch energies and thus to combine the total dose from more energy layers. Switching between energies would reduce the dose rate).

At first, the range of the DS mode was set to 23.9 g/cm^2 and the modulation to 1.5 g/cm^2, where modulation corresponds to the distance between the proximal and distal 90% dose on the percentage depth dose curve (PDD). The chosen modulation is the smallest and the chosen range is the highest available in clinics at PTC (there is a limitation in the combinations of range and modulation. It is not possible to use any range and any modulation, but the combination has to be selected from predefined options). No collimators or compensators at the nozzle exit were used to mitigate the low beam contamination by secondary particles.

The selection of the high range allowed us to choose a high-energy beam for the PBS mode (190.6 MeV) and thus to obtain a wide Bragg peak, ensuring the reduced influence of potential sample positioning uncertainties. The PBS energy has been selected in order to ensure that both modes had the same range. The PDDs for both modes are shown in Figure 6.

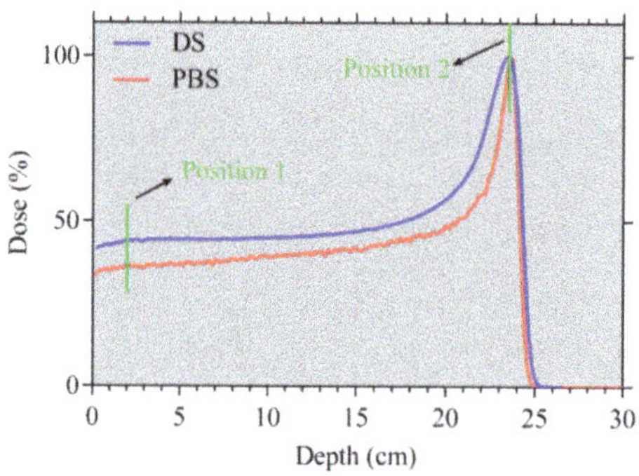

Figure 6. The percentage depth–dose curves for the selected beam energies for PBS and DS. The grey vertical lines symbolize the two irradiation positions of the cell cultures.

Monte Carlo simulations have been performed to clarify especially secondary particle properties along the primary beam (in water phantom). The Monte Carlo code Fluka v. 2011.2x.6 [24,25] has been used for the estimation of the LET spectra of protons, light ions, electrons, and positrons, as well as the energy spectra of neutrons. All the simulations have been executed for the two modes in the simplified geometry of the treatment nozzles of these modalities.

The treatment nozzles, the water phantom, as well as the source term (e.g., proton beam) have been described by the appropriate tools of Fluka code. Firstly, the depth–dose distributions of the dose deposited in the water phantom for both modalities were simulated and compared with the experimental ones. The second step consisted of estimating the LET spectra of the charged particles and energy spectra of neutrons in sample irradiation positions along the beam path in the water phantom.

4.3. Analysis of the Monte Carlo Simulations

From the Monte Carlo simulations, we have obtained numbers of charged particles with a specific LET value. The dose-averaged LET (LET_d) values were calculated from the raw data of the simulations using the following equation:

$$LET_d = \frac{\sum_i \left(\hat{LET}_i \cdot D_i \right)}{\sum_i D_i} \quad (1)$$

where LET_d is the calculated dose-averaged LET in the specific position of the Bragg peak, $\hat{LET}_i$ is the LET value, and D_i corresponds to the proportion of particles with this LET value in the specific position.

In addition to the charged particles, the production of neutrons was also simulated. Since neutrons are uncharged particles, they do not have LET values. However, it was possible to generate the energy spectra of neutrons in each position of interest for each mode. Due to the fact that neutrons have, based on their energies, different biological consequences, they were further categorized based on the ICRP Publication 103 [22] to gain a better understanding of their biological consequences.

4.4. Cell Sample Irradiations

All the irradiations and sample processing were repeated twice (two independent experiments). The irradiations took place in two different positions along the Bragg curves, corresponding to 2 cm (Position 1) and 23.6 cm (Position 2) in water adjusted using water-equivalent RW3 and PMMA plates. These positions have been selected so that one of the series of samples has been in the plateau region and the second one in the peak. In the figures and tables in this paper, the two irradiation positions are marked as PBS1 and PBS2 or DS1 and DS2, respectively. The PBS and DS stand for the used mode, while 1 stands for the plateau (Position 1) and 2 for the peak position (Position 2).

The same doses were applied at each position and for each mode (DS, PBS), which means that the irradiations were executed in four parts (DS1, DS2, PBS1, PBS2). From Figure 6, it is evident that the PDD differs in the two modes and that the peak for DS mode is wider than that for PBS. The proximal position (Position 1, Figure 6) was implemented in the study to decrease the possibility of false positive results that could have been obtained from the measurements in the peak region, where small positioning uncertainties could have led to misleading results. Another advantage of analyzing two positions is the difference in the beam quality, which should be reflected in their different biological effects.

The doses for PBS have been established in the treatment planning system XiO by ELEKTA using plans with a spot spacing of 4 mm. The irradiation maps were constructed to enable the irradiation by 1, 3, and 5 Gy in the plateau and the peak region. The doses have been verified by a PPC05 ionization chamber (IBA Dosimetry GmbH, Schwarzenbruck, Germany) at the cell sample irradiation positions.

As for the DS mode, the dose calibration and verification were performed using the same PPC05 ionization chamber to obtain the number of monitor units needed for each dose applied. The irradiation maps were created for both positions to get doses of 1, 3, and 5 Gy. The beam homogeneity was verified using the 2D LYNX detector (IBA Dosimetry GmbH, Schwarzenbruck, Germany) for all the created irradiation maps for both modes.

The irradiations were carried out in the gantry rooms of PTC, which provided the possibility of irradiating the samples through the irradiation table. This enabled us to place the samples horizontally, without the need to fill them completely with medium and thus to lower the stress of the cells.

As for the cell survival, the micronuclei assay and the apoptosis assay the cells were irradiated in T25 flasks. In case of the γH2AX assay, the irradiation was performed in glass bottom dishes (SPL Life Sciences Co., Ltd., Pocheon-si, Korea). The cells were irradiated by doses of 1, 3, and 5 Gy for the cell survival and apoptosis assays. The samples for the γH2AX assay were irradiated by 1 Gy only to obtain a reasonable number of foci per cell for analysis (about 25–35 double-strand breaks, DSB, per cell nucleus [26]). In the case of the micronuclei assay, the cells were irradiated by doses of 3 and 5 Gy to obtain a number of micronuclei sufficiently higher than the micronuclei level of the control samples.

The second set of samples was irradiated by doses of 1, 2, 3, 4, and 5 Gy using a ^{60}Co γ-ray source at the Authorized Metrology Center of the Nuclear Physics Institute of the Czech Academy of Sciences (Prague, Czech Republic) to get reference cell survival curves. Due to the horizontal beam orientation, the cells were irradiated in a vertical position in T25 flasks fully filled with medium to avoid beam inhomogeneities and back scattering. The cell irradiations took place in a water depth of 5 cm ensured by a special holder inside a water tank placed at a source-surface distance of 75 cm. The irradiation field was restricted to an area of 10×10 cm^2 by a primary and secondary collimator. The dose calibration measurements were performed in exactly the same geometry by a cylindrical PTW 30,013 ionization chamber (PTW, Freiburg, Germany). The dose rate was recalculated each time according to the irradiation date (around 0.01 Gy/s at the irradiation position), and the applied doses were achieved by increasing the time of irradiation according to the actual dose rate. The maximal positioning error was 2 mm, which corresponds to a dose uncertainty of less than 2% (calculated from the depth–dose curve obtained in the same geometry).

4.5. Post-Irradiation Sample Processing and Data Analysis

4.5.1. Clonogenic Cell Survival Assay

Immediately after the irradiation, the cells were harvested (0.05% Trypsin by GIBCO or BIOSERA) and counted by the MUSE cell analyzer (EMD Millipore, Burlington, MA, USA) using the Muse Count and Viability Assay Kit (Merck Millipore-MCH100102), following the manufacturer's instructions. The cells were then reseeded into 6-well plates for a clonogenic cell survival analysis. The seeded cell number was established according to the expected cell survival given by the delivered dose. The colonies formed nine days after irradiation were fixed and colored using crystal violet (Sigma-Aldrich, St. Louis, MO, USA) diluted in 95% methanol.

The number of colonies in each well was counted manually, and the averages and standard deviations were calculated from every 6-well plate. Since there were two experimental runs, the averages and standard deviations were established from both runs for further processing. The survival data were fitted by the Linear-Quadratic (LQ) model [27] using the least square fitting procedure in the GraphPad Prism software (Version 7.0 for Windows, GraphPad Software, La Jolla, CA, USA, www.graphpad.com).

A statistical analysis of the data was performed as follows. To estimate if the variance of the two associated data sets was significantly different, the F-test from the one-way ANOVA was used. In cases where the F-value was lower than 0.05, the data sets were recognized to have different variances. According to the result of the F-test, the two tailed Student's *t*-test with the same or different variance was used to estimate the statistical significance of the difference between the evaluated datasets. The data were considered to be significantly different if the *p*-value from the *t*-test was lower than 0.05, as visualized in the graphs using an asterisk (*). The same statistical analysis was performed for the other biological endpoints as well.

4.5.2. Apoptosis Measurement

The cell medium was replaced immediately after the irradiation with a fresh one. The level of apoptosis was established in samples 24, 48, and 96 h post irradiation based on the activation of Caspase 3/7 using the MUSE™ Caspase-3/7 Kit (Millipore-MCH100108) according to the manufacturer's instructions. Duplicates were prepared for each examined situation. Since there were two experimental runs, the averages and standard deviations were then established from the two runs.

4.5.3. Micronuclei Assay

The irradiated cells were reseeded at a 30% confluence in dishes with glass coverslips. Twenty-four hous later, cytochalasin B (Sigma-Aldrich, St. Louis, MO, USA) in a concentration of 1 μg/mL of media was added. After another 24 h, the cells were fixed with 4% paraformaldehyde and stained by 4,6-diamidine-2-phenylindole dihydrochloride (DAPI) (Sigma-Aldrich, St. Louis, MO, USA). The images were captured at the CZ-OPENSCREEN center (Institute of Molecular Genetics of the CAS, Prague, Czech Republic) by the Operetta High-Content Imaging System (Version HH12940101, PerkinElmer, Waltham, MA, USA) at 40× magnification using the Harmony software (Version 4.1, PerkinElmer, Waltham, MA, USA).

A minimum of 50 fields per slide were analyzed using in-house software following the criteria of the HUMN project [28] written in PYTHON. The binuclear cells (BNC) containing micronuclei (MN) and the total number of MN were scored. The MN frequencies were calculated as the total number of MN divided by the total number of BNC in each sample. Duplicates were prepared for each dose point. The averages and standard deviations were established from the two experimental runs.

4.5.4. γ H2AX Assay

DSBs are the most serious lesions induced in the DNA molecule by ionizing radiation [29]. DSBs contribute mostly to cell death and/or mutagenesis. The detection of γH2AX foci and/or other DSB repair foci currently represents the most sensitive method for their scoring [30,31].

Briefly, the cells were grown and irradiated in dishes with glass coverslips and fixed two hours post irradiation by 4% paraformaldehyde for 10 min at room temperature. The cells were permeabilized using 0.1% Triton X-100. Consequently, they were treated by 5% bovine serum albumin (BSA) solution (Sigma-Aldrich, St. Louis, MO, USA) before being incubated with the primary antibodies. The primary mouse anti-H2AX antibody (Millipore) diluted in 1% BSA was added and the cells were incubated overnight at 4 °C. On the next day, the samples were washed three times in 0.1% BSA and the staining was visualized by goat anti-mouse IgG ALEXA Fluor 488-conjugated secondary antibody (Invitrogen, Carlsbad, CA, USA) diluted in 1% BSA. After a repeated washing with 0.1% BSA, the nuclei were stained with DAPI.

For this "screening" analysis with large numbers of analyzed cells, the time point of two hours post irradiation was selected as a result of a compromise. The maximum number of foci is found in irradiated cells 15–30 min post irradiation [25,32]; however, after the two hours the number of foci is still high enough to allow statistically relevant comparisons with lower time expenses required for focus counting.

The images were captured at the CZ-OPENSCREEN center (Institute of Molecular Genetics of the CAS) by the Operetta High-Content Imaging System (Version HH12940101, PerkinElmer, Waltham, MA, USA) at 40× magnification using the Harmony software (Version 4.1, PerkinElmer, Waltham, MA, USA). A minimum of 100 fields per slide were analyzed by the Columbus image analysis software (Version 2.7.1, PerkinElmer, Waltham, MA, USA), which enables the detection and counting of fluorescently stained γH2AX foci in the DAPI-stained nuclei of cells. Duplicates were prepared for each time point. The averages and standard deviations were established from the two experimental runs.

5. Conclusions

Our results demonstrate that the dose rate does not significantly influence the biological effectiveness of clinical proton beams (including the effect of secondary particles produced) for doses of above 1 Gy. We have found that the PBS and DS modes in proton therapy can be considered as mutually equivalent concerning the biological responses of irradiated cells.

Author Contributions: A.M. designed the study. T.U. performed the Monte Carlo simulation of the proton beams. J.V. assured the cell cultivation. K.P.B., J.K., M.F., I.F., D.D., and S.K. participated in the analysis of the studied biological endpoints in irradiated cells. V.V. and A.M. prepared the irradiation plans and assured the dosimetry measurements and irradiation of samples. A.M. wrote the manuscript. M.D. supervised the study. All authors have read and agreed to the published version of the manuscript.

Funding: The work was supported by the INSPIRE project funded from the European Union's Horizon 2020 Research and Innovation Programme under Grant Agreement n. 730983. M.F., I.F., and D.D. acknowledge the support of the Czech Science Foundation (project n. 20-04109J) and of the German Academic Exchange Service (project n. DAAD-19-03). We acknowledge the CZ-OPENSCREEN center: National Infrastructure for Chemical Biology (IMG CAS, Prague, Czech Republic), supported by the Ministry of Education, Youth and Sports grants n. RVO: 68378050-KAV-NPUI, LM2015063 and IMG institutional support RVO: 68378050, for providing the microscopy imaging and image analysis presented herein.

Conflicts of Interest: The authors declare no conflict of interest.

References

1. Wilson, R.R. Radiological Use of Fast Protons. *Radiolgy* **1946**, *47*, 487–491. [CrossRef] [PubMed]
2. Lawrence, J.H. Proton irradiation of the pituitary. *Cancer* **1957**, *10*, 795–798. [CrossRef]
3. Paganetti, H. *Proton Therapy Physics (Series in Medical Physics and Biomedical Engineering)*; Taylor & Francis: London, UK, 2012.
4. Schneider, U.; Agosteo, S.; Pedroni, E.; Besserer, J. Secondary neutron dose during proton therapy using spot scanning. *Int. J. Radiat. Oncol.* **2002**, *53*, 244–251. [CrossRef]
5. Farah, J.; Mareš, V.; Romero-Exposito, M.; Trinkl, S.; Domingo, C.; Dufek, V.; Klodowska, M.; Kubancak, J.; Knežević, Z.; Liszka, M.; et al. Measurement of stray radiation within a scanning proton therapy facility: EURADOS WG9 intercomparison exercise of active dosimetry systems. *Med. Phys.* **2015**, *42*, 2572–2584. [CrossRef]
6. Grassberger, C.; Paganetti, H. Elevated LET components in clinical proton beams. *Phys. Med. Boil.* **2011**, *56*, 6677–6691. [CrossRef]
7. Berry, R.J. Effects of radiation dose-rate: From Protracted, Continuous Irradiation to Ultra-High Dose-Rates from Pulsed Accelerators. *Br. Med. Bull.* **1973**, *29*, 44–47. [CrossRef]
8. Epp, E.R.; Weiss, H.; Ling, C.C. Irradiation of cells by single and double pulses of high intensity radiation: Oxygen sensitization and diffusion kinetics. *Curr. Top. Radiat. Res.* **1976**, *11*, 201–250.
9. Durante, M.; Bräuer-Krisch, E.; Hill, M.A. Faster and safer? FLASH ultra-high dose rate in radiotherapy. *Br. J. Radiol.* **2017**, *91*, 20170628. [CrossRef]
10. McGarry, C.K.; Butterworth, K.T.; Trainor, C.; O'Sullivan, J.M.; Prise, K.M.; Hounsell, A.R. Temporal characterization andin vitrocomparison of cell survival following the delivery of 3D-conformal, intensity-modulated radiation therapy (IMRT) and volumetric modulated arc therapy (VMAT). *Phys. Med. Boil.* **2011**, *56*, 2445–2457. [CrossRef]
11. Lohse, I.; Tanadini-Lang, S.; Hrbacek, J.; Scheidegger, S.; Bodis, S.; Macedo, N.S.; Feng, J.; Lütolf, U.M.; Zaugg, K. Effect of high dose per pulse flattening filter-free beams on cancer cell survival. *Radiother. Oncol.* **2011**, *101*, 226–232. [CrossRef]
12. Favaudon, V.; Caplier, L.; Monceau, V.; Pouzoulet, F.; Sayarath, M.; Fouillade, C.; Poupon, M.-F.; Brito, I.; Hupé, P.; Bourhis, J.; et al. Ultrahigh dose-rate FLASH irradiation increases the differential response between normal and tumor tissue in mice. *Sci. Transl. Med.* **2014**, *6*, 245ra93. [CrossRef]
13. Vozenin, M.-C.; De Fornel, P.; Petersson, K.; Favaudon, V.; Jaccard, M.; Germond, J.-F.; Petit, B.; Burki, M.; Ferrand, G.; Patin, D.; et al. The Advantage of FLASH Radiotherapy Confirmed in Mini-pig and Cat-cancer Patients. *Clin. Cancer Res.* **2019**, *25*, 35–42. [CrossRef] [PubMed]
14. Montay-Gruel, P.-G.; Petersson, K.; Jaccard, M.; Boivin, G.; Germond, J.-F.; Petit, B.; Doenlen, R.; Favaudon, V.; Bochud, F.; Bailat, C.; et al. Irradiation in a flash: Unique sparing of memory in mice after whole brain irradiation with dose rates above 100 Gy/s. *Radiother. Oncol.* **2017**, *124*, 365–369. [CrossRef] [PubMed]
15. Iwata, H.; Ogino, H.; Hashimoto, S.; Yamada, M.; Shibata, H.; Yasui, K.; Toshito, T.; Omachi, C.; Tatekawa, K.; Manabe, Y.; et al. Spot Scanning and Passive Scattering Proton Therapy: Relative Biological Effectiveness and Oxygen Enhancement Ratio in Cultured Cells. *Int. J. Radiat. Oncol. Biol. Phys.* **2016**, *95*, 95–102. [CrossRef]

16. Auer, S.; Hable, V.; Greubel, C.; Drexler, G.A.; Schmid, T.E.; Belka, C.; Dollinger, G.; Friedl, A.A. Survival of tumor cells after proton irradiation with ultra-high dose rates. *Radiat. Oncol.* **2011**, *6*, 139. [CrossRef] [PubMed]
17. Gridley, D.S.; Pecaut, M.; Mao, X.W.; Wroe, A.J.; Luo-Owen, X. Biological Effects of Passive Versus Active Scanning Proton Beams on Human Lung Epithelial Cells. *Technol. Cancer Res. Treat.* **2015**, *14*, 81–98. [CrossRef]
18. Steel, G.G. The ESTRO Breur lecture cellular sensitivity to low dose-rate irradiation focuses the problem of tumour radioresistance. *Radiother. Oncol.* **1991**, *20*, 71–83. [CrossRef]
19. Hamilton, J.; Bernhard, E.J. Cell signalling and radiation survival: The impact of protein phosphatases. *Int. J. Radiat. Boil.* **2009**, *85*, 937–942. [CrossRef]
20. Jeggo, P.A.; Lavin, M.F. Cellular radiosensitivity: How much better do we understand it? *Int. J. Radiat. Boil.* **2009**, *85*, 1061–1081. [CrossRef]
21. Rodemann, H.P. Molecular radiation biology: Perspectives for radiation oncology. *Radiother. Oncol.* **2009**, *92*, 293–298. [CrossRef]
22. ICRP. The 2007 Recommendations of the International Commission on Radiological Protection. ICRP Publication 103. *Ann. ICRP* **2007**, *37*. [CrossRef]
23. Michaelidesová, A.; Vachelová, J.; Puchalska, M.; Brabcová, K.P.; Vondráček, V.; Sihver, L.; Davídková, M. Relative biological effectiveness in a proton spread-out Bragg peak formed by pencil beam scanning mode. *Australas. Phys. Eng. Sci. Med.* **2017**, *47*, 359–368. [CrossRef] [PubMed]
24. Ferrari, A.; Sala, P.R.; Fasso, A.; Ranft, J. *FLUKA: A Multi-Particle Transport Code (Program Version 2005)*; Stanford Linear Accelerator Center: Stanford, CA, USA, 2005.
25. Böhlen, T.; Cerutti, F.; Chin, M.; Fassò, A.; Ferrari, A.; Ortega, P.G.; Mairani, A.; Sala, P.; Smirnov, G.; Vlachoudis, V. The FLUKA Code: Developments and Challenges for High Energy and Medical Applications. *Nucl. Data Sheets* **2014**, *120*, 211–214. [CrossRef]
26. Falk, M.; Emilie, L.; Kozubek, S. Higher-order chromatin structure in DSB induction, repair and misrepair. *Mutat. Res. Mutat. Res.* **2010**, *704*, 88–100. [CrossRef] [PubMed]
27. Dale, R.G. Use of the Linear-Quadratic Radiobiological Model for Quantifying Kidney Response in Targeted Radiotherapy. *Cancer Biother. Radiopharm.* **2004**, *19*, 363–370. [CrossRef]
28. Fenech, M.; Chang, W.P.; Kirsch-Volders, M.; Holland, N.; Bonassi, S.; Zeiger, E. HUMN project: Detailed description of the scoring criteria for the cytokinesis-block micronucleus assay using isolated human lymphocyte cultures. *Mutat. Res. Genet. Toxicol. Environ. Mutagen.* **2003**, *534*, 65–75. [CrossRef]
29. Falk, M.; Hausmann, M.; Lukášová, E.; Biswas, A.; Hildenbrand, G.; Davídková, M.; Krasavin, E.; Kleibl, Z.; Falková, I.; Ježková, L.; et al. Determining Omics spatiotemporal dimensions using exciting new nanoscopy techniques to assess complex cell responses to DNA damage: Part A–Radiomics. *Crit. Rev. Eukaryot. Gene Expr.* **2014**, *24*, 205–223. [CrossRef]
30. Stefancikova, L.; Lacombe, S.; Salado, D.; Porcel, E.; Pagáčová, E.; Tillement, O.; Lux, F.; Depeš, D.; Kozubek, M.; Falk, M. Effect of gadolinium-based nanoparticles on nuclear DNA damage and repair in glioblastoma tumor cells. *J. Nanobiotechnol.* **2016**, *14*, 63. [CrossRef]
31. Jezkova, L.; Falk, M.; Falkova, I.; Davidkova, M.; Bačíková, A.; Stefancikova, L.; Vachelová, J.; Michaelidesová, A.; Lukášová, E.; Boreyko, A.; et al. Function of chromatin structure and dynamics in DNA damage, repair and misrepair: γ-rays and protons in action. *Appl. Radiat. Isot.* **2014**, *83*, 128–136. [CrossRef]
32. Sharma, N.K. Modulation of radiation-induced and mitomycin C-induced chromosome damage by apigenin in human lymphocytes in vitro. *J. Radiat. Res.* **2013**, *54*, 789–797. [CrossRef]

International Journal of
Molecular Sciences

MDPI

Article

In Vivo Validation of the BIANCA Biophysical Model: Benchmarking against Rat Spinal Cord RBE Data

Mario P. Carante [1], Giulia Aricò [2], Alfredo Ferrari [2], Christian P. Karger [3], Wioletta Kozlowska [2,4], Andrea Mairani [5], Paola Sala [6] and Francesca Ballarini [1,7,*]

1 INFN (Italian National Institute for Nuclear Physics), Sezione di Pavia, via Bassi 6, I-27100 Pavia, Italy; mario.carante@pv.infn.it
2 CERN—European Organization for Nuclear Research, 1211 Geneva, Switzerland; giulia.arico@cern.ch (G.A.); alfredo.ferrari@cern.ch (A.F.); wioletta.kozlowska@cern.ch (W.K.)
3 Department of Medical Physics in Radiation Oncology, German Cancer Research Center (DKFZ), 69120 Heidelberg, Germany; c.karger@dkfz-heidelberg.de
4 Medical University of Vienna, 1090 Vienna, Austria
5 HIT (Heidelberg Ion-Beam Therapy Center), 69120 Heidelberg, Germany; andrea.mairani@med.uni-heidelberg.de
6 INFN (Italian National Institute for Nuclear Physics), Sezione di Milano, via Celoria 16, I-20133 Milano, Italy; paola.sala@mi.infn.it
7 Physics Department, University of Pavia, via Bassi 6, I-27100 Pavia, Italy
* Correspondence: francesca.ballarini@unipv.it

Received: 24 April 2020; Accepted: 29 May 2020; Published: 1 June 2020

Abstract: (1) Background: Cancer ion therapy is constantly growing thanks to its increased precision and, for heavy ions, its increased biological effectiveness (RBE) with respect to conventional photon therapy. The complex dependence of RBE on many factors demands biophysical modeling. Up to now, only the Local Effect Model (LEM), the Microdosimetric Kinetic Model (MKM), and the "mixed-beam" model are used in clinics. (2) Methods: In this work, the BIANCA biophysical model, after extensive benchmarking in vitro, was applied to develop a database predicting cell survival for different ions, energies, and doses. Following interface with the FLUKA Monte Carlo transport code, for the first time, BIANCA was benchmarked against in vivo data obtained by C-ion or proton irradiation of the rat spinal cord. The latter is a well-established model for CNS (central nervous system) late effects, which, in turn, are the main dose-limiting factors for head-and-neck tumors. Furthermore, these data have been considered to validate the LEM version applied in clinics. (3) Results: Although further benchmarking is desirable, the agreement between simulations and data suggests that BIANCA can predict RBE for C-ion or proton treatment of head-and-neck tumors. In particular, the agreement with proton data may be relevant if the current assumption of a constant proton RBE of 1.1 is revised. (4) Conclusions: This work provides the basis for future benchmarking against patient data, as well as the development of other databases for specific tumor types and/or normal tissues.

Keywords: head-and-neck tumors; ion beam radiotherapy; relative biological effectiveness (RBE); RBE models; carbon ions; protons; Monte Carlo simulations

1. Introduction

Cancer ion beam therapy is spreading more and more worldwide, with more than 220,000 patients treated and more than 100 facilities in operation by the end of 2019. While most patients have been irradiated with protons, about 28,000 have been treated with C-ions (www.ptcog.ch).

Due to the high dose localization in the Spread-Out Bragg Peak (SOBP) region, radiotherapy treatments with charged particles like protons or carbon ions are characterized by a highly conformal dose distribution in the tumor. Furthermore, heavy ions like carbon possess a higher biological effectiveness within the SOBP. This is thought to be due to a higher level of DNA damage clustering e.g., [1–3], which is less likely to be repaired. This is an advantage especially for patients with radioresistant tumors, like head-and-neck tumors [4].

As a consequence, the RBE variation along the beam needs to be known as precisely as possible. This is especially true if the dose is delivered by an active beam scanning technique as employed at CNAO in Pavia (Italy) and HIT in Heidelberg (Germany) because the RBE has to be predicted for each point in the treatment field.

The RBE is a complex quantity depending not only on radiation quality (that is, particle type and Linear Energy Transfer, or LET), but also on the considered effect level and thus fractional dose: in general, higher survival levels imply higher RBE values, and *vice versa*. In addition, more radioresistant cells, which in general have a lower α/β ratio, tend to show higher RBE values with respect to more radiosensitive cells, which are characterized by a higher α/β ratio.

Therefore, predicting RBE for heavy ions requires appropriate models. Currently, only three models are applied in clinics: the Local Effect Model (LEM) in Europe and Shanghai (China) e.g., [5,6], the Microdosimetric Kinetic Model (MKM) e.g., [7,8] and the "mixed-beam" model e.g., [9] in Japan. The LEM was first applied to tumors of the skull base, considering late reactions of the central nervous system (CNS) as the main endpoint; indeed, late effects in the CNS are the main dose-limiting factor for head-and-neck tumors. Since data on CNS radiation tolerance in patients are scarce, animal studies are necessary.

Although absolute dose values obtained in such studies may not be directly transferable to patient cases, relative values such as the RBE and its dependence on treatment parameters may be more applicable [10]. The rat spinal cord is known as a well-established model to investigate CNS late effects [11], and it has been applied to measure RBE values for C-ion therapy e.g., [10,12–14] and protons [15]. These are the main in vivo data that have been used to benchmark the clinically employed LEM version (LEM I), but also the more recent version LEM IV [6].

In the present work, the BIANCA (BIophysical ANalysis of Cell death and chromosome Aberrations) biophysical model, after extensive in vitro validation e.g., [16–18], was applied to establish a radiobiological database (i.e., alpha and beta cell survival parameters as a function of particle type and energy). The model was then tested against the available rat spinal cord data for carbon ion and proton irradiations [15]. Using the FLUKA Monte Carlo transport code [19–22], which is used at HIT in Heidelberg (Germany) and CNAO in Pavia (Italy) for treatment plan verification/re-calculation, the RBE is predicted for the experimental settings and compared to experimental data.

2. Results

2.1. RBE-LET Relationship

Figure 1 reports the RBE-LET relationship calculated by BIANCA for chordoma cell survival following irradiation with different monochromatic carbon beams at 1 Gy Carbon dose, which is of interest for the entrance channel in a typical therapeutic fraction, and 2 Gy Carbon dose, of interest for the SOBP. The calculations were performed as explained in Section 4.2. Although a monochromatic irradiation is different from the mixed-field scenario that is typically used in patients, these plots can be useful for interpreting the results that will be subsequently shown. Both at 1 Gy and 2 Gy, the RBE shows a steep increase up to a maximum LET, followed by a moderate decrease in the so-called "overkilling" region. As expected, the RBE at 1 Gy is higher than that at 2 Gy.

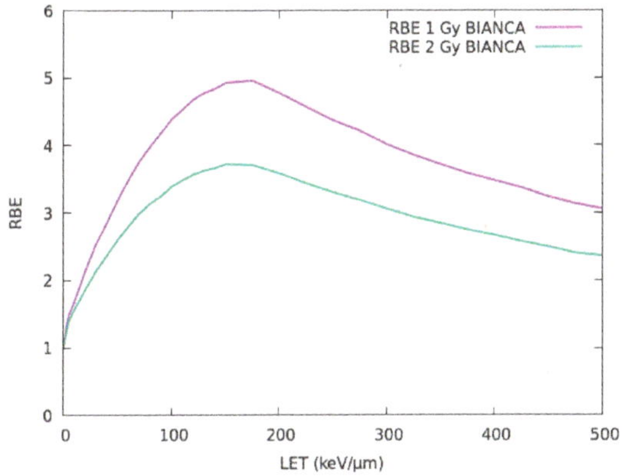

Figure 1. Relative biological effectiveness–Linear Energy Transfer (RBE-LET) relationship calculated by BIANCA for chordoma cell survival following irradiation with monochromatic carbon beams at 1 Gy (upper curve) or 2 Gy (lower curve) Carbon dose.

2.2. Carbon Ions: Dose Dependence

In Figure 2, RBE values calculated in the present work by BIANCA are compared with rat spinal cord experimental data [10,23]. In these experiments, the spinal cord of the animals was positioned either in the entrance plateau of a 270 MeV/u beam (dose-averaged LET: 13 keV/μm), or in the middle of a homogeneous 1-cm SOBP of a 140 MeV/u beam (dose-averaged LET: 125 ± 25 keV/μm). The animals were irradiated with 1, 2, 6, or 18 fractions, and RBE values with respect to 15 MV photons ranged from 1.33–1.44 (plateau) or 1.77–5.04 (peak). Further details on the experimental study can be found in [10,23], whereas the methods adopted to perform the calculations are described in Section 4.2 (RBE prediction by BIANCA for monochromatic beams) and 4.3 (interface between BIANCA and FLUKA to predict RBE for mixed fields).

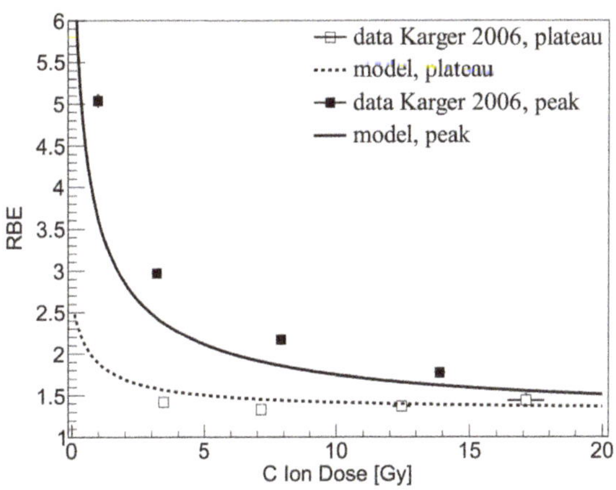

Figure 2. RBE as a function of carbon ion dose for irradiation of the rat spinal cord in the SOBP (upper line and filled data points) or in the entrance plateau (lower line and open data points). The lines are predictions by BIANCA, the points are experimental data taken from [10].

2.3. Carbon Ions: LET Dependence

BIANCA calculations, performed as described in Sections 4.2 and 4.3, were additionally compared to more recent data [12–14], where the spinal cord was located at six positions of a 6-cm SOBP (LET values: 16, 21, 36, 45, 66 and 99 keV/μm). The animals were irradiated either by a single fraction [12,14], or by two fractions [13]. RBE values relative to 15 MV photons were obtained.

In Figure 3a,b, the predictions of BIANCA are compared with these experimental data. Both for the single fraction and for the split dose (i.e., two-fraction) case, the predicted slope is smaller than for the experimental data. As a consequence, the RBE at lower LET is slightly overestimated, while at higher LET the RBE is underestimated, especially for split-doses.

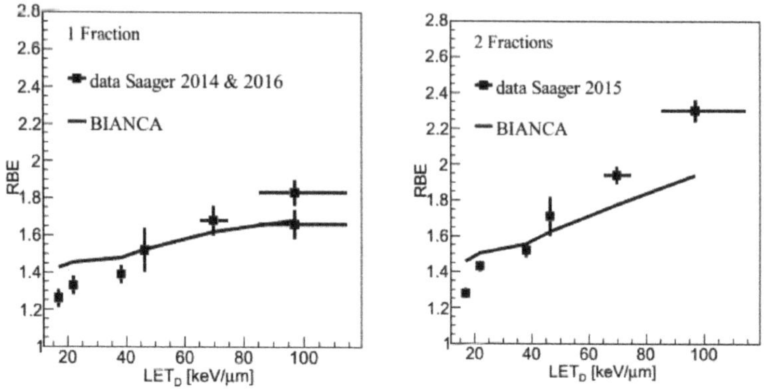

(a) (b)

Figure 3. RBE as a function of dose-averaged LET for single-fraction (panel **a**) or two-fraction (panel **b**) irradiation of the rat spinal cord at different positions within the carbon-ion Spread-Out Bragg Peak (SOBP). The lines are predictions obtained by BIANCA, the points are experimental data taken from [12,14] (panel **a**), or [13] (panel **b**). In panel a, at 99 keV/μm the higher experimental value is from a repetition experiment [14] and is considered as more reliable.

2.4. Protons

Figure 4 shows predictions performed by BIANCA (following the methods described in Sections 4.2 and 4.3) compared with RBE data for the rat spinal cord reported in [15], in the framework of the ongoing discussion whether the fixed RBE of 1.1 endorsed by ICRU [24] is still appropriate e.g., [25,26]. Analogous to the C-ion experiments, the spinal cord was positioned at four positions (35, 100, 120 and 127 mm, corresponding to LET values of 1.4, 2.7, 3.9 and 5.5 keV/μm) of a 6-cm proton SOBP ranging from 70–130 mm, and was irradiated either with single or split doses. Further details can be found in the original paper [15].

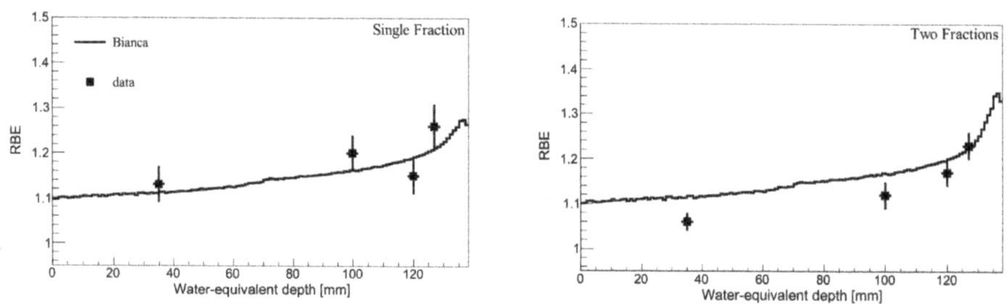

(a) (b)

Figure 4. RBE as a function of depth for single-fraction (panel **a**) or two-fraction (panel **b**) proton irradiation of the rat spinal cord. The lines are predictions by BIANCA, the points are experimental data taken from [15].

3. Discussion

Following interface with the FLUKA Monte Carlo transport code (as described in Section 4.3), for the first time the BIANCA biophysical model, described in Section 4.1, was benchmarked against in vivo data, obtained by C-ion or proton irradiation of the rat spinal cord. The C-ion data are those that have been used in the past to validate the LEM version currently used in clinics. More specifically, two C-ion experiments were considered: in the first one, the RBE dose-dependence was investigated by locating the animals in the entrance plateau or in the middle of a 1-cm SOBP, whereas in the second one, the LET-dependence was analyzed by irradiating the rat spinal cord at six positions of a 6-cm SOBP, with LET values in the range 16–99 keV/µm. In the LET experiment, the animals were irradiated either by a single fraction, or by two fractions.

Concerning the dose-dependence study, in the entrance plateau (Figure 2, lower line) the simulations show a very good agreement with the data, both qualitatively and quantitatively. This suggests that BIANCA may be applied for the evaluation of late side effects in the CNS. Concerning the SOBP (Figure 2, upper line), the simulations reproduce well the trend of the RBE dose-dependence. Quantitatively, however, there is a tendency to underestimate the data, although the degree of underestimation decreases with increasing dose. This suggests that BIANCA may underestimate the beam effectiveness in the distal SOBP region, where the LET is highest. It is worth mentioning that in the experimental study reported in [10] only the photon doses referred to measurements, whereas the carbon doses were predicted by the TPS. In a subsequent work [12], the authors found that the measured doses were lower than the TPS doses by a factor up to about 10%. This may imply that the discrepancy between simulations and data might be smaller than the one shown in Figure 2, since the data points reported in Figure 2 should be shifted towards lower doses. However, this is not automatic because the corresponding RBE values would increase as well. In addition to the experimental data, also LEM predictions are reported in [10]. According to the authors, in the peak, LEM shows a systematic underestimation of the experimental values by about 25%, whereas, in the plateau, LEM shows a (slight) RBE decrease with increasing the dose. As a consequence, LEM overestimates the point at the lowest dose (3.45 Gy) and underestimates that at the highest dose (17.1 Gy), both by about 20% [10].

In the LET-dependence study (Figure 3), the RBE at lower LET is slightly overestimated, while at higher LET the RBE is underestimated, especially for split-doses. The overestimation at the lower LET values may seem in contrast with the results of the dose-dependence investigation, where BIANCA shows a very good agreement with the data at low LET. However, it has to be taken into account that a comparison between these two scenarios is problematic as the two beam configurations result in different RBE predictions, which are then compared with data of two independent experiments. Furthermore, the two sets of experimental data use two different dose ranges (i.e., about 15–27 Gy for the LET study, 1–17 Gy for the dose study), and the LET validation is done only for relatively high doses. This suggests that the overestimation shown by BIANCA at the lower LET values might be only related to these very high doses, whereas at therapeutic doses, like those considered in the dose-dependence study, BIANCA agrees very well with the data. Concerning the higher LET values, this work shows that at high doses, like those used in the single fraction experiment (Figure 3a), BIANCA agrees well with the data, whereas at lower (fractional) doses, down to therapeutic values like those reported in the left region of Figure 2, BIANCA tends to underestimate the beam effectiveness, although the underestimation factor might be smaller than that shown by Figure 2, as already mentioned above. In the future, we will investigate whether this underestimation at high LET also holds for patient cases.

Concerning possible comparisons with LEM, while BIANCA tends to overestimate the data at low LET and to underestimate them at high LET, the LEM clinical version, also called LEM I, underestimates the data for all considered LET values; comparisons between LEM I and these data can be found in [13]. However, from a quantitative point of view, the relative discrepancy between calculations and data tends to be lower for BIANCA than for LEM I. More specifically, BIANCA shows a higher discrepancy with the data only at the lowest LET (16 keV/µm), where the discrepancy is about 13% (single fraction) or 14% (two fractions), whereas the LEM I discrepancy is 9% (1 fx) or 5% (2 fx). On the contrary, for all other five LET values the discrepancy found in this work with BIANCA is lower than that reported in [13] for LEM I. For instance, at mid-SOBP (45 keV/μm) the BIANCA discrepancy is 0% (1 fx) or 4% (2 fx), whereas that of LEM I is 19% (1 fx) or 18% (2 fx); in the distal end (99 keV/μm), these numbers become 8% (1 fx) or 15% (2 fx) for BIANCA, to be compared with 29% (1 fx) or 28% (2 fx) for LEM I. In [13], comparisons between the experimental data and the subsequent LEM IV version are also reported. Although at the moment LEM IV is not used in C-ion clinical practice, it may be worth mentioning that only at 99 keV/μm (both 1 fx and 2 fx) and 66 keV/μm (2 fx) LEM IV shows a smaller discrepancy with the data with respect to BIANCA, whereas for all other cases the reverse holds.

The C-ion study was then extended to protons, comparing BIANCA simulations with RBE data for the rat spinal cord obtained by locating the animals at four different depth positions of a 6-cm proton SOBP, and performing the irradiations either by a single or by a split dose [15]. Qualitatively, the simulations reproduced well the RBE increase with LET, and thus, with depth. The quantitative agreement with the single fraction data was also very good, since the simulation outcomes were always within the experimental error bars. While the simulation of the split-dose experiment revealed similar results as the single-fraction study, the experimental data tend to be lower than that of the single-fraction data, and thus also lower than the split-dose simulations. However, according to the authors, the split-dose experimental results should rather be regarded as a confirmation of the single-dose study. While one would expect that the split-dose RBE is not lower than the single-fraction RBE, the difference (up to 5–7%) is in the limit of the biological experimental accuracy [15]. Overall, the comparison with the proton data suggests that BIANCA may be used to predict late effects in the CNS not only for C-ions but also for protons. This may be important if it is decided that the current approach based on a constant RBE of 1.1 needs to be revised in the future.

4. Materials and Methods

4.1. The BIANCA Model: Assumptions, Parameters and Structure

The radiobiological simulations performed in this work were based on the BIANCA biophysical model, which predicts chromosome damage and cell death by different radiation types. BIANCA, which is implemented as a Monte Carlo simulation code, is based on the following main assumptions: i) ionizing radiation can induce "critical lesions" (CLs) of the DNA, where a CL is defined as a lesion that breaks the chromatin giving rise to two independent chromosome fragments; ii) distance-dependent mis-rejoining of such fragments, or fragment un-rejoining, produces chromosomal aberrations; iii) certain aberration categories (that is dicentrics, rings, and large deletions, where "large" means visible when chromatin is condensed) lead to clonogenic cell death.

Since these assumptions have been discussed in detail in previous works e.g., [27,28], only a few key issues will be addressed here. In particular, the CL yield (i.e., mean number of critical lesions per unit dose and per unit DNA mass) is an adjustable parameter of the model, because the features of the critical DNA lesions leading to chromosomal aberrations and cell death are still an open question. The value of this parameter is strongly influenced by radiation quality (that is particle type and energy), although it also depends on the characteristics of the target cell. In general, the yield of CLs yield increases with LET (excluding very high LET values, where an overkill effect may occur) as well as with the cell radiosensitivity. As in our previous work, the dependence of the rejoining probability on the (initial) fragment-end distance was modeled by a step function; the threshold distance was fixed to the average distance between the centers of two adjacent chromosome territories. Although according to recent works an exponentially-decreasing function may be better for modeling of some specific intra-chromosome aberration types [29,30], a step function has been shown to be adequate to model the main aspects of cell death [16–18,27,31,32]. As in previous works, we assumed that each chromosome fragment has a certain probability, f, to remain un-rejoined, even when possible "partners" are available within the threshold distance. In BIANCA, f is the second, and last, adjustable parameter. The assumed link between chromosomal aberrations and cell death arises from experimental studies showing a one-to-one relationship between the logarithm of the cell surviving fraction and the average number of dicentrics, rings, and deletions per cell that were scored in metaphase e.g., [33,34].

As described previously e.g., [16,27,35], the main input information for running a simulation with BIANCA consist of the radiation type (photons, light ions or heavy ions), the yield of CLs and the f value, the shape of the cell nucleus (that can be either spherical or cylindrical) and its size, as well as the treatment parameters LET and absorbed dose. While for photon irradiation the CLs induced at a given dose are uniformly distributed within the cell nucleus, for light ions like protons and α-particles they are placed along (parallel) straight lines representing the various primary particles traversing the nucleus. In case of irradiation by heavy ions like Carbon, each CL induced by a given primary ion has a 0.5 probability to be placed along the primary-ion path, and a 0.5 probability to be created at a certain radial distance with respect to the primary ion path, to take into account the effects of high-energy secondary electrons, or "delta rays". This 0.5 value derives from the consideration that 50% of the total energy deposition derives by interactions mainly occurring within the particle track core, whereas the remaining 50% derives from ionizations induced by secondary electrons, mainly occurring in the "penumbra" [36]. Afterwards, BIANCA identifies the chromosomes and the chromosome-arms that have been hit by each CL, simulates the end-joining process of the various chromosome fragments, and reproduces the scoring of dicentrics, rings, and deletions that are visible in metaphase (also called "lethal aberrations"). A cell without any lethal aberration is scored as a surviving cell, otherwise it is counted as a dead cell. For each considered dose level, the process is repeated until one obtains the desired statistical significance. The repetition for different dose levels allows for simulating a dose-response curve for lethal aberrations, and thus for cell survival. Further details on the irradiation procedure can be found in [18], whereas a detailed description of the simulation of interphase chromosome territories and arm domains is reported in [29,30].

4.2. Construction of a Radiobiological Database and RBE Prediction

As mentioned in Section 4.1, in BIANCA the CL yield depends not only on radiation quality but also on the target cell characteristics. Therefore, in principle, the CL yield for a given radiation quality should be separately adjusted for each cell line. However, we have recently developed an approach to predict the ion-survival of the cell line of interest starting from the ion-survival of a reference cell line, as well as the photon response of both lines [16]. According to this approach, the CL yield (mean number of CLs per unit length of the primary-ion path, expressed as CL/μm) to be used as an input to predict the survival of the cell line of interest for a given radiation quality, can be derived as follows:

$$\frac{CL}{\mu m} = \left(\frac{CL}{\mu m}\right)_{ref} \cdot \left[\left(\frac{CL}{Gy \cdot cell}\right) / \left(\frac{CL}{Gy \cdot cell}\right)_{ref} \right] \cdot \frac{V_{ref}}{V} \tag{1}$$

In Equation (1), $(CL/\mu m)_{ref}$ is the CL yield used for the reference cell line irradiated with the same radiation quality, whereas $(CL \cdot Gy^{-1} \cdot cell^{-1})$ and $(CL \cdot Gy^{-1} \cdot cell^{-1})_{ref}$ are the CL yields (mean number of CLs per unit dose and per cell) used to simulate photon irradiation of the cell line of interest and the reference cell line, respectively. Finally, V_{ref} and V represent the nucleus volume of the reference cell line and the cell line of interest, respectively. In a previous study, V79 cells have been chosen as a reference, and V79 cell survival curves have been simulated for a wide range of particle types and LET values. Afterwards, the aforementioned formula has been successfully applied to predict the survival of CHO cells irradiated by two opposing fields of C-ions or protons at HIT [18].

In the present work, we used the approach described above to derive a radiobiological database for chordoma patients, both for the tumor and for the adjacent normal tissue. Like in previous works, the cell nucleus was modeled by a cylinder with 6-μm height and 6-μm radius. Since chordoma is a rare tumor and basically no radiobiological data were available when LEM was introduced clinically, the LEM authors decided to use $\alpha_X = 0.10$ Gy^{-1} and $\beta_X = 0.05$ Gy^{-2}, leading to a reasonable α_X/β_X ratio of 2 Gy. Nowadays, additional data are available, both for in vitro [37,38] and for in vivo [39] settings. In particular, Henderson and co-workers irradiated chordoma patients (with a tumor volume of about 100 cm^3) with a total dose D = 35 Gy and a fractional dose d = 7 Gy getting a TCP (Tumor Control Probability) of about 60%, and they estimated an α_X/β_X ratio of 2.45 Gy:

$$TCP = \exp[-N \cdot \exp(-\alpha_X D - \beta_X Dd)] \tag{2}$$

N was calculated assuming a cell density of 10^7 cells/cm^3, as reported in the literature [40,41]. From this calculation, we obtained α_X = 0.159 and β_X = 0.065, which appears to be a reasonable estimation for photon irradiation of chordoma in vivo. Importantly, the α_X/β_X ratio of 2.45 Gy is still consistent with the value α_X/β_X=2 Gy adopted in clinical LEM applications to minimize normal tissue damage.

Afterwards, we adjusted the CL yield to reproduce this photon survival curve by our simulations, and we included this (photon) CL yield in Equation (1) to obtain ion CL yields. This allowed us to derive the CL yields to predict cell survival for many different particle types and energies. Finally, fitting of each of these survival curves by the Linear-Quadratic model $S(D) = \exp(-\alpha D - \beta D^2)$ was performed to produce a radiobiological database, i.e., a table of alpha and beta coefficients as a function of particle type and energy.

The corresponding RBE values were calculated based on the LQ-model as follows e.g., [42]:

$$RBE = 2\beta_i\left[-\alpha_X + \sqrt{\alpha_X^2 - 4\beta_X \ln S}\right]/2\beta_X\left[-\alpha_i + \sqrt{\alpha_i^2 - 4\beta_i \ln S}\right] \tag{3}$$

In Equation (3), α_X and β_X are the photon coefficients, α_i and β_i are the ion coefficients (at a given ion energy, and thus LET), and S is the considered survival level.

4.3. *Interface between BIANCA and the FLUKA Monte Carlo Transport Code*

Analogous to our previous work on in vitro irradiation of CHO cells by two opposing beams [18], the radiobiological table obtained as described in Section 4.2 was used by the FLUKA Monte Carlo transport code (version 2020.0, www.fluka.org).

More specifically, FLUKA calculated the necessary information (that is particle type and energy, and absorbed dose) based on a voxel-by-voxel approach: when a certain amount of energy was absorbed in a voxel due to a given particle type of given energy (and thus LET), FLUKA used the corresponding α_i and β_i values from the radiobiological table and calculated the respective values for the mixed beam conditions based on the TDRA (Theory of Dual Radiation Action) [43] as described in [42], i.e.,

$$\alpha = \frac{\sum_{i=1}^{n} \alpha_i D_i}{\sum_{i=1}^{n} D_i} \tag{4}$$

$$\sqrt{\beta} = \frac{\sum_{i=1}^{n} \sqrt{\beta_i} D_i}{\sum_{i=1}^{n} D_i} \tag{5}$$

Here, D_i is the absorbed dose (in the considered voxel) due to the i-th particle calculated by FLUKA, α_i and β_i are the corresponding radiobiological parameters provided by BIANCA, and α and β are their values under mixed-beam conditions. The corresponding cell survival fraction, RBE-weighted dose (D_{RBE}) and RBE were then calculated as follows:

$$-\ln S = \alpha D + \beta D^2 \tag{6}$$

$$D_{RBE} = \left[\sqrt{\alpha_X^2 - 4\beta_X \ln S}\right]/2\beta_X \tag{7}$$

$$RBE = \frac{D_{RBE}}{D} \tag{8}$$

In Equations (7) and (8), α_X and β_X are the photon parameters provided by BIANCA.

5. Conclusions

Based on the BIANCA biophysical model, a radiobiological database was constructed to predict RBE values for head-and-neck tumor irradiation with C-ions or protons. Establishing an interface with the FLUKA Monte Carlo transport code, this allowed benchmarking BIANCA against in vivo data for the first time. The RBE values predicted by BIANCA were in line with experimental RBE values obtained by irradiating the rat spinal cord at different depth positions along C-ion or proton SOBPs. Although further benchmarking is also necessary to investigate whether this approach can be extended to patient cases, this work suggests that BIANCA may be applied to predict RBE values of normal CNS tissue response in carbon ion or proton treatments. The agreement with the proton data may be especially relevant if the current assumption of a constant proton RBE of 1.1 is revised in the future. Further experimental in vivo data at lower fractional doses, comparable to those used for patient treatment, are needed.

Author Contributions: Conceptualization, F.B. and M.P.C.; methodology, F.B. and M.P.C.; software, A.F., P.S., F.B., G.A., W.K. and M.P.C.; validation, M.P.C., A.M., G.A., W.K. and C.P.K.; writing—original draft preparation, F.B.; writing—review and editing, F.B., M.P.C., G.A., W.K., A.M. and C.P.K. All authors have read and agreed to the published version of the manuscript.

Funding: This research was funded by the Italian National Institute for Nuclear Physics, project MC-INFN/FLUKA. G.A. wishes to thank the European Union's Horizon 2020 research and innovation program under the Marie Sklodowska-Curie grant agreement No 675265, Optimization of Medical Accelerators (OMA) project, for funding her research.

Conflicts of Interest: The authors declare no conflict of interest. The funders had no role in the design of the study; in the collection, analyses, or interpretation of data; in the writing of the manuscript, or in the decision to publish the results.

References

1. Georgakilas, A.G.; O'Neill, P.; Stewart, R.D. Induction and repair of clustered DNA lesions: What do we know so far? *Radiat. Res.* **2013**, *180*, 100–109. [CrossRef]
2. Karger, C.P.; Peschke, P. RBE and related modeling in carbon-ion therapy. *Phys. Med. Biol.* **2017**, *63*, 01TR02. [CrossRef]
3. Schuemann, J.; McNamara, A.L.; Warmenhoven, J.W.; Henthorn, N.T.; Kirkby, K.J.; Merchant, M.J.; Ingram, S.; Paganetti, H.; Held, K.D.; Faddegon, B.; et al. A new standard DNA damage data format (SDD). *Radiat. Res.* **2019**, *191*, 76–92. [CrossRef]
4. Suit, H.; DeLaney, T.; Goldberg, S.; Paganetti, H.; Clasie, B.; Gerweck, L.; Niemierko, A.; Hall, E.; Flanz, J.; Hallman, J.; et al. Proton vs carbon ion beams in the definitive radiation treatment of cancer patients. *Radiother. Oncol.* **2010**, *95*, 3–22. [CrossRef]
5. Scholz, M.; Kellerer, A.M.; Kraft-Weyrather, W.; Kraft, G. Computation of cell survival in heavy ion beams for therapy. The model and its approximation. *Radiat. Environ. Biophys.* **1997**, *36*, 59–66. [CrossRef]
6. Elsässer, T.; Weyrather, W.K.; Friedrich, T.; Durante, M.; Iancu, G.; Krämer, M.; Kragl, G.; Brons, S.; Winter, M.; Weber, K.J.; et al. Quantification of the relative biological effectiveness for ion beam radiotherapy: Direct experimental comparison of proton and carbon ion beams and a novel approach for treatment planning. *Int. J. Radiat. Oncol. Biol. Phys.* **2010**, *78*, 1177–1183. [CrossRef]
7. Inaniwa, T.; Furukawa, T.; Kase, Y.; Matsufuji, N.; Toshito, T.; Matsumoto, Y.; Furusawa, Y.; Noda, K. Treatment planning for a scanned carbon beam with a modified microdosimetric kinetic model. *Phys. Med. Biol.* **2010**, *55*, 6721–6737. [CrossRef]
8. Inaniwa, T.; Kanematsu, N.; Matsufuji, N.; Kanai, T.; Shirai, T.; Noda, K.; Tsuji, H.; Kamada, T.; Tsujii, H. Reformulation of a clinical-dose system for carbon-ion radiotherapy treatment planning at the National Institute of Radiological Sciences, Japan. *Phys. Med. Biol.* **2015**, *60*, 3271–3286. [CrossRef]
9. Kanai, T.; Furusawa, Y.; Fukutsu, K.; Itsukaichi, H.; Eguchi-Kasai, K.; Ohara, H. Irradiation of mixed beam and design of spread-out Bragg peak for heavy-ion radiotherapy. *Radiat. Res.* **1997**, *147*, 78–85. [CrossRef]
10. Karger, C.P.; Peschke, P.; Sanchez-Brandelik, R.; Scholz, M.; Debus, J. Radiation tolerance of the rat spinal cord after 6 and 18 fractions of photons and carbon ions: Experimental results and clinical implications. *Int. J. Radiat. Oncol. Biol. Phys.* **2006**, *66*, 1488–1497. [CrossRef]

11. van der Kogel, A.J.; Barendsen, G.W. Late effects of spinal cord irradiation with 300 kV X rays and 15 MeV neutrons. *Br. J. Radiol.* **1974**, *47*, 393–398. [CrossRef] [PubMed]
12. Saager, M.; Glowa, C.; Peschke, P.; Brons, S.; Scholz, M.; Huber, P.E.; Debus, J.; Karger, C.P. Carbon ion irradiation of the rat spinal cord: Dependence of the relative biological effectiveness on linear energy transfer. *Int. J. Radiat. Oncol. Biol. Phys.* **2014**, *90*, 63–70. [CrossRef] [PubMed]
13. Saager, M.; Glowa, C.; Peschke, P.; Brons, S.; Grün, R.; Scholz, M.; Huber, P.E.; Debus, J.; Karger, C.P. Split dose carbon ion irradiation of the rat spinal cord: Dependence of the relative biological effectiveness on dose and linear energy transfer. *Radiother. Oncol.* **2015**, *117*, 358–363. [CrossRef]
14. Saager, M.; Glowa, C.; Peschke, P.; Brons, S.; Grün, R.; Scholz, M.; Huber, P.E.; Debus, J.; Karger, C.P. The relative biological effectiveness of carbon ion irradiations of the rat spinal cord increases linearly with LET up to 99 keV/μm. *Acta Oncol.* **2016**, *55*, 1512–1515. [CrossRef]
15. Saager, M.; Peschke, P.; Brons, S.; Debus, J.; Karger, C.P. Determination of the proton RBE in the rat spinal cord: Is there an increase towards the end of the spread-out Bragg peak? *Radiother. Oncol.* **2018**, *128*, 115–120. [CrossRef]
16. Carante, M.P.; Aimè, C.; Tello Cajiao, J.J.; Ballarini, F. BIANCA, a biophysical model of cell survival and chromosome damage by protons, C-ions and He-ions at energies and doses used in hadrontherapy. *Phys. Med. Biol.* **2018**, *63*, 075007. [CrossRef]
17. Carante, M.P.; Tello Cajiao, J.J.; Ballarini, F. Predicting biological effects along hadrontherapy dose profiles by the BIANCA biophysical model. *Radiat. Prot. Dosim.* **2019**, *183*, 111–115. [CrossRef]
18. Carante, M.P.; Aricò, G.; Ferrari, A.; Kozlowska, K.; Mairani, A.; Ballarini, F. First benchmarking of the BIANCA model for cell survival predictions in a clinical hadron therapy scenario. *Phys. Med. Biol.* **2019**, *64*, 215008. [CrossRef]
19. Ferrari, A.; Sala, P.; Fassò, A.; Ranft, J. *FLUKA: A Multi-Particle Transport Code Technical Report CERN-2005-10, INFN/TC 05/11, SLAC-R-773 CERN, INFN*; SLAC: Menlo Park, CA, USA, 2005.
20. Bohlen, T.T.; Cerutti, F.; Chin, M.; Fassò, A.; Ferrari, A.; Ortega, P.; Mairani, A.; Sala, P.; Smirnov, G.; Vlachoudis, V. The FLUKA code: Developments and challenges for high energy and medical applications. *Nucl. Data Sheets* **2014**, *120*, 211–214. [CrossRef]
21. Battistoni, G.; Bauer, J.; Boehlen, T.T.; Cerutti, F.; Chin, M.P.W.; Dos Santos Augusto, R.; Ferrari, A.; Ortega, P.G.; Kozłowska, W.; Magro, G.; et al. The FLUKA Code: An Accurate Simulation Tool for Particle Therapy. *Front. Oncol.* **2016**, *6*, 116. [CrossRef]
22. Kozłowska, W.S.; Böhlen, T.T.; Cuccagna, C.; Ferrari, A.; Fracchiolla, F.; Magro, G.; Mairani, A.; Schwarz, M.; Vlachoudis, V.; Georg, D. FLUKA particle therapy tool for Monte Carlo independent calculation of scanned proton and carbon ion beam therapy. *Phys. Med. Biol.* **2019**, *64*, 075012. [CrossRef]
23. Debus, J.; Scholz, M.; Haberer, T.; Peschke, P.; Jäkel, O.; Karger, C.P.; Wannenmacher, M. Radiation Tolerance of the Rat Spinal Cord after Single and Split Doses of Photons and Carbon Ions. *Radiat. Res.* **2003**, *160*, 536–542. [CrossRef]
24. International Commission on Radiation Units and Measurements. *Prescribing, Recording, and Reporting Proton-Beam Therapy*; Report 78; International Commission on Radiation Units and Measurements: Bethesda, MD, USA, 2007.
25. Paganetti, H. Relative biological effectiveness (RBE) values for proton beam therapy. Variations as a function of biological endpoint, dose, and linear energy transfer. *Phys. Med. Biol.* **2014**, *59*, R419–R472. [CrossRef]
26. Mohan, R.; Grosshans, D. Proton therapy—Present and future. *Adv. Drug Deliv. Rev.* **2017**, *109*, 26–44. [CrossRef]
27. Carante, M.P.; Ballarini, F. Calculating Variations in Biological Effectiveness for a 62 MeV Proton Beam. *Front. Oncol.* **2016**, *6*, 76.
28. Ballarini, F.; Carante, M.P. Chromosome aberrations and cell death by ionizing radiation: Evolution of a biophysical model. *Radiat. Phys. Chem.* **2016**, *128*, 18–25. [CrossRef]
29. Tello, J.J.; Carante, M.P.; Bernal, M.; Ballarini, F. Proximity effects in chromosome aberration induction by low-LET ionizing radiation. *DNA Repair* **2017**, *58*, 38–46. [CrossRef]
30. Tello, J.J.; Carante, M.P.; Bernal, M.; Ballarini, F. Proximity effects in chromosome aberration induction: Dependence on radiation quality, cell type and dose. *DNA Repair* **2018**, *64*, 45–52. [CrossRef]

31. Ballarini, F.; Altieri, S.; Bortolussi, S.; Carante, M.P.; Giroletti, E.; Protti, N. The BIANCA model/code of radiation-induced cell death: Application to human cells exposed to different radiation types. *Radiat. Environ. Biophys.* **2014**, *53*, 525–533. [CrossRef]
32. Carante, M.P.; Altieri, S.; Bortolussi, S.; Postuma, I.; Protti, N.; Ballarini, F. Modelling radiation-induced cell death: Role of different levels of DNA damage clustering. *Radiat. Environ. Biophys.* **2015**, *54*, 305–316. [CrossRef]
33. Carrano, A.V. Chromosome aberrations and radiation-induced cell death. I. Transmission and survival parameters of aberrations. *Mutat. Res.* **1973**, *17*, 341–353. [CrossRef]
34. Cornforth, M.; Bedford, J. A quantitative comparison of potentially lethal damage repair and the rejoining of interphase chromosome breaks in low passage normal human fibroblasts. *Radiat. Res.* **1987**, *111*, 385–405. [CrossRef]
35. Carante, M.P.; Ballarini, F. Modelling cell death for cancer hadrontherapy. *AIMS Biophys.* **2017**, *4*, 465–490. [CrossRef]
36. Chatterjee, A.; Schaefer, H.J. Microdosimetric structure of heavy ion tracks in tissue. *Radiat. Environ. Biophys.* **1976**, *13*, 215–227. [CrossRef]
37. Kato, T.A.; Tsuda, A.; Uesaka, M.; Fujimori, A.; Kamada, T.; Tsujii, H.; Okayasu, R. In vitro characterization of cells derived from chordoma cell line U-CH1 following treatment with X-rays, heavy ions and chemotherapeutic drugs. *Radiat. Oncol.* **2011**, *6*, 116. [CrossRef]
38. Fujisawa, H.; Genik, P.C.; Kitamura, H.; Fujimori, A.; Uesaka, M.; Kato, T.A. Comparison of human chordoma cell-kill for 290 MeV/n carbon ions versus 70 MeV protons in vitro. *Radiat. Oncol.* **2013**, *8*, 91. [CrossRef]
39. Henderson, F.C.; McCool, K.; Seigle, J.; Jean, W.; Harter, W.; Gagnon, G.J. Treatment of chordomas with CyberKnife: Georgetown University experience and treatment recommendations. *Neurosurgery* **2009**, *64*, A44–A53. [CrossRef]
40. Stavreva, N.; Stavrev, P.; Warkentin, B.; Fallone, B.G. Derivation of the expressions for γ_{50} and D_{50} for different individual TCP and NTCP models. *Phys. Med. Biol.* **2002**, *47*, 3591–3604. [CrossRef]
41. Nahum, A.E.; Tait, D.M. Maximizing Local Control by Customized Dose Prescription for Pelvic Tumours. In *Tumor Response Monitoring and Treatment Planning*; Breit, A., Heuck, A., Lukas, P., Kneschaurek, P., Mayr, M., Eds.; Springer: Berlin/Heidelberg, Germany, 1992.
42. Mairani, A.; Brons, S.; Cerutti, F.; Fassò, A.; Ferrari, A.; Kramer, A.; Parodi, K.; Scholz, M.; Sommerer, F. The FLUKA Monte Carlo code coupled with the local effect model for biological calculations in carbon ion therapy. *Phys. Med. Biol.* **2010**, *55*, 4273–4289. [CrossRef]
43. Zaider, M.; Rossi, H.H. The Synergistic Effects of Different Radiations. *Radiat. Res.* **1980**, *83*, 732–739. [CrossRef]

International Journal of
Molecular Sciences

MDPI

Article

Radiation-Stimulated Translocation of CD166 and CRYAB to the Endothelial Surface Provides Potential Vascular Targets on Irradiated Brain Arteriovenous Malformations

Lucinda S. McRobb [1,*], Matthew J. McKay [2], Andrew J. Gauden [1], Vivienne S. Lee [1], Sinduja Subramanian [1], Santhosh George Thomas [1], Markus K. H. Wiedmann [1], Vaughan Moutrie [3], Michael Grace [3], Zhenjun Zhao [1], Mark P. Molloy [2,4] and Marcus A. Stoodley [1]

1 Department of Clinical Medicine, Macquarie University, Sydney 2109, Australia; andrew.gauden@gmail.com (A.J.G.); vivienne.s.lee@gmail.com (V.S.L.); Sinduja.subramanian@mq.edu.au (S.S.); Santhoshgeorgethomas@gmail.com (S.G.T.); markus.kh.wiedmann@gmail.com (M.K.H.W.); zhenjun.zhao@mq.edu.au (Z.Z.); marcus.stoodley@mq.edu.au (M.A.S.)
2 Australian Proteome Analysis Facility, Department of Molecular Sciences, Macquarie University, Sydney 2109, Australia; matthew.mckay@mq.edu.au (M.J.M.); m.molloy@sydney.edu.au (M.P.M.)
3 Genesis Cancer Care, Macquarie University Hospital, Sydney 2109, Australia; Vaughan.moutrie@health.nsw.gov.au (V.M.); michael.grace@genesiscare.com (M.G.)
4 Kolling Institute, Northern Clinical School, University of Sydney, Sydney 2065, Australia
* Correspondence: Lucinda.mcrobb@mq.edu.au; Tel.: +61-2-9850-2707

Received: 1 November 2019; Accepted: 19 November 2019; Published: 20 November 2019

Abstract: Vascular targeting with pro-thrombotic antibody-conjugates is a promising biological treatment for brain arteriovenous malformations (bAVMs). However, targeted drug delivery relies on the identification of unique or overexpressed markers on the surface of a target cell. In the absence of inherent biological markers, stereotactic radiosurgery may be used to prime induction of site-specific and targetable molecular changes on the endothelial surface. To investigate lumen-accessible, endothelial targets induced by radiation, we combined Gamma knife surgery in an AVM animal model with in vivo biotin-labeling and comparative proteomics. Two proteins, αB-crystallin (CRYAB)—a small heat shock protein that normally acts as an intracellular chaperone to misfolded proteins—and activated leukocyte cell adhesion molecule CD166, were further validated for endothelial surface expression after irradiation. Immunostaining of endothelial cells in vitro and rat AVM tissue ex vivo confirmed de novo induction of CRYAB following irradiation (20 Gy). Western analysis demonstrated that CRYAB accumulated intracellularly as a 20 kDa monomer, but, at the cell surface, a novel 65 kDa protein was observed, suggesting radiation stimulates translocation of an atypical CRYAB isoform. In contrast, CD166 had relatively high expression in non-irradiated cells, localized predominantly to the lateral surfaces. Radiation increased CD166 surface exposure by inducing translocation from intercellular junctions to the apical surface without significantly altering total protein levels. These findings reinforce the dynamic molecular changes induced by radiation exposure, particularly at the cell surface, and support further investigation of radiation as a priming mechanism and these molecules as putative targets for focused drug delivery in irradiated tissue.

Keywords: brain arteriovenous malformation; endothelial cells; ionizing radiation; proteomics; stereotactic radiosurgery; vascular targeting

1. Introduction

Brain arteriovenous malformations (AVMs) are vascular abnormalities characterized by direct artery to vein connections that shunt high-flow blood across the brain [1]. They are a significant cause of intracerebral hemorrhage, disability and death, in children and young adults [2]. Traditional approaches to AVM treatment include surgical removal, endovascular embolization, or radiosurgery, however, all have limitations and, currently, approximately one third of AVM patients remain untreatable. We have proposed that vascular targeting may be useful as an alternative treatment approach for brain AVMs. Vascular targeting was first described in cancer therapy to deliver thrombotic agents for tumor vessel destruction that capitalizes on the inherent molecular differences between tumor endothelium and normal, healthy endothelium that can be used to target agents selectively to the site of disease [3,4]. To date, no inherent, discriminatory vascular markers have been identified on human AVMs, hence we proposed that a combination of focused radiosurgical priming and vascular targeting may be an effective approach for AVM cure [5–11]. Radiation has previously been used to enhance target expression on tumor vasculature [12,13].

We recently utilized a technique of biotin-labeling to specifically tag and enrich endothelial surface proteins using a modified biotin that is impermeable to the cell membrane. Streptavidin enrichment of the biotinylated surface proteins prior to mass spectrometry and proteomic analysis allowed us to examine the surface proteome of irradiated endothelial cells in vitro and in vivo in order to discover putative molecular targets [5,14]. Several typically intracellular proteins that undergo an unconventional surface translocation in response to radiation were identified using this approach. The mechanisms and biological functions of these translocations remain unclear, but these findings contribute to our understanding of radiation effects on the vasculature, and, further, provide a set of novel radiation biomarkers for further investigation as putative target molecules for drug delivery. In this report, we performed further proteomic screening of radiation-stimulated surface molecules in an AVM animal model [15,16] and chose two proteins for additional examination and validation: activated leukocyte adhesion molecule (ALCAM or CD166) and αB-crystallin (CRYAB).

CD166 is a membrane glycoprotein of the immunoglobulin family that forms adhesive intercellular complexes in epithelial or endothelial layers, but also plays a role in transendothelial monocyte migration [17,18]. It has previously been investigated as a putative target in prostate cancer but has not previously been explored in any context as a radiation marker [19]. In contrast, CRYAB is a member of the small heat shock family of proteins that can act as an intracellular chaperone to misfolded proteins in response to a variety of stresses, including radiation [20,21], but can also play a protective role when excreted in exosomes [22]. Here, we further characterized the association of these targets with the luminal endothelial surface after radiation. Our findings support the observation that radiation drives both canonical and non-canonical transport pathways regulating subcellular protein localization and that focused radiosurgery may be a source of "neo-antigens" suitable for ligand-directed vascular targeting.

2. Results

2.1. LC-MS/MS and SWATH Analysis

Using a previously established method, we treated a rat model of AVM with Gamma Knife surgery and used in vivo biotin-labeling with a membrane-impermeable form of biotin to specifically label the luminal surface proteins [5]. Proteins were then extracted from the AVM tissue and the biotin-tagged proteins enriched by streptavidin pull-down. Due to the limiting protein amounts, LC-MS/MS was performed on enriched extracts from pooled samples ($n = 4$) at each time point. This enabled detection of 280 proteins, which were used to form a spectral library for SWATH-MS analysis and the quantitative comparisons of irradiated and non-irradiated AVM extracts [23]. Proteins were sorted in order of highest to lowest fold-change for day 21, and the corresponding fold-change values for day 3 and day 7 were assembled alongside (Table S1). Due to the need to pool samples, analysis focused on proteins that demonstrated a consistent fold-change of at least 1.4 across all three time points to increase the confidence in each (Table 1), rather than assessing those with the highest total expression at a single time point. The presence of the mitochondrial protein PDCE2 (pyruvate dehydrogenase complex subunit E2) in the surface extracts was consistent with our previous findings [5], further validating the utility of this method, however, this was not further examined in this study. Suitable commercial antibodies were not available for the two solute carrier proteins. Hence, CD166 and CRYAB were selected for further study to validate their regulation at the cell surface in response to radiation.

Table 1. Proteins from the proteomic datasets that increased in response to radiation at all time points.

Uniprot Accession Number|Protein Symbol	Fold-Change at Time Point (Days)			Protein Name/Alternate Name
	3	7	21	
O35112|CD166	3.5	1.4	4.0	Activated leukocyte cell adhesion molecule (ALCAM)
Q09073|ADT2	1.8	1.4	2.7	ADP/ATP translocase 2 (SLC25a5)
B0BNL4|HRG1	1.6	8.1	1.6	Heme responsive gene 1 (SLC48a1)
P23928|CRYAB	2.0	1.5	1.6	αB-crystallin/heat shock protein β5 (HSPB5)
P08461|ODP2	2.5	2.7	1.6	Dihydrolipoamide S-acetyltransferase (DLAT) or pyruvate dehydrogenase complex subunit E2 (PDCE2)

2.2. *Radiation Stimulates Up-Regulation of CRYAB, but not CD166 Expression, in the Vascular Endothelium*

Total levels of CRYAB and CD166 expression were examined in the vascular wall of the model AVM using immunostaining of ex vivo sections of rat AVM vessels taken 3 days after sham or Gamma Knife treatment (Figure 1). CD166 was localized specifically to the endothelium of both small and large vessels, co-localizing with the CD31 endothelial marker (Figure 1a). High basal levels of expression were observed for CD166 in the endothelium of both non-irradiated and irradiated vessels. There was no significant increase observed in response to Gamma Knife treatment. In contrast, CRYAB was absent at the endothelium in non-irradiated AVMs but, in response to radiation, there was a consistent and significant increase in the intensity of endothelial CRYAB staining (2-fold, $p < 0.05$) (Figure 1b,c).

2.3. *Radiation Alters Subcellular Localization of CD166 and CRYAB*

We further investigated the effect of radiation on CD166 and CRYAB expression using cultured brain microvascular endothelial cells to gain a better understanding of subcellular localization. Cells were examined after irradiation or sham treatment over a period of 6 days by fixing briefly in para-formaldehyde without permeation, to examine surface expression by immunostaining and fluorescence microscopy (Figure 2). Cells that remained adherent after radiation progressively developed a characteristic senescent morphology and phenotype, including flattening and enlargement, as previously described and characterized in this cell type [14]. Consistent with the in vivo findings, CD166 was present on non-irradiated cells, where it localized predominantly at the intercellular junctions between cells (Figure 2). In response to radiation, CD166 translocated from the intercellular junctions and accumulated in small immune-positive puncta across the surface of the enlarged, senescent cells.

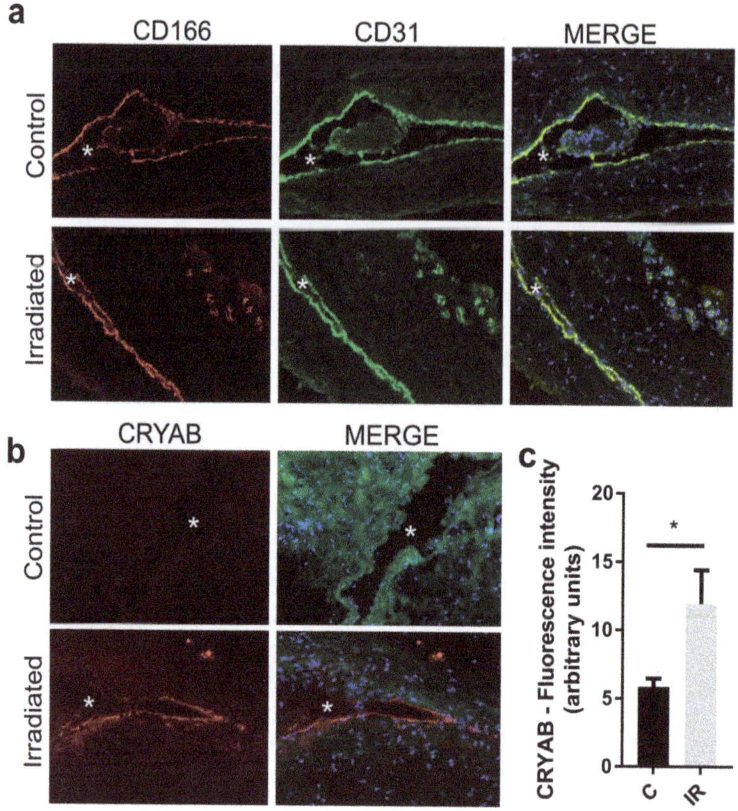

Figure 1. Immunohistochemical localization of CD166 and CRYAB in rat arteriovenous malformations (AVM). The surgically created AVM was excised 3 days after Gamma Knife or sham treatment and frozen for cryosectioning. Representative images of immunostained AVM vessels in the central nidus stained with antibodies targeting CD166 (**a**) or CRYAB (**b**). Target protein (AF647, red); CD166 sections were co-stained with endothelial marker CD31 (AF488, green); CRYAB sections show elevated background autofluorescence (green, 488 nm em) to outline the vessel wall; nuclei were stained with DAPI (blue). Original magnification, 200×. Asterisk indicates vessel lumen. (**c**) Mean intensity of fluorescence for CRYAB-stained sections, sham control (C) or irradiated (IR). Data represent mean fluorescence intensity ± SEM of CRYAB-stained sections from three independent animals per treatment group. Unpaired Student's t-test, * $p < 0.05$.

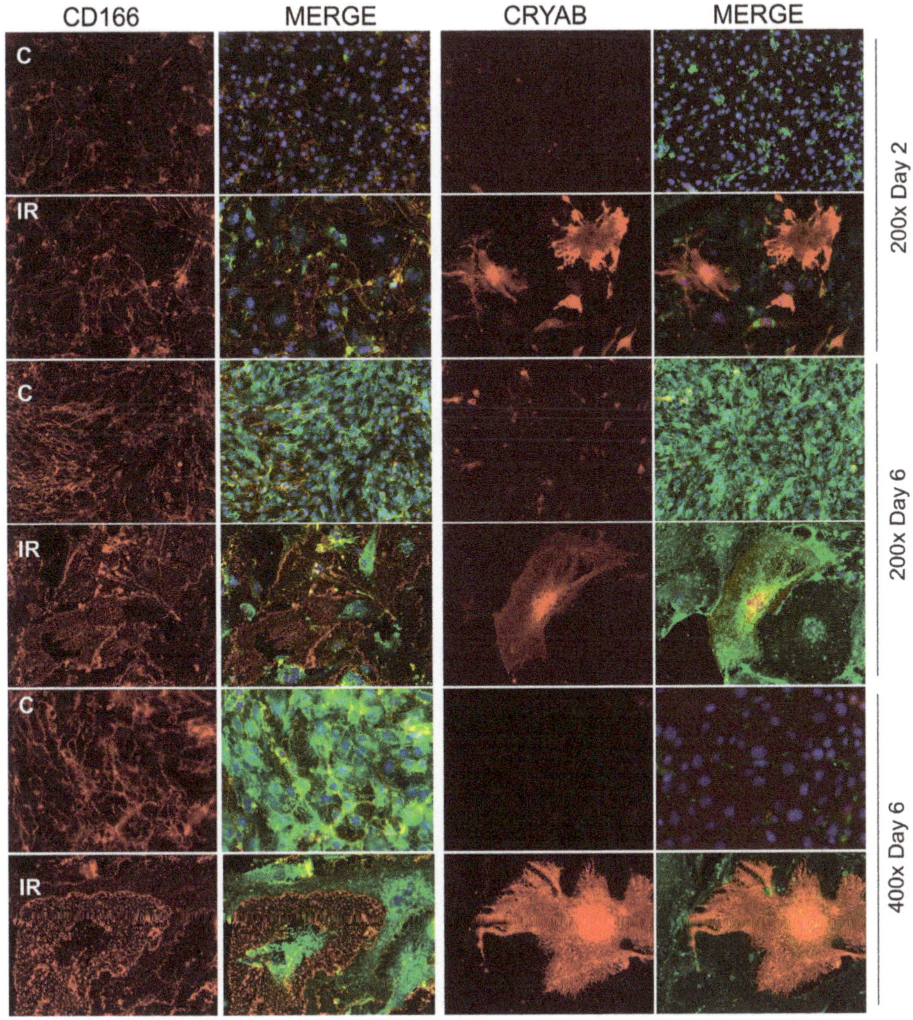

Figure 2. Immunofluorescent localization of target proteins in brain microvascular endothelial cells. Microvascular bEnd.3 cells were exposed to ionizing radiation (20 Gy) or sham treatment in 8-well chamber slides and immunostaining performed after 2 or 6 days on fixed (2% PFA, 5 min) but non-permeabilized cells. Representative immunofluorescent images of bEnd.3 cells probed with anti-CD166 or anti-CRYAB antibodies. Irradiated cells were typically senescent, showing a considerably enlarged, hypertrophic cell type with multi-lobed or fragmented nuclei. Target protein (AF647, red); wheat germ agglutinin surface marker (AF488, green); nuclei were stained with DAPI (blue). Original magnification, 200× or 400×, as indicted.

For CRYAB, expression was scarcely detected in the population of non-irradiated endothelial cells (0.5% ± 0.2% were CRYAB-positive at day 2; 0.6% ± 0.2% CRYAB-positive at day 6), but after irradiation the number of CRYAB-positive cells increased significantly (21.1% ± 9.2% by day 2; 30.8% ± 2.5% by day 6; $p < 0.001$, $n = 3$) (Figure 2). This de novo induction of CRYAB in response to radiation was consistent with the immunostaining of AVM tissue sections and demonstrated surface exposure occurring in a sub-population of the senescent-like cells that became enlarged and hypertrophic.

2.4. Radiation Increases CD166 and CRYAB Expression at the Apical Surface

To assess CD166 and CRYAB expression changes more quantitatively, Western analysis was used to compare total protein levels in whole cell lysates taken from brain endothelial cells with biotin-tagged, streptavidin-enriched extracts that represent only the cell surface fraction.

In whole cell lysates, significant basal levels of CD166 protein (105–110 kDa) were evident in non-irradiated cells, consistent with previous findings; radiation had minimal effects on total protein levels in the first 6 days (Figure 3a,b). In contrast, very low basal levels of CRYAB monomer (20 kDa) were detectable in whole cell extracts from non-irradiated cells but this increased significantly in response to radiation in a time-dependent manner (Figure 3a,b). These patterns of expression were consistent with those demonstrated by immunostaining.

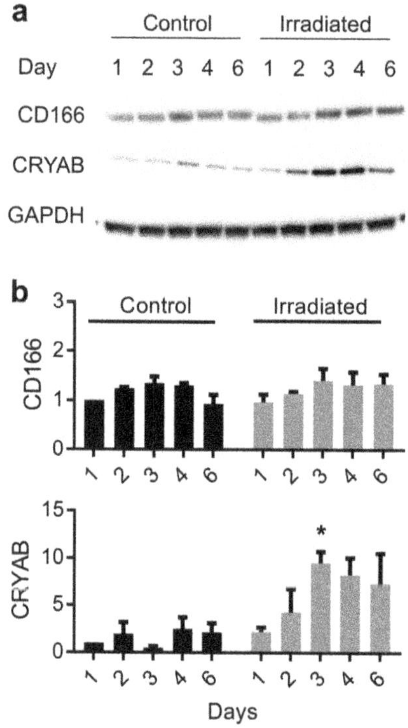

Figure 3. Time course of radiation-stimulated target protein expression in vitro. Representative Western immunoblots probed for CD166 and CRYAB expression in whole cell lysates (15 µg) extracted 1–6 days after sham or radiation treatment (20 Gy) of bEnd.3 cells by linear accelerator (**a**). Western blot images were quantitated using Image J and intensity, compared to respective day 1, non-irradiated controls for each blot. GAPDH was used to ensure equal loading (**b**). Data represent mean ± SEM of three independent experiments. Two-way ANOVA with Sidak's multiple comparison test, * $p < 0.05$.

Relative quantitation of surface expression was investigated by biotin-tagging proteins at the day 6 time point post-irradiation or sham, with subsequent streptavidin-enrichment of biotinylated proteins prior to Western blotting. Proteins in the biotin-tagged extracts (biotin-bound or "BB" extract) from non-irradiated and irradiated cell cultures were compared to the total protein input before streptavidin enrichment (IN), and the protein output remaining after removal of streptavidin-enriched, biotin-tagged surface proteins (OUT) (Figure 4a–e). In the biotin-bound extracts, a very different pattern of expression was observed between non-irradiated and irradiated extracts for each target relative to that observed in whole cell extracts. Consistent with the pattern of CD166 immunostaining, a significant level of CD166 expression was apparent at the surface, even in the non-irradiated, biotin-tagged extracts, however, expression increased 2-fold ($p < 0.05$) in irradiated cells in the biotin-enriched, surface fractions (Figure 4a,e). For CRYAB, whole cell extracts demonstrated the presence of strong bands consistent with the predicted size of the monomeric CRYAB protein (20 kDa) and there was significant induction in response to radiation (Figure 4b,e). A second weak band was also detected, with an approximate molecular weight of 60 kDa. However, the intensity of this band did not change with radiation treatment. In the streptavidin-enriched extracts, the 20 kDa and 60 kDa bands were completely absent, suggesting these proteins were intracellular and not present at the cell surface. Instead, a third protein band of approximately 65 kDa was repeatedly identified in the irradiated extracts ($n = 3$). This third band was not present in the INPUT samples representing total protein, suggesting the surface-localized form represents only a small fraction of the total protein pool. This band was not evident in non-irradiated extracts but increased approximately 2-fold ($p < 0.05$) in response to radiation (Figure 4e). These findings suggest that radiation induces a small pool of atypical CRYAB that is able to translocate to the cell surface.

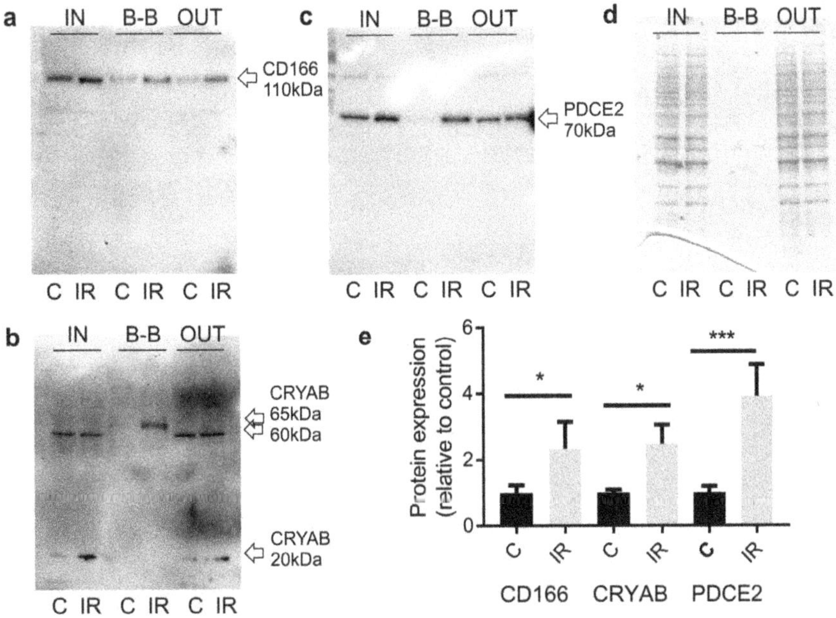

Figure 4. Radiation alters target expression in biotin-bound protein fractions. Representative Western immunoblot images of fractionated extracts from biotin-labeled bEnd.3 cells before and after streptavidin enrichment. Membranes were probed for CD166 (**a**), CRYAB (**b**), or PDCE2 (**c**). Representative coomassie-stained SDS-PAGE gel showing protein loadings (**d**). IN, total cell extract prior to streptavidin enrichment, 15 µg loaded; B-B, biotin-bound fraction, 30 µg loaded; OUT, non-biotinylated fraction eluted after streptavidin binding, 15 µg loaded; C, control; IR irradiated. Images are representative of 3–5 independent experiments. (**e**) Proteins in biotin-bound extracts were quantitated using Image J. Band intensities in irradiated samples were compared to band intensities in non-irradiated control samples for each blot. Paired Student's t-test, * $p < 0.05$, *** $p < 0.001$.

As a positive control, the same protein extracts were analyzed for PDCE2 expression. PDCE2 is a mitochondrial protein we previously validated as undergoing an atypical surface translocation in response to radiation [5]. No band was evident in the non-irradiated, biotin-labelled extracts, but a strong band was induced approximately 4-fold ($p < 0.001$) in the irradiated, biotin-tagged protein extract (Figure 4c,e), in line with our earlier findings.

3. Discussion

The ability of stereotactic radiosurgery to deliver a highly focused dose of radiation to a designated region provides a potential priming mechanism for the induction of novel biomarkers. Radiation is known to induce a myriad of cellular and molecular changes, depending on dose and cell type. Studying changes that occur specifically at the cell surface increases our understanding of how radiation-damaged cells may interact with their local environment after exposure, but could also provide a pool of targets for ligand or antibody-directed drug delivery in pathologies amenable to stereotactic radiosurgery, such as brain AVMs.

For target discovery, we employed a combined approach of surface protein biotinylation with proteomic analysis, as we require target localization specifically at the luminal endothelial surface. Using this approach, we previously identified and validated several novel proteins that demonstrate non-canonical surface translocations in response to radiation [5,14]. Here, we extended our proteomic discovery approach and identified and validated two additional proteins for their endothelial expression and surface translocation in response to radiation.

CD166 demonstrated a mean 3-fold increase in response to radiation in the proteomic datasets, but was also of interest as it has previously been investigated as a vascular target in prostate cancer and an internalizing, single-chain variable fragment (scFV) antibody to CD166 developed [19]. Using Western blotting, we observed a steady level of expression of the CD166 protein in non-irradiated cells both in vitro and in vivo that did not change significantly in response to radiation, although there was a small increase over time in both irradiated and non-irradiated cells, most likely due to ongoing adjustments in cell confluence and the nature of cell–cell contacts. However, there was a significant increase in the fraction at the apical surface that could be identified after biotin-labeling and enrichment. Most notably, a significant redistribution of CD166 across the apical surface of the enlarged senescent cells was observed using immunofluorescence, consistent with this finding. This effect of radiation on CD166 was quite different to those observed for other inflammatory adhesion molecules, such as intercellular adhesion molecule 1 (ICAM-1) and vascular cell adhesion molecule 1 (VCAM-1), where surface exposure was regulated primarily at the level of transcription [7]. Like ICAM-1 and VCAM-1, CD166 can play a role at the apical surface in transendothelial monocyte migration but is primarily localized on epithelial and endothelial cells at intercellular junctions, forming an adhesive complex between cells that helps to maintain tissue architecture [18]. Thus, radiation-induced changes to CD166 localization may contribute to the pro-inflammatory milieu by increasing monocyte binding, as well as extravasation through disrupted junctions. As a molecular target, further experiments are required to assess whether the level of induction in response to radiation can overcome the relatively high basal levels of exposure of CD166 on the vasculature to give the necessary specificity required for drug delivery, but the results suggest that radiosurgery could potentially be investigated in other contexts, such as prostate cancer, for further enhancement of CD166 expression and improved targeting specificity [12].

CRYAB was chosen for further investigation in this study, due to its presence in all three proteomic datasets analyzed here with the animal model, but also because independent biotinylation and proteomic studies in cultured cells showed a 7-fold ($p < 0.001$) increase in surface expression after radiation, although it was not independently validated at the time [14]. CRYAB is a member of the small heat shock family of proteins that acts as a molecular chaperone to misfolded proteins, maintaining them in large soluble aggregates. It is well-known to be up-regulated in neurodegenerative proteinopathies such as Alzheimer's disease and Parkinson's disease [24] and has been identified as a candidate autoantigen in multiple sclerosis [22,25]. CRYAB is up-regulated in response to various stresses, including radiation, and plays a role in cellular survival and radioresistance [20,21,26]. In the vasculature, CRYAB is expressed in a subset of human tumor vessels, but not in normal capillaries, and is thought to assist tumor angiogenesis as it promotes survival [27,28]. In accordance with the literature, we observed negligible CRYAB expression in the normal vasculature but demonstrated a significant induction at the endothelium in response to radiation. Western blotting demonstrated that the majority of the protein appeared to remain intracellular, however, a novel 65 kDa immunopositive band was detectable in the biotin-labeled fraction, that was consistently identified in three independent experiments. The absence of this band in the whole cell extracts suggests this form of CRYAB may represent a very small proportion of the total CRYAB pool, but its presence only in response to radiation suggests a potentially unique radiation target.

It is unclear at this time what form of CRYAB this band may represent or whether it has a unique function. CRYAB is typically cytoplasmic but can translocate to the mitochondria during heat shock, where it acts as an inhibitor of apoptosis [27], and there is evidence for its secretion in exosomes in certain cell types [26,29,30]. Cubedo et al., identified the presence of both monomeric (20 kDa) and oligomeric (36 and 50 kDa) CRYAB forms in ischemic cardiomyoctes but found only the larger oligomeric forms could be isolated in the insoluble membrane fraction associated with exosomal translocation and secretion [26]. Fujii et al., demonstrated that very high doses (> 1000 Gy) of gamma radiation delivered directly to α-crystallin could induce cross-linking and molecular weight modifications [31]. While this dose is far higher than the one used in the current study, the relatively small fraction of modified CRYAB could fit with the hypothesis that radiation produces a cross-linked multimeric form, but further investigation is required to fully analyze the nature of this isoform and its uniqueness in the vasculature for targeting.

As per our previous study, the proteomic aspect of this study was limited by the necessity to combine multiple animal AVM extracts into one for analysis at each time point, given the unanticipated low level of proteins that were obtained after enrichment [5]. This was not ideal, and future experiments of this nature would be advised to increase animal numbers in anticipation of this. However, we consider it a viable discovery approach, assuming downstream validation is performed as for any "omics" approach. The advantage of the present study was the availability of data from multiple time points, *in lieu* of animal replicates, which gave us greater confidence in the preliminary "hits" as potential targets. Inclusion of a non-AVM vessel extraction would also have been of value to assess basal levels of protein relative to the sham AVM and irradiated AVM samples, as the surgical creation of the AVM may also have altered basal levels of protein expression. Despite this, considering that this is our third proteomic study to independently identify the mitochondrial protein PDCE2 as a radiation-stimulated surface protein [5,14], this further supports the approach as a consistent and valid method for investigation of the surface proteome. Validation of other identified proteins in the proteomic datasets may be of interest to clarify any functional role in radiation injury or potential utility for vascular targeting. Further assessment of these molecules with respect to their internalization or maintenance at the cell surface would also be of value for assessing potential downstream applications. For vascular targeting with thrombin conjugates, prolonged surface exposure is desirable to allow contact with blood components for coagulation; in contrast, internalization and lysosomal activation is useful for cancer targeting.

Future studies will use in vivo imaging approaches to further investigate CD166, CRYAB and PDCE2 biodistribution at the AVM vascular wall, however the current findings support the observations that radiation drives multiple non-canonical transport pathways regulating subcellular protein localization and luminal exposure of non-classical surface proteins and that further investigation of the transport mechanisms and functional outcomes of translocation is warranted. In addition, the findings support our hypothesis that focused radiosurgery may be a source of "neo-antigens", suitable for ligand-directed vascular targeting in AVMs.

4. Materials and Methods

4.1. Animal Model and Gamma Knife Surgery (GKS)

All experiments involving animals were approved by the Macquarie University Animal Care and Ethics Committee (Sydney, Australia) under approval numbers 2014/013 (1 April 2014) and 2014/053 (1 January 2015). Animal experiments were performed in accordance with committee guidelines and the Australian Code of Practice for the Care and Use of Animals for Scientific Purposes. The rat AVM model is formed in Sprague–Dawley rats by surgically anastomosing the rostral end of the left jugular vein to the center of the left common carotid artery, as has been described in detail previously [5,15,16]. The arteriovenous fistula and the arterialized veins (of the model AVM) are irradiated six weeks after creation in anesthetized animals using a single-fraction stereotactic radiosurgical dose, which delivers a minimum of 20 Gy to the margin of the defined AVM, to a peak value of 40 Gy at the center. This was administered using a Leksell Gamma Knife Perfexion (Elekta Instruments, Stockholm, Sweden) at Macquarie University Hospital (Sydney, Australia), as described previously [5,15]. Sham-irradiated animals were treated identically but did not receive any radiation.

4.2. *Tissue Biotinylation and Proteomic Analysis*

A cohort of 24 animals with AVM creation (n = 12 sham, n = 12 GKS) were used for in vivo biotin-labeling of luminal surface proteins. In vivo biotinylation with membrane-impermeable EZ-link Sulfo-NHS-LC Biotin (Thermoscientific, Waltham, MA, USA), and subsequent decellularization and protein extraction, was performed as described previously [5]; however, for this study the concentration of sodium dodecyl sulphate in the extraction buffer was reduced 10-fold from 1% to 0.1% to reduce carryover and interference in downstream mass spectrometry. Biotin labeling was performed either at day 3 (n = 8), day 7 (n = 8) or day 21 (n = 8) post-irradiation or sham treatment (n = 4 per treatment group, per time point). However, as initial extracts obtained from single AVM tissues contained only limited amounts of protein, it was necessary to pool the four available tissue samples per treatment group and time point to increase the detection limit. Enrichment of the biotin-bound proteins in the final six pooled samples with Streptavidin–Sepharose HP (GE Healthcare, Pittsburgh, PA, USA) and downstream proteomic sample analysis, using data-dependent and SWATH LC/MS/MS procedures [32], was as described [5,14].

4.3. *Cell Culture and Western Analysis*

Cultured mouse brain microvascular endothelial cells (bEnd.3; ATCC CRL-2299) were irradiated by linear accelerator (Elekta Synergy, Crawley, UK) (20 Gy) at Macquarie University Hospital (Sydney, Australia), as previously described [5,14]. At designated time points, cells were immunostained to examine protein localization, or whole cell lysates were extracted for western analysis of total protein content [5,14]. Alternatively, bEnd.3 cells were biotinylated in vitro and whole cell extracts subjected to streptavidin-enrichment of surface-accessible proteins before SDS-PAGE and Western analysis as per published methods [5,14].

4.4. *Immunohistochemistry and Immunocytochemistry*

A second cohort of rats received AVM creation surgery and irradiation (n = 3) or treatment (n = 3) as described above but without biotinylation. At 3 days post-irradiation or sham treatment, these animals were perfused with 500 mL phosphate-buffered saline (PBS, pH 7.4) and the excised AVM tissue snap frozen in OCT freezing compound (ProSciTech, Thuringowa Central, Australia) and used for immunohistochemistry. Cryosections (10 μm) of non-biotinylated AVM tissues were fixed with 4% paraformaldehyde for 20 min and processed as described previously [5]. Immunocytochemistry was performed on irradiated bEnd.3 cells fixed with 4% paraformaldehyde for 5 min without cell permeabilization to enable assessment of surface localization [5,14].

Antibodies against CRYAB (mouse monoclonal, ab13496, Abcam, Cambridge, MA, USA), and CD166 (goat polyclonal, PA5-47083, Thermo Fisher Scientific) were used with species-specific AlexaFluor647-conjugated secondary antibodies (Life Technologies, Grand Island, NY, USA). Ex vivo sections stained with CD166 were co-stained with anti-CD31 mouse monoclonal antibody (#550025 PECAM-1; BD Biosciences, San Jose, CA, USA) and species-specific AlexaFluor488-conjugated secondary antibody (Life Technologies) to visualize the endothelium. Control sections incubated with rabbit IgG (Santa Cruz Biotechnologies, Dallas, TX, USA) or mouse IgG (BD Biosciences) showed no reactivity. Antibodies against PDCE2 (rabbit polyclonal, sc-32925, Santa Cruz Biotechnologies) were used as a positive control in Western analyses. For immunofluorescence, cells were co-stained with an AF488-labelled lectin, wheat germ agglutinin (Life Technologies), to highlight cell surface area. Nuclei were counterstained with 4′,6-diamidino-2-phenylindole dihydrochloride (DAPI, 5 μg/mL, Life Technologies). Digital images were captured under fixed parameters using a Zeiss microscope with AxioCam HRc camera and Zen 2012 software (Carl Zeiss Microscopy, Jena, Germany).

4.5. Statistical Analysis

All statistical analyses were performed with Prism version 7.02 (GraphPad Software, La Jolla, CA, USA). For Western blotting, a two-sided, paired Student's t-test was used to compare relative protein expression between biotin-enriched extracts; multiple comparisons were performed using two-way ANOVA with Sidak's post-hoc analysis. Two-sided, unpaired Student's *t*-tests were used to compare immunostaining data. Data are shown as mean ± SEM. Statistical significance was set at $p < 0.05$.

Supplementary Materials: Supplementary materials can be found at http://www.mdpi.com/1422-0067/20/23/5830/s1.

Author Contributions: Conceptualization, L.S.M., M.G., Z.Z., M.P.M. and M.A.S.; Data curation, L.S.M.; Formal analysis, L.S.M.; Funding acquisition, Z.Z., M.P.M. and M.A.S.; Investigation, L.S.M., M.J.M., A.J.G., V.S.L., S.S. and S.G.T.; Methodology, L.S.M., M.J.M., M.K.H.W., V.M. and M.G.; Supervision, M.P.M. and M.A.S.; Writing—original draft, L.S.M.; Writing—review and editing, L.S.M., M.J.M., A.J.G., V.S.L., S.S., S.G.T., M.K.H.W., V.M., M.G., Z.Z., M.P.M. and M.A.S.

Funding: This research was funded by the National Health and Medical Research Council of Australia, grant number APP1047302. Aspects of this work were facilitated by access to the Australian Proteome Analysis Facility supported by the Australian Government's National Collaborative Research Infrastructure Scheme.

Conflicts of Interest: The authors declare no conflict of interest. The funders had no role in the design of the study; in the collection, analyses, or interpretation of data; in the writing of the manuscript, or in the decision to publish the results.

References

1. Lawton, M.T.; Rutledge, W.C.; Kim, H.; Stapf, C.; Whitehead, K.J.; Li, D.Y.; Krings, T.; Terbrugge, K.; Kondziolka, D.; Morgan, M.K.; et al. Brain arteriovenous malformations. *Nat. Rev. Dis. Primers* **2015**, *1*, 15008. [CrossRef]
2. Gross, B.A.; Du, R. Natural history of cerebral arteriovenous malformations: A meta-analysis. *J. Neurosurg.* **2013**, *118*, 437–443. [CrossRef]
3. Huang, X.; Molema, G.; King, S.; Watkins, L.; Edgington, T.S.; Thorpe, P.E. Tumor infarction in mice by antibody-directed targeting of tissue factor to tumor vasculature. *Science* **1997**, *275*, 547–550. [CrossRef]
4. Thorpe, P.E. Vascular targeting agents as cancer therapeutics. *Clin. Cancer Res.* **2004**, *10*, 415–427. [CrossRef] [PubMed]
5. McRobb, L.S.; Lee, V.S.; Simonian, M.; Zhao, Z.; Thomas, S.G.; Wiedmann, M.; Raj, J.V.; Grace, M.; Moutrie, V.; McKay, M.J.; et al. Radiosurgery Alters the Endothelial Surface Proteome: Externalized Intracellular Molecules as Potential Vascular Targets in Irradiated Brain Arteriovenous Malformations. *Radiat. Res.* **2017**, *187*, 66–78. [CrossRef] [PubMed]
6. Raoufi Rad, N.; McRobb, L.S.; Zhao, Z.; Lee, V.S.; Patel, N.J.; Qureshi, A.S.; Grace, M.; McHattan, J.J.; Amal Raj, J.V.; Duong, H.; et al. Phosphatidylserine Translocation after Radiosurgery in an Animal Model of Arteriovenous Malformation. *Radiat. Res.* **2017**, *187*, 701–707. [CrossRef] [PubMed]
7. Raoufi Rad, N.; McRobb, L.S.; Lee, V.S.; Bervini, D.; Grace, M.; Ukath, J.; McHattan, J.; Sreenivasan, V.K.A.; Duong, T.T.H.; Zhao, Z.; et al. In vivo imaging of endothelial cell adhesion molecule expression after radiosurgery in an animal model of arteriovenous malformation. *PLoS ONE* **2017**, *12*, e0185393. [CrossRef]
8. Reddy, R.; Duong, T.T.; Fairhall, J.M.; Smee, R.I.; Stoodley, M.A. Durable thrombosis in a rat model of arteriovenous malformation treated with radiosurgery and vascular targeting. *J. Neurosurg.* **2014**, *120*, 113–119. [CrossRef]

9. Storer, K.; Tu, J.; Karunanayaka, A.; Smee, R.; Short, R.; Thorpe, P.; Stoodley, M. Coadministration of low-dose lipopolysaccharide and soluble tissue factor induces thrombosis after radiosurgery in an animal arteriovenous malformation model. *Neurosurgery* **2007**, *61*, 604–611. [CrossRef]
10. Storer, K.P.; Tu, J.; Stoodley, M.A.; Smee, R.I. Expression of endothelial adhesion molecules after radiosurgery in an animal model of arteriovenous malformation. *Neurosurgery* **2010**, *67*, 976–983. [CrossRef]
11. Zhao, Z.; Johnson, M.S.; Chen, B.; Grace, M.; Ukath, J.; Lee, V.S.; McRobb, L.S.; Sedger, L.M.; Stoodley, M.A. Live-cell imaging to detect phosphatidylserine externalization in brain endothelial cells exposed to ionizing radiation: Implications for the treatment of brain arteriovenous malformations. *J. Neurosurg.* **2016**, *124*, 1780–1787. [CrossRef] [PubMed]
12. He, J.; Luster, T.A.; Thorpe, P.E. Radiation-enhanced vascular targeting of human lung cancers in mice with a monoclonal antibody that binds anionic phospholipids. *Clin. Cancer Res.* **2007**, *13*, 5211–5218. [CrossRef] [PubMed]
13. He, J.; Yin, Y.; Luster, T.A.; Watkins, L.; Thorpe, P.E. Antiphosphatidylserine antibody combined with irradiation damages tumor blood vessels and induces tumor immunity in a rat model of glioblastoma. *Clin. Cancer Res.* **2009**, *15*, 6871–6880. [CrossRef] [PubMed]
14. McRobb, L.S.; McKay, M.J.; Gamble, J.R.; Grace, M.; Moutrie, V.; Santos, E.D.; Lee, V.S.; Zhao, Z.; Molloy, M.P.; Stoodley, M.A. Ionizing radiation reduces ADAM10 expression in brain microvascular endothelial cells undergoing stress-induced senescence. *Aging* **2017**, *9*, 1248–1268. [CrossRef]
15. Kashba, S.R.; Patel, N.J.; Grace, M.; Lee, V.S.; Raoufi-Rad, N.; Amal Raj, J.V.; Duong, T.T.; Stoodley, M. Angiographic, hemodynamic, and histological changes in an animal model of brain arteriovenous malformations treated with Gamma Knife radiosurgery. *J. Neurosurg.* **2015**, *123*, 954–960. [CrossRef]
16. Yassari, R.; Sayama, T.; Jahromi, B.S.; Aihara, Y.; Stoodley, M.; Macdonald, R.L. Angiographic, hemodynamic and histological characterization of an arteriovenous fistula in rats. *Acta Neurochir.* **2004**, *146*, 495–504. [CrossRef]
17. Bowen, M.A.; Patel, D.D.; Li, X.; Modrell, B.; Malacko, A.R.; Wang, W.C.; Marquardt, H.; Neubauer, M.; Pesando, J.M.; Francke, U.; et al. Cloning, mapping, and characterization of activated leukocyte-cell adhesion molecule (ALCAM), a CD6 ligand. *J. Exp. Med.* **1995**, *181*, 2213–2220. [CrossRef]
18. Masedunskas, A.; King, J.A.; Tan, F.; Cochran, R.; Stevens, T.; Sviridov, D.; Ofori-Acquah, S.F. Activated leukocyte cell adhesion molecule is a component of the endothelial junction involved in transendothelial monocyte migration. *FEBS Lett.* **2006**, *580*, 2637–2645. [CrossRef]
19. Roth, A.; Drummond, D.C.; Conrad, F.; Hayes, M.E.; Kirpotin, D.B.; Benz, C.C.; Marks, J.D.; Liu, B. Anti-CD166 single chain antibody-mediated intracellular delivery of liposomal drugs to prostate cancer cells. *Mol. Cancer Ther.* **2007**, *6*, 2737–2746. [CrossRef]
20. Golenhofen, N.; Ness, W.; Wawrousek, E.F.; Drenckhahn, D. Expression and induction of the stress protein alpha-B-crystallin in vascular endothelial cells. *Histochem. Cell Biol.* **2002**, *117*, 203–209. [CrossRef]
21. Xu, Q.Y.; Gao, Y.; Liu, Y.; Yang, W.Z.; Xu, X.Y. Identification of differential gene expression profiles of radioresistant lung cancer cell line established by fractionated ionizing radiation in vitro. *Chin. Med. J.* **2008**, *121*, 1830–1837. [CrossRef] [PubMed]
22. Ousman, S.S.; Tomooka, B.H.; van Noort, J.M.; Wawrousek, E.F.; O'Connor, K.C.; Hafler, D.A.; Sobel, R.A.; Robinson, W.H.; Steinman, L. Protective and therapeutic role for alphaB-crystallin in autoimmune demyelination. *Nature* **2007**, *448*, 474–479. [CrossRef] [PubMed]
23. Gillet, L.C.; Navarro, P.; Tate, S.; Rost, H.; Selevsek, N.; Reiter, L.; Bonner, R.; Aebersold, R. Targeted data extraction of the MS/MS spectra generated by data-independent acquisition: A new concept for consistent and accurate proteome analysis. *Mol. Cell. Proteomics* **2012**, *11*. [CrossRef]
24. Liu, Y.; Zhou, Q.; Tang, M.; Fu, N.; Shao, W.; Zhang, S.; Yin, Y.; Zeng, R.; Wang, X.; Hu, G.; et al. Upregulation of alphaB-crystallin expression in the substantia nigra of patients with Parkinson's disease. *Neurobiol. Aging* **2015**, *36*, 1686–1691. [CrossRef]
25. Van Noort, J.M.; van Sechel, A.C.; Bajramovic, J.J.; Ouagmiri, M.; Polman, C.H.; Lassmann, H.; Ravid, R. The small heat-shock protein alpha B-crystallin as candidate autoantigen in multiple sclerosis. *Nature* **1995**, *375*, 798–801. [CrossRef]
26. Cubedo, J.; Vilahur, G.; Casani, L.; Mendieta, G.; Gomez-Jabalera, E.; Juan-Babot, O.; Padro, T.; Badimon, L. Targeting the molecular mechanisms of ischemic damage: Protective effects of alpha-crystallin-B. *Int. J. Cardiol.* **2016**, *215*, 406–416. [CrossRef]

27. Dimberg, A.; Rylova, S.; Dieterich, L.C.; Olsson, A.K.; Schiller, P.; Wikner, C.; Bohman, S.; Botling, J.; Lukinius, A.; Wawrousek, E.F.; et al. alphaB-crystallin promotes tumor angiogenesis by increasing vascular survival during tube morphogenesis. *Blood* **2008**, *111*, 2015–2023. [CrossRef]
28. Ruan, Q.; Han, S.; Jiang, W.G.; Boulton, M.E.; Chen, Z.J.; Law, B.K.; Cai, J. alphaB-crystallin, an effector of unfolded protein response, confers anti-VEGF resistance to breast cancer via maintenance of intracrine VEGF in endothelial cells. *Mol. Cancer Res.* **2011**, *9*, 1632–1643. [CrossRef]
29. Kore, R.A.; Abraham, E.C. Phosphorylation negatively regulates exosome mediated secretion of cryAB in glioma cells. *Biochim. Biophys. Acta* **2016**, *1863*, 368–377. [CrossRef]
30. Kore, R.A.; Abraham, E.C. Inflammatory cytokines, interleukin-1 beta and tumor necrosis factor-alpha, upregulated in glioblastoma multiforme, raise the levels of CRYAB in exosomes secreted by U373 glioma cells. *Biochem. Biophys. Res. Commun.* **2014**, *453*, 326–331. [CrossRef]
31. Fujii, N.; Hiroki, K.; Matsumoto, S.; Masuda, K.; Inoue, M.; Tanaka, Y.; Awakura, M.; Akaboshi, M. Correlation between the loss of the chaperone-like activity and the oxidation, isomerization and racemization of gamma-irradiated alpha-crystallin. *Photochem. Photobiol.* **2001**, *74*, 477–482. [CrossRef]
32. Krisp, C.; Yang, H.; van Soest, R.; Molloy, M.P. Online Peptide fractionation using a multiphasic microfluidic liquid chromatography chip improves reproducibility and detection limits for quantitation in discovery and targeted proteomics. *Mol. Cell. Proteomics* **2015**, *14*, 1708–1719. [CrossRef] [PubMed]

International Journal of
Molecular Sciences

Article

Proton Therapy and Src Family Kinase Inhibitor Combined Treatments on U87 Human Glioblastoma Multiforme Cell Line

Francesco P Cammarata [1,2,†], Filippo Torrisi [2,3,†], Giusi I Forte [1,2], Luigi Minafra [1,2,*], Valentina Bravatà [1,2], Pietro Pisciotta [2,4], Gaetano Savoca [1], Marco Calvaruso [1,2], Giada Petringa [2,3], Giuseppe A. P. Cirrone [2], Anna L Fallacara [5,6], Laura Maccari [5], Maurizio Botta [5,6], Silvia Schenone [7], Rosalba Parenti [3], Giacomo Cuttone [2] and Giorgio Russo [1,2]

1 Institute of Molecular Bioimaging and Physiology, National Research Council, IBFM-CNR, 90015 Cefalù, Italy; francesco.cammarata@ibfm.cnr.it (F.P.C.); giusi.forte@ibfm.cnr.it (G.I.F.); valentina.bravata@ibfm.cnr.it (V.B.); savoca.gaetano@gmail.com (G.S.); marco.calvaruso@ibfm.cnr.it (M.C.); giorgio.russo@ibfm.cnr.it (G.R.)
2 National Institute for Nuclear Physics, Laboratori Nazionali del Sud, INFN-LNS, 95123 Catania, Italy; filippo.torrisi@unict.it (F.T.); pietro.pisciotta@lns.infn.it (P.P.); giada.petringa@lns.infn.it (G.P.); pablo.cirrone@lns.infn.it (G.A.P.C.); cuttone@lns.infn.it (G.C.)
3 Department of Biomedical and Biotechnological Sciences (BIOMETEC), University of Catania, 95123 Catania, Italy; parenti@unict.it
4 Departments of Physics and Astronomy, University of Catania, 95123 Catania, Italy
5 Lead Discovery Siena s.r.l. (LDS), 53100 Siena, Italy; al.fallacara@gmail.com (A.L.F.); l.maccari@leaddiscoverysiena.it (L.M.); maurizio.botta@unisi.it (M.B.)
6 Department of Biotechnology, Chemistry and Pharmacy, Università degli Studi di Siena, 53100 Siena, Italy
7 Department of Pharmacy, Università degli Studi di Genova, 16126 Genova, Italy; schenone@difar.unige.it
* Correspondence: luigi.minafra@ibfm.cnr.it
† These authors contributed equally to this work.

Received: 3 September 2019; Accepted: 24 September 2019; Published: 24 September 2019

Abstract: Glioblastoma Multiforme (GBM) is the most common of malignant gliomas in adults with an exiguous life expectancy. Standard treatments are not curative and the resistance to both chemotherapy and conventional radiotherapy (RT) plans is the main cause of GBM care failures. Proton therapy (PT) shows a ballistic precision and a higher dose conformity than conventional RT. In this study we investigated the radiosensitive effects of a new targeted compound, SRC inhibitor, named Si306, in combination with PT on the U87 glioblastoma cell line. Clonogenic survival assay, dose modifying factor calculation and linear-quadratic model were performed to evaluate radiosensitizing effects mediated by combination of the Si306 with PT. Gene expression profiling by microarray was also conducted after PT treatments alone or combined, to identify gene signatures as biomarkers of response to treatments. Our results indicate that the Si306 compound exhibits a radiosensitizing action on the U87 cells causing a synergic cytotoxic effect with PT. In addition, microarray data confirm the SRC role as the main Si306 target and highlights new genes modulated by the combined action of Si306 and PT. We suggest, the Si306 as a new candidate to treat GBM in combination with PT, overcoming resistance to conventional treatments.

Keywords: glioblastoma multiforme; proton therapy; combined treatments; gene signatures

1. Introduction

Glioblastoma multiforme (GBM) is a central nervous system tumor classified as grade IV of high-grade malignant gliomas (HGG), according to the World Health Organization (WHO) guidelines [1]. GBM belongs to the group of diffuse astrocytic and oligodendroglial tumor, joining oligodendrocytomas, ependymomas, and mixed gliomas, under the glioma classification [2]. According to the ASTRO guidelines statements, the current standard care for GBM is surgical resection to the feasible extent, followed by conventional radiotherapy (RT) of 60 Gy delivered by fractions of 2 Gy, up to seven weeks. Moreover, chemotherapy is concurrent to RT with daily temozolomide (TMZ) administration [3–5]. These treatment modalities are not currently curative and the resistance to both chemotherapy and RT plans is the main cause of GBM care failures (the median survival time is 14.6 months) [6]. Moreover, the percentage of relapses and side effects post TMZ and RT treatments is more than 90% [7]. More precisely, even if the application of TMZ has significantly improved clinical GBM outcomes, cases of drug resistance related to the activity of the enzyme methyl guanine methyl transferase (MGMT) have been observed [8]. The hypermethylation of its promoter, is indeed associated with a better survival rate in patients receiving TMZ with or without RT [9]. In addition, the dose release onto healthy brain tissue or surrounding organs at risk during irradiation may, substantially, contribute to late tissue toxicities, such as radionecrosis and neurocognitive dysfunction, because of their limited dose tolerance.

In recent years, different dose fractionation schedules have been improved to have a better prognosis, avoiding the large side effect even in case of focal re-irradiation of recurrences. In this scenario, proton therapy (PT) could be used as a successful strategy for GBM treatment, being able to regulate the balance between tumor control and the normal tissue tolerance [10–14]. In particular, when heavy particles cross the tissues, they deposit a minimal radiation dose on their track to the tumor. The depth-dose distribution, described by the Bragg peak trend, gradually increases as a function of the depth. So, the so-called spread-out Bragg peak (SOBP) lead to a complete irradiation of the target volume and a more conformal dose distribution, sparing the surrounding healthy tissues from damage [15,16]. This specific dose distribution curve represents a key topic for GBM tumor treatments in which the sparing of healthy tissue is a key factor for the patient's quality of life. Therefore, there is a robust scientific rationale motivating the need to enlarge studies that guide towards new clinical trials for PT combined with targeted therapy rather than conventional RT with photons or electrons [17,18].

Today, in the context of personalized medicine, prognostic and predictive molecular biomarkers are useful to select cancer therapeutic planning [19,20]. A critical point in RT success is the prediction of cancer radiosensitivity. At the molecular level, the idea that genes may behave as biomarkers of a disease response represents the base for the development of gene signatures, to predict response to cancer radiation treatments [21]. Several genes have been shown to be responsive to radiation exposure and thanks to the use of high-throughput technologies, such as gene expression profiling (GEP) by microarray, radiosensitivity assays have been developed with gene signatures predicting radioresponse in many cancer types, including GBM [22]. However, the response to radiation is highly cell-line dependent and some specific genes and pathways may be linked both to tumor subtypes and dose delivered [23–25].

Actually, few published studies have evaluated the effectiveness of radiosensitizing agents combined with PT in GBM and none of them consider genes and response pathways induced by RT. Most studies have demonstrated that different genetic pathways and molecular features can provide reliable prognostic biomarkers, overlooking the treatment responses and predictive outcomes. However, according to WHO guidelines, IDH1/IDH2 gene status distinguishes a more radioresistant tumor type (primary GBM, IDH-wild type) from a more sensitive one (secondary GBM, *IDH-mutant*). IDH mutation is correlated with epigenetic modifications of the MGMT gene and assumes a prognostic value together with other biomarkers such as, the presence of LOH 10q, epidermal growth factor receptor (EGFR) amplification, $p16^{INK4a}$ deletion, TP53 mutation, PTEN mutation, and the co-deletion of 1p/19q [26–28].

Based on this evidence, a large group of molecularly targeted agents have been designed, but none of them seem to overcome tumor radioresistance [29]. Previous studies support an involvement of the SRC-family protein kinases in the irradiation induction of radioresistance mechanisms. SRC protein is a non-receptor tyrosine kinase that interacts with many intracellular proteins involved in GBM carcinogenesis and progression. In addition, in vitro and in vivo studies confirmed the correlation between SRC activity and GBM carcinogenesis. [30].

In this work we analyzed the GEP on the U87 MG human glioblastoma cell line after treatment with PT alone or in combination with a new targeted compound, named Si306 (Lead Discovery Siena, Siena, Italy), inhibiting SRC proteins. The Si306 molecule is a new TKI, chosen among the family of pyrazolo[3,4-d] pyrimidines, that exhibited the ability to specifically bind the ATP site of SRC protein, making it inactive. Furthermore, previous in vitro and in vivo studies have shown that the Si306 determines a significant reduction of the β-PDGFR active phosphorylated form and a greater loss of the migratory ability in GBM cells stimulated by Epidermal Growth Factor (EGF). In addition, the antiproliferative effect of Si306 has been tested in association with conventional RT treatments both in vitro and in vivo [31].

Here, in order to clarify the Si306 activity in GBM cells exposed to PT, we firstly evaluated radiosensitive effects of different amounts of the Si306 compound on the U87 cell line in combination with PT exposed at the doses of 1, 2, 3, 4, 10, and 21 Gy. Clonogenic assay and dose modifying factor (DMF) calculations were performed. We also analyzed the U87 cell radiosensitivity by applying the radiobiological linear-quadratic (LQ) model and calculated the α, β, and $\frac{\alpha}{\beta}$ ratio, commonly used to predict radiosensitivity of normal and tumor cells [32].

In addition, at molecular level we selected 2 and 10 Gy of proton radiation doses combined with the Si306 to evaluate GEP induced responses, by using whole genome cDNA microarray. We described networks and specific gene signatures of response to both treatments, highlighting for the first time, the cell pathways induced by Si306.

2. Results

2.1. IC50 Determination

In order to evaluate cytotoxicity ability of Si306 in term of concentration that determined the 50% of growth inhibition (IC_{50}), U87 cells were incubated with Si306 at increasing concentrations of 0.1, 1.0, 10, and 100 μM for 24, 48, and 72 h under normal cell culture conditions. Cell numbers and viability were evaluated and the IC_{50} values calculated at each time points (Table 1).

Table 1. IC_{50} values calculated after 24, 48, and 72 h of treatment with Si306 on U87 glioblastoma cell line.

IC50	IC50	IC50
24 h 17.3 μM	48 h 6.8 μM	72 h 1.98 μM

2.2. Cell Radiosensitization Following Combined Treatments with Protons and Si306

To evaluate the radiosensitizing ability of Si306 compound, we investigated the combined effects of this molecule on U87 cells exposed to different proton doses (1, 2, 3, 4, 10, and 21 Gy). Surviving fraction values, obtained by clonogenic assay, after irradiation with protons alone or after pretreatment with 10 and 20 μM Si306, are shown in Table 2. These surviving fraction (SF) values were plotted to obtain dose-response curves with the exception of the 10 Gy and 21 Gy doses because of the lack of LQ model validity at high doses (Figure 1). We then calculated the DMF, which represents the relative reduction of dose to be delivered following a combined treatment with Si306 to get the isoeffect of SF = 0.5 compared to radiation treatment without modification. The DMF values were 1.09 (10 μM of Si306) and 1.21 (20 μM of Si306), showing a radiosensitive effect at both concentrations (Table 3).

Table 2. Surviving fraction (SF) values of U87 cells after irradiation with only protons and after combined treatments with 10 and 20 μM of Si306.

Dose (Gy)	SF (Only Protons)	SF (Protons + 10 μM Si306)	SF (Protons + 20 μM Si306)
0	1.000 ± 0.185	1.000 ± 0.121	1.000 ± 0.127
1	0.756 ± 0.126	0.722 ± 0.107	0.694 ± 0.104
2	0.516 ± 0.066	0.509 ± 0.088	0.474 ± 0.078
3	0.409 ± 0.069	0.342 ± 0.057	0.305 ± 0.051
4	0.257 ± 0.050	0.239 ± 0.050	0.216 ± 0.044
10	0.109 ± 0.022	0.072 ± 0.018	0.064 ± 0.018
21	0.056 ± 0.015	0.039 ± 0.009	0.035 ± 0.012

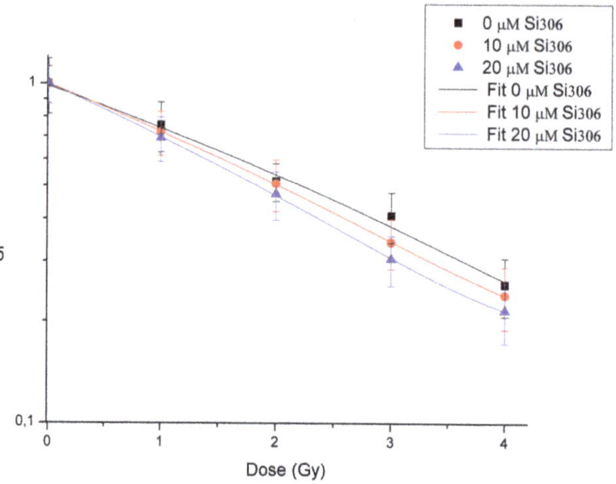

Figure 1. Cell survival curves of U87 cells. Cells treated with protons only (black line), protons plus 10 μM of Si306 (red line), and protons plus 20 μM of Si306 (blue line).

Table 3. Dose modifying factor (DMF) values calculated as isoeffective dose at surviving fraction of 0.5.

Treatment	SF 50% (Gy)	DMF
Protons	2.22	1
Protons + 10 μM Si306	2.03	1.09
Protons + 20 μM Si306	1.84	1.21

2.3. LQ Model

We calculated LQ parameters α and β of U87 cells, which provided information about the intrinsic cell radiosensitivity. Together with $\frac{\alpha}{\beta}$ ratio they have a pivotal role for a reliable estimation of radiation response, although most of the studies reported a large heterogeneity in LQ parameters and limited data is published about PT [33,34]. The U87 fitted survival curve, generated after only protons administration, gives us the values of 0.292 Gy^{-1} for α and of 0.010 Gy^{-2} for β, that result in an $\frac{\alpha}{\beta}$ ratio of 28.6 Gy (Table 4).

Table 4. Values of the α and β parameters estimated by fitting the cell survival to the linear-quadratic (LQ) model.

Treatment	α (Gy-1)	β (Gy-2)	α/β (Gy)
Proton	0.292 ± 0.036	0.010 ± 0.003	28.6
Proton + 10 μM Si306	0.322 ± 0.011	0.010 ± 0.003	32.2
Proton + 20 μM Si306	0.372 ± 0.018	0.004 ± 0.001	93.0

The higher $\frac{\alpha}{\beta}$ ratio showed, when the Si306 is added, especially at higher concentrations, determines a more linear cell survival as reasonably expected and demonstrates the molecule radiosensitivity role. Moreover, the shape variations at the origin of survival curves are linked with the DMF values. Other points are evident for the relationship between the LQ parameters and survival curve. Si306 affects substantially the linear component (α), whereas the quadratic component (β) is slightly decreased at higher concentrations. These results can be interpreted according to the LQ model, in which the cell death is lead, in our case, to the greater accumulation of lethal lesions. The use of Si306, both at concentrations of 10 and 20 μM, combined with PT contributes to sensitize GBM cells to protons exposure with an increase in cell killing.

2.4. *Gene Expression Profiles (GEP) Experiments*

As a second aim of this work, here we have reported GEP data obtained applying a Two-Color cDNA Microarray-Based Gene Expression Analysis (Agilent technologies) on U87 cells exposed to PT, with or without 10 μM Si306 compound. Comparative differential gene-expression analysis revealed that multiple deregulated genes (DEG) were significantly altered, by 2-fold or greater according to the specific experimental configuration reported as follows.

In addition, as described by several authors and also by our group [35,36], we have studied GEP lists, using PubMatrix, a tool for multiplex literature mining, in order to confirm our assumptions and to test their involvement in selected queries, radiation related, to draw assumptions described in the "Discussion" section. In this way, lists of terms, such as gene names, were assigned to a genetic, biological, or clinical relevance in a flexible systematic fashion in order to confirm our hypothesis, highlighting the involvement of known and lesser known genes able to drive cell radiation responses (Table S1).

2.4.1. GEP Induced by Proton Irradiation in U87 Glioblastoma Cells

Firstly, we analyzed the gene expression changes uniquely induced by protons irradiation with 2 and 10 Gy of IR doses. It should be remembered that 2 Gy is the daily dose delivered in fractionated RT treatments, so it is a dose of clinical interest, while 10 Gy represents a high dose of interest for comparisons with high-dose GEP studies of our research group [36].

In particular, U87 cell line treated with 2 Gy changed the expression levels of 936 genes (215 down regulated and 721 up regulated). On the other hand, 1018 DEGs were selected in U87 cells treated with 10 Gy and, among these, 251 were down regulated while 767 up regulated (Table 5).

Table 5. Number of genes significantly deregulated by 2-fold or greater in all the configuration modalities assayed in this work.

Configuration	Number of Genes	Down	Up
U87 2 Gy versus U87 n.t	936	215	721
U87 10 Gy versus U87 n.t	1018	767	251
U87 +Si306 + 2 Gy versus U87 2 Gy	1419	563	856
U87 + Si306 + 10 Gy versus U87 10 Gy	969	353	616

Deregulated transcripts obtained were grouped by using the DAVID tool [37,38] according to pathway analysis and the top-five molecular pathways selected are reported in Table 6. The analysis on DEGs induced by PT treatment with 2 Gy revealed the involvement of a set of factors controlling cellular processes, such as Hippo signaling pathway, cAMP signaling pathway, antigen processing and presentation, Wnt signaling pathway, and cell adhesion molecules (CAMs).

Table 6. Top-five statistically relevant pathways activated in U87 cells exposed to proton therapy (PT).

		Pathway Name	Genes Count	%	*p* Value	Genes
2 Gy	1	Hippo signaling pathway	19	0.016	0.000255	*WNT5A, DVL3, WNT10A, NF2, FZD3, TCF7L2, LLGL1, LATS2, TP73, DVL1, CTNNB1, PPP1CA, CCND3, CSNK1E, CCND2, DLG4, PARD6G, WNT6, BMP8B*
	2	cAMP signaling pathway	18	0.015	0.012333	*FXYD2, HCN2, VAV3, MAP2K2, GRIN1, GRIN2A, ATP1A4, VIPR2, ADORA1, AKT1, ATP2B2, PPP1CA, GRIN2D, ABCC4, CALML6, HCN4, PIK3R3, HTR1D*
	3	Antigen processing and presentation	9	0.007	0.026474	*CIITA, KLRC2, HLA-A, NFYC, HLA-C, HSPA1A, HLA-B, CTSB, HLA-E*
	4	Wnt signaling pathway	13	0.011	0.029905	*WNT5A, WNT10A, DVL3, FZD3, TCF7L2, DVL1, CTNNB1, SFRP1, CCND3, CSNK1E, CCND2, NFATC2, WNT6*
	5	Cell adhesion molecules (CAMs)	13	0.011	0.036193	*PVR, LRRC4, ITGAL, CD276, HLA-A, HLA-C, HLA-B, HLA-E, SDC4, NRCAM, SDC1, ITGB8, CLDN1*
10 Gy	1	PI3K-Akt signaling pathway	31	0.025	0.000968	*CSH1, PHLPP1, FGF7, PGF, KITLG, RPS6KB2, BCL2L1, GNG8, AKT1, COL6A5, COL6A3, TEK, COL6A2, COL6A1, PRKAA2, INSR, GHR, AKT2, FN1, TNXB, PKN2, HSP90B1, CDKN1A, EIF4E, CCND3, GNB2, CCND2, ITGA5, VEGFA, MDM2, LAMC2*
	2	p53 signaling pathway	11	0.008	0.001175	*PPM1D, CDKN1A, CCND3, CCND2, BBC3, BAX, MDM2, FAS, GADD45B, SESN1, TP73*
	3	Proteoglycans in cancer	21	0.017	0.001320	*ERBB2, RPS6KB2, IGF2, FLNC, FLNA, PXN, CTNNB1, AKT1, WNT7B, SDC1, PPP1CA, CDKN1A, MAPK12, ITGA5, VEGFA, MDM2, FAS, MSN, WNT6, FN1, AKT2*
	4	Hippo signaling pathway	15	0.012	0.012836	*NF2, TEAD1, TCF7L2, LATS2, TP73, DVL1, CTNNB1, WNT7B, PPP1CA, CCND3, BBC3, CCND2, PARD6G, WNT6, BMP8B*
	5	cAMP signaling pathway	18	0.014	0.013410	*FXYD2, HCN2, VAV3, GRIN1, HTR4, ATP1A4, VIPR2, ADORA1, AKT1, ATP2B2, FOS, PPP1CA, SSTR1, GRIN2D, HTR6, ABCC4, HCN4, AKT2*
Common between 2 and 10 Gy	1	Hippo signaling pathway	12	0.018	0.001636	*PPP1CA, CCND3, NF2, CCND2, PARD6G, WNT6, TCF7L2, BMP8B, LATS2, TP73, CTNNB1, DVL1*
	2	cAMP signaling pathway	13	0.019	0.004726	*HCN2, FXYD2, VAV3, GRIN1, ATP1A4, VIPR2, ADORA1, AKT1, ATP2B2, PPP1CA, GRIN2D, ABCC4, HCN4*
	3	Proteoglycans in cancer	12	0.018	0.013466	*AKT1, PPP1CA, SDC1, MAPK12, ERBB2, IGF2, MSN, FLNC, WNT6, FLNA, PXN, CTNNB1*
	4	Neuroactive ligand-receptor interaction	14	0.021	0.025160	*CSH1, PRLHR, GRIN1, DRD4, ADORA1, VIPR2, NTSR2, CRHR2, CHRM3, GALR3, GRIN2D, GALR2, UTS2R, CHRNA1*
	5	Antigen processing and presentation	6	0.009	0.044750	*NFYC, HLA-C, HSPA1A, HLA-B, CTSB, HLA-E*

On the other hand, U87 glioblastoma cells exposed to 10 Gy of proton irradiation activate specific cell pathways as displayed in Table 6: PI3K-Akt signaling pathway, p53 signaling pathway, proteoglycans in cancer, Hippo signaling pathway, and cAMP signaling pathway. Finally, the GEP lists were analyzed by Venn diagrams in order to identify the overlapping deregulated genes (537 DEGs), between the two configurations of 2 and 10 Gy assayed (Figure 2A). Some genes were specifically deregulated following the dose provided, showing a dose-dependent transcriptional response. Moreover, cells respond to radiation treatment also in a common manner with activation of common genes and pathways, as displayed in Table 6 and listed as follows: Hippo signaling pathway; cAMP signaling pathway; proteoglycans in cancer; neuroactive ligand-receptor interaction; and antigen processing and presentation. Except for the neuroactive ligand-receptor interaction pathway, formed overall by molecules driving neuronal cell signaling, the involvement of these cellular processes in U87 cells proton exposed, has described above.

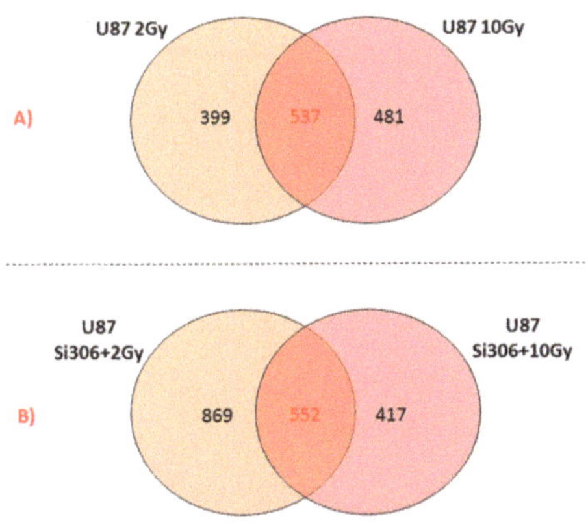

Figure 2. Venn diagrams showing the number of unique and shared differentially expressed genes after exposure to: (**A**) PT and (**B**) Si306 + PT combined treatments.

2.4.2. GEP Induced by Si306 and Proton Combined Treatments in U87 Glioblastoma Cells

In a second step, we have evaluated the effect on GEPs after a combined administration of 10 μM Si306 compound and PT using the doses of 2 and 10 Gy, hereafter named as follows: U87 Si306 + 2 Gy and U87 Si306 + 10 Gy, which were analyzed in comparison to the respective samples treated with PT alone (U87 2 Gy and U87 10 Gy). We selected a large amount of deregulated genes, caused by the Si306 compound addition to PT treatment: 1419 DEGs (563 down and 856 up regulated) in U87 Si306 + 2 Gy, while 969 DEGs (353 down and 616 up regulated) changed their expression levels in U87 Si306 + 10 Gy (Table 5). Thus, also for these experimental configurations, up and down regulated transcripts were grouped according to their involvement in specific biological pathways using DAVID tool [38]. The top-five statistically relevant molecular pathways of deregulated gene datasets are reported in Table 7. In particular, the Si306 + 2 Gy combined treatments deregulated the expression levels of genes controlling: Phagosome, antigen processing and presentation, cell adhesion molecules, inflammatory disease, and calcium signaling pathway.

Some of the above described pathways were also deregulated in U87 cells exposed to Si306 + 10 Gy and following reported and listed in Table 7: Proteoglycans in cancer, leukocyte transendothelial migration, phagosome, cell adhesion molecules, and autoimmune disease. Three out of the five pathways selected in U87 Si306 + 10 Gy (proteoglycans in cancer, phagosome, and cell adhesion molecules), were also deregulated in the other configurations analyzed, underling once again their interesting role in U87 cells response to radiation and/or to the Si306 molecule.

Finally, the Venn diagram shown in Figure 2B displays 552 deregulated common genes between the two configurations: U87 Si306 + 2 Gy and U87 Si306 + 10 Gy. The top-five statistically relevant pathways selected by DAVID tool using the 552 common gene list, are displayed in Table 7, and following listed: Autoimmune disease, antigen processing and presentation, proteoglycans in cancer, apoptosis, and inflammatory bowel disease.

Table 7. Top-five Statistically relevant pathways activated in U87 cells pretreated with Si306 and exposed to PT.

		Pathway Name	Genes Count	%	*p* Value	Genes
2 Gy	1	Phagosome	23	0.013	0.00014	*HLA-DQB1, NOS1, HLA-DRB1, MRC2, HLA-A, HLA-C, HLA-B, ITGB3, SFTPA1, HLA-E, CLEC4M, FCAR, CD209, COMP, TUBAL3, HLA-DPA1, SCARB1, HLA-DPB1, HLA-DOA, TUBB1, ATP6V0D2, TUBB4A, HLA-DRA*
	2	Antigen processing and presentation	15	0.009	0.00017	*CIITA, HLA-DQB1, HLA-DRB1, HLA-A, HLA-C, HSPA1A, HLA-B, HLA-E, CD74, KIR3DL3, HSPA6, HLA-DPA1, HLA-DPB1, HLA-DOA, HLA-DRA*
	3	Cell adhesion molecules (CAMs)	21	0.012	0.00036	*PVR, HLA-DQB1, HLA-DRB1, SELL, CLDN5, HLA-A, NLGN1, CTLA4, HLA-C, HLA-B, HLA-E, CLDN15, ALCAM, NCAM2, SDC1, CD2, MADCAM1, HLA-DPA1, HLA-DPB1, HLA-DOA, HLA-DRA*
	4	Inflammatory bowel disease (IBD)	13	0.007	0.00041	*HLA-DQB1, HLA-DRB1, TBX21, RORC, STAT1, STAT3, IL12RB2, IL17A, IL1B, HLA-DPA1, HLA-DPB1, HLA-DOA, HLA-DRA*
	5	Calcium signaling pathway	19	0.011	0.02473	*ORAI2, PTGER1, NOS1, ERBB4, TNNC1, ERBB3, ERBB2, STIM2, OXTR, EDNRA, ATP2B2, P2RX1, CHRM3, LTB4R2, GRPR, CHRNA7, CALML6, PLCB2, CACNA1B*
10 Gy	1	Proteoglycans in cancer	22	0.019	0.000094	*NANOG, ERBB4, ROCK2, HCLS1, ERBB2, FASLG, IGF2, FZD3, HGF, DCN, ITGB3, MMP2, PXN, KDR, CTNNB1, SMO, MAPK13, HPSE, PLCG2, HSPB2, PRKACA, TWIST1*
	2	Leukocyte transendothelial migration	12	0.010	0.01064	*ITGAL, ROCK2, MAPK13, PLCG2, CLDN5, CTNND1, MYLPF, RAPGEF3, JAM2, MMP2, PXN, CTNNB1*
	3	Phagosome	14	0.012	0.01214	*HLA-DQB1, HLA-DRB1, SFTPA1, ITGB3, COLEC11, TUBA8, CD36, FCGR2B, PIKFYVE, TUBAL3, HLA-DPA1, HLA-DPB1, TUBB1, TUBB4A*
	4	Cell adhesion molecules (CAMs)	12	0.010	0.03671	*HLA-DQB1, NRCAM, ITGAL, HLA-DRB1, CLDN5, NLGN1, CTLA4, HLA-DPA1, HLA-DPB1, JAM2, SELE, PDCD1LG2*
	5	Autoimmune disease	6	0.005	0.06648	*HLA-DQB1, HLA-DRB1, CTLA4, FASLG, HLA-DPA1, HLA-DPB1*
Common between 2 and 10 Gy	1	Autoimmune disease	6	0.009	0.00768	*HLA-DQB1, HLA-DRB1, CTLA4, FASLG, HLA-DPA1, HLA-DPB1*
	2	Antigen processing and presentation	6	0.009	0.03468	*HLA-DQB1, HLA-DRB1, KIR3DL3, HLA-DPA1, HLA-DPB1, CD74*
	3	Proteoglycans in cancer	10	0.015	0.04961	*ERBB4, MAPK13, ERBB2, FASLG, FZD3, HGF, ITGB3, MMP2, KDR, TWIST1*
	4	Apoptosis	5	0.007	0.06011	*DFFB, CYCS, CASP8, FASLG, IL3RA*
	5	Inflammatory bowel disease (IBD)	5	0.007	0.06604	*HLA-DQB1, HLA-DRB1, TBX21, HLA-DPA1, HLA-DPB1*

3. Discussion

The first purpose of this study was to evaluate the radiosensitizing effects mediated by combination of the new compound, the Si306 targeting SRC proteins, with PT on the U87 human glioblastoma cell line. The IC_{50} evaluation showed that this cell line is sensitive to treatment with the Si306 compound. Based on the IC_{50} values, we tested the radiosensitizing effect of Si306, used at concentrations of 10 and 20 μM, in combination with proton irradiation at increasing doses of 1,2, 3, 4, 10, and 21 Gy, in order to generate dose/response curves for the dose configurations tested.

The radiosensitizing effect was evaluated by calculating the DMF, obtained at the SF of 50%, in order to highlight the combined treatment capacity of enhancing tumor cells killing in respect of irradiation only [39]. Our data show that pretreatment with Si306 at both concentrations leads to a synergic cytotoxic effect with PT on the U87 cell line, further suggesting this compound as a new possible candidate to treat GBM in combination with PT. Indeed, the possibility to use drug/IR combined treatments, permits to increase the tumor control probability (TCP) even for radioresistant tumors, such as GBM. In addition, we also analyzed the U87 cell radiosensitivity by applying the radiobiological LQ model calculating the α, β parameters, and $\frac{\alpha}{\beta}$ ratio, which predict the radiosensitivity of normal and tumor cells [32]. The LQ model is considered to be the best-fitting model to describe cell survival and, therefore, is of great interest in radiation oncology to highlight the link existing between the $\frac{\alpha}{\beta}$ ratio and the following RT-induced tissue reactions [34,40,41]. The $\frac{\alpha}{\beta}$ ratio obtained on U87 cell line is in line with the $\frac{\alpha}{\beta}$ ratio calculated for a population of glioma cells reported by Barazzuol et al., who used a mathematical model to extract radiobiological information from clinical GBM patients data [42]. In addition, our results showed a higher $\frac{\alpha}{\beta}$ ratio by using combined treatments of Si306 and protons. Therefore, we speculate that the clinical effect of using combined treatments of PT/Si306 administration, with an optimized Si306 pharmacological quantity for the patients, could be translated into the possibility of modifying the PT schedule treatment. Thus, all of this gains an efficacy in TCP, by using a more tolerable fractionated PT treatment plan and a reduced total dose delivered to the tumor [43,44].

As a second aim of this work, we carried out a transcriptomic study in order to define gene signatures as biomarkers of treatment response. GEP by whole genome cDNA microarray was firstly performed to analyze the gene expression changes uniquely induced by proton irradiation with 2 and 10 Gy of IR doses, which represent two clinical doses of interest and also for comparison with high-dose GEP studies of our research group [23,24,36,45].

In particular, the treatment of U87 with 2 Gy revealed that a large number of genes were deregulated and involved in the regulation of specific cellular processes (Table 6). One of the activated pathways was the Hippo signaling pathway, an emerging growth control and tumor suppressor pathway that regulates cell proliferation and stem cell functions; the hyperactivation of its downstream effectors (such as TAZ protein, up regulated in U87 2 Gy with a fold change of 1,89) contributes to the development of cancer including GBM, suggesting that pharmacological inhibition of these factors may be an effective anticancer strategy [46,47]. In turn, in GBM cells Yang et al. recently reported that the Hippo transducer TAZ promotes cell proliferation and tumor formation through the EGFR pathway [48]. In addition, Hippo and Wnt signaling, up regulated in U87 2 Gy cells, reciprocally regulate each other's activity through a variety of mechanisms that needs to be better clarified in GBM cells [49]. As known, Wnt/β-catenin signaling plays important roles in maintaining the stemness of cancer stem cells in various cancer types and in promoting cellular invasiveness. Multimodality in vivo and in vitro studies revealed a key role of Wnt activation in GBM radiation resistance. In turn, literature data report a pivotal role of the Wnt/β-catenin signaling pathway in IR-induced invasion of U87 GBM cells, indicating that β-catenin is a potential therapeutic target for overcoming evasive radioresistance [50,51].

In U87 2 Gy the involvement of cAMP signaling pathway was also observed. Existing evidence suggests that intracellular cAMP level and signaling may affect the survival of cancer cells, including resistant cancer cells to standard chemotherapeutic drugs. Suppression of the cAMP pathway is a common feature across different cancers including GBM. [52,53]. In addition, IR is known to be able to activate the transcription of genes, through the presence of cAMP responsive elements (CREs) in their promoters, in order to guide cell response and survival after radiation exposure [54].

Moreover, the activation of antigen processing and presentation pathway after proton exposure with dose of 2 Gy in GBM cells is sustained by an up regulation of genes belonging to the human leukocyte antigen (HLA) class family (probably activated by β-catenin), factors involved in antigen presentation. As reported by Ghosh et al., HLA genes increasing level, often caused by a hypoxic tumor microenvironment, is associated with evasion of immune responses in cancer cells [55]. Finally, an overall activation of several cell adhesion molecules was highlighted in U87 2 Gy cells, involved in the activation of inflammation process and in the regulation of cancer invasiveness.

On the other hand, U87 cells exposed to 10 Gy of proton irradiation activate specific cell pathways, including the phosphatidylinositol-3-kinase (PI3K)-protein kinase B (Akt) signaling pathway (Table 6). As known, the PI3K/AKT pathway is commonly activated in cancer initiation and progression, including GBM, as it regulates different processes, such as proliferation, apoptosis, and migration [56], therefore representing a key target for cancer therapeutics. Moreover, the activation of TP53 pathway was observed in U87 10 Gy and driven by TP53 gene that was significantly altered by 1.77-fold. As described, TP53, exerts a crucial role following IR-induced DNA damage because it is able to cause cell cycle arrest, DNA repair, and apoptosis processes. Moreover, the influence of TP53 status on DNA damage repair after cell irradiation has been studied in several malignancies and also reported by our group in breast cancer cells after a high dose of electron irradiation [45,57]. Finally, in U87 10 Gy, an activation of proteoglycan signaling was observed. Proteoglycans are known to have many roles in tumor progression and are the main extracellular matrix (ECM) components of normal brain tissue, playing an important role in brain development; an overproduction of different molecules of this family were found in GBM cells [58,59].

Interestingly, in U87 10 Gy Hippo and cAMP signaling pathways were activated, as above described in U87 2 Gy configuration, underling once again the important role of these processes in GBM cells after proton exposure.

In a second step, we evaluated the GEPs induced by Si306 molecule in U87 cells irradiated with 2 and 10 Gy of proton doses and we selected a large number of deregulated genes, grouped according to their involvement in specific biological pathways (Table 7). In particular, in U87 Si306 + 2 Gy combined treatments a deregulated expression level of genes controlling phagosome was observed.

In GBM an intensive autophagic activity regulated by several signaling pathways was described [60]. As recently reported by Yasui et al., an altered autophagic flux was described in GBM cell lines exposed to 10 Gy of γ-rays. Our data also confirms this trend after proton exposure. These altered fluxes represent a useful biomarker of metabolic stress induced by IR and provide a metabolic context for radiation sensitization [61]. Here the Si306 radiosensitization effect seems to act by stressing this molecular mechanism. In addition, in U87 Si306 + 2 Gy configuration the involvement of antigen processing and presentation and cell adhesion molecules pathways were observed, similarly to that shown in U87 cells proton treated with only 2 Gy. Therefore, the Si306 treatment seems to cause an overall down regulation of HLA molecules (up regulated in U87 2 Gy), suggesting the activation of immune surveillance escaping mechanism induced by Si306 [55,62].

The latest two pathways deregulated in U87 Si306 + 2 Gy were linked to inflammation and calcium signaling. As known, the inflammation process is often activated in cell exposed to radiation, affecting cell fate by the activation of key transcription factors (TFs), such as NF-KB and STATs (i.e., STAT1 and STAT3) [63]. Interestingly, the combined Si306 + 2 Gy treatment induced a down regulation of STAT1 and STAT3 proteins. Thus, we speculate that this inhibition could promote radiation sensitivity decreasing angiogenesis and cell survival as hypothesized in other malignancies by several authors [64,65]. Indeed, a number of studies confirm that selective inhibitors of these proinflammatory pathways driven by STAT TFs, could be combined to conventional radiation or chemotherapy to increase their effectiveness [66,67].

On the other hand, the combined treatment with Si306 and 2 Gy PT seem to affect survival/death balance by modulating the intracellular calcium levels, a mechanism known to be involved in regulating IR-induced cell cycle arrest, apoptosis, and chromatin structure modifications [45,68,69].

Some of these pathways were also deregulated in U87 cells exposed to Si306 + 10 Gy, such as: Proteoglycans in cancer, leukocyte transendothelial migration, phagosome, cell adhesion molecules, and autoimmune disease. Three of the five pathways (proteoglycans in cancer, phagosome, and cell adhesion molecules), were also deregulated in the other configurations analyzed, suggesting once again their important role in U87 cells in response to radiation and/or to Si306 molecule. The other two selected pathways in U87 Si306 + 10 Gy (i.e., leukocyte transendothelial migration and autoimmune disease), highlight the involvement of a complex immunological response induced by IR, as known from the literature, and by the Si306 compound addition, as observed in this study.

Finally, we reported the number of overlapping deregulated genes between the two configurations of the combined treatments, such as U87 Si306 + 2 Gy and U87 Si306 + 10 Gy (Figure 2B). The top-five statistically relevant pathways selected and displayed in Table 7, were previously described.

Summarizing, our GEP results show that combined treatments on U87 cells can activate multiple signal transduction pathways described, to our knowledge, for the first time, to be new targets of

Si306. Finally, considering that the main target of Si306 is the tyrosine kinase SRC, we analyzed the known cellular target downstream to this transducer, in order to better clarify its role as molecular radiosensitizing. Thus, we observed that the combined treatment Si306 + protons (with 2 and 10 Gy) in U87 cells, is able to inhibit several signal transduction pathways, normally regulated by SRC as shown in Figure 3.

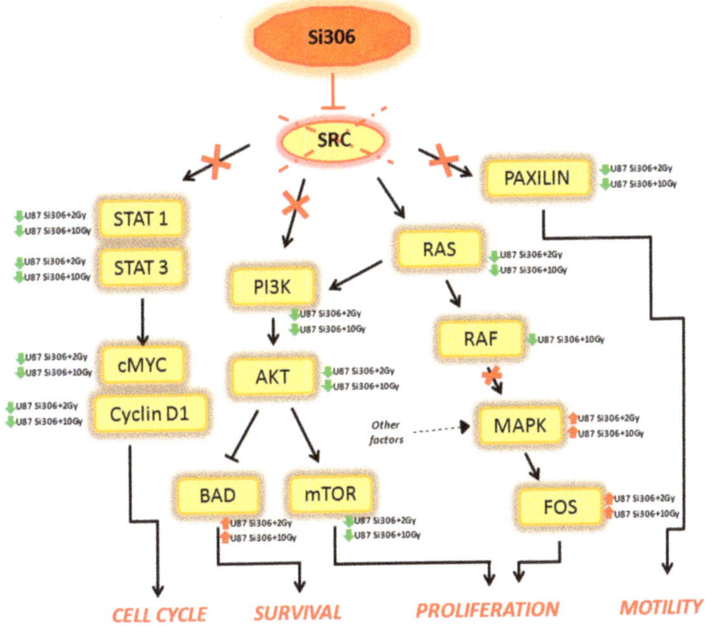

Figure 3. The figure displays the main targets of Si306 compound observed. The arrows define an activation and the T bars the inhibition. Red arrows define gene upregulation and green arrows gene downregulation.

In particular, the STAT1, STAT3, c-MYC, and Cyclin D1 genes, which are able to control the cell cycle, were downregulated in our analysis. Cell survival was negatively regulated by the downstream PI3K, AKT, and mTOR downregulation and by the BAD upregulation. In addition, Si306 is able to cause a partial inhibition of cell proliferation, downregulating RAS and RAF gene expression. However, the MAPK and FOS genes were not targets of Si306, so these factors (up regulated in our data), were probably activated by other cellular pathways. Finally, Si306 is also able to negatively regulate cell motility, through the downregulation of the paxilin gene.

These data confirm the SRC role as a main target of Si306 compound and highlight the transcriptional events occurring downstream of SRC inhibition by the combined treatments. The SRC blockage observed after Si306 and PT combined treatments seems to increase the single treatments effectiveness, thus promoting a radiosensitizing effect.

Today, very little data is available regarding the combination of molecularly targeted drugs and PT. Indeed, many studies debate about chemotherapeutic agents combined with high-linear energy transfer (LET) particle beams or protons for GBM treatment, overlooking the clinical perspective of target therapy [70,71].

The results obtained from this work have highlighted the radiosensitizing capacity of the Si306 targeted compound on U87 GBM cell line, acting in tandem with PT. Taking into account previously in vivo pharmacokinetic data, demonstrating that Si306 was able to reach the brain, overcoming the hurdle represented by the blood–brain barrier [31], this compound can be considered a new candidate for combined treatments of GBM. In addition, our GEP results confirm the important role of SRC as the main Si306 target and highlight new genes and pathways modulated by the combined action of Si306 and PT, which can be further explored as new radiosensitizing therapeutic targets in GBM.

4. Materials and Methods

4.1. Proton Irradiation Configuration and Cell Irradiation

The proton beam irradiation was performed at the CATANA (Centro di Adroterapia ed Applicazioni Nucleari Avanzate) facility of INFN-LNS (Catania, Italy) [72], using 62 MeV of proton beams accelerated by a cyclotron superconducting. The beamline is composed of several passive elements optimized for the clinical application: Scattering foils to spread the beam laterally, collimators to define beam profile in accordance to the tumor shape, and monitoring chambers to measure the dose delivered [73]. In order to irradiate the entire 25 cm^2 (T25) standard tissue culture flasks, a motorized system for biological samples irradiation was used. Radiochromic film detectors were adopted to check the delivered dose and the lateral dose distribution during each irradiation. The dosimetric system was calibrated under reference conditions according to the International Atomic Energy Agency Technical Reports Series No. 398 "Absorbed Dose Determination in External Beam Radiotherapy" [74].

For combined treatments with Si306, U87 irradiations were carried out using six dose values of 1, 2, 3, 4, 10, and 21 Gy. Cell irradiations were conducted placing the cell at the middle spread-out Bragg peak, to simulate a clinical condition, with a dose rate of 15 Gy/min.

4.2. Cell Culture and IC50 Determination

In vitro experiments were carried out using the U87 MG human glioblastoma cell line. Cells were purchased from European Collection of Authenticated Cell Cultures (ECACC, Public Health England, Porton Down Salisbury, UK) and cultured as previously described [31]. Cells were maintained in an exponentially growing culture condition in an incubator at 37 °C in a humidified atmosphere (95% air and 5% CO_2) and were routinely sub-cultured in T25 standard tissue culture flasks.

To calculate IC_{50} (drug concentration that determined the 50% of growth inhibition), 2.5×10^4 U87 cells were plated in 12-well plates and incubated with Si306 dissolved in DMSO at increasing concentrations (0.1, 1.0, 10, and 100 μM) for 24, 48, and 72 h under normal cell culture conditions. Cell numbers and viability were evaluated using Z2 Coulter Counter (Beckman Coulter, Indianapolis, United States). IC_{50} was calculated by GraphPad Prism 6.0 software (GraphPad Software, San Diego, CA, USA) using the best fitting sigmoid curve.

4.3. Clonogenic Survival Assay

Forty-eight hours before irradiations U87 cells were seeded in T25 flasks and the day after were incubated with the concentrations of 10 and 20 μM of Si306, chosen on the base of IC_{50} results, for 24 h prior to radiation treatment. Cells were irradiated at subconfluence. Combined effects of Si306 and protons were evaluated by clonogenic survival assay, performed as previously described [45,57]. Briefly, after irradiation, U87 cells were detached, counted by hemocytometer and seeded into a six-well plate in triplicate at a density of 50–2000 cells per well, by plating an increasing cell quantity according to the dose delivered raising, in order to assay the SF. Colonies were allowed to grow under normal cell culture conditions for two weeks and then were fixed with 50% methanol and stained 0.5% crystal violet (both from Sigma-Aldrich, St. Louis, MO, USA). Colonies with more than 50 cells were counted manually under Olympus CK30 phase-contrast microscope (Olympus, Tokyo, Japan) and also automatically with a computer-assisted methodology [75]. The calculation of SFs in U87 cells irradiated with protons and pre-treated with Si306 were determined taking into consideration the plating efficiency (PE) for all treatment modalities based on three independent experiments.

55. Ghosh, S.; Paul, A.; Sen, E. Tumor necrosis factor α-induced hypoxia-inducible factor 1α-β-catenin axis regulates major histocompatibility complex class I gene activation through chromatin remodeling. *Mol. Cell Biol.* **2013**, *33*, 2718–2731. [CrossRef]
56. Lino, M.M.; Merlo, A. PI3Kinase signaling in glioblastoma. *J. Neurooncol.* **2011**, *103*, 417–427. [CrossRef]
57. Minafra, L.; Bravata, V.; Russo, G.; Forte, G.I.; Cammarata, F.P.; Ripamonti, M.; Messa, C.; Candiano, G.; Cervello, M.; Giallongo, A.; et al. Gene expression profiling of MCF10A breast epithelial cells exposed to IOERT. *Anticancer Res.* **2015**, *35*, 3223–3234.
58. Iozzo, R.V.; Sanderson, R.D. Proteoglycans in cancer biology, tumour microenvironment and angiogenesis. *J. Cell Mol. Med.* **2011**, *15*, 1013–1031. [CrossRef]
59. Kazanskaya, G.M.; Tsidulko, A.Y.; Volkov, A.M.; Kiselev, R.S.; Suhovskih, A.V.; Kobozev, V.V.; Grigorieva, E.V.; Gaytan, A.S.; Aidagulova, S.V.; Krivoshapkin, A.L.; et al. Heparan sulfate accumulation and perlecan/HSPG2 up-regulation in tumour tissue predict low relapse-free survival for patients with glioblastoma. *Histochem. Cell Biol.* **2018**, *149*, 235–244. [CrossRef]
60. Giatromanolaki, A.; Sivridis, E.; Mitrakas, A.; Kalamida, D.; Zois, C.E.; Haider, S.; Koukourakis, M.I.; Piperidou, C.; Pappa, A.; Gatter, K.C.; et al. Autophagy and lysosomal related protein expression patterns in human glioblastoma. *Cancer Biol. Ther.* **2014**, *15*, 1468–1478. [CrossRef]
61. Yasui, L.S.; Duran, M.; Andorf, C.; Kroc, T.; Owens, K.; Allen-Durdan, K.; Becker, R.; Schuck, A.; Grayburn, S. Autophagic flux in glioblastoma cells. *Int. J. Radiat. Biol.* **2016**, *92*, 665–678. [CrossRef]
62. Yaghi, L.; Poras, I.; Simoes, R.T.; Donadi, E.A.; Tost, J.; Daunay, A.; Moreau, P.; de Almeida, B.S.; Carosella, E.D. Hypoxia inducible factor-1 mediates the expression of the immune checkpoint HLA-G in glioma cells through hypoxia response element located in exon 2. *Oncotarget* **2016**, *7*, 63690–63707. [CrossRef]
63. Di Maggio, F.M.; Minafra, L.; Forte, G.I.; Cammarata, F.P.; Lio, D.; Messa, C.; Bravatà, V.; Gilardi, M.C. Portrait of inflammatory response to ionizing radiation treatment. *J. Inflamm.* **2015**, *12*, 14. [CrossRef]
64. Kim, K.W.; Mutter, R.W.; Cao, C.; Albert, J.M.; Shinohara, E.T.; Sekhar, K.R.; Lu, B. Inhibition of signal transducer and activator of transcription 3 activity results in down-regulation of Survivin following irradiation. *Mol. Cancer Ther.* **2006**, *5*, 2659–2665. [CrossRef]
65. Yu, H.; Pardoll, D.; Jove, R. STATs in cancer inflammation and immunity: A leading role for STAT3. *Nat. Rev. Cancer.* **2009**, *9*, 798–809. [CrossRef]
66. Sun, Y.; Cheng, M.K.; Thomas, R.L.G.; Kilian, M.J.; Kai, B.; Kriajevska, M.; Manson, M. Inhibition of STAT signalling in bladder cancer by diindolylmethane: Relevance to cell adhesion, migration and proliferation. *Curr. Cancer Drug Targets.* **2013**, *13*, 57–68. [CrossRef]
67. Du, Y.C.; Gu, S.; Zhou, J.; Wang, T.; Cai, H.; MacInnes, M.A.; Chen, X.; Bradbury, E.M. The dynamic alterations of H2AX complex during DNA repair detected by a proteomic approach reveal the critical roles of Ca(2+)/calmodulin in the ionizing radiationinduced cell cycle arrest. *Mol. Cell. Proteom.* **2006**, *5*, 1033–1044. [CrossRef]
68. Lao, Y.; Chang, D.C. Mobilization of Ca2+ from endoplasmic reticulum to mitochondria plays a positive role in the early stage of UV-or TNFα-induced apoptosis. *Biochem. Biophys. Res. Commun.* **2008**, *373*, 42–47. [CrossRef]
69. Combs, S.E.; Bohl, J.; Elsässer, T.; Weber, K.J.; Schulz-Ertner, D.; Debus, J.; Weyrather, W.K. Radiobiological evaluation and correlation with the local effect model (LEM) of carbon ion radiation therapy and temozolomide in glioblastoma cell lines. *Int. J. Radiat. Biol.* **2009**, *85*, 126–137. [CrossRef]
70. Combs, S.E.; Zipp, L.; Rieken, S.; Habermehl, D.; Brons, S.; Winter, M.; Weber, K.J.; Haberer, T.; Debus, J. In vitro evaluation of photon and carbon ion radiotherapy in combination with chemotherapy in glioblastoma cells. *Radiat. Oncol.* **2012**, *7*, 9. [CrossRef]
71. Barazzuol, L.; Jena, R.; Burnet, N.G.; Jeynes, J.C.; Merchant, M.J.; Kirkby, K.J.; Kirkby, N.F. In Vitro Evaluation of Combined Temozolomide and Radiotherapy Using X Rays and High-Linear Energy Transfer Radiation for Glioblastoma. *Radiat. Res.* **2012**, *177*, 651–662. [CrossRef]
72. Cirrone, G.A.P.; Cuttone, G.; Lojacono, P.A.; Lo Nigro, S.; Mongelli, V.; Patti, I.V.; Privitera, G.; Raffaele, L.; Rifuggiato, D.; Sabini, M.G.; et al. A 62-MeV proton beam for the treatment of ocular melanoma at laboratori nazionali del sud-INFN. *IEEE Transact. Nuclear Sci.* **2004**, *51*, 860–865. [CrossRef]
73. Cirrone, G.A.P.; Cuttone, G.; Lo Nigro, S.; Mongelli, V.; Raffaele, L.; Sabini, M.G. Dosimetric characterization of CVD diamonds in photon, electron and proton beams. *Nuclear Physics B (Proc. Suppl.)* **2006**, *150*, 330–333. [CrossRef]

74. Sartini, L.; Simeone, F.; Pani, P.; Lo Bue, N.; Marinaro, G.; Grubich, A.; Gasparoni, F.; Lobko, A.; Etiope, G.; Gapone, A.; et al. Nuclear Instruments and Methods in Physics Research Section A: Accelerators, Spectrometers, Detectors and Associated Equipment. *Nucl. Instrum. Methods Phys. Res. A* **2017**, *846*. [CrossRef]
75. Militello, C.; Rundo, L.; Conti, V.; Minafra, L.; Cammarata, F.P.; Mauri, G.; Porcino, N.; Gilardi, M.C. Area-based cell colony surviving fraction evaluation: A novel fully automatic approach using general-purpose acquisition hardware. *Comput. Biol. Med.* **2017**, *89*, 454–465. [CrossRef]
76. Minafra, L.; Porcino, N.; Bravatà, V.; Gaglio, D.; Bonanomi, M.; Amore, E.; Baglio, M.; Cammarata, F.P.; Russo, G.; Militello, C.; et al. Radiosensitizing effect of curcumin-loaded lipid nanoparticles in breast cancer cells. *Sci. Rep.* **2019**, *9*, 11134. [CrossRef]

International Journal of
Molecular Sciences

Review

FLASH Radiotherapy: Current Knowledge and Future Insights Using Proton-Beam Therapy

Jonathan R. Hughes [1] and Jason L. Parsons [1,2,*]

[1] Cancer Research Centre, Department of Molecular and Clinical Cancer Medicine, University of Liverpool, 200 London Road, Liverpool L3 9TA, UK; jonathan.hughes@liverpool.ac.uk
[2] Clatterbridge Cancer Centre NHS Foundation Trust, Clatterbridge Road, Bebington CH63 4JY, UK
* Correspondence: j.parsons@liverpool.ac.uk; Tel.: +44-151-794-8848

Received: 31 July 2020; Accepted: 2 September 2020; Published: 5 September 2020

Abstract: FLASH radiotherapy is the delivery of ultra-high dose rate radiation several orders of magnitude higher than what is currently used in conventional clinical radiotherapy, and has the potential to revolutionize the future of cancer treatment. FLASH radiotherapy induces a phenomenon known as the FLASH effect, whereby the ultra-high dose rate radiation reduces the normal tissue toxicities commonly associated with conventional radiotherapy, while still maintaining local tumor control. The underlying mechanism(s) responsible for the FLASH effect are yet to be fully elucidated, but a prominent role for oxygen tension and reactive oxygen species production is the most current valid hypothesis. The FLASH effect has been confirmed in many studies in recent years, both *in vitro* and *in vivo*, with even the first patient with T-cell cutaneous lymphoma being treated using FLASH radiotherapy. However, most of the studies into FLASH radiotherapy have used electron beams that have low tissue penetration, which presents a limitation for translation into clinical practice. A promising alternate FLASH delivery method is via proton beam therapy, as the dose can be deposited deeper within the tissue. However, studies into FLASH protons are currently sparse. This review will summarize FLASH radiotherapy research conducted to date and the current theories explaining the FLASH effect, with an emphasis on the future potential for FLASH proton beam therapy.

Keywords: FLASH; ionizing radiation; proton beam therapy; radiotherapy; radiobiology

---

1. Introduction

Radiotherapy using X-rays (photons) is a main treatment strategy employed to combat human tumors, with ~50% of all cancer patients receiving radiotherapy. However, the major drawback of radiotherapy treatment is that in order to deliver a lethal dose to cancerous cells, short- and long-term adverse side-effects are evident due to the irradiation of the surrounding normal healthy tissues that can severely impact the health and quality of life of the cancer patient [1–3]. This occurs because radiotherapy uses an external radiation beam where the dose decreases exponentially but which can deposit energy within a certain depth of the patient tissue [4]. Therefore, in the cases of deep-seated tumors, the healthy normal tissue in front of the tumor receives a large dose of ionizing radiation relative to the tumor. Furthermore, it is possible that healthy normal tissue located behind the tumor can receive an exit dose of radiation if the beam passes through the tumor. This can present significant challenges to sensitive tissues and organs at risk, such as the brain and spinal cord. Advancements in modern radiotherapy deliverance and imaging techniques such as image-guided radiotherapy, intensity-modulated radiotherapy, and volumetric modulated arc therapy, along with targeted combinatorial drug therapies and immunotherapy, have increased the therapeutic index of radiotherapy [5–10]. Furthermore, the increased use of proton beam therapy (PBT) which displays a lower entrance dose compared to conventional radiotherapy and where the majority of the radiation dose can be specifically targeted at the tumor, can also limit the unnecessary irradiation of surrounding normal tissues leading to reduced adverse side-effects [11]. Despite this, many tumors remain intrinsically radioresistant and therefore further discovery and research into novel treatment strategies is critical to maximize the tumor-killing effect of radiotherapy, while simultaneously minimizing the toxic impact to surrounding normal tissues.

Excitingly, a recent series of research studies examining "FLASH" irradiation, a term first coined in 2014 by Favaudon et al. and the Vozenin group in Lausanne, has demonstrated that it possesses a normal tissue sparing capability while maintaining tumor cytotoxicity when compared to conventional radiotherapy in several *in vivo* models [12–15]. FLASH irradiation is the deliverance of dose at ultra-high dose rates (>40 Gy/s) that are several orders of magnitude higher than conventional dose rates (~5 Gy/min) that are used clinically. The normal tissue sparing phenotype is consequently a phenomenon called the "FLASH effect". Despite the spike in FLASH interest in recent years, the advantages of using ultra-high dose rate radiotherapy and the FLASH effect was originally reported as far back as 1960–1970, although further investigations were largely halted due to logistical difficulties translating the findings into clinical practice [16,17]. However, with the improvements in modern-day technology and a greater understanding of radiobiology, FLASH is demonstrating potential as a key tool in the future of clinical radiotherapy. Before this can happen though, it is critical that the underlying biological mechanisms and optimal beam delivery parameters are realized, as these currently remain largely uncovered.

2. The FLASH Effect

2.1. Normal Tissue Sparing

The FLASH effect is defined as the decrease in radiation-induced normal tissue toxicities with dose delivery at ultra-high dose rates (FLASH), compared to conventional dose rates used clinically. The FLASH effect has now been observed in several *in vitro* cellular models, and multiple *in vivo* animal models (summarized in Table 1). The earliest report of the FLASH effect was described in 1966 where it was discovered that mice irradiated at ultra-high dose rates had a greater survival than those irradiated at conventional dose rates [17]. A renewed interest in FLASH peaked more recently in 2014 in which C57BL/6J mice were comparatively treated with 17 Gy of either FLASH dose rates (60 Gy/s, 4.5 MeV electrons) or conventional dose rates (0.03 Gy/s, γ-rays or 4.5 MeV electrons) and the presence of lung fibrogenesis was investigated [12]. Here, mice were observed for up to 36 weeks following bilateral thorax irradiation and for the conventional dose-rate-treated mice, pulmonary fibrosis developed as early as 8 weeks and progressively worsened up to 36 weeks. By contrast, mice treated with 17 Gy FLASH dose rates were relatively free of pulmonary fibrosis, and doses of 30 Gy FLASH irradiation were required to induce significant fibrosis which was seen with 17 Gy at conventional dose rates [12]. In addition to the lungs, the FLASH effect has also been investigated and confirmed in several other organs using mouse models, including brain, skin and gut [13,18–21].

Recent data investigating the neurocognitive development of juvenile (3-week old) mice showed a radioprotective FLASH effect following 8 Gy whole brain irradiation with ultra-high FLASH dose rates (4.4×10^6 Gy/s, 6 MeV electrons) when compared to conventional dose rates (0.077 Gy/s, 6 MeV electrons) [22]. Mice were subjected to several neurocognitive tests following irradiation and in all cases the performance of the FLASH-irradiated animals was indistinguishable from the control group, whereas conventional irradiation caused a significant detriment. It was suggested that the neurocognitive benefits of FLASH irradiation was due to a preservation of the neurogenic niche and neurogenesis in the FLASH treated mice, with conventional dose-rate-irradiated mice showing considerably lower levels of immature and mature neurons four months post-irradiation. Furthermore, the long-term benefits of FLASH on pituitary function was also investigated and it was found that 8 Gy conventional dose-rate-irradiated mice had a two-fold reduction in levels of plasma growth hormone levels one-week post-treatment compared to the non-irradiated controls, whereas no significant decrease was observed in the FLASH-irradiated animals [22].

Table 1. Evidence of normal tissue sparing from FLASH irradiation.

Model (Site of Irradiation)	Assay/Endpoint	Dose (Gy)	Dose Rate (Gy/s)	Radiation Source	Reference
Mice (WBI) [1]	Memory tests, neurogenesis	10	>100	Electron	[13]
Mice (WBI) [1]	Neurocognitive tests, mature/immature neurons, growth hormone levels	8	4.4×10^6	Electron	[22]
Mice (WBI) [1]	Neurocognitive tests, dendritic spine density, microglial activation, inflammation	30	200/300	Electron	[20]
Mice (WBI) [1]	Neurocognitive tests, neuroinflammation, neuronal morphology	10	>100	Electron	[23]
Mice (WBI) [1]	Neurocognitive tests, hippocampal cell division, astrogliosis	10	37	X-ray	[21]
Mice (thorax)	Survival, dermatitis, breathing function, lung pathology	15/17.5/20	40	Proton	[24]
Mice (thorax)	Lung fibrosis, skin dermatitis, survival	15/17.5/20	40	Proton	[25]
Mice (thorax)	Lung fibrosis, TGF-β signaling, apoptosis	17	40–60	Electron	[12]
Mice (thorax)	Cellular proliferation, pro-inflammatory gene expression, DNA damage (53BP1/γH2AX foci), senescence	17	40–60	Electron	[26]
Mice (abdomen)	Survival	10–22	70–210	Electron	[19]
Mice (abdomen)	Survival, stool production, crypt cell regeneration, apoptosis, DNA damage	12–16	216	Electron	[18]
Mice (abdomen)	Intestinal crypt cell proliferation	15 Gy	78	Proton	[27]
Mice (local intestinal)	Fibrosis	18 Gy	78	Proton	[27]
Mini-pig (skin)	Skin toxicity/injury	22–34	300	Electron	[14]
Zebrafish Embryo	Morphology	8	>100	Electron	[23]

[1] WBI refers to whole brain irradiation.

Aside from mice, the FLASH effect has also been confirmed in mini-pigs and cats, higher animal models that are more similar to humans [14]. Pig skin irradiated at the same time with either FLASH (300 Gy/s) or conventional dose-rate (0.083 Gy/s) radiation were used to comparatively investigate the difference in cutaneous lesions formed. The experiment was performed by irradiating multiple 26 mm diameter circular patches on the skin of a single mini-pig with single doses ranging from 22–34 Gy. Over the course of 48 weeks, the clinical pathologies following skin irradiation including, depilation/destruction of hair follicles, fibronecrosis, epithelial ulceration, and inflammation, were observed with the conventional dose-rate treatments. However, the results observed following FLASH remained comparable to that of non-irradiated skin, showing only minor depilation and pigmentation and therefore starkly different to conventional dose rates. Furthermore, it was suggested that the dose modifying factor was >1.36 for FLASH compared to conventional dose rates using the absence of late stage necrosis at 9 months as an endpoint, whereby similar results were obtained for 34 Gy FLASH and 25 Gy at conventional dose rates [28].

2.2. Tumor Control

An important attribute of FLASH that has been reported in only a limited number of studies, is the ability to generate a similar anti-tumor response as the equivalent conventional dose-rate radiation (summarized in Table 2). This potentially means that larger doses could be administered to radioresistant tumors using FLASH radiotherapy due to the increased therapeutic index. For example, in breast (HBCx-12A) and head and neck cancer (Hep-2) xenograft models, FLASH was found to be as efficient at controlling tumor growth as conventional radiotherapy [12]. In parallel, an orthotopic lung tumor model using luciferase-positive TC-1 cells injected into C57BL/6J mice, revealed no observed difference in anti-tumor efficiency when mice were exposed to either FLASH or conventional radiotherapy. In a subsequent dose-escalation experiment, it was observed that at 8–9 weeks post-irradiation, only 20% of 15 Gy conventional dose-rate radiotherapy treated mice were tumor free, whereas 70% of the 28 Gy FLASH treated mice were free of tumors. Furthermore, the conventional radiotherapy treated mice displayed inflammatory and fibrotic remodeling, whereas the FLASH treated mice did not [12].

Table 2. Evidence of tumor control from FLASH irradiation.

Model	Assay/Endpoint	Dose (Gy)	Dose Rate (Gy/s)	Radiation Source	Reference
Mice, HBCx-12A, and Hep-2 human xenografts (local)	Tumor growth	17–25	60	Electron	[12]
Mice, orthotopic engrafted lung carcinoma luciferase+ TC-1 cells (thorax)	Tumor growth	15–28	60	Electron	[12]
Mice, ID8 syngeneic ovarian cancer (thorax)	Tumor number/weight	14	216	Electron	[18]
Mice, orthotopic engrafted Lewis lung carcinoma (thorax)	Tumor size	18	40	Proton	[29]
Mice, pancreatic MH641905 flank tumor	Tumor growth	12/15	78	Proton	[27]
Cat, nasal planum SCC (local)	Tumor growth	25–41	130–390	Electron	[14]
Human, CD30+ T-cell cutaneous lymphoma	Tumor response	15	167	Electron	[15]

In the same study described above investigating mini-pig skin, the impact of FLASH electrons on six cats with advanced squamous cell carcinoma of the nasal planum were examined, although conventional dose rates were not comparatively used [14]. Each of the six cats were given a single individual dose ranging from 25–41 Gy, and it was reported that the cats responded very well, with only mild dermatitis/mucositis observed and no late stage toxicities. In terms of the tumor, 5 of the 6 cats achieved complete remission by 16 months, with one cat experiencing a local recurrence at 21 months. Although the images presented within this study are striking, and the overall results are promising, it is limited by the lack of a control using conventional dose-rate radiotherapy, so comparisons analyzing the anti-tumor control of FLASH versus conventional dose rates was not reported.

Interestingly, the first patient to be treated with FLASH radiotherapy has been performed at the Lausanne University Hospital [15]. The patient was a 75-year old male that presented with CD30+ T-cell cutaneous lymphoma that he was diagnosed with in 1999, and over the course of a ten-year period (2008–2018), the patient had received localized radiotherapy that generally controlled the lymphoma but he experienced severed acute toxicity to the surrounding skin. For the FLASH treatment, the 3.5 cm tumor was treated with a 15 Gy total dose delivered over 10×1 μs pulses ($\geq 10^6$ Gy/s, 1.5 Gy per pulse) with a total treatment time of 90 ms. Initial tumor shrinkage began after 10 days, and a complete tumor response was achieved at 36 days that was preserved for 5 months. In terms of toxicity to the surrounding skin, redness, mild epithelitis and edema (grade 1) was observed at 10–12 days post-irradiation with a maximal reaction at 3 weeks, which was deemed mild and healed much quicker compared to the patients previous localized radiotherapy treatments. Although this report shows promising data demonstrating the feasibility of FLASH radiotherapy in the clinic, as well as the observed FLASH effect and positive patient outcome, FLASH is not yet ready to be fully translated for cancer patient treatment. Larger patient trials comparing conventional dose-rate radiotherapy with

FLASH still need to be performed, along with investigations into the appropriate radiation sources and equipment that can treat tumors other than superficial skin tumors.

3. Mechanisms Contributing to the FLASH Effect

3.1. Oxygen Depletion

The exact biochemical mechanisms that result in the FLASH effect are yet to be fully elucidated, although the current theory gaining the most ground implicates oxygen as a critical molecule in the biological response to FLASH irradiation [30]. Generally in response to ionizing radiation, indirect DNA damage occurs through the radiolysis of water and subsequent generation of reactive oxygen species (ROS), such as hydroxyl radicals, that attack the DNA [31]. It is estimated that for low linear energy transfer (LET) radiation, such as photons and electrons, 60–70% of the DNA damage induced is through generation of ROS whereas 30–40% is via direct interaction of the radiation with DNA [32,33]. The oxygen fixation hypothesis suggests that if this indirect DNA damage is a result of reaction with a free radical (e.g., hydroxyl radical), the damage is fixed due to the presence of molecular oxygen through the formation of a more damaging peroxyl radical [34]. Indeed, this is a major contributor as to why hypoxic tumors are more radioresistant than well-oxygenated tumors that display an oxygen enhancement ratio of ~2–3 [35,36]. In terms of FLASH, the oxygen depletion hypothesis suggests that the ultra-high dose rate modulates the immediate radiochemical events that occur in the irradiated tissue [37]. In this short exposure time frame, local oxygen is depleted faster than reoxygenation can occur, leading to a transient state of radiation-induced hypoxia, and therefore radioresistance and protection of the normal tissues to the FLASH irradiation [38].

The relationship between increasing dose rates and the oxygen depletion hypothesis was realized in early bacterial studies, whereby irradiation at ultra-high doses produced a survival curve indicative of those irradiated in an anaerobic (hypoxic) environment [39–42]. A subsequent study also suggested oxygen depletion as the reason why there was resistance in the tails of mice irradiated at high dose rates to epithelial necrosis [43]. However, *in vitro* evidence using mammalian cell lines to observe the FLASH effect have been lacking, with mixed reports as to whether this phenomenon was observed or not [16,44–48]. This can be explained, in part, due to FLASH studies using cells cultured in atmospheric oxygen concentrations (~20%), whereas the normal tissue sparring observed *in vivo* is generally at physiological oxygen tensions from 3–7%. This means that the FLASH doses used in these *in vitro* studies were not sufficient enough to significantly reduce the oxygen tension [28,49,50]. In support of this, a recent study using prostate cancer cells irradiated at 600 Gy/s (10 MeV electrons) showed significant survival versus conventional dose-rate irradiation (14 Gy/min) at oxygen concentrations of 1.6%, 2.7% and 4.4%, but no significant difference was seen at higher oxygen concentrations of 8.3% and 20% [45]. The oxygen depletion hypothesis does raise an important issue whether FLASH can be translated clinically. This is because for tumors that contain a heterogonous population of cells at different oxygen concentrations, the FLASH effect may be therapeutically detrimental by actually increasing tumor radioresistance. Therefore, the role of oxygen tension and the impact on FLASH radiotherapy must be explored in more detail experimentally.

3.2. ROS

Other oxygen-related products, including ROS and free radicals, have been theorized to have an altered biochemistry between normal tissue and tumors, thus contributing to the FLASH effect. In an experiment conducted using zebrafish embryos following conventional dose rate (0.1 Gy/s) or FLASH (1 pulse of 1.8×10^{-6} s) electron irradiation, it was concluded that FLASH led to less of an effect on zebrafish morphology 5 days post-fertilization due to a lower production in ROS [23]. However, if the zebrafish were incubated with the ROS scavengers, amifostine, or *N*-acetyl-cysteine 1 h prior to irradiation with conventional or FLASH radiotherapy, body length measurements 5 days post-fertilization revealed no significant difference between the FLASH or conventional radiotherapy treated zebrafish. Overall, this study demonstrated that FLASH offers radioresistance in normal tissue due to reduced ROS levels. Differences in redox chemistry and free radical production have recently been used to explain the contrasting biological effects between normal and cancer tissue following FLASH [33]. It has been hypothesized that due to normal cells having lower pro-oxidant burdens during normal redox metabolism and an increased ability to sequester labile iron compared to cancerous cells, normal cells can more effectively reduce the levels of free radicals and hydroperoxides generated from peroxidation chain reaction and Fenton type chain reactions following FLASH, therefore increasing the oxidative burden in cancer cells [33].

3.3. Immune Response

The inflammatory and immune responses have also been suggested as underlying mechanisms that contribute to the FLASH effect. Transforming growth factor beta (TGF-β), an important pro-inflammatory cytokine, has particularly been implicated to alter the effects of FLASH compared to conventional dose-rate radiotherapy. In an *in vitro* study using proton irradiation, the induction of TGF-β levels in human lung fibroblasts were significantly reduced following 20 Gy FLASH (1000 Gy/s) versus conventional dose rates (0.2 Gy/s) [47]. For the FLASH dose rate, a ~1.8-fold induction in TGF-β levels was observed 24 h post-irradiation, while a ~6.5-fold increase was observed following conventional dose rates, suggesting that FLASH may have the potential to reduce radiation-induced chronic inflammation. A reduction in TGF-β signaling was also previously reported for FLASH-irradiated mice versus conventional dose rates [12]. In support of a shifting balance from a pro-inflammatory towards an anti-inflammatory phenotype, a study investigating whole brain irradiation of C57BL/6J mice showed a reduction in hippocampal pro-inflammatory cytokine levels following FLASH compared to conventional dose-rate irradiation [20]. It was reported that at 10 weeks post-irradiation, there was a statistically significant increase in five out of ten cytokines tested following conventional dose rates, whereas FLASH generated an increase in only three cytokines.

In general, the role of TGF-β and its associated signaling pathway is known to be involved in the anti-tumor immune response following conventional radiotherapy, although the precise effects are still debated [51]. One study has reported that TGF-β is key to the radioresistance of tumor infiltrating T-cells [52], while others suggest TGF-β signaling suppresses the immune system and promotes cancer progression, pushing the need for the use of TGF-β pathway inhibitors [53]. Consequently, the alterations observed in TGF-β signaling and immune system activation following FLASH irradiation need to be carefully considered for clinical translation of FLASH, especially when radiotherapy is combined with immunotherapy. It has also been suggested that FLASH may offer an improved immune response due to the fast exposure time leading to less irradiation of circulating immune cells, although this effect may be reduced for fractionated FLASH radiotherapy [54]. Finally, it has been reported that proton irradiation of mice at FLASH dose rates showed an increased T-lymphocyte recruitment into the tumor microenvironment compared to conventional dose rates, further supporting the notation that changes in the immune response may contribute to the FLASH effect [29].

4. The Potential for FLASH Proton-Beam Therapy

Although in the simplest terms FLASH is the use of radiation dose rates multiple orders of magnitude higher than conventional dose rates, several other factors need to be taken into consideration to elicit the FLASH effect. Along with dose rate, these factors include total dose delivered, pulse rate/duration/width/number and total delivery time. Another important parameter is the irradiation source, with many of the current FLASH investigations using electron linear accelerators [12,14,15,55,56]. However, these experimental electron beams are currently limited to treatment of superficial cancers and intraoperative radiation therapy due to the low tissue penetration and limited field size of these beams (~4–20 MeV) [8]. On the other hand, clinical PBT offers a much greater tissue penetration and allow the irradiation of more deep-seated tumors. The significant advantage of PBT over conventional photon radiotherapy is that the majority of the beam energy is deposited in a narrow range called the Bragg peak following a low entrance dose (Figure 1), allowing the precise targeting of the tumor volume while sparing normal healthy tissue and organs at risk [11]. As a result, there has been an increase in the clinical use of PBT with ~150,000 cancer patients being treated to date. However, there are still significant biological uncertainties following proton irradiation largely due to the increases in LET at and around the Bragg peak, leading to changes in the DNA damage spectrum and increases in the radiobiological effectiveness [11]. Radiobiological research has also been impeded by the lack of accessible proton facilities for *in vitro* and *in vivo* experimentation. Despite this, the promise of proton FLASH has been invested in by multiple companies such as Varian, IBA and Mevion who are funding both the development of FLASH PBT machinery and research [57].

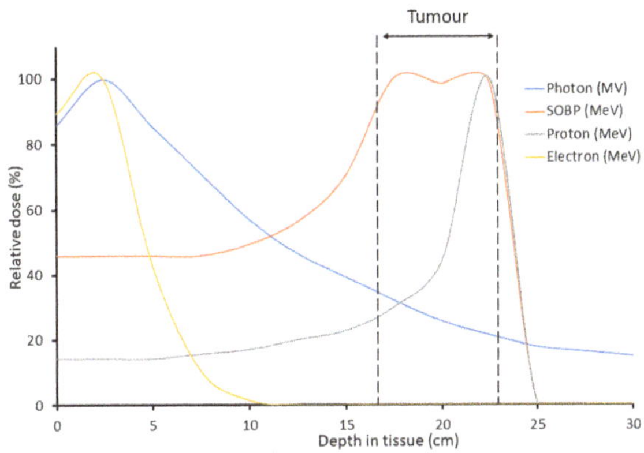

Figure 1. Comparison of depth–dose distribution of photons, protons, and electrons relative to a target tumor. A spread-out Bragg peak (SOBP) from several modulated proton beams is also shown, but which demonstrates the precise targeting of the tumor using PBT.

Studies Investigating FLASH Protons

Current research on FLASH protons conducted *in vitro* and *in vivo* has revealed mixed information as to whether the FLASH effect was induced or not. In general, *in vitro* studies investigating FLASH protons have produced a lack of positive results observing the FLASH effect, particularly in terms of acute endpoints, such as clonogenic survival, γH2AX foci formation and cell cycle arrest. Data from such studies has been recently reviewed, and only one of ten studies demonstrated evidence of a FLASH effect [58]. Interestingly, though, all these studies were performed at aerobic oxygen levels (21%) and it is likely that this is the major reason for the absence of the FLASH effect. It is therefore clear that in order to investigate the FLASH effect *in vitro* with protons, experiments at varying oxygen tensions need to be performed, similar to those performed with electrons [45,50]. Regarding the one study reporting positive results *in vitro*, this was conducted using normal human lung fibroblasts (IMR90) and comparing conventional dose rate (0.05 Gy/s) and FLASH (100 or 1000 Gy/s) proton irradiation (4.5 MeV). It was observed that the increasing dose rate reduced the number of prematurely senescent cells (measured using β-galactosidase positive cells) and also reduced the induction of TGF-β expression, suggesting a long-term role of the FLASH effect particularly on chronic inflammation [47]. However, the difference in proton dose rate was concluded to have little effect on acute biological outcomes, including clonogenic survival and γH2AX foci formation. In fact, there was an indication of decreased clonogenic survival at both the FLASH dose rates compared to the conventional dose rate, albeit the data was not statistically significant. Interestingly, significantly less γH2AX foci formation was observed only following 20 Gy of 1000 Gy/s FLASH compared to 100 Gy/s FLASH and the conventional dose rates, potentially suggesting a role for reduced yields of double strand breaks (DSBs) and an altered DNA repair capacity following FLASH protons but only at very high doses [47]. It is worth noting that changes in the DNA damage response following FLASH in general are surprisingly understudied considering this is a major factor in radiobiology. Indeed, impacts on endpoints such as cell cycle progression, chromosomal aberrations, ROS levels, as well as DNA damage signaling and DNA damage foci relating to DSB formation following FLASH could be key to further understanding the underlying mechanisms that cause the FLASH effect. Another important consideration, which has not been investigated up to now, is whether the FLASH effect is still observed with increasing LET at and around the Bragg peak, and whether the profile of DSBs and complex DNA damage induced is altered [11]. We have recently demonstrated that complex DNA damage, containing multiple DNA lesions including oxidative DNA base damage and DNA single strand breaks within close proximity (1–2 helical turns of the DNA), along with DSBs, are a critical factor in radiation-induced cell killing and which triggers a specific cellular DNA damage response [59,60]. Therefore, it is important to determine the DNA damage profile with FLASH protons at higher LET.

Regarding *in vivo* studies, one report has investigated morphological changes in zebrafish embryos with either conventional dose rates (5 Gy/min) or FLASH (100 Gy/s) [61] but could not replicate FLASH sparing with protons that was previously observed with electrons [23]. A possible reason for not observing the FLASH effect was suggested due to the proton-beam pulse characteristics that delivered a lower maximum dose rate per pulse. Micro-pulse dose rates delivered by the cyclotron were ~10^3 Gy/s, whereas electron macro-pulse dose rates have been reported as ~10^7 Gy/s. Furthermore, the zebrafish embryos in this study were irradiated at a later developmental stage post-fertilization compared to the previous electron-focused study (~24 hpf vs. 4 hpf), potentially making the zebrafish less sensitive to FLASH PBT irradiation and contributing to the lack of an observed FLASH effect. However in general, more recent *in vivo* studies investigating FLASH PBT have yielded much more positive findings by observing the FLASH effect, and associated tumor control compared to using conventional dose rates (summarized in Table 3). In an innovative study, a clinical 230 MeV proton accelerator using double-scattered protons under CT guidance was designed to deliver FLASH dose rates of 60–100 Gy/s and conventional dose rates of 0.5–1 Gy/s [27]. Here, 8–10-week-old C57BL/6J mice were subjected to whole abdominal irradiation with 15 Gy of either FLASH (78 Gy/s) or conventional dose-rate (0.9 Gy/s) protons and intestinal segments were harvested 3.5 days post-irradiation. It was found that FLASH significantly reduced the loss of proliferating intestinal crypt cells versus conventional dose-rate radiotherapy. In addition, mice irradiated with 18 Gy protons focused on the intestines and harvested 8 weeks post-irradiation, revealed that conventional dose-rate-irradiated mice had considerably increased fibrosis compared to the FLASH-irradiated mice. The degree of fibrosis following FLASH PBT treatment was actually comparable to that of unirradiated mice. Finally, MH641905 pancreatic tumor cells injected into the mice to generate flank tumors were irradiated with 12 or 18 Gy FLASH alongside conventional dose-rate irradiation, and no significant difference between tumor growth delay was observed between the treatments. Therefore, these results demonstrate normal tissue sparring along with effective tumor control with FLASH PBT, at least for gastrointestinal tumors [27].

Several other mouse model studies have shown the benefits of FLASH PBT. In the first, whole thorax irradiation (15–20 Gy) was delivered to C57BL/6J mice using FLASH (40 Gy/s) or conventional (1 Gy/s) dose rates, and responses analyzed at 8–34 weeks post-irradiation [25]. FLASH tissue sparring was observed through a 30% reduction in lung fibrosis, reduced skin dermatitis, and improved overall survival in the FLASH PBT treated mice. In addition, genome-wide microarray analysis was performed in order to uncover the underlying mechanisms involved in the FLASH effect, with this demonstrating that DNA repair, inflammation, and the immune response as the major pathways differentially regulated between the two PBT dose rates. In a related study, the whole thorax region of C57BL/6J mice were irradiated with 15–20 Gy using a clinical pencil-beam scanning PBT system with either FLASH (40 Gy/s) or conventional dose-rate protons (0.5 Gy/s), and the mice analyzed at 8–36 weeks post-irradiation. Surprisingly, gender-specific differences were observed with only the female mice cohort showing improved outcomes following FLASH [24]. Nevertheless, these mice displayed better breathing function, reduced dermatitis, altered lung pathology, and greater overall survival with FLASH radiotherapy compared to using conventional dose rates. Finally, a study injected Lewis lung carcinoma (LLC) cells into the left lung of C57BL/6J mice and the whole lungs were irradiated with an 18 Gy dose of protons delivered using a clinical pencil-beam scanning PBT system at either FLASH or conventional dose rates. Tumor sizes were measured 7 days post-irradiation by imaging, and then at 10 days post-irradiation when the mice were sacrificed. Remarkably, it was observed that the lung tumors in the FLASH PBT irradiated mice were significantly smaller in comparison to conventional dose rates, suggesting in this case that FLASH protons have an increased tumor control capability compared to protons used at conventional dose rates [29]. Nevertheless, additional and more comprehensive *in vivo* studies examining FLASH PBT using the appropriate tumor models need to be conducted.

Table 3. Summary of outcomes in in vivo studies comparing FLASH and conventional dose-rate PBT.

Model	Dose (Gy)	FLASH Dose-Rate (Gy/s)	Outcome	Reference
Zebrafish embryo	0–43	100	No survival difference	[61]
Mice (thorax)	15/17.5/20	40	Normal tissue protection with FLASH	[24]
Mice (thorax)	15/17.5/20	40	Normal tissue protection with FLASH	[25]
Mice (abdomen)	15	78	Normal tissue protection with FLASH	[27]
Mice (local intestinal)	18	78	Normal tissue protection with FLASH	[27]
Mice, orthotopic engrafted Lewis lung carcinoma (thorax)	18	40	Improved tumor control with FLASH, increased T-lymphocyte tumor infiltration	[29]
Mice, pancreatic MH641905 flank tumor	12/15	78	No difference in tumor control	[27]
Mice, FaDu head, and neck squamous cell carcinoma transplantation	17.4	>10^9	No difference in tumor control	[62]

5. Conclusions

FLASH radiotherapy is an exciting new treatment strategy that has the potential to change the future of clinical cancer treatment. The use of ultra-high dose rates several orders of magnitude higher than conventional dose rates generates a phenomenon known as the "FLASH effect", through which sparing of normal healthy tissue is observed, while maintaining equivalent tumor control properties compared to conventional dose-rate radiotherapy. Current radiotherapy regimes are limited by the tolerance of surrounding normal tissues to radiation-induced toxicities, meaning that some radioresistant tumors may not receive the required dose of radiation for the treatment to be effective. However, FLASH radiotherapy has the potential to overcome this and allow an increased radiation dose delivered to tumors while keeping the toxicity to surrounding healthy tissues low. Remarkably, the first patient with CD30+ T-cell cutaneous lymphoma has recently been treated using FLASH radiotherapy. It is clear that oxygen plays a key role in the underlying biological mechanism resulting in the FLASH effect. In fact, multiple studies have found that the ultra-high dose rate radiation is able to deplete local oxygen and induce a short-lived protective hypoxic environment within the normal healthy tissues that increases radioresistance. Furthermore, theories have suggested changes in ROS and redox chemistry between normal and tumor cells following FLASH dose rates. Although the oxygen depletion hypothesis is the most popular current explanation for the FLASH effect, other phenomena may play an important role, including the immune response and tumor microenvironment that require further examination. Despite this, an area that has been surprisingly understudied is whether there are any differences in the DNA damage profile (e.g., actual numbers and ratios of DNA base damage, DNA single strand breaks and DSBs) and the subsequent DNA damage response following FLASH irradiation (photons/electrons and PBT), in comparison to conventional dose rates. Therefore, future studies should focus on quantifying the levels and persistence of particularly DSBs and complex DNA damage (measured directly or using DNA damage foci) that are the key drivers contributing to the therapeutic effect of radiotherapy, in the appropriate 3D *in vitro* (spheroids/organoids) and/or *in vivo* models. Additionally, the DNA repair pathways responsive to FLASH-induced DNA damage, particularly non-homologous end-joining or homologous recombination involved in DSB repair, should be monitored. It is important to consider both the FLASH effect on sparing of normal cells/tissues, but also its impact in tumor cell killing, as well as appreciation of the oxygen levels at which the experiments are conducted. Nevertheless, it is likely that a myriad of biological changes are observed following FLASH irradiation.

Although the FLASH effect in theory appears revolutionary, translation into the clinic is still difficult at this early stage. This is because several factors contribute to the FLASH effect, including total dose, pulse rate, pulse duration, pulse width, pulse number, and total delivery time. Based on *in vitro* and *in vivo* reported data, doses upwards of tens of Gy are required to induce FLASH radioprotection, which can be too high to treat a significant number of patients clinically. Furthermore, questions arise as to whether a fractionation regime for FLASH to deliver higher doses will be able to induce a FLASH effect. Another question that needs to be answered is which source of radiation is best to deliver FLASH radiotherapy. Much of the current data has used electron sources, however this is currently limited to treatment of superficial cancers or intraoperative radiation therapy. PBT may offer the best solution to be able to treat some deep-seated tumors, and there are several high-energy clinical PBT facilities already in place that can be modified to generate FLASH dose rates [63]. Furthermore, several innovative set-ups are already being tested using modified clinically available PBT beams [27,64]. However, implementation of FLASH PBT still has its technical limitations. To deliver protons to a large tumor volume, the proton beam must be scattered which may cause particle loss and decrease the total dose delivered. Pencil-beam scanning enables the delivery of ultra-high dose rates per individual spot, however the time taken to perform this is extended, therefore reducing the total dose rate which may not be enough to induce the FLASH effect [37]. Furthermore, research output using protons has produced largely mixed results, and it is also still unknown how the increasing LET at and around the Bragg peak will impact on the FLASH effect. Therefore, significantly more research into FLASH PBT is required, particularly investigations at physiological oxygen concentrations, for this to be potentially translated into the clinic for the benefit of cancer patients.

Author Contributions: Conceptualization, J.L.P.; writing—original draft preparation, J.R.H.; writing—review and editing, J.R.H. and J.L.P.; supervision, J.L.P.; funding acquisition, J.L.P. All authors have read and agreed to the published version of the manuscript.

Funding: This research was funded by the Science and Technology Facilities Council (ST/T002158/1) and by the Clatterbridge Cancer Centre NHS Foundation Trust, awarded to J.L.P.

Conflicts of Interest: The authors declare no conflict of interest.

Abbreviations

DSB	DNA double strand break
LET	Linear energy transfer
PBT	Proton-beam therapy
ROS	Reactive oxygen species
TGF-β	Transforming growth factor beta

References

1. Moding, E.J.; Kastan, M.B.; Kirsch, D.G. Strategies for optimizing the response of cancer and normal tissues to radiation. *Nat. Rev. Drug. Discov.* **2013**, *12*, 526–542. [CrossRef] [PubMed]
2. Berkey, F.J. Managing the adverse effects of radiation therapy. *Am. Fam. Physician* **2010**, *82*, 381–388, 394.
3. Siddiqui, F.; Movsas, B. Management of Radiation Toxicity in Head and Neck Cancers. *Semin. Radiat. Oncol.* **2017**, *27*, 340–349. [CrossRef]

4. Kurup, A.; Pasternak, J.; Taylor, R.; Murgatroyd, L.; Ettlinger, O.; Shields, W.; Nevay, L.; Gruber, S.; Pozimski, J.; Lau, H.T.; et al. Simulation of a radiobiology facility for the Centre for the Clinical Application of Particles. *Phys. Med.* **2019**, *65*, 21–28. [CrossRef] [PubMed]
5. Kruger, S.; Ilmer, M.; Kobold, S.; Cadilha, B.L.; Endres, S.; Ormanns, S.; Schuebbe, G.; Renz, B.W.; D'Haese, J.G.; Schloesser, H.; et al. Advances in cancer immunotherapy 2019 - latest trends. *J. Exp. Clin. Cancer Res.* **2019**, *38*, 268. [CrossRef] [PubMed]
6. Rosenberg, S.A. IL-2: The first effective immunotherapy for human cancer. *J. Immunol.* **2014**, *192*, 5451–5458. [CrossRef]
7. Al-Lazikani, B.; Banerji, U.; Workman, P. Combinatorial drug therapy for cancer in the post-genomic era. *Nat. Biotechnol.* **2012**, *30*, 679–692. [CrossRef]
8. Montay-Gruel, P.; Meziani, L.; Yakkala, C.; Vozenin, M.C. Expanding the therapeutic index of radiation therapy by normal tissue protection. *Br. J. Radiol.* **2019**, *92*, 20180008. [CrossRef]
9. Garibaldi, C.; Jereczek-Fossa, B.A.; Marvaso, G.; Dicuonzo, S.; Rojas, D.P.; Cattani, F.; Starzynska, A.; Ciardo, D.; Surgo, A.; Leonardi, M.C.; et al. Recent advances in radiation oncology. *Ecancermedicalscience* **2017**, *11*, 785. [CrossRef]
10. Beaton, L.; Bandula, S.; Gaze, M.N.; Sharma, R.A. How rapid advances in imaging are defining the future of precision radiation oncology. *Br. J. Cancer* **2019**, *120*, 779–790. [CrossRef]
11. Vitti, E.T.; Parsons, J.L. The Radiobiological Effects of Proton Beam Therapy: Impact on DNA Damage and Repair. *Cancers* **2019**, *11*, 946. [CrossRef] [PubMed]
12. Favaudon, V.; Caplier, L.; Monceau, V.; Pouzoulet, F.; Sayarath, M.; Fouillade, C.; Poupon, M.F.; Brito, I.; Hupe, P.; Bourhis, J.; et al. Ultrahigh dose-rate FLASH irradiation increases the differential response between normal and tumor tissue in mice. *Sci. Transl. Med.* **2014**, *6*, 245ra293. [CrossRef]
13. Montay-Gruel, P.; Petersson, K.; Jaccard, M.; Boivin, G.; Germond, J.F.; Petit, B.; Doenlen, R.; Favaudon, V.; Bochud, F.; Bailat, C.; et al. Irradiation in a flash: Unique sparing of memory in mice after whole brain irradiation with dose rates above 100Gy/s. *Radiother. Oncol.* **2017**, *124*, 365–369. [CrossRef] [PubMed]
14. Vozenin, M.C.; De Fornel, P.; Petersson, K.; Favaudon, V.; Jaccard, M.; Germond, J.F.; Petit, B.; Burki, M.; Ferrand, G.; Patin, D.; et al. The Advantage of FLASH Radiotherapy Confirmed in Mini-pig and Cat-cancer Patients. *Clin. Cancer Res.* **2019**, *25*, 35–42. [CrossRef] [PubMed]
15. Bourhis, J.; Sozzi, W.J.; Jorge, P.G.; Gaide, O.; Bailat, C.; Duclos, F.; Patin, D.; Ozsahin, M.; Bochud, F.; Germond, J.F.; et al. Treatment of a first patient with FLASH-radiotherapy. *Radiother. Oncol.* **2019**, *139*, 18–22. [CrossRef]
16. Berry, R.J. Effects of radiation dose-rate from protracted, continuous irradiation to ultra-high dose-rates from pulsed accelerators. *Br. Med. Bull.* **1973**, *29*, 44–47. [CrossRef]
17. Hornsey, S.; Alper, T. Unexpected dose-rate effect in the killing of mice by radiation. *Nature* **1966**, *210*, 212–213. [CrossRef]
18. Levy, K.; Natarajan, S.; Wang, J.; Chow, S.; Eggold, J.; Loo, P.; Manjappa, R.; Lartey, F.; Schüler, E.; Skinner, L.; et al. FLASH irradiation enhances the therapeutic index of abdominal radiotherapy in mice. *bioRxiv [Preprint]* **2020**, 1–35. [CrossRef]
19. Loo, B.W.; Schuler, E.; Lartey, F.M.; Rafat, M.; King, G.J.; Trovati, S.; Koong, A.C.; Maxim, P.G. (P003) Delivery of Ultra-Rapid Flash Radiation Therapy and Demonstration of Normal Tissue Sparing After Abdominal Irradiation of Mice. *Int. J. Radiat. Oncol. Biol. Phys.* **2017**, *98*. [CrossRef]
20. Simmons, D.A.; Lartey, F.M.; Schuler, E.; Rafat, M.; King, G.; Kim, A.; Ko, R.; Semaan, S.; Gonzalez, S.; Jenkins, M.; et al. Reduced cognitive deficits after FLASH irradiation of whole mouse brain are associated with less hippocampal dendritic spine loss and neuroinflammation. *Radiother. Oncol.* **2019**, *139*, 4–10. [CrossRef]
21. Montay-Gruel, P.; Bouchet, A.; Jaccard, M.; Patin, D.; Serduc, R.; Aim, W.; Petersson, K.; Petit, B.; Bailat, C.; Bourhis, J.; et al. X-rays can trigger the FLASH effect: Ultra-high dose-rate synchrotron light source prevents normal brain injury after whole brain irradiation in mice. *Radiother. Oncol.* **2018**, *129*, 582–588. [CrossRef] [PubMed]
22. Alaghband, Y.; Cheeks, S.N.; Allen, B.D.; Montay-Gruel, P.; Doan, N.L.; Petit, B.; Jorge, P.G.; Giedzinski, E.; Acharya, M.M.; Vozenin, M.C.; et al. Neuroprotection of Radiosensitive Juvenile Mice by Ultra-High Dose Rate FLASH Irradiation. *Cancers* **2020**, *12*, 1671. [CrossRef]

23. Montay-Gruel, P.; Acharya, M.M.; Petersson, K.; Alikhani, L.; Yakkala, C.; Allen, B.D.; Ollivier, J.; Petit, B.; Jorge, P.G.; Syage, A.R.; et al. Long-term neurocognitive benefits of FLASH radiotherapy driven by reduced reactive oxygen species. *Proc. Natl. Acad. Sci. USA* **2019**, *116*, 10943–10951. [CrossRef] [PubMed]
24. Abel, E.; Girdhani, S.; Jackson, I.; Eley, J.; Katsis, A.; Marshall, A.; Rodriguez, A.; Senapati, S.; Bentzen, S.M.; Vujaskovic, Z.; et al. Characterization of Radiation-Induced Lung Fibrosis and Mode of Cell Death Using Single and Multi-Pulsed Proton Flash Irradiation. *Int. J. Radiat. Oncol. Biol. Phys.* **2019**, *105*, E652–E653. [CrossRef]
25. Girdhani, S.; Abel, E.; Katsis, A.; Rodriquez, A.; Senapati, S.; KuVillanueva, A.; Jackson, I.; Eley, J.; Vujaskovic, Z.; Parry, R. Abstract LB-280: FLASH: A novel paradigm changing tumor irradiation platform that enhances therapeutic ratio by reducing normal tissue toxicity and activating immune pathways. *Cancer Res.* **2019**, *79*, LB-280. [CrossRef]
26. Fouillade, C.; Curras-Alonso, S.; Giuranno, L.; Quelennec, E.; Heinrich, S.; Bonnet-Boissinot, S.; Beddok, A.; Leboucher, S.; Karakurt, H.U.; Bohec, M.; et al. FLASH Irradiation Spares Lung Progenitor Cells and Limits the Incidence of Radio-induced Senescence. *Clin. Cancer Res.* **2020**, *26*, 1497–1506. [CrossRef] [PubMed]
27. Diffenderfer, E.S.; Verginadis, I.I.; Kim, M.M.; Shoniyozov, K.; Velalopoulou, A.; Goia, D.; Putt, M.; Hagan, S.; Avery, S.; Teo, K.; et al. Design, Implementation, and in Vivo Validation of a Novel Proton FLASH Radiation Therapy System. *Int J. Radiat. Oncol. Biol. Phys.* **2020**, *106*, 440–448. [CrossRef]
28. Vozenin, M.C.; Hendry, J.H.; Limoli, C.L. Biological Benefits of Ultra-high Dose Rate FLASH Radiotherapy: Sleeping Beauty Awoken. *Clin. Oncol.* **2019**, *31*, 407–415. [CrossRef]
29. Rama, N.; Saha, T.; Shukla, S.; Goda, C.; Milewski, D.; Mascia, A.E.; Vatner, R.E.; Sengupta, D.; Katsis, A.; Abel, E.; et al. Improved Tumor Control Through T-cell Infiltration Modulated by Ultra-High Dose Rate Proton FLASH Using a Clinical Pencil Beam Scanning Proton System. *Int. J. Radiat. Oncol. Biol. Phys.* **2019**, *105*, S164–S165. [CrossRef]
30. Wilson, J.D.; Hammond, E.M.; Higgins, G.S.; Petersson, K. Ultra-High Dose Rate (FLASH) Radiotherapy: Silver Bullet or Fool's Gold? *Front. Oncol.* **2019**, *9*, 1563. [CrossRef]
31. Morgan, W.F.; Sowa, M.B. Effects of ionizing radiation in nonirradiated cells. *Proc. Natl. Acad. Sci. USA* **2005**, *102*, 14127–14128. [CrossRef] [PubMed]
32. Santivasi, W.L.; Xia, F. Ionizing radiation-induced DNA damage, response, and repair. *Antioxid. Redox. Signal.* **2014**, *21*, 251–259. [CrossRef] [PubMed]
33. Spitz, D.R.; Buettner, G.R.; Petronek, M.S.; St-Aubin, J.J.; Flynn, R.T.; Waldron, T.J.; Limoli, C.L. An integrated physico-chemical approach for explaining the differential impact of FLASH versus conventional dose rate irradiation on cancer and normal tissue responses. *Radiother. Oncol.* **2019**, *139*, 23–27. [CrossRef] [PubMed]
34. Grimes, D.R.; Partridge, M. A mechanistic investigation of the oxygen fixation hypothesis and oxygen enhancement ratio. *Biomed. Phys. Eng. Express* **2015**, *1*, 045209. [CrossRef]
35. Antonovic, L.; Lindblom, E.; Dasu, A.; Bassler, N.; Furusawa, Y.; Toma-Dasu, I. Clinical oxygen enhancement ratio of tumors in carbon ion radiotherapy: The influence of local oxygenation changes. *J. Radiat. Res.* **2014**, *55*, 902–911. [CrossRef] [PubMed]
36. Bristow, R.G.; Hill, R.P. Hypoxia and metabolism. Hypoxia, DNA repair and genetic instability. *Nat. Rev. Cancer* **2008**, *8*, 180–192. [CrossRef]
37. Bourhis, J.; Montay-Gruel, P.; Goncalves Jorge, P.; Bailat, C.; Petit, B.; Ollivier, J.; Jeanneret-Sozzi, W.; Ozsahin, M.; Bochud, F.; Moeckli, R.; et al. Clinical translation of FLASH radiotherapy: Why and how? *Radiother. Oncol.* **2019**, *139*, 11–17. [CrossRef]
38. Pratx, G.; Kapp, D.S. A computational model of radiolytic oxygen depletion during FLASH irradiation and its effect on the oxygen enhancement ratio. *Phys. Med. Biol.* **2019**, *64*, 185005. [CrossRef]
39. Dewey, D.L.; Boag, J.W. Modification of the oxygen effect when bacteria are given large pulses of radiation. *Nature* **1959**, *183*, 1450–1451. [CrossRef]
40. Dewey, D.L. An oxygen-dependent X-ray dose-rate effect in Serratia marcescens. *Radiat. Res.* **1969**, *38*, 467–474. [CrossRef]
41. Epp, E.R.; Weiss, H.; Santomasso, A. The oxygen effect in bacterial cells irradiated with high-intensity pulsed electrons. *Radiat. Res.* **1968**, *34*, 320–325. [CrossRef] [PubMed]
42. Phillips, T.L.; Worsnop, B.R. Ultra-high dose-rate effects in radiosensitive bacteria. *Int. J. Radiat. Biol. Relat. Stud. Phys. Chem. Med.* **1969**, *14*, 573–575. [CrossRef] [PubMed]

43. Hendry, J.H.; Moore, J.V.; Hodgson, B.W.; Keene, J.P. The constant low oxygen concentration in all the target cells for mouse tail radionecrosis. *Radiat. Res.* **1982**, *92*, 172–181. [CrossRef]
44. Epp, E.R.; Weiss, H.; Djordjevic, B.; Santomasso, A. The radiosensitivity of cultured mammalian cells exposed to single high intensity pulses of electrons in various concentrations of oxygen. *Radiat. Res.* **1972**, *52*, 324–332. [CrossRef]
45. Adrian, G.; Konradsson, E.; Lempart, M.; Back, S.; Ceberg, C.; Petersson, K. The FLASH effect depends on oxygen concentration. *Br. J. Radiol.* **2020**, *93*, 20190702. [CrossRef] [PubMed]
46. Cygler, J.; Klassen, N.V.; Ross, C.K.; Bichay, T.J.; Raaphorst, G.P. The survival of aerobic and anoxic human glioma and melanoma cells after irradiation at ultrahigh and clinical dose rates. *Radiat. Res.* **1994**, *140*, 79–84. [CrossRef]
47. Buonanno, M.; Grilj, V.; Brenner, D.J. Biological effects in normal cells exposed to FLASH dose rate protons. *Radiother. Oncol.* **2019**, *139*, 51–55. [CrossRef]
48. Town, C.D. Radiobiology. Effect of high dose rates on survival of mammalian cells. *Nature* **1967**, *215*, 847–848. [CrossRef]
49. McKeown, S.R. Defining normoxia, physoxia and hypoxia in tumours-implications for treatment response. *Br. J. Radiol.* **2014**, *87*, 20130676. [CrossRef]
50. Petersson, K.; Adrian, G.; Butterworth, K.; McMahon, S.J. A Quantitative Analysis of the Role of Oxygen Tension in FLASH Radiation Therapy. *Int. J. Radiat. Oncol. Biol. Phys.* **2020**, *107*, 539–547. [CrossRef]
51. Vanpouille-Box, C.; Diamond, J.M.; Pilones, K.A.; Zavadil, J.; Babb, J.S.; Formenti, S.C.; Barcellos-Hoff, M.H.; Demaria, S. TGFbeta Is a Master Regulator of Radiation Therapy-Induced Antitumor Immunity. *Cancer Res.* **2015**, *75*, 2232–2242. [CrossRef] [PubMed]
52. Arina, A.; Beckett, M.; Fernandez, C.; Zheng, W.; Pitroda, S.; Chmura, S.J.; Luke, J.J.; Forde, M.; Hou, Y.; Burnette, B.; et al. Tumor-reprogrammed resident T cells resist radiation to control tumors. *Nat. Commun.* **2019**, *10*, 3959. [CrossRef] [PubMed]
53. Holmgaard, R.B.; Schaer, D.A.; Li, Y.; Castaneda, S.P.; Murphy, M.Y.; Xu, X.; Inigo, I.; Dobkin, J.; Manro, J.R.; Iversen, P.W.; et al. Targeting the TGFbeta pathway with galunisertib, a TGFbetaRI small molecule inhibitor, promotes anti-tumor immunity leading to durable, complete responses, as monotherapy and in combination with checkpoint blockade. *J. Immunother. Cancer* **2018**, *6*, 47. [CrossRef] [PubMed]
54. Durante, M.; Brauer-Krisch, E.; Hill, M. Faster and safer? FLASH ultra-high dose rate in radiotherapy. *Br. J. Radiol.* **2018**, *91*, 20170628. [CrossRef]
55. Schuler, E.; Trovati, S.; King, G.; Lartey, F.; Rafat, M.; Villegas, M.; Praxel, A.J.; Loo, B.W., Jr.; Maxim, P.G. Experimental Platform for Ultra-high Dose Rate FLASH Irradiation of Small Animals Using a Clinical Linear Accelerator. *Int. J. Radiat. Oncol. Biol. Phys.* **2017**, *97*, 195–203. [CrossRef]
56. Lempart, M.; Blad, B.; Adrian, G.; Back, S.; Knoos, T.; Ceberg, C.; Petersson, K. Modifying a clinical linear accelerator for delivery of ultra-high dose rate irradiation. *Radiother. Oncol.* **2019**, *139*, 40–45. [CrossRef]
57. Van Marlen, P.; Dahele, M.; Folkerts, M.; Abel, E.; Slotman, B.J.; Verbakel, W. Bringing FLASH to the Clinic: Treatment Planning Considerations for Ultrahigh Dose-Rate Proton Beams. *Int. J. Radiat. Oncol. Biol. Phys.* **2020**, *106*, 621–629. [CrossRef]
58. Colangelo, N.W.; Azzam, E.I. The Importance and Clinical Implications of FLASH Ultra-High Dose-Rate Studies for Proton and Heavy Ion Radiotherapy. *Radiat. Res.* **2020**, *193*, 1–4. [CrossRef]
59. Carter, R.J.; Nickson, C.M.; Thompson, J.M.; Kacperek, A.; Hill, M.A.; Parsons, J.L. Complex DNA Damage Induced by High Linear Energy Transfer Alpha-Particles and Protons Triggers a Specific Cellular DNA Damage Response. *Int. J. Radiat. Oncol. Biol. Phys.* **2018**, *100*, 776–784. [CrossRef]
60. Carter, R.J.; Nickson, C.M.; Thompson, J.M.; Kacperek, A.; Hill, M.A.; Parsons, J.L. Characterisation of Deubiquitylating Enzymes in the Cellular Response to High-LET Ionizing Radiation and Complex DNA Damage. *Int. J. Radiat. Oncol. Biol. Phys.* **2019**, *104*, 656–665. [CrossRef]
61. Beyreuther, E.; Brand, M.; Hans, S.; Hideghety, K.; Karsch, L.; Lessmann, E.; Schurer, M.; Szabo, E.R.; Pawelke, J. Feasibility of proton FLASH effect tested by zebrafish embryo irradiation. *Radiother. Oncol.* **2019**, *139*, 46–50. [CrossRef] [PubMed]
62. Zlobinskaya, O.; Siebenwirth, C.; Greubel, C.; Hable, V.; Hertenberger, R.; Humble, N.; Reinhardt, S.; Michalski, D.; Roper, B.; Multhoff, G.; et al. The effects of ultra-high dose rate proton irradiation on growth delay in the treatment of human tumor xenografts in nude mice. *Radiat. Res.* **2014**, *181*, 177–183. [CrossRef] [PubMed]

63. Van de Water, S.; Safai, S.; Schippers, J.M.; Weber, D.C.; Lomax, A.J. Towards FLASH proton therapy: The impact of treatment planning and machine characteristics on achievable dose rates. *Acta Oncol.* **2019**, *58*, 1463–1469. [CrossRef] [PubMed]
64. Patriarca, A.; Fouillade, C.; Auger, M.; Martin, F.; Pouzoulet, F.; Nauraye, C.; Heinrich, S.; Favaudon, V.; Meyroneinc, S.; Dendale, R.; et al. Experimental Set-up for FLASH Proton Irradiation of Small Animals Using a Clinical System. *Int. J. Radiat. Oncol. Biol. Phys.* **2018**, *102*, 619–626. [CrossRef]

International Journal of
Molecular Sciences

MDPI

Review

Stage-Specific Effects of Ionizing Radiation during Early Development

Yasuko Honjo * and Tatsuo Ichinohe *

Department of Hematology and Oncology, Research Institute for Radiation Biology and Medicine (RIRBM), Hiroshima University, Hiroshima 734-8553, Japan
* Correspondence: yhonjo@hiroshima-u.ac.jp (Y.H.); nohe@hiroshima-u.ac.jp (T.I.)

Received: 7 May 2020; Accepted: 30 May 2020; Published: 1 June 2020

Abstract: Early embryonic cells are sensitive to genotoxic stressors such as ionizing radiation. However, sensitivity to these stressors varies depending on the embryonic stage. Recently, the sensitivity and response to ionizing radiation were found to differ during the preimplantation period. The cellular and molecular mechanisms underlying the change during this period are beginning to be elucidated. In this review, we focus on the changes in radio-sensitivity and responses to ionizing radiation during the early developmental stages of the preimplantation (before gastrulation) period in mammals, *Xenopus*, and fish. Furthermore, we discuss the underlying cellular and molecular mechanisms and the similarities and differences between species.

Keywords: embryo; gastrulation; preimplantation; DNA repair; apoptosis

1. Introduction

Since the discovery of X-ray radiation by Röntgen in 1895, ionizing radiation has been used extensively in medical applications and in crystallography. However, after its discovery, there were increasing studies indicating that ionizing radiation could be very harmful to living organisms. We now have a basic understanding of how ionizing radiation affects animal cells and know that not all cells show the same sensitivity to radiation [1].

Ionizing radiation induces DNA lesions directly or indirectly by formation of hydroxyl free radicals. The type of induced DNA lesions are double strand breaks (DSBs) and single strand breaks (SSBs) as well as cluster of DNA lesions, i.e., two or more individual lesions within one or two helical turns of the DNA [1–4]. Both DSBs and clusters of DNA lesions are considered significantly damaging for all cell type. DNA breaks by irradiation typically raise the cellular responses beginning by detection of DNA damages and the cell cycle arrest. The inefficient repair of DNA lesions is widely considered a critical initiation event for mutagenesis, genomic instability and cell death.

Bergonie–Tribondeau's law states that the radio-sensitivity of a tissue is directly proportional to the number of undifferentiated cells in the tissue, their mitotic activity, and the length of time that they spend in proliferation. This fundamental law of radiation biology also applies to embryos during development. Early-stage embryos are often extremely sensitive to the effects of ionizing radiation [5,6]. Although embryos are generally very sensitive to radiation, sensitivity varies across different stages of embryonic development. Russell and Russell (1954) concisely reviewed the effects of radiation in human and rodent embryos and fetuses [1,7] and showed that there are three different periods of differential radiation sensitivity during mouse and human development: preimplantation, organogenesis, and fetus. Each period can be identified by the morphological effects associated with exposure to radiation. Embryos irradiated during the preimplantation period, namely, the period before implantation, follow an "all-or-none" rule: some irradiated embryos grow perfectly normally if they survive the gestation period, but others die or fail to implant. Embryos irradiated during organogenesis, during which the major organs develop, show the most distinctive phenotypes that include abnormal head formation and temporary small body size. Depending on the dosage of irradiation, some cases show microcephaly and mental retardation. The fetus period is marked by growth, and exposure to ionizing radiation during this period causes permanent growth retardation in mice and rats. In humans, some fetuses exposed to irradiation have shown mental retardation.

Recently, the specific cellular and molecular mechanisms underlying the response to ionizing radiation during the preimplantation period have been analyzed [8–11] and may explain the differences in sensitivity and response to ionizing radiation between each developmental period. Given that there are detailed reviews about the mechanisms of the response to ionizing radiation during the organogenesis and fetus periods [12–14], we will focus on the variation in sensitivity and response to radiation during the preimplantation (before gastrulation) period and discuss the underlying cellular and molecular mechanisms in this review. Embryos of this period have totipotency and the lack of cellular responses is closely related to this characteristic. It is also important to note here that temporal differences among species exist in what cellular mechanisms responds to ionizing radiation. Reviewing the cellular mechanisms during this period of each model organism shed lights on the critical mechanisms of totipotency or pluripotency.

2. Overview of the Effects of Radiation in Early Development

Different species have different responses to radiation during the preimplantation period (before gastrulation). We chose mammals, xenopus, and fish as model organisms since there are a relatively higher number of previous studies available for this period. We will discuss the stage-specific effects of each species separately.

2.1. Mammals

The knowledge of the effects of radiation during the preimplantation stage primarily stems from mice experiments. In mammals, radiation during this period is typically lethal [5]. Only certain strains of mice exhibit abnormalities, which are described below. The highest sensitivity is seen 2–6 h post-fertilization (hpf) during the early pronuclear stage [15], and embryos become more resistant during the S phase of the first cell cycle [16,17]. Sensitivity decreases rapidly over the days following this period [17–19]. The LD_{50} of this period changes from 1.5 Gy (immediately following sperm entry) to 0.3 Gy (4–6 hpf), and then gradually increases to approximately 3.5 Gy by the 5th day post fertilization (dpf). However, sensitivity seems to fluctuate, presumably depending on the timing of irradiation relative to the stage of the cell cycle [16–20]. Malformations caused by ionizing radiation are observed in the 'Heiligenberger Stamm' (HLG) and CF1 mouse strains [5]. In these strains, embryos irradiated during the preimplantation stage show gastroschisis, exencephaly, and polydactyly [21,22].

2.2. Xenopus

Although few experiments have been performed in *Xenopus*, or in amphibians at all, at the early stage prior to gastrulation, the most significant effect of irradiation on *Xenopus* embryos is lethality, similar to that in mice. However, compared to mice, *Xenopus* embryos are more resistant to ionizing radiation and some embryos show abnormalities. Hamilton [23] described *Xenopus* embryos irradiated 40 min after fertilization (the stage of pronuclei fusion, before the 2-cell stage) with 2.39–56.36 Gy. The survival rates of embryos to stage 28 (tailbud, 32.5 hpf) after being irradiated with 6.94, 11.95, 19.12, or 23.9 Gy were 80%, 89.7%, 62.7%, and 43.4%, respectively, while the average survival rate was 94.6% for control embryos. Although the LD_{50} at stage 28 was 25.5 Gy, many surviving embryos showed abnormalities, and it is possible that they will die at a later stage. Abnormalities included developmental delays, small heads, and short axes. Similar trends were observed when *Xenopus* embryos were irradiated at stage 9 (blastula, 7 hpf) with 6–24 Gy [24]. Survival rates at stage 29/30 (tailbud, 35 hpf) were 65%, 58%, 61%, and 45% when irradiated with 6, 12, 18, and 24 Gy, respectively, compared to control embryos with 81% survival. At stage 44 (tadpole, 92 hpf), the survival rate decreased to 52.1%, 8.8%, 1.2%, and 0%, respectively, compared to control embryos with 79% survival. Only 6.4% of the embryos irradiated with 6 Gy survived to stage 47 (tadpole) while all embryos irradiated with higher doses (12, 18, and 24 Gy) demonstrated 100% fatality at this stage. Although these results should be interpreted with caution because these experiments were performed independently, exposure to ionizing radiation appears to be lethal for *Xenopus* embryos. This is particularly true for embryos subjected to high doses of radiation that exhibit abnormalities including developmental delays, short axes, lordosis, and microcephaly [24], eventually resulting in death at later stages.

2.3. Fish

Fish embryos are more resistant to ionizing radiation than mice. In zebrafish, the highest sensitivity is observed 4-cell to 64-cell stage (1–2 hpf) [8,25], and the LD_{50} is 5–10 Gy. The effects of radiation on fish embryos include lethality and abnormality. Zebrafish embryos irradiated at 2 hpf with 5 Gy showed 100% body shape abnormality at 24 hpf, embryos irradiated at 4 hpf (blastula) with 5 or 10 Gy showed 31% and 100% body shape abnormality, respectively, and embryos irradiated at 6 hpf (gastrula) with 5 or 10 Gy showed 2.7% and 57.2% body shape abnormality, respectively. The primary abnormality observed at 24 hpf (pharyngula period) was a short axis phenotype [8]. Regarding low dose irradiation, Miyachi et al. [26] reported that irradiation with 0.025 Gy at 3.5 hpf (blastula) accelerated hatching time without affecting body length. Similarly, zebrafish embryos irradiated from 3 hpf to 20 dpf with a low dose (1 mGy/day) were resistant to starvation and showed accelerated hatching times. It is difficult to interpret these results in the absence of further morphological and cellular analyses.

In medaka *Oryzias latipes*, the day of fertilization is the most sensitive with respect to hatchability. Sensitivity decreases at 1 dpf, increases slightly at 2 dpf, and then gradually decreases as development progresses [27,28]. Abnormalities were also observed in embryos irradiated at 32-cell stage (5 hpf), germ-ring (30 hpf), beginning of blood circulation (80 hpf), and enlargement of tail (6 dpf). It appears that irradiation at earlier stages requires a smaller dose to induce abnormalities [27]. With respect to lethality, medaka are resistant to radiation. Hyodo-Taguchi and Egami [29] showed that the survival time of irradiated embryos steadily decreased as the radiation dose increased. However, high doses of radiation were required to shorten the survival time, and the age of the fish with the use of such doses was not relevant. For example, medaka embryos at 8–12 hpf (morulae) irradiated with 20, 40, or 80 Gy survived for approximately 15.3, 10.9, and 5.8 days, respectively, compared to control embryos that survived for 26.6 days. Medaka irradiated at the 1-year adult stage with 20, 40, or 80 Gy survived for 27.5, 12.3, and 12.0 days, respectively, compared to 29.6 days in control groups.

3. Cellular and Molecular Mechanisms Underlying Stage-Specific Effects

DNA damage is known to be the most significant threat for living organisms. A total of 1 Gy of radiation induces DNA lesions, including as many as 40 double strand breaks (DSBs) and 1000 single strand breaks (SSBs). DSBs are considered significantly damaging for all cell types. Embryonic cells are constantly dividing and embryos at the preimplantation (before gastrulation) stage are particularly active with rapid cell cycles. The events of the cell cycle are regulated by successive waves of cyclin/cyclin-dependent kinase (CDK) and ubiquitin ligase activity. Upon detection of DNA breaks caused by irradiation, cellular responses begin by recruiting the Mre11-Rad50-Nbs1 complex and ataxia-telangiectasia mutated kinase (ATM) to the site of DSBs [16,17]. ATM accumulation at the site of the DSB activates Chk1/Chk2, then auto-activates TP53. Activated TP53 upregulates p21 mRNA and P21/WAF1 protein levels [30–32]. P21 is also a CDK inhibitor; therefore, accumulation of P21/WAF1 causes cell cycle arrest, known as a cell cycle checkpoint, and promotes DNA repair [33–36]. ATM is primarily responsible for the DSB response whereas ataxia-telangiectasia and Rad3-related (ATR) is responsible for the SSB response. During the cell cycle, cells do not normally arrest at any point. There are two major checkpoints, namely, G1/S and G2/M, that can trigger cell cycle arrest [1,37,38]. If DNA repair fails and DNA remains damaged, embryonic cells progress to cell death [30,33]. Divergent responses to irradiation among species or developmental stages are primarily caused by the differences in the ability of these cellular or molecular mechanisms controlling DNA repair and the cell cycle. Specifically, cell cycle checkpoints, DNA repair capability, and the ability to undergo apoptosis are crucial in determining the effects of irradiation. Embryos at the preimplantation (before gastrulation) stage are often limited DNA damage responses. We will describe the status of these mechanisms in each organism during the preimplantation stage and discuss possible explanations of the morphological outcomes and genome integrity.

3.1. Mammals

As described above, the highest sensitivity to ionizing radiation occurs at the pronuclear stage in mammals. It is very rare that irradiated embryos show malformations, and these phenotypes are restricted to certain strains. Rather, most irradiated embryos die at or before the implantation stage. There is no midblastula transition (MBT) during mammalian development although this process occurs in many other phyla. The MBT is a timepoint of dramatic developmental changes, including slowing of the cell cycle, asynchronized cell divisions, and initiation of major zygotic transcription. In mouse embryos, cellular responses during early development differ from somatic cells in their cell cycle checkpoints and transcription ability. Transcription begins with minor activation before the 2-cell stage, followed by major activation at the 2-cell stage, which is extremely early compared to that in other organisms. This is a clear advantage for maintaining genome integrity during early development; however, it may be a disadvantage for the rapid development that is presumed to be important for many organisms.

3.1.1. Cell Cycle Checkpoints

As described above, changes in the sensitivity to ionizing radiation correlate with the stages of the cell cycle. This trend has also been reported in other cell types [37,38]. Generally, the most sensitive phase of the cell cycle to irradiation is during mitosis and in G2. Cells are less sensitive in G1 and the least sensitive during the latter part of the S phase. Early embryos are well known to have no or very short G1 phases. In the early mouse embryo, the G1/S checkpoint is defective [39–42], but embryos retain the ability to undergo G2/M cell cycle arrest at the one- or two-cell stage [43]. Mouse embryonic stem cells (mESCs) also lack a G1 checkpoint in response to DNA damage. After irradiation, the p53 protein is phosphorylated and accumulates in the nucleus [9,44,45]. This accumulation of activated p53 induces upregulation of the cell cycle inhibitor p21 mRNA. However, if the levels of P21 protein are insufficient [9,44,45], the cell cycle checkpoint will fail. Human embryonic stem cells (hESCs) do not display a G1/S checkpoint in response to any form of DNA damage, including that induced by ionizing radiation [46,47]. Similar to hESCs, induced pluripotent stem cells (iPSCs) also appear to lack a radiation-induced G1 checkpoint and instead arrest in the G2 phase of the cell cycle [47].

3.1.2. DNA Repair Capacity

DNA repair capacity has primarily been investigated using mESCs or hESCs (reviewed in [48,49]). There are two major repair pathways for DSBs. One is homologous recombination (HR), a high-fidelity repair system that relies on homologous regions of the sister chromatid. The other is non-homologous end joining (NHEJ), an error-prone system that simply joins broken DNA ends. Tichy et al. [50] demonstrated that HR-related proteins (RAD51, RAD52, and RAD54) and NHEJ-related proteins (Ku70/Ku80) are upregulated in mESCs compared to mouse embryonic fibroblasts. When mESCs were functionally tested for the preferred pathway of DSB repair, the high-fidelity HR pathway was predominantly utilized [50,51]. Moreover, the kinetics of the repair system in mESCs are efficient compared to that in NIH3T3 cells [52]. hESCs also have efficient DSB repair that is largely HR-mediated; however, hESCs rely on ATR, rather than ATM, for regulating DSB repair, and this relationship is dynamically changed as cells differentiate [53]. In addition, it was demonstrated that repair at a targeted DSB is relatively highly precise in hESCs compared to human somatic cells or murine embryonic stem cells, while differentiating hESCs harboring the targeted reporter into astrocytes reduces both the efficiency and precision of repair [54].

3.1.3. Apoptosis

An apoptotic response to irradiation arises in two-cell stage mouse embryos [55]. This suggests that embryonic cells with damaged chromosomes quickly become apoptotic from an early developmental stage. Several studies have shown that embryonic cells are hypersensitive to DNA damaging agents. For example, treatment with the topoisomerase II inhibitor etoposide causes cell death in mESCs at a dose 10 times lower than that in 3T3 cells [50,56]. Similarly, massive cell death was induced when mESCs were treated with ultraviolet irradiation or methylating agents [57,58]. He et al. [10] reported that p53 and p73 play critical roles in apoptosis, as opposed to cell cycle arrest, after DNA damage. hESCs and iPSCs also appear to be very sensitive to DNA damaging agents, such as ionizing radiation, and undergo caspase 3-dependent apoptosis after exposure to such agents [46,47,59].

3.1.4. Genome Integrity

Although the G2 checkpoint prevents the segregation of damaged chromosomes during M phase, several reports have described damaged chromosomes remaining after irradiation. After 0.4 Gy irradiation, 20.6% of embryos irradiated at the 1-cell stage were found to have chromosomal abnormalities, which was significantly higher than in irradiated sperm (2.9%) or irradiated unfertilized eggs (11.0%) [60]. In addition to structural chromosomal aberrations, chromosome loss was also observed regardless of the cell cycle phase [61]. Most chromosomal damage was expressed at later developmental stages after the embryonic cells had undergone several cell cycles after irradiation [61,62]. This genomic instability caused by irradiation in early embryos may be partially owing to the lack of a G1/S checkpoint.

Although *Atm* or *Chk2* are dispensable for embryonic development [63,64], embryos lacking *Atr* or *Chk1* die shortly after implantation and exhibit high degrees of chromosomal fragmentation [65–68]. Furthermore, embryos lacking many DNA repair-related genes, such as *Rad50* or *Nbs1*, show embryonic lethality [69,70]. These data strongly suggest that DNA repair machinery and cell cycle regulation correlate with genome integrity.

3.2. *Xenopus*

Xenopus embryo seem to be highly sensitive to ionizing radiation at the tadpole stage in terms of lethality. Irradiated *Xenopus* embryos exhibit lethality and malformation. During *Xenopus* development, embryos undergo a major transition called the MBT. At the MBT, the cell cycle begins to slow, the G1 phase of cell the cycle is lengthened, the zygotic genome is transcriptionally activated, and asynchronous cell division begins.

3.2.1. Cell Cycle Checkpoints

Cellular responses to ionizing radiation differ before and after MBT. In *Xenopus laevis* embryos, irradiation during early developmental stages (before MBT; stage 8, 7 hpf) leads to apoptosis of all embryonic cells [71]. In later stages (after MBT), cells develop the ability to induce cell cycle arrest to prevent apoptosis in the embryos [71]. In several studies, double-stranded DNA have been injected into embryos to mimic DSBs. These data revealed that the ATM/ATR-mediated pathway stimulates Chk1-dependent degradation of Cdc25A in pre-MBT embryos to result in negative regulation of CDK activity [72]. Lysates from embryos treated with double-stranded DNA ends revealed that of Cdc25C was phosphorylated by Chk1 in response to DNA damage [73–75]. This treatment not only mimics DSBs in early embryos but also changes the DNA-to-cytoplasmic ratio. The DNA-to-cytoplasmic ratio is an important factor controlling the onset of MBT. Titration of double-stranded DNA and/or co-injection with vector DNA revealed that the activation of Chk1 correlates with an appropriate total DNA-to-cytoplasmic ratio. Amodeo et al. [76] reported that a reduction of histone H3 protein in embryos induced premature transcriptional activation and cell cycle lengthening in *Xenopus* embryos. The addition of histone H3/H4 shortened post-MBT cell cycles. This suggests that the MBT is regulated by a DNA-based titration against free histones in the cytoplasm. From these experiments, it was demonstrated that pre-MBT *Xenopus* embryos are able to activate a DNA damage response; however, DNA damage response signaling is blocked at some steps, most likely at the detection of DNA damage [73,77].

3.2.2. DNA Repair Capacity

There are insufficient studies to compare HR and NHEJ kinetics in early *Xenopus* embryos. Anderson et al. [71] stated that pre- and post-MBT *Xenopus* embryos appear to have a similar DNA repair capacity.

3.2.3. Apoptosis

Apoptosis only occurs after the onset of gastrulation in *Xenopus* embryos [78]. Embryos exposed to ionizing radiation before MBT undergo apoptosis at the time of gastrulation. There are no studies directly comparing survival rates between embryos irradiated before and after MBT. From a comparison of the data from Hamilton [23] and Ijiri [24], it cannot be suggested that the changes at MBT correlate with the difference in survival rate, although this might be biased by the difference in dose and timing of scoring. It is possible that if embryos were scored at an earlier stage, immediately after gastrulation when apoptosis occurs, cellular differences might have been observed. Alternatively, lethality of the whole embryo does not necessarily correlate with apoptosis in individual cells.

3.2.4. Genome Integrity

Surprisingly, there is insufficient knowledge of genome integrity in *Xenopus* embryos following irradiation or DNA damage. Irradiation of embryos before MBT results in apoptosis at gastrulation, and irradiation after MBT causes cell cycle arrest, presumably due to repair DNA damage. This can explain why embryos die at later stages rather than at the time of DNA damage. *Xenopus* embryos show abnormalities at a high frequency after irradiation. More information regarding the capabilities of DNA repair in early embryos is needed to explain and evaluate these results. Specifically, a comparison with mammalian embryonic cells and their usage of DSB repair machinery would provide considerable insights into the unique mechanisms operating in *Xenopus* embryos to maintain their genome integrity.

3.3. *Fish*

The highest sensitivity to irradiation in fish embryos occurs immediately after fertilization in zebrafish or 1 dpf (late gastrula to early neurula stage) in medaka. The phenotype of irradiated embryos is typically abnormality rather than lethality in fish. Fish are generally thought to have an MBT; however, medaka were recently found not to have an MBT [79]. In medaka embryos, cell division becomes asynchronous and asymmetric DNA cleavage occurs before the blastula stage and zygotic transcription is extensively activated at the cleavage stage. This is in contrast to the findings of Aizawa et al. [80] who showed that zygotic genome activation and presumably MBT begins at stage 11 (late blastula stage) based on the expression of expressed sequence tag markers. Additional analysis is needed to draw a complete picture of medaka development; for example, neither study analyzed cell cycle checkpoints.

3.3.1. Cell Cycle Checkpoints

Similar to *Xenopus*, the transition to somatic adaptive responses occurs during MBT (3 hpf, blastula) in zebrafish embryos. During zebrafish development, cell cycle arrest does not occur prior to MBT [8]. Upregulation of *p21* mRNA, but not P21 protein accumulation, was observed in embryos irradiated before MBT. This is the same result as in mESC and is probably the reason underlying the lack of cell cycle arrest. γH2AX foci were observed only in 4 hpf or older embryos, although *p21* mRNA was upregulated, suggesting that the DNA damage response occurred to some extent before MBT [8]. Abnormalities occur more frequently in younger zebrafish embryos. Even though cell cycle arrest and P21 accumulation are observed in embryos irradiated at 4 and 6 hpf, P21 accumulation is only delayed in the 4 hpf group. This could indicate inferior function of the cellular response at this stage. Treatment with a topoisomerase I inhibitor after MBT, but not before MBT, has been shown to cause cell cycle arrest in zebrafish embryos [81,82].

3.3.2. DNA Repair Capacity

HR and NHEJ activity have been observed in the early developmental stages of zebrafish (reviewed in [83,84]). HR has also been observed in zebrafish embryonic stem cells [85]. Sussman [86] demonstrated that zebrafish embryos have a much higher DNA repair capability for DNA damage from UV irradiation compared to that by human lymphoblast cells. In medaka, DNA lesions caused by gamma-irradiation were repaired within 2 h, and the level of DSBs decreased until reaching the control level within 30 min after irradiation [87,88].

3.3.3. Apoptosis

Zebrafish embryos do not undergo apoptosis until the mid-gastrulation stage [81,89]. Ikegami et al. [81,82] showed that embryos treated with a topoisomerase I inhibitor before MBT do not show apoptosis immediately; however, cells eventually become apoptotic at the late gastrulation stage. Zebrafish embryos irradiated at 2 hpf with 10 Gy undergo cell division until the mid-gastrula stage and cells abruptly initiate apoptosis during late gastrulation [90].

3.3.4. Genome Integrity

Zebrafish embryos treated with a topoisomerase I inhibitor both before and after MBT show distinctive fragmented chromosomes. Earlier treatment more rapidly results in a severe phenotype [82]. Irradiation of zebrafish embryos before MBT also caused similarly fragmented chromosomes (Honjo, unpublished observation). These observations suggest that cell cycle checkpoints are a key factor for preserving genome integrity. Lindelman et al. [11] reported that exposing zebrafish embryos to radiation for 3 h beginning at 2.5 hpf (256-cell stage) induced locus-specific changes in the enrichment of histone modifications. Similar results were observed when Atlantic salmon embryos were exposed to radiation from the 1-cell stage to the early gastrula stage [11]. To assess whether this change affects chromatin structure or organization, additional experiments must be carried out. Before MBT, zebrafish embryos lack cell cycle checkpoints and the ability to undergo apoptosis. This likely causes the most sensitized period to DNA damage. After MBT, zebrafish embryos become quite resistant to irradiation through more strict cell cycle checkpoints, likely together with a high DNA repair capability and regenerative potential. There is insufficient information regarding cell cycle checkpoints in early medaka embryos, and it is challenging to analyze the mechanisms activated after DNA damage in early-stage medaka embryos. Even if medaka embryos do not undergo MBT, the onset of transcription and apoptotic ability appear to occur later than that in mammalian embryos. This may explain why the phenotype caused by irradiation is more similar to that of *Xenopus* and zebrafish than that of mammals.

4. Perspectives

During early development, embryos inactivate some cellular mechanisms to maintain totipotency or pluripotency, making them extremely sensitive to genotoxic stress. We discussed the morphological effects of ionizing radiation and the resulting cellular responses in several species (also see Supplementary Table S1). Each species has divergent cellular mechanisms and responses. Mammals are quite different from *Xenopus* and fish in several points. First, mammals show a much higher sensitivity, with an LD_{50} as low as 0.3 Gy in mouse embryos, whereas LD_{50} is 25.5 Gy or between 5 and 10 Gy in *Xenopus* and zebrafish, respectively. Second, generally, no malformations are observed in mouse embryos irradiated at the preimplantation stage, whereas malformations are frequently observed in *Xenopus* and zebrafish. Currently, we have no explanation as to why *Xenopus* and fish are more resistant to ionizing radiation than mammals. One possible explanation could be the differences in their cellular mechanisms of early embryogenesis. In mammals, many mechanisms, such as G2/M checkpoints, zygotic genome activation, and apoptosis, start at the 2-cell stage. This is relatively early considering that those mechanisms do not occur until MBT or the gastrula stage in *Xenopus* and zebrafish. These mechanisms might prevent malformed embryos by forcing damaged cells to quickly undergo apoptosis. Another possible explanation is the capability of DNA repair. Certain mouse strains that exhibit malformation rather than lethality seem to have reduced DNA repair activity. DNA repair capability is an important factor for recovering from the DNA damage induced by ionizing radiation. Together with the prolonged period of suspended cellular mechanisms, DNA repair capabilities or mechanisms in *Xenopus* and fish differ from those of mammals. The selection of DNA repair machinery, either HR or NHEJ, might also make a difference. Mouse or human embryos appear preferentially utilize HR over NHEJ. This enables precise DNA repair, although it might take more time than NHEJ. There have been no studies identifying the preferential use of either repair pathway in *Xenopus* or fish embryos. If NHEJ is utilized more frequently than in mammalian embryos, this might explain the higher occurrence of abnormalities and later lethality. Increasing data have shown that genomic reprogramming occurs immediately after fertilization and epigenetic marks are dramatically reset and/or re-modified on the maternal and paternal genomes. This process is a consequence of reprogramming and is strongly associated with genome integrity in many organisms. These epigenetic marks are also affected by genotoxic stressors such as radiation. The effects of epigenetic marks on the response to genotoxic stressors during early embryo stages are still unknown.

Supplementary Materials: Supplementary materials can be found at http://www.mdpi.com/1422-0067/21/11/3975/s1.

Author Contributions: Writing—original draft preparation, Y.H.; writing—review and editing, T.I.; funding acquisition, Y.H., T.I. All authors have read and agreed to the published version of the manuscript.

Funding: This work was supported in part by a Grant-in-Aid from the Ministry of Education, Culture, Sports, Science, and Technology in Japan (#15K09453 and #26293277 to T.I., #18K116440 to Y.H.). This work was also supported by the Program of the network-type joint Usage/Research Disaster Medical Science of Hiroshima University, Nagasaki University, and Fukushima Medical University (Y.H. and T.I.) and Hiroshima University Research Promotion Award for Women Scientists (Y.H.). We also thank the Low-Dose Radiation Effects Advanced Research Program of the Research Institute for Radiation Biology and Medicine, Hiroshima University, for funding and supporting this study.

Acknowledgments: We thank Ryoko Matsumoto, Nanae Nakaju, Sachiko Fukumoto, Sajeda Chowdhury and Masako Ninomiya for their excellent technical and secretarial assistance.

Conflicts of Interest: The authors declare no conflict of interest.

Abbreviations

hpf	Hour post fertilization
MBT	Midblastula transition

References

1. Hall, J.E.; Giaccia, J.A. *Radiation Biology for Radiation Biologist*; Lippincott Williams & Wilkins: Philadelphia, PA, USA; ISBN 1-4963-3541-4.
2. Georgakilas, A.G.; O'Neill, P.; Stewart, R.D. Induction and Repair of Clustered DNA Lesions: What Do We Know So Far? *Rare* **2013**, *180*, 100–109. [CrossRef]
3. Ward, J.F. Some Biochemical Consequences of the Spatial Distribution of Ionizing Radiation-Produced Free Radicals. *Radiat. Res.* **1981**, *86*, 185. [CrossRef]
4. Goodhead, D.T.; Thacker, J.; Cox, R. Effects of Radiations of Different Qualities on Cells: Molecular Mechanisms of Damage and Repair. *Int. J. Radiat. Biol.* **1993**, *63*, 543–556. [CrossRef]

5. Valentin, J. Biological effects after prenatal irradiation (embryo and fetus) ICRP Publication 90 Approved by the Commission in October 2002. *Ann. ICRP* **2003**, *33*, 1–206. [CrossRef]
6. Heyer, B.S.; MacAuley, A.; Behrendtsen, O.; Werb, Z. Hypersensitivity to DNA damage leads to increased apoptosis during early mouse development. *Genes Dev.* **2000**, *14*, 2072–2084. [CrossRef]
7. Russell, L.B.; Russell, W.L. An analysis of the changing radiation response of the developing mouse embryo. *J. Cell. Physiol.* **1954**, *43*, 103–149. [CrossRef]
8. Honjo, Y.; Ichinohe, T. Cellular responses to ionizing radiation change quickly over time during early development in zebrafish. *Cell Biol. Int.* **2019**, *43*, 516–527. [CrossRef]
9. Suvorova, I.I.; Grigorash, B.B.; Chuykin, I.A.; Pospelova, T.V.; Pospelov, V.A. G1 checkpoint is compromised in mouse ESCs due to functional uncoupling of p53-p21Waf1 signaling. *Cell Cycle* **2015**, *15*, 52–63. [CrossRef]
10. He, H.; Wang, C.; Dai, Q.; Li, F.; Bergholz, J.; Li, Z.; Li, Q.; Xiao, Z.-X. p53 and p73 Regulate Apoptosis but Not Cell-Cycle Progression in Mouse Embryonic Stem Cells upon DNA Damage and Differentiation. *Stem Cell Rep.* **2016**, *7*, 1087–1098. [CrossRef]
11. Lindeman, L.C.; Kamstra, J.H.; Ballangby, J.; Hurem, S.; Martín, L.M.; Brede, D.A.; Teien, H.C.; Oughton, D.H.; Salbu, B.; Lyche, J.L.; et al. Gamma radiation induces locus specific changes to histone modification enrichment in zebrafish and Atlantic salmon. *PLoS ONE* **2019**, *14*, e0212123. [CrossRef]
12. De Santis, M.; Cesari, E.; Nobili, E.; Straface, G.; Cavaliere, A.F.; Caruso, A. Radiation effects on development. *Birth Defects Res. C Embryo Today* **2007**, *81*, 177–182. [CrossRef]
13. Hulse, E.V. The effects of ionising radiation on the embryo and foetus: A review of experimental data. *Clin. Radiol.* **1964**, *15*, 312–319. [CrossRef]
14. Yang, B.; Ren, B.X.; Tang, F.R. Prenatal irradiation–induced brain neuropathology and cognitive impairment. *Brain Dev.* **2017**, *39*, 10–22. [CrossRef]
15. Russell, L.B.; Montgomery, C.S. Radiation-sensitivity Differences within Cell-division Cycles during Mouse Cleavage. *Int. J. Radiat. Biol. Relat. Stud. Phys. Chem. Med.* **1966**, *10*, 151–164. [CrossRef]
16. Domon, M. Cell Cycle-Dependent Radiosensitivity in Two-Cell Mouse Embryos in Culture. *Radiat. Res.* **1980**, *81*, 236. [CrossRef]
17. Domon, M. Radiosensitivity variation during the cell cycle in pronuclear mouse embryos in vitro. *Cell Prolif.* **1982**, *15*, 89–98. [CrossRef]
18. Rugh, R.; Duhamel, L.; Somogyi, C.; Chandler, A.; Cooper, W.R.; Smith, R.; Stanford, G. Sequelae Of The Ld/50 X-Ray Exposure of the Pre-Implantation Mouse Embryo: Days 0.0 To 5.0. *Biol. bull.* **1966**, *131*, 145–154. [CrossRef]
19. Müller, W.-U.; Streffer, C. Lethal and teratogenic effects after exposure to X-rays at various times of early murine gestation. *Teratology* **1990**, *42*, 643–650. [CrossRef]
20. Gu, Y.; Kai, M.; Kusama, T. The Embryonic and Fatal Effects in ICR Mice Irradiated in the Various Stages of the Preimplantation Period. *Radiat. Res.* **1997**, *147*, 735–740. [CrossRef]
21. Jacquet, P.; de Saint-Georges, L.; Vankerkom, J.; Baugnet-Mahieu, L. Embryonic death, dwarfism and fetal malformations after irradiation of embryos at the zygote stage: Studies on two mouse strains. *Mutat. Res.* **1995**, *332*, 73–87. [CrossRef]
22. Streffer, C.; Muller, W.U. Malformations after radiation exposure of preimplantation stages. *Int. J. Dev. Biol.* **2003**, *40*, 355–360. [CrossRef]
23. Hamilton, L. A Comparison of the X-Ray Sensitivity of Haploid and Diploid Zygotes of Xenopus Iaevis. *Radiat. Res.* **1967**, *30*, 248. [CrossRef]
24. Ijiri, K.-I. X-Ray Effect on the Development of Xenopus laevis Embryos—With Special Reference to Primordial Germ Cells. *J. Radiat. Res.* **1979**, *20*, 133–136. [CrossRef] [PubMed]
25. McAleer, M.F.; Davidson, C.; Davidson, W.R.; Yentzer, B.; Farber, S.A.; Rodeck, U.; Dicker, A.P. Novel use of zebrafish as a vertebrate model to screen radiation protectors and sensitizers. *Int. J. Radiat. Oncol. Biol. Phys.* **2005**, *61*, 10–13. [CrossRef]
26. Miyachi, Y.; Kanao, T.; Okamoto, T. Marked depression of time interval between fertilization period and hatching period following exposure to low-dose X-rays in zebrafish. *Environ. Res.* **2003**, *93*, 216–219. [CrossRef]
27. Hyodo-Taguchi, Y.; Etoh, H.; Egami, N. RBE of Fast Neutrons for Inhibition of Hatchability in Fish Embryos Irradiated at Different Developmental Stages. *Radiat. Res.* **1973**, *53*, 385. [CrossRef]

28. Shimada, Y.; Egami, N.; Shima, A. Effect of Heat on Radiosensitivity at Different Developmental Stages of Embryos of the Fish *Oryzias Latipes*. *Int. J. Radiat. Biol. Relat. Stud. Phys. Chem. Med.* **1985**, *48*, 505–512. [CrossRef]
29. Hyodo-Taguchi, Y.; Egami, N. Change in Dose-survival Time Relationship after X-irradiation during Embryonic Development in the Fish, *Oryzias latipes*. *J. Radiat. Res.* **1969**, *10*, 121–125. [CrossRef]
30. Hirao, A.; Kong, Y.-Y.; Matsuoka, S.; Wakeham, A.; Ruland, J.; Yoshida, H.; Liu, D.; Elledge, S.J.; Mak, T.W. DNA Damage-Induced Activation of p53 by the Checkpoint Kinase Chk2. *Science* **2000**, *287*, 1824–1827. [CrossRef]
31. Lossaint, G.; Besnard, E.; Fisher, D.; Piette, J.; Dulić, V. Chk1 is dispensable for G2 arrest in response to sustained DNA damage when the ATM/p53/p21 pathway is functional. *Oncogene* **2011**, *30*, 4261–4274. [CrossRef]
32. Bartek, J.; Lukas, J. Chk1 and Chk2 kinases in checkpoint control and cancer. *Cancer Cell* **2003**, *3*, 421–429. [CrossRef]
33. Wang, Y.A.; Elson, A.; Leder, P. Loss of p21 increases sensitivity to ionizing radiation and delays the onset of lymphoma in atm-deficient mice. *Proc. Natl. Acad. Sci. USA* **1997**, *94*, 14590–14595. [CrossRef]
34. Gartel, A.L.; Serfas, M.S.; Tyner, A.L. p21–negative regulator of the cell cycle. *Proc. Soc. Exp. Biol. Med.* **1996**, *213*, 138–149. [CrossRef]
35. Brugarolas, J.; Moberg, K.; Boyd, S.D.; Taya, Y.; Jacks, T.; Lees, J.A. Inhibition of cyclin-dependent kinase 2 by p21 is necessary for retinoblastoma protein-mediated G1 arrest after gamma-irradiation. *Proc. Natl. Acad. Sci. USA* **1999**, *96*, 1002–1007. [CrossRef]
36. Roque, T.; Haton, C.; Etienne, O.; Chicheportiche, A.; Rousseau, L.; Martin, L.; Mouthon, M.-A.; Boussin, F.D. Lack of a *p21* $^{waf1/cip}$ -Dependent G1/S Checkpoint in Neural Stem and Progenitor Cells After DNA Damage In Vivo. *Stem Cells* **2012**, *30*, 537–547. [CrossRef]
37. Pawlik, T.M.; Keyomarsi, K. Role of cell cycle in mediating sensitivity to radiotherapy. *Int. J. Radiat. Oncol. Biol. Phys.* **2004**, *59*, 928–942. [CrossRef]
38. Sinclair, W.K.; Morton, R.A. X-Ray Sensitivity during the Cell Generation Cycle of Cultured Chinese Hamster Cells. *Radiat. Res.* **1966**, *29*, 450–474. [CrossRef]
39. Aladjem, M.I.; Spike, B.T.; Rodewald, L.W.; Hope, T.J.; Klemm, M.; Jaenisch, R.; Wahl, G.M. ES cells do not activate p53-dependent stress responses and undergo p53-independent apoptosis in response to DNA damage. *Curr. Biol.* **1998**, *8*, 145–155. [CrossRef]
40. Orford, K.W.; Scadden, D.T. Deconstructing stem cell self-renewal: Genetic insights into cell-cycle regulation. *Nat. Rev. Genet.* **2008**, *9*, 115–128. [CrossRef]
41. Singh, A.M.; Dalton, S. The Cell Cycle and Myc Intersect with Mechanisms that Regulate Pluripotency and Reprogramming. *Cell Stem Cell* **2009**, *5*, 141–149. [CrossRef]
42. Lange, C.; Calegari, F. Cdks and cyclins link G_1 length and differentiation of embryonic, neural and hematopoietic stem cells. *Cell Cycle* **2010**, *9*, 1893–1900. [CrossRef]
43. Yukawa, M.; Oda, S.; Mitani, H.; Nagata, M.; Aoki, F. Deficiency in the response to DNA double-strand breaks in mouse early preimplantation embryos. *Biochem. Biophys. Res. Commun.* **2007**, *358*, 578–584. [CrossRef]
44. Solozobova, V.; Rolletschek, A.; Blattner, C. Nuclear accumulation and activation of p53 in embryonic stem cells after DNA damage. *BMC Cell Biol.* **2009**, *10*, 46. [CrossRef]
45. Hong, Y.; Stambrook, P.J. Restoration of an absent G1 arrest and protection from apoptosis in embryonic stem cells after ionizing radiation. *Proc. Natl. Acad. Sci. USA* **2004**, *101*, 14443–14448. [CrossRef]
46. Filion, T.M.; Qiao, M.; Ghule, P.N.; Mandeville, M.; van Wijnen, A.J.; Stein, J.L.; Lian, J.B.; Altieri, D.C.; Stein, G.S. Survival responses of human embryonic stem cells to DNA damage. *J. Cell. Physiol.* **2009**, *220*, 586–592. [CrossRef]
47. Momcilovic, O.; Knobloch, L.; Fornsaglio, J.; Varum, S.; Easley, C.; Schatten, G. DNA Damage Responses in Human Induced Pluripotent Stem Cells and Embryonic Stem Cells. *PLoS ONE* **2010**, *5*, e13410. [CrossRef]
48. Giachino, C.; Orlando, L.; Turinetto, V. Maintenance of Genomic Stability in Mouse Embryonic Stem Cells: Relevance in Aging and Disease. *Int. J. Mol. Sci.* **2013**, *14*, 2617–2636. [CrossRef]
49. Nagaria, P.; Robert, C.; Rassool, F.V. DNA double-strand break response in stem cells: Mechanisms to maintain genomic integrity. *Biochim. Biophys. Acta, Gen. Subj.* **2013**, *1830*, 2345–2353. [CrossRef]

50. Tichy, E.D.; Pillai, R.; Deng, L.; Liang, L.; Tischfield, J.; Schwemberger, S.J.; Babcock, G.F.; Stambrook, P.J. Mouse Embryonic Stem Cells, but Not Somatic Cells, Predominantly Use Homologous Recombination to Repair Double-Strand DNA Breaks. *Stem Cells Dev.* **2010**, *19*, 1699–1711. [CrossRef]
51. Serrano, L.; Liang, L.; Chang, Y.; Deng, L.; Maulion, C.; Nguyen, S.; Tischfield, J.A. Homologous Recombination Conserves DNA Sequence Integrity Throughout the Cell Cycle in Embryonic Stem Cells. *Stem Cells Dev.* **2011**, *20*, 363–374. [CrossRef]
52. Saretzki, G.; Armstrong, L.; Leake, A.; Lako, M.; Zglinicki, T. von Stress Defense in Murine Embryonic Stem Cells Is Superior to That of Various Differentiated Murine Cells. *Stem Cells* **2004**, *22*, 962–971. [CrossRef]
53. Adams, B.R.; Golding, S.E.; Rao, R.R.; Valerie, K. Dynamic Dependence on ATR and ATM for Double-Strand Break Repair in Human Embryonic Stem Cells and Neural Descendants. *PLoS ONE* **2010**, *5*, e10001. [CrossRef]
54. Fung, H.; Weinstock, D.M. Repair at Single Targeted DNA Double-Strand Breaks in Pluripotent and Differentiated Human Cells. *PLoS ONE* **2011**, *6*, e20514. [CrossRef]
55. Adiga, S.K.; Toyoshima, M.; Shimura, T.; Takeda, J.; Uematsu, N.; Niwa, O. Delayed and stage specific phosphorylation of H2AX during preimplantation development of γ-irradiated mouse embryos. *Reproduction* **2007**, *133*, 415–422. [CrossRef]
56. Tichy, E.D. Mechanisms maintaining genomic integrity in embryonic stem cells and induced pluripotent stem cells. *Exp. Biol. Med.* **2011**, *236*, 987–996. [CrossRef]
57. Van Sloun, P.P.H.; Jansen, J.G.; Weeda, G.; Mullenders, L.H.F.; van Zeeland, A.A.; Lohman, P.H.M.; Vrieling, H. The role of nucleotide excision repair in protecting embryonic stem cells from genotoxic effects of UV-induced DNA damage. *Nucleic Acids Res.* **1999**, *27*, 3276–3282. [CrossRef]
58. Roos, W.P.; Christmann, M.; Fraser, S.T.; Kaina, B. Mouse embryonic stem cells are hypersensitive to apoptosis triggered by the DNA damage O6-methylguanine due to high E2F1 regulated mismatch repair. *Cell Death Differ.* **2007**, *14*, 1422–1432. [CrossRef]
59. Momčilović, O.; Choi, S.; Varum, S.; Bakkenist, C.; Schatten, G.; Navara, C. Ionizing Radiation Induces Ataxia Telangiectasia Mutated-Dependent Checkpoint Signaling and G2 But Not G1 Cell Cycle Arrest in Pluripotent Human Embryonic Stem Cells. *Stem Cells* **2009**, *27*, 1822–1835. [CrossRef]
60. Matsuda, Y.; Yamada, T.; Tobari, I.; Ohkawa, A. Preliminary study on chromosomal aberrations in eggs of mice fertilized in vitro after X-irradiation. *Mut. Res. Let.* **1983**, *121*, 125–130. [CrossRef]
61. Weissenborn, U.; Streffer, C. The One-cell Mouse Embryo: Cell Cycle-dependent Radiosensitivity and Development of Chromosomal Anomalies in Postradiation Cell Cycles. *Int. J. Radiat. Biol.* **1988**, *54*, 659–674. [CrossRef]
62. Weissenborn, U.; Streffer, C. Analysis of Structural and Numerical Chromosomal Anomalies at the First, Second, and Third Mitosis after Irradiation of One-cell Mouse Embryos with X-rays or Neutrons. *Int. J. Radiat. Biol.* **1988**, *54*, 381–394. [CrossRef] [PubMed]
63. Elson, A.; Wang, Y.; Daugherty, C.J.; Morton, C.C.; Zhou, F.; Campos-Torres, J.; Leder, P. Pleiotropic defects in ataxia-telangiectasia protein-deficient mice. *Proc. Natl. Acad. Sci. USA* **1996**, *93*, 13084–13089. [CrossRef]
64. Takai, H.; Naka, K.; Okada, Y.; Watanabe, M.; Harada, N.; Saito, S.; Anderson, C.W.; Appella, E.; Nakanishi, M.; Suzuki, H.; et al. Chk2-deficient mice exhibit radioresistance and defective p53-mediated transcription. *EMBO J.* **2002**, *21*, 5195–5205. [CrossRef] [PubMed]
65. Brown, E.J.; Baltimore, D. ATR disruption leads to chromosomal fragmentation and early embryonic lethality. *Genes Dev.* **2000**, *14*, 397–402. [CrossRef] [PubMed]
66. Liu, Q.; Guntuku, S.; Cui, X.-S.; Matsuoka, S.; Cortez, D.; Tamai, K.; Luo, G.; Carattini-Rivera, S.; DeMayo, F.; Bradley, A.; et al. Chk1 is an essential kinase that is regulated by Atr and required for the G2/M DNA damage checkpoint. *Genes Dev.* **2000**, *14*, 1448–1459. [CrossRef]
67. Takai, H.; Tominaga, K.; Motoyama, N.; Minamishima, Y.A.; Nagahama, H.; Tsukiyama, T.; Ikeda, K.; Nakayama, K.; Nakanishi, M.; Nakayama, K. Aberrant cell cycle checkpoint function and early embryonic death in $Chk1^{-/-}$ mice. *Genes Dev.* **2000**, *14*, 1439–1447. [CrossRef]
68. Artus, J.; Cohen-Tannoudji, M. Cell cycle regulation during early mouse embryogenesis. *Mol. Cell. Endocrinol.* **2008**, *282*, 78–86. [CrossRef]
69. Luo, G.; Yao, M.S.; Bender, C.F.; Mills, M.; Bladl, A.R.; Bradley, A.; Petrini, J.H.J. Disruption of mRad50 causes embryonic stem cell lethality, abnormal embryonic development, and sensitivity to ionizing radiation. *Proc. Natl. Acad. Sci. USA* **1999**, *96*, 7376–7381. [CrossRef]

70. Dumon-Jones, V.; Frappart, P.-O.; Tong, W.-M.; Sajithlal, G.; Hulla, W.; Schmid, G.; Herceg, Z.; Digweed, M.; Wang, Z.-Q. Nbn Heterozygosity Renders Mice Susceptible to Tumor Formation and Ionizing Radiation-Induced Tumorigenesis. *Cancer Res.* **2003**, *63*, 7263–7269.
71. Anderson, J.A.; Lewellyn, A.L.; Maller, J.L. Ionizing radiation induces apoptosis and elevates cyclin A1-Cdk2 activity before but not after the midblastula transition in Xenopus. *Mol. Biol. Cell* **1997**, *8*, 1195–1206. [CrossRef]
72. Conn, C.W.; Lewellyn, A.L.; Maller, J.L. The DNA Damage Checkpoint in Embryonic Cell Cycles Is Dependent on the DNA-to-Cytoplasmic Ratio. *Dev. Cell* **2004**, *7*, 275–281. [CrossRef]
73. Peng, A.; Lewellyn, A.L.; Maller, J.L. Undamaged DNA Transmits and Enhances DNA Damage Checkpoint Signals in Early Embryos. *Mol. Cell. Biol.* **2007**, *27*, 6852–6862. [CrossRef]
74. Peng, C.-Y.; Graves, P.R.; Thoma, R.S.; Wu, Z.; Shaw, A.S.; Piwnica-Worms, H. Mitotic and G2 Checkpoint Control: Regulation of 14-3-3 Protein Binding by Phosphorylation of Cdc25C on Serine-216. *Science* **1997**, *277*, 1501–1505. [CrossRef]
75. Kumagai, A.; Guo, Z.; Emami, K.H.; Wang, S.X.; Dunphy, W.G. The Xenopus Chk1 Protein Kinase Mediates a Caffeine-sensitive Pathway of Checkpoint Control in Cell-free Extracts. *J. Cell Biol.* **1998**, *142*, 1559–1569. [CrossRef]
76. Amodeo, A.A.; Jukam, D.; Straight, A.F.; Skotheim, J.M. Histone titration against the genome sets the DNA-to-cytoplasm threshold for the Xenopus midblastula transition. *Proc. Natl. Acad. Sci. USA* **2015**, *112*, E1086–E1095. [CrossRef]
77. Peng, A.; Lewellyn, A.L.; Maller, J.L. DNA damage signaling in early Xenopus embryos. *Cell Cycle* **2008**, *7*, 3–6. [CrossRef]
78. Hensey, C.; Gautier, J. A developmental timer that regulates apoptosis at the onset of gastrulation. *Mech. Dev.* **1997**, *69*, 183–195. [CrossRef]
79. Kraeussling, M.; Wagner, T.U.; Schartl, M. Highly asynchronous and asymmetric cleavage divisions accompany early transcriptional activity in pre-blastula medaka embryos. *PLoS ONE* **2011**, *6*. [CrossRef]
80. Aizawa, K.; Shimada, A.; Naruse, K.; Mitani, H.; Shima, A. The medaka midblastula transition as revealed by the expression of the paternal genome. *Gene Expr. Patterns* **2003**, *3*, 43–47. [CrossRef]
81. Ikegami, R.; Rivera-Bennetts, A.K.; Brooker, D.L.; Yager, T.D. Effect of inhibitors of DNA replication on early zebrafish embryos: Evidence for coordinate activation of multiple intrinsic cell-cycle checkpoints at the mid-blastula transition. *Zygote* **1997**, *5*, 153–175. [CrossRef]
82. Ikegami, R.; Hunter, P.; Yager, T.D. Developmental Activation of the Capability to Undergo Checkpoint-Induced Apoptosis in the Early Zebrafish Embryo. *Dev. Biol.* **1999**, *209*, 409–433. [CrossRef]
83. Kienzler, A.; Bony, S.; Devaux, A. DNA repair activity in fish and interest in ecotoxicology: A review. *Aquat. Toxicol.* **2013**, *134–135*, 47–56. [CrossRef] [PubMed]
84. Hagmann, M.; Bruggmann, R.; Xue, L.; Georgiev, O.; Schaffner, W.; Rungger, D.; Spaniol, P.; Gerster, T. Homologous Recombination and DNA-End Joining Reactions in Zygotes and Early Embryos of Zebrafish (Danio rerio) and Drosophila melanogaster. *Biol. Chem.* **1998**, *379*, 673–682. [CrossRef] [PubMed]
85. Fan, L.; Moon, J.; Crodian, J.; Collodi, P. Homologous Recombination in Zebrafish ES Cells. *Transgenic Res.* **2006**, *15*, 21–30. [CrossRef]
86. Sussman, R. DNA repair capacity of zebrafish. *Proc. Natl. Acad. Sci. USA* **2007**, *104*, 13379–13383. [CrossRef]
87. Aizawa, K.; Mitani, H.; Kogure, N.; Shimada, A.; Hirose, Y.; Sasado, T.; Morinaga, C.; Yasuoka, A.; Yoda, H.; Watanabe, T.; et al. Identification of radiation-sensitive mutants in the Medaka, Oryzias latipes. *Mech. Dev.* **2004**, *121*, 895–902. [CrossRef]
88. Hidaka, M.; Oda, S.; Kuwahara, Y.; Fukumoto, M.; Mitani, H. Cell Lines Derived from a Medaka Radiation-Sensitive Mutant have Defects in DNA Double-Strand Break Responses. *J. Radiat. Res.* **2010**, *51*, 165–171. [CrossRef]
89. Negron, J.F.; Lockshin, R.A. Activation of apoptosis and caspase-3 in zebrafish early gastrulae. *Dev. Dyn.* **2004**, *231*, 161–170. [CrossRef]
90. Honjo, Y.; Hiroshima University, Higashihiroshima, Japan. Personal communication, 2020.

International Journal of
Molecular Sciences

MDPI

Review

Ionizing Radiation-Induced Epigenetic Modifications and Their Relevance to Radiation Protection

Mauro Belli [1] **and Maria Antonella Tabocchini** [2,*]

[1] Independent Researcher, formerly: Istituto Superiore di Sanità, 00161 Rome, Italy; mau.belli1@gmail.com
[2] National Center for Innovative Technologies in Public Health, Istituto Superiore di Sanità, Viale Regima Elena 299, 00161 Rome, Italy
* Correspondence: antonella.tabocchini@iss.it

Received: 26 July 2020; Accepted: 17 August 2020; Published: 20 August 2020

Abstract: The present system of radiation protection assumes that exposure at low doses and/or low dose-rates leads to health risks linearly related to the dose. They are evaluated by a combination of epidemiological data and radiobiological models. The latter imply that radiation induces deleterious effects via genetic mutation caused by DNA damage with a linear dose-dependence. This picture is challenged by the observation of radiation-induced epigenetic effects (changes in gene expression without altering the DNA sequence) and of non-linear responses, such as non-targeted and adaptive responses, that in turn can be controlled by gene expression networks. Here, we review important aspects of the biological response to ionizing radiation in which epigenetic mechanisms are, or could be, involved, focusing on the possible implications to the low dose issue in radiation protection. We examine in particular radiation-induced cancer, non-cancer diseases and transgenerational (hereditary) effects. We conclude that more realistic models of radiation-induced cancer should include epigenetic contribution, particularly in the initiation and progression phases, while the impact on hereditary risk evaluation is expected to be low. Epigenetic effects are also relevant in the dispute about possible "beneficial" effects at low dose and/or low dose-rate exposures, including those given by the natural background radiation.

Keywords: ionizing radiation; radiation biology; radiation protection; health effects; epigenetics; low dose radiation; DNA methylation; non-targeted effects

---

1. Introduction

There is increasing interest in assessing the robustness of the present system of radiation protection at low doses and/or low dose-rates, typical of those exposures encountered in the workplace, in the environment and in diagnostic medicine (also irradiation of normal tissues in radiotherapy may fall in this type of exposure).

Quantitative evaluation of health risks at these levels of exposure is currently obtained by a combination of epidemiological and radiobiological data and models. Even though no comprehensive and "universal" model of radiation action on living systems, i.e., a model capable of describing all aspects at the different scale involved (molecular, cellular, tissue, organ, organisms), has been developed yet, nevertheless, radiobiology research, after just over a century of existence, has provided a wealth of information on biological response to ionizing radiation. Some important general notions are currently used by international bodies, such as the United Nations Scientific Committee on the Effects of Atomic Radiation (UNSCEAR) and the International Commission on Radiological Protection (ICRP), to extrapolate to low doses and low dose rates the health risk derived from epidemiological data at higher acute doses. These notions are essentially the harmful mutagenic potential of ionizing radiation and its linear dose-dependence at low levels of exposure [1,2]. In particular, the fundamental role of radiation-induced DNA damage in the induction of mutations and chromosome aberrations is currently assumed to provide a framework for the analysis of risks at low radiation doses and low dose-rate exposures [2,3]. Additionally, for the induction of cancer and heritable disease at low doses/low dose-rates, the use of a linear relationship between increments of dose and increased risk is considered a scientifically plausible assumption, even if uncertainties regarding this judgement are recognized [1,2,4,5].

However, non-linear radiobiological responses that can be relevant at low level exposures have been observed for many years, such as the so-called "non-targeted effects" (NTEs), and the (radio) adaptive response (AR). Moreover, it is now well established that ionizing radiation, besides genetic mutations, may also cause epigenetic alterations. In effect, epigenetic events are known to regulate gene activity and expression not only during development and differentiation, but also in response to environmental stimuli, such as ionizing radiation [6,7]. Interestingly, there is evidence that NTEs and AR are inter-related and even more interesting is the possibility that epigenetic mechanisms may have a role in them.

Evidence that such biological phenomena do not fit the classical paradigm of radiobiology, on which the internationally agreed system of radiation protection is currently based [1], has led to much discussion on if and how this paradigm should be modified [8–10].

Some excellent reviews have been reported on the historical and methodological aspects of radiobiology paradigm evolution [11] and on the effects of ionizing radiation on DNA methylation [12]. The present review encompasses many important aspects of the biological response to ionizing radiation in which epigenetic mechanisms are shown to be, or could likely be, involved with a focus on the possible implications in health risk assessment at low doses, a key issue in radiation protection.

2. The Role of Radiation Biology in Radiation Protection

2.1. The Current Paradigm of Radiation Biology

Ionizing radiation is capable of inducing a wide spectrum of DNA alterations, such as: base damage, sugar damage, single strand breaks (SSBs), double strand breaks (DSBs), DNA–DNA and DNA–protein cross-links. Clustered DNA lesions (defined as two or more lesions within one or two helical turns of DNA), such as complex DSBs and non-DSB clustered lesions [13] are considered to be the most biologically relevant form of radiation-induced DNA damage [14–18]. They are expected to be less readily repaired as compared to other radiation-induced damage and to endogenous or metabolism-related cellular damage. Indeed, ionizing radiation is uniquely very efficient at inducing clustered DNA lesions [19]. At low doses, even the passage of a single particle can produce clustered DNA lesions [15,17,20].

The frequency and degree of clustering of DNA damage depend on radiation quality [21]. There is evidence that clustered DNA damage, such as multiple DSB as well as non-DSB lesions close together [22] is the most challenging to repair and that the proportion of clustered damage increases with Linear Energy Transfer (LET), reaching ~70% or more for high-LET radiation (see the review in [23]).

High-LET charged particles typically induce complex chromosome aberrations [24,25] (defined as those aberrations involving three or more breaks in two or more chromosomes [26], although they can also be observed less frequently after exposure to γ-rays. In particular, high-LET heavy ions induce a high fraction of complex-type exchanges, and possibly unique chromosome rearrangements [27,28].

Un-repaired or mis-repaired DNA lesions cause changes in the DNA sequence, i.e., (genetic) mutations, that in turn are considered as the main event leading to deleterious biological effects, resulting, even at low doses, in an increase in both the probability of developing cancers and the rates of hereditary diseases naturally occurring in the population [2].

The association of genetic mutation to detrimental effects dates back to 1926 with Muller's discovery of mutagenic effects of X-rays or γ-rays on the fruit fly *Drosophila melanogaster*, although they were observed after high doses [29]. Muller, who for this discovery was awarded the 1946 Nobel Prize in medicine and physiology, became convinced that the vast majority of mutations were deleterious and consequently that exposure to radiation should be strictly controlled.

Indeed, it is now generally assumed that a vast majority of mutations are neutral or detrimental, as in many cases gene mutation is a process which burdens the population with a load of harmful genes. On the other hand, mutations may occur that, despite their rarity, increase the fitness of the biological system. However, considering the low likelihood of these favorable mutation events, radiation-induced mutations in humans, even at low doses, are generally assumed to be detrimental for radiation protection purposes [30].

A schematic and rationalized picture of the radiobiological knowledge for radiation protection purposes can be summarized by the following statements, forming what is sometimes referred to as the "conventional paradigm of radiobiology" [8], still considered as an useful reference framework:

(i) The DNA damage in directly exposed cells is the main event for biological effects;
(ii) the DNA damage occurs during, or very shortly after, irradiation of the nuclei in targeted cells;
(iii) the potential for biological consequences can be expressed within one or two cell generations;
(iv) at low doses, the biological effect is in direct proportion to the energy deposited in nuclear DNA.

The present internationally agreed system for radiation protection has used this paradigm, although with many simplifications and assumptions [1]. It forms the rational basis for assuming a linear relationship between risk and dose in radiation protection, known as the "Linear No-Threshold" (LNT) assumption.

2.2. Challenges to the Current Paradigm

Awareness is presently growing that a number of observations challenges the conventional paradigm, based on the target theory of radiation-induced effects. The most relevant are the occurrence of: (i) radiation induced epigenetic effects, i.e., changes in gene expression, for example through alteration of DNA and chromatin organization without altering DNA sequence; (ii) non-linear responses, such as non-targeted effects, i.e., effects observed in cells not directly traversed by radiation (bystander effects, BE) or occurring in the genome of the progeny of irradiated or bystander cells (genomic instability, GI), and (radio)adaptive responses (AR); all these NTEs can be described as the expression of inter- or intra-cellular signaling and are deemed to be particularly relevant to cell response to low doses.

3. Ionizing Radiation Induces Epigenetic Changes

3.1. The Main Epigenetic Modifications

By the second half of the last century, it was recognized that DNA by itself does not determine all characteristics of an organism, including the human one. The role emerged of those characteristics that crucially determine which genes are expressed in each cell type ("epigenetic" traits). The term "epigenetics" was coined in 1942 but its contemporary usage is quite recent, and for some years it has been used with variable meanings [31]. The modern definition of epigenetics is "the study of mitotically and/or meiotically heritable changes in gene function that cannot be explained by changes in DNA sequence" and the epigenetic trait (epigenome) of an organism is intended as the "stably heritable phenotype resulting from changes in a chromosome without alterations in the DNA sequence" [32].

Epigenetic events are known to regulate gene activity and expression during development and differentiation. In particular, epigenetic mechanisms regulate the gene expression in our body's cells to create all the different cell types, although they have the same genome. However, they also affect gene expression in response to environmental stimuli, including ionizing radiation (see the reviews in [6,7,33]). Epigenetics is thus considered to be a bridge between genotype and phenotype. Genetic mechanisms, such as mutations, are heritable, but not very susceptible to, or driven by, environmental influence (even if mutations can be induced by the environmental radiation, they are relatively rare events). At the other extreme, there are the metabolic pathways, responsive to environmental changes through interactions of chemical agents or other stressors with proteins involved in gene expression, that are not heritable. Epigenetic modifications, instead, are susceptible to environmental change and heritable at the same time. An interesting aspect is that they can persist after the stressor is removed, but they can also be reversible [34]. The main epigenetic changes currently considered are DNA methylation, histone modification, and modulation of non-coding RNAs (ncRNAs) (Figure 1) [35].

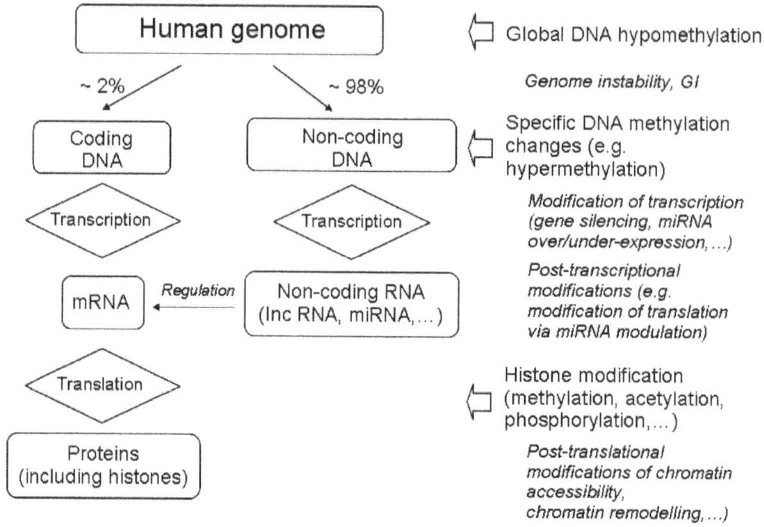

Figure 1. Epigenetic mechanisms involve different levels of gene expression (transcriptional, post-transcriptional, and post-translational). Only a small fraction of the human genome (2% or even less) accounts for protein-coding genes while the majority is associated with non-coding sequences, notably non-coding RNA genes. Of the non-coding DNA, only the regulatory part, giving rise to non-coding RNAs, is considered here. Epigenetic mechanisms can involve both protein-coding and non-coding RNA genes, with interplay between DNA methylation, histone modification and miRNA expression.

DNA methylation, i.e., the addition of methyl groups to the DNA. In mammals, DNA methylation is mostly at CpG sites to give 5-methylcytosine (5-mC). These sites are concentrated in specific regions called CpG islands, i.e., DNA sequences with high level of CpG sites (typically 300–3000 bp with C + G content > 50%), sometimes located consecutively. In humans, CpG islands occupy about 70% of human gene promoter regions [36]. In transcriptionally active regions of the genome, GpC islands are normally hypomethylated, allowing that gene to be expressed. Therefore, the methylation of CpG sites is a critical factor affecting gene transcription because of its ability to directly silence gene expression. DNA methylation was one of the first identified and the most widely studied epigenetic alteration [37]. It is now a consolidated notion that hypermethylation of genomic DNA is linked to transcriptional silencing and hypomethylation to chromosomal instability [38,39]. DNA methylation is considered a heritable epigenetic mark since methylation modifications that regulate gene expression are usually heritable in mitotically dividing cells. In contrast, it shows dynamic changes during development and cell differentiation, even if some methylation patterns may be retained as a form of epigenetic memory [40]. In mammals, DNA methylation patterns are maintained or established by a family of enzymes, the DNA methyltransferases (DNMTs), notably DNMT1 (maintenance methylation) and DNMT3 (de novo methylation), while other proteins can achieve active de-methylation ([38] and refs therein).

Histone modifications, including, inter alia, acetylation, methylation, phosphorylation and ubiquitination. For years, histones were regarded as merely structural proteins, but now they are recognized to control the organization of chromatin and hence transcriptional responses [41]. Post-translational modifications on histones can change gene transcription by changing DNA accessibility, but also by recruiting other proteins. Histone acetylation, the first epigenetic modification shown to be connected with biological activity [42], neutralizes histone positive charges and reduces its interaction with the negatively charged DNA, thereby inducing chromatin structure relaxation and a marked increase in gene expression. On the contrary, histone methylation does not alter the charge of the modified residue and can either repress or activate transcription depending on the methylation site [43].

Modulation of non-coding RNAs (ncRNAs). Among these RNAs, much attention has been paid to microRNAs (miRNAs), which are small RNA molecules (usually 21–23 nucleotides) discovered in 1993 [44]. Countless microRNAs have been discovered and described in the past years [45,46]. They play an important role in animals and plants in regulating gene expression by transiently inhibiting the translation of a messenger RNA molecule or by inducing its degradation [47,48]. In addition, long non-coding (lnc) RNA molecules may have an epigenetic role [49], since they bind to the transcripts in the nucleus as they emerge from the DNA. miRNAs are involved in multiple biological processes, including cell proliferation, differentiation, and programmed cell death. Since the dysregulation of these processes is a hallmark of cancer, miRNAs can be viewed as important contributors to the pathogenesis of cancer, including initiation and progression [50,51]. They are estimated to regulate the expression of up to 60% of the human protein coding genes [52,53] by means of mRNA degradation or translational repression, acting through a multitude of interconnecting regulatory pathways [51,54–56].

3.2. Radiation-Induced Changes in DNA Methylation

Early findings obtained at the end of the 1980s indicate that exposure to ^{60}Co γ-radiation causes a dose-dependent decrease in DNA methylation, in terms of levels of 5-mC, in several cultured cell lines [57]. Since then, considerably amount of research carried out both in vitro and in vivo showed that X- or γ-rays can change the DNA methylation pattern (see the reviews in [12] and [58]).

Studies on cultured human cells showed that low-LET radiation induces DNA hypomethylation that displayed different profiles in radioresistant and radiosensitive cultured human cells [59,60]. Animal studies, in particular on rodent models, indicated that low-LET radiation induces global DNA hypomethylation that is not ubiquitous among different tissues and cells [61], that occurs in a dose-dependent, sex-, and tissue-specific manner [62,63], and that can be persistent [64,65].

Overall, these data indicate that low-LET radiation exposure results in global DNA hypomethylation. However, it is important to identify whether or not hypomethylation is uniformly distributed throughout the genome, and whether there is also specific locus hypermethylation, which is known to be associated with inactive chromatin state and in most cases with repressed gene expression activity [66–68]. It should be considered that the majority of the eukaryotic genome is composed of repetitive elements (REs), while only less than 2% is occupied by protein-coding genes [69]. Non-coding REs, in particular the so-called transposable elements (TE), provide a rich source of gene regulation. Their hypomethylation, especially in the regions called "Long Interspersed Nucleotide Element 1" (LINE-1), has been observed in virtually all human cancers and is frequently associated with a poor prognosis [70]. Loss of DNA methylation in the TEs enhances transcriptional activity so that reactivation of TEs potentially leads to GI [58,71–74], considered as a major hallmark of many cancer ([75] and refs therein). Many lines of evidence clearly demonstrate that alterations in methylation and expression of TEs are caused by exposure to environmental stressors, many of which are carcinogens or suspected carcinogens so that it has been proposed that TEs can serve as biomarkers of exposure to environmental stressors [72]. However, hypermethylation of TEs has also been detected in some in vitro experiments, suggesting that alterations in the methylation status of TEs is tissue-, dose-, and radiation quality-dependent (see [72] for a review).

Specific-gene hypermethylation often involves normally unmethylated CpG islands, and can be associated with transcriptional silencing of the corresponding gene. If it is a suppressor gene, its loss of function may be a key event contributing to the oncogenic process [76–79]. In effect, some studies demonstrated significant DNA hypermethylation of tumor-suppressor genes in workers exposed to ionizing radiation [80,81]. Also, gene-specific DNA methylation changes was found in human breast cancer cells irradiated with X-rays [82]. Interestingly, this differential methylation changes correlate with already known biological responses to radiation, such as those on cell cycle, DNA repair, and apoptosis.

3.3. Radiation-Induced Histone Modifications

Cell exposure to ionizing radiation results in a wide variety of histone modifications. A well-known radiation-induced histone modification is phosphorylation of histone H2AX, which is crucially important for the repair of DNA double strand breaks and for the maintenance of genome stability. Phosphorylation of this histone at serine 139 (γ-H2AX) is an early cellular response to ionizing radiation and is used as a measure of DSBs [83,84].

In an in vivo murine model, low-dose X-ray irradiation resulted in decreased tri-methylation of histone H4 in the thymus accompanied by an overall reduction in chromatin compactness, a significant increase in global DNA hypomethylation as well as an accumulation of DNA damage, and was associated to a reduced expression of DNMTs [85]. Similar histone modifications were found in human breast cancers [86]. These findings demonstrate that radiation-induced changes in DNA methylation and histone modifications result in overall GI (see [43] for a review).

Furthermore, it has been shown that chromatin modification by histone acetylation is also crucial for DNA repair [87], and that chromatin acetylation is involved in several important steps such as chromatin remodelling and tagging of DSBs, activation of repair regulators, cell cycle regulation, and apoptosis [88].

3.4. Radiation-Induced Modulation of Non-Coding RNA Expression

Another type of epigenetic radiation-induced modification involves ncRNAs, in particular miRNAs that have specific roles in the regulation of gene expression. Since their discovery in 1993 [44], miRNAs have emerged as important modulators in many cellular pathways, including cell proliferation, differentiation, and programmed cell death, and the roles of specific miRNAs have begun to be elucidated.

A number of studies have examined the general and specific effects of miRNA perturbation in different cell types exposed to low-LET ionizing radiation (see [89] for a review). miRNAs have been shown to be involved in the response of irradiated cultured human cells [90]. In particular, it was shown that ionizing radiation affects miRNA levels in human endothelial cells [91]. Overall, these studies revealed that the expression levels of several miRNAs change significantly upon irradiation and indicated a specific role of various miRNAs on cellular radiosensitivity [92]. miRNAs have also shown to have a fundamental role in several radiation-induced cell signaling events, such as those involving cell cycle arrest and cell death (reviewed in [93]). Many studies demonstrated that miRNA expression levels change in response to radiation, and that certain miRNAs alter radiation sensitivity, suggesting they are good potential targets for enhancing the efficacy of cancer radiation therapy [89,94–97]. Expression levels of a variety of miRNAs after low-LET ionizing radiation were reviewed and listed in [98].

3.5. Radiation Quality May Affect Epigenetic Changes

Most research on the impact of radiation exposure on the epigenome has focused on the effects of low-LET X- or γ-rays. In contrast, few studies have assessed the effects of high-LET radiation on the epigenome. Increased interest in the mechanisms underlying biological effects of high-LET radiation was triggered quite recently by the investigation on the health risk posed by the space radiation during manned space missions and to the introduction of high-LET radiation into clinical practice (hadrontherapy). Comparison between epigenetic effects induced by low- and high-LET radiation was addressed in particular by Morgan's group [58,90]. It is expected that high-LET radiation has the potential for unique effects on the epigenome, given the unique characteristics of its track structure. Indeed, there are now a number of studies showing that exposure to high-LET radiation can result in lasting changes in the total levels of DNA methylation and in the miRNA expression that may be different from those induced by equivalent doses of low-LET radiation [58,90,99–102].

Some of these studies focused on the effect of high energy and charge (HZE) particles, such as high energy Fe-ions (usually 600–1000 MeV/u, LET 180–150 keV/μm), as they are representative of the most detrimental component of space radiation associated to health risks encountered by astronauts in deep space [103,104]. Comparison between X-ray and high-LET Fe-ions exposures of cultured cells showed that Fe-ions elicited more chromosomal damage and cell killing than X-rays do [90]. Global DNA methylation was affected in a different way, as hypermethylation was found in cultured cells 16–20 doublings after exposure to protons and high-LET Fe-ions in contrast to hypomethylation for cells exposed to X-rays [58,101]. Global DNA hypermethylation was also confirmed after exposure to Fe-ions in a mouse model [100]. Interestingly, high energy protons of relatively low-LET gave an effect similar to that caused by high-LET Fe-ions, suggesting that epigenetic responses to radiation may be based on radiation quality rather than LET [58].

A possible explanation for the difference between sparsely and densely ionizing radiation comes from the possible difference in oxidative stress [58] or from the observations [102] that stable DNA methylation can result at the sites of DNA break repair [105], likely produced with higher yield by densely ionizing radiation.

However, after exposure to high-LET Fe-ions, TE hypomethylation was detected in the same cultured cells that displayed global hypermethylation [58]. In vivo experiments performed on mouse models irradiated with Fe-ions showed a complex picture: hypo- or hyper-methylation in TEs depended on the organ analyzed and on the observation time (see [12] for a review). It was also clarified that DNA hypermethylation of LINE-1 elements found in the lungs of mice irradiated with Fe-ions depended on their evolutionary age [106].

Presently, little information is available on the effect of high-LET radiation on methylation at specific genes. In vitro experiments showed hyper- or hypo-methylation or no changes at promoters of specific loci that are used as biomarkers for the early detection of carcinogenesis [12,58,90,99]. The observed differences are likely related to differences in cell types, doses/dose rates, time of observation, or assay used. In a mouse model irradiated with Fe-ions, an increase in 5-mC content was reported that, however, was not associated with increased DNA methylation in a panel of tumor-suppressor genes frequently hypermethylated and inactivated in lung cancer [100]. Some information has come from human data on exposed workers (as reviewed in [12]). Significant DNA hypermethylation of the cyclin-dependent kinase CDKN2A, and of the DNA methyltransferase MGMT genes was found in the sputum of uranium miners exposed to radon [80]. This analysis was also proposed to predict lung cancer. Another study found high levels of p16 hypermethylation in lung adenocarcinomas from plutonium-exposed workers at the Russian nuclear plant MAYAK [107]. However, these results should be regarded as qualitative, since it is not easy to quantify the high-LET exposure in these cases.

Dependence on radiation quality was also found for effects on miRNA expression. In cultured cells, Fe-ions irradiation caused a lower incidence of alteration of miRNA expression levels than X-rays do [90], a quite surprising result given the higher effectiveness of Fe-ions for chromosomal damage and cell killing. Irradiation with high energy protons, γ-rays, or Fe-ions in mouse blood resulted in a radiation type- and dose-specific modifications of a panel of 31 miRNAs [108], so that the extent of miRNA expression signatures derived from mouse blood was proposed as a biomarker for exposure to high-energy protons [109].

4. Basic Mechanisms of Radiation-Induced Epigenetic Changes

It is well known that ionizing radiation can cause DNA lesions by direct deposition of energy in the DNA as well as by the indirect action of reactive chemical species formed near the DNA [15,110] and that the spectrum of lesions depends on radiation quality [17,18]. Indirect DNA damage from water free radicals is the most frequent mechanism for low-LET radiation, while direct DNA damage is predominant for high-LET radiation [111,112]. These radicals are formed through the radiolysis of water, the hydroxyl radicals being considered the most damaging among them. In aerobic conditions, these free radicals are converted to reactive oxygen species (ROS) that include free radicals as well as non-free radicals. Organic radicals are also formed, giving rise to peroxyl radicals (strong oxidizing species) and hydroperoxydes (see e.g., [113]). Ionizing radiation can also generate reactive nitrogen species (RNS) through the upregulation of several enzymes, including inducible nitric oxide synthase. Nitric oxide reacts with superoxide radical, generating peroxynitrite, a strong oxidant radical [43]. The yield and spatial distribution of ROS and RNS are strongly modulated by radiation quality as a consequence of the specific track structure of each quality [113]. ROS and RNS can attack DNA resulting in several alterations, including DNA breaks, base damage and destruction of sugars. These lesions, if unrepaired or mis-repaired, may lead to genetic mutations in surviving cells. In this context, particularly relevant are the DNA clustered lesions [22], since they appear to be "highly resistant" to faithful repair (see the review in [23]).

The mechanisms by which ROS are generated by ionizing radiation were studied in some detail in fibroblasts, where it was shown that ROS can be directly generated by radiation exposure and indirectly through the damage of mitochondria. This leads to the activation of the signaling pathway, which sustains an increase in ROS levels by increasing oxidase expression, thereby setting up a cycle of high oxidative stress, i.e., excess of ROS/RNS not compensated by the scavenging mechanisms of the cell [114].

Besides the mutagenic action of ROS and RNS, there is also evidence that oxidative stress has a fundamental role in epigenetic modifications [115–117]. Oxidative stress can modify the epigenome by multiple mechanisms, the most important of which involve oxidation of DNA bases and/or mitochondria-mediated changes, with the main target being the CpG sites, especially in the CpG islands.

Among the mechanisms leading to global DNA de-methylation/hypomethylation, an important one is the oxidation of 5 mC to 5-hydroxymethylcytosine (5 hmC), which serves as an intermediate in active DNA demethylation [118,119]. In addition, oxidation of guanine to 8-Oxo-2′-deoxyguanosine (8-oxo-dG) can create mismatches via pairing with A, thus leading to G > T transversion. In addition, 8-oxo-dG can also affect dC methylation by interfering with the ability of DNA to function as a substrate for the DNMTs, inhibiting DNA methylation at nearby C bases [120]. A complete understanding of the effect of 8-oxo-dG is still a matter of study, since it may alter gene expression in multiple ways [121].

Mitochondria also appear to have an important role in radiation-induced global DNA hypomethylation. Dysfunction of mitochondria can affect epigenetic regulation [122]. Mitochondria constitute a major intracellular source of reactive species, as they generate almost 90% of the total number of cellular ROS [123]. High intra-mitochondrial ROS level can damage the mitochondrial DNA, causing global DNA hypomethylation, by decreasing the activity of DNMTs and these changes are transmitted to the progeny of the irradiated cells [124]. These observations suggest that mitochondrial dysfunction can cause oxidative DNA damage and contributes to an altered epigenetic landscape to perpetuate radiation-induced instability [125].

In addition to hypomethylation, ROS can also induce site-specific hypermethylation by different mechanisms, such as catalysis of DNA methylation or upregulation of DNMTs levels, thereby leading to gene silencing [126] (Figure 2).

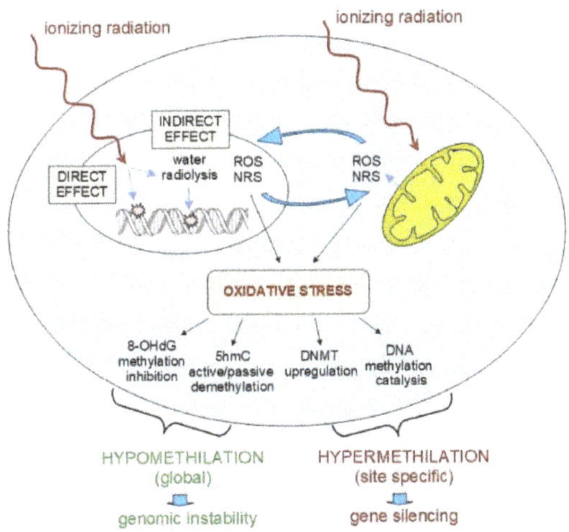

Figure 2. Simplified representation of the role of reactive oxygen species (ROS) and reactive nitrogen species (RNS) in the epigenetic response (DNA methylation) to ionizing radiation. Ionizing radiation can cause DNA lesions by direct deposition of energy in the DNA as well as by the indirect action of reactive chemical species formed near the DNA. Indirect DNA damage arises from free radicals formed through the radiolysis of water molecules. In aerobic conditions, these free radicals are converted to reactive oxygen species (ROS). Ionizing radiation can also generate reactive nitrogen species (RNS) through the upregulation of several enzymes. The yield and spatial distribution of ROS and RNS are strongly modulated by radiation quality because of the specific track structure of each quality [113]. ROS and RNS can attack DNA, resulting in several alterations that, if unrepaired or mis-repaired, may lead to genetic mutations in surviving cells. In addition, they can drive various epigenetic modifications through several mechanisms. It was demonstrated, especially in fibroblasts [114], that ROS can be directly generated by radiation exposure and indirectly through the radiation damage of mitochondria, leading to the activation of signaling pathways, which in turn sustains an increase in ROS levels. Oxidative stress results when excess of ROS/RNS are not compensated by the scavenging mechanisms of the cell. DNA hypomethylation can be achieved by oxidation of guanine to 8-OHdG that inhibits DNA methylation at nearby cytosine bases, and by hydroxylation of 5 mC to 5 hmC that causes active DNA demethylation processes. ROS can also induce site-specific DNA hypermethylation by up-regulation of expression of DNMTs or by acting as catalysts of DNA methylation [126]. Oxidative stress can influence the epigenetic landscape of the cell on other levels, such as by histone modifications and miRNA expression (not shown here).

Oxidative stress also contributes to epigenetic changes by altering the action of ncRNAs, in particular miRNA. However, the interactions between ROS metabolism and miRNA levels appear to be complex. There is evidence that miRNAs are critical regulators of the cellular stress response and thus are responsive to ROS, some of them being themselves able to regulate ROS levels ([127], and refs therein). Analysis of ROS-mediated miRNA expression patterns revealed that the gene locations for epigenetic changes correspond to fragile sites known to be targets of oxidative damage [43].

It is important to note that research on radiation-induced epigenetic mechanisms was initially addressed to DNA methylation as a process capable of modulating gene expression by changing chromatin organization, and subsequently integrated with the roles of histone modifications and changes in miRNA expression as they would act independently. However, there is accumulating evidence of interactions between these different types of epigenetic changes.

Histone modifications can regulate the DNA methylation [128,129]. DNA methylation and histone methylation are tied together in a reinforcing loop [130–134]. Cross-talks between DNA methylation and histone modification were shown at specific gene loci and are present in eukaryotic organisms, although they vary widely, in fungi, plants and animals [135]. DNA methylation can also be regulated by miRNAs targeting DNMTs or critical methylation-related proteins, whereas DNA methylation regulates miRNA expression via hypermethylation or hypomethylation of the promoter-associated CpG islands, thereby achieving a sort of mutual regulation [136,137].

In summary, it appears that radiation-induced oxidative stress is an important player in shaping the epigenetic landscape of the entire genome [127], that is a result of a cross-link between DNA methylation, histone modification and ncRNA (in particular miRNA) expression [137,138].

Interestingly, it appears that the production of oxidizing species that are responsible for inducing DNA damage via indirect effects can also have a role in the damage repair processes via epigenetic changes that enable DNA accessibility to repair enzymes. For example, this can be accomplished through histone modification or replacing canonical histones with histone variants, thereby inducing the needed changes in chromatin structure and dynamics [139].

5. Epigenetic Changes Have a Role in Radiation-Induced NTE and AR

In the last three decades, a wealth of investigations have been carried out on NTE, namely BE and GI, discovered between the end of 1980 and the beginning of the 1990s [140–146], and on radiation-induced AR, discovered even earlier [147], that are phenomena that do not fit into the conventional paradigm.

Radiation-induced GI is an encompassing term which is used to describe the acquisition of an increased rate of alterations within the genome, manifested in the unexposed progeny of irradiated cells [145,146] (Figure 3). Radiation-induced BE describes the ability of cells affected by radiation to convey manifestations of damage to other cells not directly targeted [141,148]. Abscopal, or out-of-field, effects, defined in radiotherapy as radiation-induced effects observed outside the irradiated volume, are currently considered as a special type of BE [149]. Abscopal effects were seen in rodents, such as the induction of profound epigenetic dysregulation in spleen tissue after localized cranial radiation exposure [150], and the increased induction of malignancy in the shielded head (specifically in the brain) of radio-sensitive mice after exposure of the remainder of the body to X-rays [151].

These phenomena challenge the concepts on which the conventional paradigm of radiation biology is based and are potentially relevant for radiation risk assessment, especially at low doses [149,152,153]. They have been seen in many in vitro and in vivo experiments, including experiments with blood samples from irradiated humans [145,146,152].

There is evidence that all these phenomena are inter-related and that they may share some common mechanistic pathways (see, e.g., [153–155]). For example, the radiation-induced intercellular signaling cascades, including cytokine production, nitric oxide production and persistent free radical production have the potential to mediate both GI and BE. Indeed, GI was observed in the progeny of unirradiated neighbors of irradiated cells [156].

Most NTE have been observed in vitro, but they can also be relevant in vivo, even if the question remains whether the non-targeted effects demonstrated in vitro can be extrapolated to in vivo situations. In vitro experiments have provided some important insights into the nature of these effects, but in spite of extensive research, their mechanisms remain to be completely understood. An intriguing observation is that, even if NTE and AR have been observed in a variety of cell and tissue types, biological end-points and radiation qualities [152,157–159], they have not been universally observed [153,160–162].

Epigenetic mechanisms, encompassing DNA methylation, histone modification, and RNA-associated gene silencing, have been shown to be plausible mediators of the mentioned effects. These inter-relationships have stimulated much interest, especially for their possible impact in the risk assessment at low radiation doses and have been the subject of a number of studies [73,90,163]. It is worth noting here that in its latest recommendations, ICRP classifies GI and BE as epigenetic responses to radiation [1].

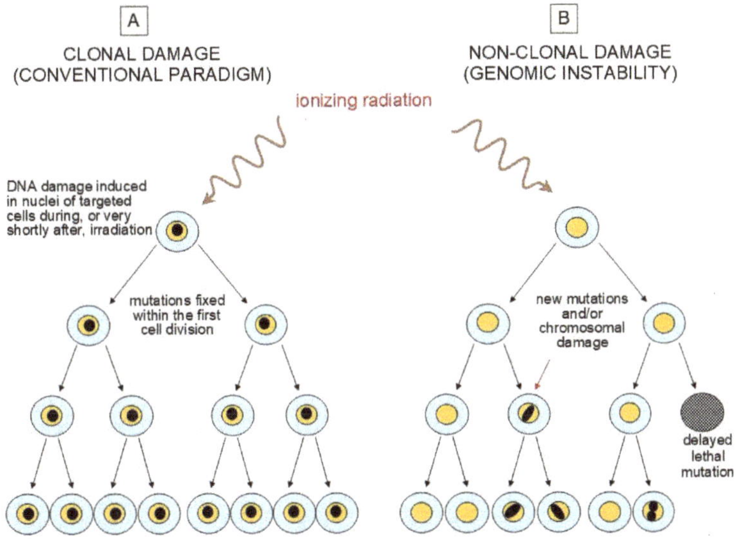

Figure 3. Schematic description of two possible ways for radiation-induced damage propagation in the progeny of irradiated cells: (**A**) according to the conventional paradigm of radiobiology, damage is induced during, or shortly after, irradiation, and clonally propagates to the progeny; (**B**) by genomic instability, a non-clonal effects that is observed as new mutations and/or new chromosomal damage in the unirradiated progeny of the irradiated cell.

Indeed, there are many lines of evidence that epigenetic mechanisms have a potential role in GI. An early observation was made on micronuclei induction in cultured cells irradiated with different fluences of alpha-particles, indicating that the target for GI is larger than the cell nucleus [164]. It was shown that GI can occur without the need for genetic alterations as an initiating or perpetuating factor [165,166]. Cells irradiated with low to medium doses exhibited a much larger proportion of GI than mutations from targeted effects, suggesting that instability might arise, rather than from a genetic mutation, through epigenetic mechanisms [152].

All the above observations give support to the idea that epigenetic alterations could be a mechanism of GI induction [167]. Indeed, experimental evidence points to a causal relationship between GI in the exposed animals and the radiation-induced global DNA hypomethylation (see the review in [168]). The fundamental role of DNA methylation in the transmission of GI is clearly demonstrated in embryonic stem cells since the disruption of specific DNMT genes completely eliminates the transmission of GI. Interestingly, this inactivation also protects neighboring cells from indirect induction of GI [169].

A recent review highlighted the link between radiation-induced ROS, DNA hypomethylation and GI and/or AR [170]. Moreover, it has been reported that mitochondrial-derived reactive species can not only cause oxidative DNA damage but also directly affect aberrant changes in 5 mC levels, suggesting a link between radiation-induced genomic instability, epigenetic mechanisms and mitochondrial dysfunction [125].

There is some evidence that, besides DNA methylation, miRNA may also have a role in radiation-induced BE and GI. Concerning BE, an experiment conducted in vivo showed that partial exposure of a mouse body induced a significant upregulation of a specific miRNA in distant lead-shielded liver tissue [150]. Various bystander end-points, such as apoptosis, cell cycle deregulation, and DNA hypomethylation, are shown to be mediated by the altered expression of miRNAs, even if they do not appear to be the primary bystander signaling molecules in the formation of bystander-induced DNA strand breaks [163]. Concerning GI, upregulation of miRNAs was found in directly exposed male mice, leading to hypomethylation of the exposed animals as well as of their unexposed offspring, demonstrating the possibility that they may play a role in the transgenerational epigenetic inheritance of GI [171].

Information about the possible role of epigenetic mechanisms in the AR to ionizing radiation is scarce. AR can be regarded as a quite general phenomenon of biological response as it has been observed in cells, tissues and organisms using various indicators of biological damage after exposure to ionizing radiation and to other stressors. Although in the literature there are a plethora of descriptions about the adapting conditions, and DNA repair and antioxidative mechanisms are among the best described pathways involved in it, the mechanisms underlying AR remain poorly understood [172]. Both intracellular and intercellular signaling (the latter being mainly related to BE) can account for the occurrence of AR. Enhanced efficiency of DSB repair through homologous recombination and a significant increase in gene expression of antioxidant enzymes appear to play a predominant role in the adaptive response (see the review in [9]). An interesting AR model has been developed accordingly [173].

However, the processes by which increasing in DNA repair efficiency and in antioxidant levels would be accomplished by exposure to the priming dose are not clear. It has been proposed that, in order to adapt the gene expression program to the stress situation, and to achieve proper functioning of DNA repair processes, epigenetic processes are involved, notably transient protein acetylation [88]. Furthermore, data on endothelial cells suggest that the radiation-induced changes in miRNAs expression modulate the intrinsic radiosensitivity of these cells in subsequent irradiations [91]. A recent review pointed to the radiation-induced oxidative stress as the source of various processes connected to AR [170], which is consistent with the occurrence of epigenetic mechanisms.

6. Epigenetics in Radiation Risk Assessment

6.1. Radiation-Induced Cancer

There is large consensus on the fact that cancer, in general, is a disease that results from both genetic and epigenetic changes and several studies pointed to the description of cancer as due to a dysregulated epigenome allowing cellular growth advantage at the expense of the host, with mechanisms involving both genetic mutations and epigenetic modification4s [174,175]. This notion applies not only to solid cancers but also to leukemia, in particular to myeloid leukemia [176]. Dramatic changes in DNA methylation are common in cancer and are considered as an early event in many of them [177,178]. DNA methylation changes appear to be even more frequent events than genetic mutations [179,180]. The global loss of DNA methylation at CpG dinucleotides was the first epigenetic abnormality identified in cancer cells [177,181]. Loss of genome-wide methylation, especially in repetitive elements [77], promotes GI, considered as a major hallmark of cancer [182,183]. For its part, gene hypermethylation, often involving normally unmethylated CpG islands, can be associated with their transcriptional silencing and, if they are suppressor genes, their loss of function may be a key event contributing to the oncogenic process [78,79]. For example, the silencing of the BRCA1 gene by promoter hypermethylation occurs in primary breast and ovarian carcinomas, supporting a role for this tumor suppressor gene in sporadic breast and ovarian tumorigenesis [184]. It has been evaluated that more than 300 genes and gene products are epigenetically altered in various human cancers [185] and a meta-analysis of the altered genes in colorectal cancer reinforces their involvement in tumorigenesis [186].

Additionally, in radiation-induced cancers, a role for tumor suppressor gene hypermethylation has been demonstrated. Silencing of suppressor genes was detected in murine models of radiation-induced lymphoma, in lung tumors of rats induced by exposure to Pu-239, and in human lung adenocarcinoma occurring in workers of the Russian MAYAK plutonium plant ([33,81] and references therein). Interestingly, a study focused on lung carcinoma in radiation-exposed MAYAK workers compared to non-worker controls showed that methylation at one gene (coding for a tumor suppressor protein) occurred more often in carcinomas found in exposed workers than in non-worker controls, with a dose-dependent prevalence [107]. Aberrant hypermethylation was observed in an appreciable fraction of patients with renal cell carcinomas living in radiocontaminated areas after the Chernobyl accident [187]. Significant DNA hypermethylation of tumor suppressor genes was detected in workers exposed to radon in uranium mines, even without detectable cancers [80].

The findings described above indicate that radiation exposure, although normally thought to be pathogenic through DNA damage such as deletions and point mutations [188], may also elevate the cancer risk through epigenetic alteration, resulting in GI increase and/or specific silencing of tumor suppressor genes.

While research in cancer epigenetics was initially focused on DNA methylation abnormalities, particularly on CpG island promoter methylation [189], other players have eventually emerged, a not unexpected result given that probably about 40% of human genes do not contain CpG islands in their promoters [190]. Indeed, besides aberrant DNA methylation, which is one of the most well studied epigenetic changes in cancer cells, it was found that histone modifications and chromatin remodeling are also involved in cancer [191,192].

Next-generation sequencing revealed that more than 50% of human cancers harbor mutations in enzymes that are involved in chromatin organization. [193]. Importantly, aberrant activity of histone-modifying factors may promote cancer development by mis-regulating chromatin structure and activity [194], as frequently found in human leukemias [195].

In recent years, there has been tremendous and growing interest in investigating the role of dysregulation of ncRNAs, notably miRNAs, in normal cellular functions as well as in disease processes. Indeed, less than 2% of the entire human genome encodes proteins, while the majority of it (at least 75%) encodes ncRNAs [69]. There is now emerging evidence that these RNAs are involved in the development and progression of leukemia and cancer [196–198].

Alterations in miRNA expression may occur following exposure to several stress-inducing anticancer agents including ionizing radiation, etoposide, and hydrogen peroxide (H_2O_2). Dysregulation of a family of miRNA was found in Ptch1 ± mice that are highly susceptible to radiation-induced medulloblastoma [199].

These findings are consistent with the general notion that typically, miRNAs involved in radiation tumorigenesis are dysregulated, and this dysregulation is believed to alter the expression of protein-coding mRNA, thereby favoring uncontrolled tumor cell growth, in some cases by decreasing tumor suppressor expression [200]. miRNA-related epigenetic changes have been proposed to be the "missing link" between radiation exposure, radiation-induced genomic instability, and radiation-induced carcinogenesis [90].

It should be noted that most of the investigations focusing on the relationship between radiation-induced cancer and miRNA changes were obtained using rodent models, while relatively fewer studies have been performed on human cancers. An interesting finding of one of these few is the upregulation of a specific miRNA in breast cancer tissue samples derived from Chernobyl radiation-exposed female clean-up workers [201].

While the occurrence is well established of a relationship between radiation-induced cancers and epigenetic changes, the question can be posed whether these changes are the cause of cell transformation, or rather the consequence of it. Indeed, it is now accepted that epigenetic abnormalities along with genetic alterations are involved in the initiation and progression of cancer [189,202]. For example, it was found that in rat mammary cells, the frequency of initiation (the first step in oncogenesis) induced by γ-rays was much higher than specific locus mutations [203–205], and that the observed frequency of radiation-induced GI is considerably higher than that observed for gene mutations at a similar dose, suggesting that the latter is highly unlikely to be the initiating mechanism for GI [145,146,188].

It has also been suggested that a crucial role in such steps is played by epigenetically disrupted stem/progenitor cells [205], a hypothesis consistent with the importance that is now given to cancer stem cells in cancer development and perpetuation [206].

Epigenetic considerations also affect individual susceptibility to radiation-induced cancer. Assessment of individual variability in cancer risk is a key area to address for radiation protection. It is recognized that differences in radiation sensitivity between individuals, or groups, may relate to gender, age at exposure, state of health, genetic and epigenetic make-up, lifestyle, and age attained [207].

6.2. Transgenerational Effects

The heritable change in gene expression that is induced by a previous stimulus, such as ionizing radiation, is often described as epigenetic memory. Epigenetic memory is a sort of "footprint" that maintains gene expression states through cell generations without changes in DNA sequence and in the absence of the initial stimulus. Epigenetic memory can be considered over different time scales: cellular and transcriptional memory (mitotically heritable) and transgenerational memory (meiotically heritable) (see, e.g., [208]). In this paper, "transgenerational epigenetic effects" are intended as those effects which arise in the offspring of the irradiated organism and that are not due to the inheritance of DNA mutations through the parental germline, according to the current use in radiation protection issues [209]. Epigenetic variation induced by environmental factors contributes to the phenotypic plasticity and adaptive capacity of various species. The molecular basis of cellular memory is a fascinating topic that has been addressed during the last few decades [40].

In many cases, epigenetic changes have been proven to be stable and can lead to transgenerational heritable changes. In plants and in some animals, such as nematodes, transgenerational epigenetic inheritance is well-documented and relatively common [210]. Many examples have been reported for transgenerational epigenetic effects in which environmental exposures, including ionizing radiation, lead to heritable phenotypic changes that pass through male, female and sometimes both germlines (reviewed in [211]). In mammals, epigenetic patterns are largely erased and then remodeled during germ cell development and early embryonic development (epigenetic reprogramming) [212,213].

The first evidence for a radiation-induced transgenerational effect was reported in 1976 by Luning et al. [214], who showed elevated rates of dominant lethal mutations following intraperitonial injection of male mice with a plutonium salt solution. Afterwards, animal models demonstrated that effects of the parental radiation exposure are transmitted through the germline to the progeny of the irradiated parent [145,146,215].

Radiation-induced transgenerational effects may involve radiation-induced genome instability. Indeed, in vitro data have shown that ionizing radiation can induce genomic instability that can manifest in the progeny of the irradiated cells for many divisions [145] and transgenerational induction of chromosomal instability has also been documented in vivo, notably in irradiated rodents [2,146,216].

Immediately relevant questions are whether the effects are common or rare, and whether they are long-lasting or transient. Indeed, in several species, transgenerational effects have been detected in many generations after the parents were exposed to ionizing radiation, (see the review in [168]. Recent results on vertebrates (zebrafish) show that ionizing radiation-related effects in offspring can be linked to DNA methylation changes, many of which could be associated to pathways involved in cancers and apoptosis, that partly can persist over generations. It has also been suggested that monitoring DNA methylation could serve as a biomarker to provide an indication of ancestral exposure to ionizing radiation [217].

A question especially relevant for radiation protection purposes is whether transgenerational radiation effects occurs in humans. While animal studies show such effects, their occurrence is highly controversial in humans. A high risk of leukemia and birth defects has been reported in the children of fathers who had been exposed to radionuclides in the nuclear reprocessing plants [218] and an increase in minisatellite mutations was found in offspring of various groups living close to the Chernobyl site, to nuclear test sites in Kazakhstan and to the Techa-river region [219–221]. However, these findings were not supported by studies in the children of atomic bomb survivors in Hiroshima and Nagasaki [222]. A review published in 2013 of these and other available data concluded that "studies of disease in the offspring of irradiated humans have not so far identified any effects on health, possibly in part a result of lack of statistical power", and that transgenerational effects of radiation, if any, "may be restricted to relatively short times post-exposure, when in humans conception is likely to be rare" [209]. A subsequent review [223] also considered more recent results from a long-term monitoring by Russian Federation of the children of residents exposed to radionuclides after the Chernobyl accident, which showed an increased prevalence of malignant neoplasms, especially childhood cancer, and other disorders. Based on these findings and on the consideration that the negative results of gene mutations in Hiroshima and Nagasaki might be caused by erroneous methodology, these authors concluded that radiation-induced persistent accumulation of genomic instability may cause various disorders in a further generation in humans [223]. Research has been undertaken using plant and animal systems to understand the mechanisms governing the epigenetic transgenerational effects in organisms exposed chronically to low- doses in Chernobyl and in Fukushima areas [224]. The results so far obtained from laboratory and field studies confirm that DNA methylation might be the key to transfer the response to ionizing radiation from one generation to the next, but more in depth studies are needed, involving other epigenetic mechanisms such as histone modifications and microRNAs, linked to responses at higher levels of biological complexity [225].

If transgenerational effects of radiation were to be demonstrated to apply to humans, it may have implications in radiation protection when estimating the hereditary risks (i.e., the risk of induction of genetic diseases expressed in future generations) of ionizing radiation in human populations. According to the current risk assessment system, they are quantified as the harmful genetic effects on the descendants of those exposed, resulting from the induction of germline mutations and their transmission over generations [1]. This implies that mutation induction in directly exposed cells is regarded as the cause of this risk for humans. Since epidemiological studies have not provided clear evidence of heritable effects of radiation exposure in humans, current estimates for radiation hereditary risk are derived from measured germline mutation frequencies in mice [226]. The underlying rationale is that "experimental studies in plants and animals have demonstrated that radiation can induce hereditary effects, and humans are unlikely to be an exception in this regard" [227]. However, if the results of animal and cellular studies on epigenetic transgenerational destabilization of the genome do apply to human populations, then the hereditary risk could be greater than currently predicted. In this case, the question remains about the magnitude and significance of such an effect in the perspective of radiation protection.

6.3. Non Cancer Effects

Manifestations of health effects other than cancer and hereditary diseases have been well known after medium/high doses of ionizing radiation. Within months of Roentgen's discovery of X-rays, severe adverse effects were reported, such as eye and skin injuries. They were historically termed as "deterministic" effects in contrast to the stochastic cancer and hereditary effects, and later referred to as "tissue reactions" [228]. In general, tissue reactions to high/moderate doses are thought to arise mainly as a consequence of cell killing or functional inactivation, but other non-cytotoxic effects, such as disturbances in molecular cell signaling, also play a crucial role in determining tissue response to radiation. For radiation protection purposes, it is currently assumed that they show a "practical" threshold, defined as the dose required to lead to 1% excess incidence [228], at doses that are well above the levels of exposure typically encountered in the public environment, at work or in diagnostic medical uses of ionizing radiation. Recent results from epidemiological and experimental studies indicate possible increased risks for circulatory diseases, cognitive/neurological effects, and cataracts, not only at high doses but also at doses around 500 mGy and, possibly, even lower. In this section, we will give a glimpse of the role of epigenetics in radiation-induced cognitive and cardiovascular effects and cataract.

6.3.1. Possible Epigenetic Role in Radiation-Induced Cognitive Effects

Clinicians have known for decades that patients subjected to cranial radiotherapy for the control of brain malignancies develop severe and progressive cognitive deficits (see, e.g., [229]). Indeed, many studies of childhood cancer survivors (mainly of leukemia) documented cognitive impairment associated with high-dose (40–50 Gy) cranial irradiation [230]. Quite surprisingly, cognitive impairment was observed in a Swedish group treated for hemangioma in infancy with much lower doses, expressed as a ~50% reduction in high school attendance associated with 100 mGy exposure [231]. *In utero* exposed Japanese atomic bomb survivor data also suggest cognitive impairment at high dose, but no cognitive impairment can be demonstrated in the 0–100 mGy dose range [232,233]. However, the obvious differences in the age-at-exposure values (infancy vs. in utero) make it difficult to draw any meaningful comparison between the two studies [234].

Investigations showing cognitive/behavioral deficits caused by charged particles (relevant for protection against space radiation) in rodent models were carried out to understand the possible limitation to human exploration of our solar system [235,236]. Interestingly, in rats a correlation was recently found between behavioral changes and epigenomic remodeling in the hippocampus [237] and between adverse effects on cognition of space relevant irradiation and epigenetic aberrations consisting in increased levels of the DNA methylating enzymes [238].

6.3.2. Possible Epigenetic Role in Radiation-Induced Cardiovascular Effects

Circulatory disease has been recognized as an important late effect of radiation exposure after the evidence arising from radiotherapeutic experience and epidemiological studies following nuclear and other radiation activities [226]. ICRP has classified circulatory disease as a tissue reaction (a generalized definition of the deterministic effects), with a threshold dose of 0.5 Gy [228]. Several studies addressed the candidate biological mechanisms for the circulatory disease effects of radiation At radiotherapeutic doses > 5 Gy, the cell-killing effect on capillaries and endothelial cells plausibly explains effects on the heart and other parts of the circulatory system [239]. At lower doses (0.5–5 Gy), in humans and in in vivo and in vitro experiments, many inflammatory markers are upregulated long after exposure to radiation, while for doses less than about 0.5 Gy, the balance shifts toward anti-inflammatory effects [240,241]. The involvement of epigenetics, namely demethylation of a gene involved in aging endothelial cells, has been reported as one of the several events that contribute to the eventual development of atherosclerotic plaques after a dose of 10 Gy [242]. Changes in DNA methylation of repetitive elements in the heart tissue have also been observed after the irradiation of mice with 0.1 Gy of protons and 0.5 Gy of ^{56}Fe-ions, which are charged particles relevant to space radiation. These changes are dynamic and may vary depending on the time after irradiation, going from early global and repetitive elements-associated DNA hypomethylation to late DNA hypermethylation [243].

6.3.3. Possible Epigenetic Role in Radiation-Induced Cataract

Cataract is a progressive opacification of the crystalline lens of the eye which can determine a decrease in central vision, and is very common in the elderly [244]. It is due to a cumulative physiological response to toxic environmental factors leading to an excessive generation of ROS in the lens epithelium cells and in the superficial lens fiber cells, as well as in the aqueous humor [245]. The main effect of ionizing radiation on the eyes is the onset of posterior cortical and subcapsular cataracts [246], while there is little evidence that nuclear cataracts are radiation-induced [234].

In 2012, ICRP indicated a value of approximately 500 mGy as the threshold for cataract induction by low-LET radiation for acute and fractionated/protracted exposure [228]. This is a value lower by a factor of 10 than that deduced in earlier studies.

Even if it remains unknown exactly how ionizing radiation exposure contributes to opacification [247], epigenetic mechanisms, mainly DNA methylation, have been shown to play an important role in the pathophysiology of numerous ocular diseases [248]. For example, a decreased level

Understanding the role of natural radiation background on living organisms is also essential to complete this scenario, given that it provides the biological background on which the response to man-made exposures overlap, and that this background shows large geographical variations [2]. It is expected that controlled long-term experiments with various model organisms, conducted in underground laboratories where conditions with extremely reduced background radiation are realized, can provide this basic information and, at the same time, can increase our knowledge about the role played by the natural radiation in life's evolution.

Clearly, the decision whether the current LNT extrapolation of health risk estimates to low doses is still appropriate or whether a new paradigm has to be developed to provide more realistic protection against low radiation doses has significant social and economic implications.

Therefore, it is not surprising that quite often, the controversy is implicitly driven by considerations that are not strictly scientific but nevertheless can have a great impact on radiation protection practice. One of these considerations is that the LNT assumption makes the present system quite well manageable since a given dose can be a direct index of health risk and different doses received by an individual in different time periods can be summed up to evaluate the overall risk. On the other hand, no real alternative model based on recent scientific achievements has yet been proposed, likely because of the complex picture that has emerged for the biological response at low doses. However, these achievements suggest some general considerations useful in radiation protection practice such as, for example, that summing up many small doses to get an indication of the total health risk is unreasonable from the biological point of view, both at the individual and at the population level. This should be taken into account at least in performing the process called "optimization of the protection", which is one of the three fundamental principles of radiation protection, so as to extend to this aspect the conclusion already expressed by the ICRP that the "collective effective dose is not intended as a tool for epidemiological studies, and it is inappropriate to use it in risk projections" [1], a reasonable conclusion when taking into account that the collective dose may be made of a sum of a large number of small individual doses.

Author Contributions: M.B. conceptualization, original draft preparation, figure drawing, reviewing and editing. M.A.T. writing, reviewing and editing. All authors have read and agreed to the published version of the manuscript.

Funding. This research received no external funding.

Acknowledgments: M.B. acknowledges the support of the Multidisciplinary European Low Dose Initiative (MELODI) for participation, as member of its Scientific Committee, in several MELODI Workshops where subjects relevant to this review were discussed. We apologize to authors whose relevant publications were not cited due to space limitation.

Conflicts of Interest: The authors declare no conflict of interest.

Abbreviations

UNSCEAR	United Nations Scientific Committee on the Effects of Atomic Radiation
ICRP	International Commission on Radiological Protection
UNEP	United Nations Environment Programme
DSB	Double strand break
LET	Linear energy transfer
LNT	Linear No-Threshold
NTE	Non-targeted effects
AR	Adaptive response
BE	Bystander effect
GI	Genomic instability
C, G	Cytosine, Guanine
CpG	5′—C—phosphate—G—3′
DNMT	DNA methyltransferase
ncRNA	Non-coding RNA

lncRNA	Long non-coding RNA
miRNA	Micro RNA
ROS	Reactive oxygen species
RNS	Reactive nitrogen species
TE	Transposable element
LINE-1	Long interspersed nucleotide element 1

References

1. ICRP. ICRP Publication 103: The 2007 Recommendations of the International Commission on Radiological Protection. *Ann. ICRP* **2007**, *37*, 1–332.
2. United Nations Scientific Committee on the Effects of Atomic Radiation (UNSCEAR). *Biological Mechanisms of Radiation Actions at Low Doses*; United Nations: New York, NY, USA, 2012.
3. ICRP. ICRP Publication 99: Low-dose Extrapolation of Radiation-related Cancer Risk. *Ann. ICRP* **2005**, *35*, 1–140.
4. National Research Council. *Health Risks from Exposure to Low Levels of Ionizing Radiation: BEIR VII Phase 2*; The National Academies Press: Washington, DC, USA, 2006. [CrossRef]
5. United Nations Environment Programme (UNEP). *Radiation Effects and Sources*; UNSCEAR: Vienna, Austria, 2016; Available online: http://hdl.handle.net/20.500.11822/7790 (accessed on 18 August 2020).
6. Feil, R.; Fraga, M.R. Epigenetics and the environment: Emerging patterns and implications. *Nat. Rev. Genet.* **2012**, *13*, 97–109. [CrossRef] [PubMed]
7. Pacchierotti, F.; Spanò, M. Environmental Impact on DNA Methylation in the Germline: State of the Art and Gaps of Knowledge. *BioMed. Res. Int.* **2015**, 1–23. [CrossRef]
8. Goodhead, D.T. New radiobiological, radiation risk and radiation protection paradigms. *Mutat. Res.* **2010**, *687*, 13–16. [CrossRef] [PubMed]
9. Averbeck, D. Non-targeted effects as a paradigm breaking evidence. *Mutat. Res.* **2010**, *687*, 7–12. [CrossRef]
10. Mothersill, C.; Seymour, C. Targets, pools, shoulders, and communication–a reflection on the evolution of low-dose radiobiology. *Int. J. Radiat. Biol.* **2019**, *95*, 851–860. [CrossRef]
11. Schofield, P.N.; Kondratowicz, M. Evolving paradigms for the biological response to low dose ionizing radiation: The role of epigenetics. *Int. J. Radiat. Biol.* **2018**, *94*, 769–781. [CrossRef]
12. Miousse, I.R.; Kutanzi, K.R.; Koturbash, I. Effects of Ionizing Radiation on DNA Methylation: From Experimental Biology to Clinical Applications. *Int. J. Radiat. Biol.* **2017**, *93*, 457–469. [CrossRef]
13. O'Neill, P.; Wardman, P. Radiation chemistry comes before radiation biology. *Int. J. Radiat. Biol.* **2009**, *85*, 9–25. [CrossRef]
14. Ward, J.F. Biochemistry of DNA Lesions. *Radiat. Res.* **1985**, *104*, S103–S111. [CrossRef]
15. Ward, J.F. The complexity of DNA damage: Relevance to biological consequences. *Int. J. Radiat. Biol.* **1994**, *66*, 427–432. [CrossRef] [PubMed]
16. Goodhead, D.T.; Thacker, J.; Cox, R. Weiss Lecture. Effects of radiations of different qualities on cells: Molecular mechanisms of damage and repair. *Int. J. Radiat. Biol.* **1993**, *63*, 543–556. [CrossRef] [PubMed]
17. Goodhead, D.T. Initial events in the cellular effects of ionising radiation: Clustered damage in DNA. *Int. J. Radiat. Biol.* **1994**, *65*, 7–17. [CrossRef] [PubMed]
18. Nikjoo, H.; O'Neill, P.; Wilson, W.E.; Goodhead, D.T. Computational Approach for Determining the Spectrum of DNA Damage Induced by Ionizing Radiation. *Radiat. Res.* **2001**, *156*, 577–583. [CrossRef]
19. Hill, M.A. Radiation Track Structure: How the Spatial Distribution of Energy Deposition Drives Biological Response. *Clin. Oncolol.* **2001**, *32*, 75–83. [CrossRef]
20. Prise, K.M.; Pinto, M.; Newman, H.C.; Michael, B.D. A review of studies of ionizing radiation-induced double-strand break clustering. *Radiat. Res.* **2001**, *156*, 572–576. [CrossRef]
21. Prise, K.M.; Folkard, M.; Newman, H.C.; Michael, B.D. Effect of Radiation Quality on Lesion Complexity in Cellular DNA. *Int. J. Radat. Biol.* **1994**, *66*, 537–542. [CrossRef]
22. Georgakilas, A.G.; O'Neill, P.; Stewart, R.D. Induction and repair of clustered DNA lesions: What do we know so far? *Radiat. Res.* **2013**, *180*, 100–109. [CrossRef]

23. Mavragani, I.V.; Nikitaki, Z.; Kalospyros, S.A.; Georgakilas, A.G. Ionizing Radiation and Complex DNA Damage: From Prediction to Detection Challenges and Biological Significance. *Cancers* **2019**, *11*, 1789. [CrossRef]
24. Anderson, R.M.; Stevens, D.L.; Goodhead, D.T. M-FISH analysis shows that complex chromosome aberrations induced by α-particle tracks are cumulative products of localized rearrangements. *Proc. Natl. Acad. Sci. USA* **2002**, *99*, 12167–12172. [CrossRef] [PubMed]
25. Anderson, R.M.; Marsden, S.J.; Paice, S.J.; Bristow, A.E.; Kadhim, M.A.; Griffin, C.S.; Goodhead, D.T. Transmissible and Non transmissible Complex Chromosome Aberrations Characterized by Three-Color and mFISH Define a Biomarker of Exposure to High-LET α Particles. *Radiat. Res.* **2003**, *159*, 40–48. [CrossRef]
26. Savage, J.R.; Simpson, P.J. FISH "painting" patterns resulting from complex exchanges. *Mutat. Res.* **1994**, *312*, 51–60. [CrossRef]
27. Ritter, S.; Durante, M. Heavy-ion induced chromosomal aberrations: A review. *Mutat. Res. Genet. Toxicol. Environ. Mut.* **2010**, *701*, 38–46. [CrossRef]
28. Loucas, B.D.; Durante, M.; Bailey, S.; Cornforth, M.N. Chromosome damage in human cells by γ-rays, α-particles and heavy ions: Track interactions in basic dose–response relationships. *Radiat. Res.* **2013**, *179*, 9–20. [CrossRef]
29. Muller, H.J. Nobel Lecture. NobelPrize.org. Nobel Media AB. 2020. Available online: https://www.nobelprize.org/prizes/medicine/1946/muller/lecture/ (accessed on 18 August 2020).
30. United Nations Scientific Committee on the Effects of Atomic Radiation (UNSCEAR). *Sources, Effects and Risks of Ionizing Radiation*; United Nations: New York, NY, USA, 1988.
31. Ledford, H. Language: Disputed definitions. *Nature* **2008**, *455*, 1023–1028. [CrossRef]
32. Berger, S.L.; Kouzarides, T.; Shiekhattar, R.; Shilatifard, A. An operational definition of epigenetics. *Genes Develop.* **2009**, *23*, 781–783. [CrossRef]
33. Christensen, B.C.; Marsit, C.J. Epigenomics in environmental health. *Front. Genet.* **2011**, *2*, 1–10. [CrossRef]
34. Head, J.A.; Dolinoy, D.C.; Basu, N. Epigenetics for Ecotoxicologists. *Environ. Toxicol. Chem.* **2012**, *31*, 221–227. [CrossRef]
35. Jaenisch, R.; Bird, A. Epigenetic regulation of gene expression: How the genome integrates intrinsic and environmental signals. *Nat. Genet.* **2003**, *33*, 245–254. [CrossRef]
36. Saxonov, S.; Berg, P.; Brutlag, D.L. A genome-wide analysis of CpG dinucleotides in the human genome distinguishes two distinct classes of promoters. *Proc. Natl. Acad. Sci. USA* **2006**, *103*, 1412–1417. [CrossRef] [PubMed]
37. Bird, A.P. DNA methylation versus gene expression. *J. Embryol. Exp. Morph.* **1984**, *83*, 31–40. Available online: https://dev.biologists.org/content/develop/83/Supplement/31.full.pdf (accessed on 18 August 2020). [PubMed]
38. Greenberg, M.V.C.; Bourc'his, D. The diverse roles of DNA methylation in mammalian development and disease. *Nat. Rev. Mol. Cell. Biol.* **2019**, *20*, 590–607. [CrossRef] [PubMed]
39. Madakashira, B.P.; Sadler, K.C. DNA Methylation, Nuclear Organization, and Cancer. *Front. Genet.* **2017**. [CrossRef]
40. Kim, M.; Costello, J. DNA methylation: An epigenetic mark of cellular memory. *Exper. Mol. Med.* **2017**, *49*, e322. [CrossRef]
41. Margueron, R.; Trojer, P.; Reinberg, D. The key to development: Interpreting the histone code? *Curr. Opin. Genet. Dev.* **2005**, *15*, 163–176. [CrossRef]
42. Allfrey, V.G.; Faulkner, R.; Mirsky, A.E. Acetylation and Methylation of Histones and Their Possible Role in the Regulation of RNA Synthesis. *Proc. Natl. Acad. Sci. USA* **1964**, *51*, 786–794. [CrossRef]
43. Tharmalingam, S.; Sreetharan, S.; Kulesza, A.V.; Boreham, D.R.; Tai, T.C. Low-Dose Ionizing Radiation Exposure, Oxidative Stress and Epigenetic Programing of Health and Disease. *Radiat. Res.* **2017**, *188*, 525–538. [CrossRef]
44. Lee, R.C.; Feinbaum, R.L.; Ambros, V. The C. elegans heterochronic gene lin-4 encodes small RNAs with antisense complementarity to lin-14. *Cell* **1993**, *75*, 843–854. [CrossRef]
45. Morozova, N.; Zinovyev, A.; Nonne, N.; Pritchard, L.-L.; Gorban, A.N.; Harel-Bellan, A. Kinetic signatures of microRNA modes of action. *RNA* **2012**, *18*, 1635–1655. [CrossRef]
46. Mao, A.; Liu, Y.; Zhang, H.; Di, C.; Sun, C. MicroRNA expression and biogenesis in cellular response to ionizing radiation. *DNA Cell Biol.* **2014**, *33*, 667–679. [CrossRef] [PubMed]

47. Carthew, R.W.; Sontheime, E.J. Origins and Mechanisms of miRNAs and siRNAs. *Cell* **2009**, *136*, 642–655. [CrossRef] [PubMed]
48. Fabian, M.R.; Sonenberg, N.; Filipowicz, W. Regulation of mRNA translation and stability by microRNAs. *Annu. Rev. Biochem.* **2010**, *79*, 351–379. [CrossRef] [PubMed]
49. Morris, K.V. Long antisense non-coding RNAs function to direct epigenetic complexes that regulate transcription in human cells. *Epigenetics* **2009**, *4*, 296–301. [CrossRef]
50. Koturbash, I.; Zemp, F.J.; Pogribny, I.; Kovalchuk, O. Small molecules with big effects: The role of the microRNAome in cancer and carcinogenesis. *Mutat. Res. Genet. Toxicol. Environ. Mut.* **2011**, *722*, 94–105. [CrossRef]
51. Esteller, M. Non-coding RNAs in human disease. *Nat. Rev. Genet.* **2011**, *12*, 861–874. [CrossRef]
52. Friedman, R.C.; Farh, K.K.H.; Burge, C.B.; Bartel, D.P. Most mammalian mRNAs are conserved targets of microRNAs. *Genome Res.* **2009**, *19*, 92–105. [CrossRef]
53. Shkumatava, A.; Stark, A.; Sive, H.; Bartel, D.P. Coherent but overlapping expression of microRNAs and their targets during vertebrate development. *Genes Develop.* **2009**, *23*, 466–481. [CrossRef]
54. Fatica, A.; Bozzoni, I. Long non-coding RNAs: New players in cell differentiation and development. *Nat. Rev. Genet.* **2014**, *15*, 7–21. [CrossRef]
55. Rinn, J.L.; Chang, H.Y. Genome regulation by long noncoding RNAs. *Ann. Rev. Biochem.* **2012**, *81*, 145–166. [CrossRef]
56. Anastasiadou, E.; Jacob, L.S.; Slack, F.J. Non-coding RNA networks in cancer. *Nat. Rev. Cancer* **2018**, *18*, 5–18. [CrossRef] [PubMed]
57. Kalinich, J.F.; Catravas, G.N.; Snyder, S.L. The effect of gamma radiation on DNA methylation. *Radiat. Res.* **1989**, *117*, 185–197. [CrossRef] [PubMed]
58. Goetz, W.; Morgan, M.N.M.; Baulch, J.E. The effect of radiation quality on genomic DNA methylation profiles in irradiated human cell lines. *Radiat. Res.* **2011**, *175*, 575–587. [CrossRef] [PubMed]
59. Kim, E.H.; Park, A.K.; Dong, S.M.; Ahn, J.H.; Park, W.Y. Global analysis of CpG methylation reveals epigenetic control of the radiosensitivity in lung cancer cell lines. *Oncogene* **2010**, *29*, 4725–4731. [CrossRef] [PubMed]
60. Chaudhry, M.A.; Omaruddin, R.A. Differential DNA Methylation Alterations in Radiation-Sensitive and Resistant Cells. *DNA Cell Biol.* **2012**, *31*, 908–916. [CrossRef] [PubMed]
61. Tawa, R.; Kimura, Y.; Komura, J.; Miyamura, Y.; Kurishita, A.; Sasaki, M.S.; Sakurai, H.; Ono, T. Effects of X-ray irradiation on genomic DNA methylation levels in mouse tissues. *J. Radiat. Res.* **1998**, *39*, 271–278. [CrossRef]
62. Pogribny, I.; Raiche, J.; Slovack, M.; Kovalchuk, O. Dose-dependence, sex- and tissue-specificity, and persistence of radiation-induced genomic DNA methylation changes. *Biochem. Biophys. Res. Commun.* **2004**, *320*, 1253–1261. [CrossRef]
63. Raiche, J.; Rodriguez-Juarez, R.; Pogribny, I.; Kovalchuk, O. Sex- and tissue-specific expression of maintenance and de novo DNA methyltransferases upon low dose X-irradiation in mice. *Biochem. Biophys. Res. Commun.* **2004**, *325*, 39–47. [CrossRef]
64. Koturbash, I.; Pogribny, I.; Kovalchuk, O. Stable loss of global DNA methylation in the radiation-target tissue-A possible mechanism contributing to radiation carcinogenesis? *Biochem. Biophys. Res. Commun.* **2005**, *337*, 526–533. [CrossRef]
65. Giotopoulos, G.; McCormick, C.; Cole, C.; Zanker, A.; Jawad, M.; Brown, R. DNA methylation during mouse hemopoietic differentiation and radiation-induced leukemia. *Exp. Hematol.* **2006**, *34*, 1462–1470. [CrossRef]
66. Klose, R.J.; Bird, A.P. Genomic DNA methylation: The mark and its mediators. *Trends Biochem. Sci.* **2006**, *31*, 89–97. [CrossRef] [PubMed]
67. Weber, M.; Schübeler, D. Genomic patterns of DNA methylation: Targets and function of an epigenetic mark. *Curr. Opin. Cell Biol.* **2007**, *19*, 273–280. [CrossRef] [PubMed]
68. Baylin, S.B.; Jones, P.A. Epigenetic determinants of cancer. *Cold Spring Harb. Perspect. Biol.* **2016**. [CrossRef] [PubMed]
69. Collins, L.J.; Penny, D. The RNA infrastructure: Dark matter of the eukaryotic cell? *Trends Genet.* **2009**, *25*, 120–128. [CrossRef]
70. Miousse, I.R.; Koturbash, I. The Fine LINE: Methylation Drawing the Cancer Landscape. *Biomed. Res. Int.* **2015**, 131547. [CrossRef]

71. Tanaka, A.; Nakatani, Y.; Hamada, N.; Jinno-Oue, A.; Shimizu, N.; Wada, S. Ionising irradiation alters the dynamics of human long interspersed nuclear elements 1 (LINE1) retrotransposon. *Mutagenesis* **2012**, *27*, 599–607. [CrossRef]
72. Miousse, I.R.; Chalbot, M.C.; Lumen, A.; Ferguson, A.; Kavouras, I.G.; Koturbash, I. Response of transposable elements to environmental stressors. *Mutat. Res. Rev. Mutat. Res.* **2015**, *765*, 19–39. [CrossRef]
73. Kovalchuk, O.; Baulch, J.E. Epigenetic changes and nontargeted radiation effects-Is there a link? *Environ. Mol. Mutagen.* **2008**, *49*, 16–25. [CrossRef]
74. Goetz, W.; Morgan, M.N.M.; Belliveau, B.J.; Baulch, J.E. Effects of high and low LET radiation exposure on DNA methylation. *Environ. Mol. Mutagen.* **2009**, *50*, 575.
75. Koturbash, I. Michael Fry Award Lecture: When DNA is Actually Not a Target: Radiation Epigenetics as a Tool to Understand and Control Cellular Response to Ionizing Radiation. *Radiat Res.* **2018**, *190*, 5–11. [CrossRef]
76. Mendonca, M.S.; Antoniono, R.J.; Redpath, J.L. Delayed Heritable Damage and Epigenetics in Radiation-Induced Neoplastic Transformation of Human Hybrid Cells. *Radiat. Res.* **1993**, *134*, 209–216. [CrossRef] [PubMed]
77. Hoffmann, M.J.; Schulz, W.A. Causes and consequences of DNA hypomethylation in human cancer. *Biochem. Cell Biol.* **2005**, *83*, 296–321. [CrossRef] [PubMed]
78. Toyota, M.; Issa, J.P. The role of DNA hypermethylation in human neoplasia. *Electrophoresis* **2000**, *21*, 329–333. [CrossRef]
79. Baylin, S.B.; Jones, P.A. A decade of exploring the cancer epigenome-biological and translational implications. *Nat. Rev. Cancer* **2012**, *11*, 726–734. [CrossRef]
80. Su, S.; Jin, Y.; Zhang, W.; Yang, L.; Shen, Y.; Cao, Y.; Tong, J. Aberrant promoter methylation of p16 (INK4a) and O(6)-methylguanine-DNA methyltransferase genes in workers at a Chinese uranium mine. *J. Occup. Health* **2006**, *48*, 261–266. [CrossRef] [PubMed]
81. Lyon, C.M.; Klinge, D.M.; Liechty, K.C.; Gentry, F.D.; March, T.H.; Kang, T.; Gilliland, F.D.; Adamova, G.; Rusinova, G.; Telnov, V.; et al. Radiation-induced lung adenocarcinoma is associated with increased frequency of genes, inactivated by promoter hypermethylation. *Radiat. Res.* **2007**, *168*, 409–414. [CrossRef]
82. Antwih, K.M.; Gabbara, W.D.; Lancaster, D.M.; Ruden, S.P.; Zielske, S.P. Radiation-induced epigenetic DNA methylation modification of radiation-response pathways. *Epigenetics* **2013**, *8*, 839–848. [CrossRef]
83. Rogakou, E.P.; Pilch, D.R.; Orr, A.H.; Ivanova, V.S.; Bonner, W.M. DNA double-stranded breaks induce histone H2AX phosphorylation on serine 139. *J. Biol. Chem.* **1998**, *273*, 5858–5868. [CrossRef]
84. Pilch, D.R.; Sedelnikova, O.A.; Redon, C.; Celeste, A.; Nussenzweig, A.; Bonner, W.M. Characteristics of gamma-H2AX foci at DNA double-strand breaks sites. *Biochem. Cell Biol.* **2003**, *81*, 123–129. [CrossRef]
85. Pogribny, I.; Koturbash, I.; Tryndyak, V.; Hudson, D.; Stevenson, S.M.; Sedelnikova, O.; Bonner, W.; Kovalchuk, O. Fractionated low-dose radiation exposure leads to accumulation of DNA damage and profound alterations in DNA and histone methylation in the murine thymus. *Mol. Cancer Res.* **2005**, *3*, 553–561. [CrossRef]
86. Tryndyak, V.P.; Kovalchuk, O.; Pogribny, I.P. Loss of DNA methylation and histone H4 lysine 20 trimethylation in human breast cancer cells is associated with aberrant expression of DNA methyltransferase 1, Suv4-20h2 histone methyltransferase and methyl-binding proteins. *Cancer Biol. Ther.* **2006**, *5*, 65–70. [CrossRef] [PubMed]
87. Mendez-Acuna, L.; Di Tomaso, M.V.; Palitti, F.; Martinez-Lopez, W. Histone posttranslational modifications in DNA damage response. *Cytogenet. Genome Res.* **2010**, *128*, 28–36. [CrossRef] [PubMed]
88. Averbeck, N.B.; Durante, M. Protein acetylation within the cellular response to radiation. *J. Cell. Physiol.* **2011**, *226*, 962–967. [CrossRef] [PubMed]
89. Metheetrairut, C.; Slack, F.J. MicroRNAs in the Ionizing Radiation Response and in Radiotherapy. *Curr. Opin. Genet. Dev.* **2013**, *23*, 12–19. [CrossRef] [PubMed]
90. Aypar, U.; Morgan, W.F.; Baulch, J.E. Radiation-induced epigenetic alterations after low and high LET irradiations. *Mutat. Res.* **2011**, *707*, 24–33. [CrossRef]
91. Wagner-Ecker, M.; Schwager, C.; Wirkner, U.; Abdollahi, A.; Huber, P.E. MicroRNA expression after ionizing radiation in human endothelial cells. *Radiat. Oncol.* **2010**, *5*, 25. [CrossRef]
92. Chaudhry, M.A.; Kreger, B.; Omaruddin, R.A. Transcriptional modulation of micro-RNA in human cells differing in radiosensitivity. *Int. J. Radiat. Biol.* **2010**, *86*, 569–583. [CrossRef]

93. Halimi, M.; Asghari, S.M.; Sariri, R.; Moslemi, D.; Parsian, H. Cellular Response to Ionizing Radiation: A MicroRNA Story. *Int. J. Mol. Cell Med.* **2012**, *1*, 178–184.
94. Gong, P.; Zhang, T.; He, D.; Hsieh, J.T. MicroRNA-145 Modulates Tumor Sensitivity to Radiation in Prostate Cancer. *Radiat. Res.* **2015**, *184*, 630–638. [CrossRef]
95. El Bezawy, R.; Tinelli, S.; Tortoreto, M.; Doldi, V.; Zuco, V.; Folini, M.; Stucchi, C.; Rancati, T.; Valdagni, R.; Gandellini, P.; et al. MiR-205 enhances radiation sensitivity of prostate cancer cells by impairing DNA damage repair through PKCε and ZEB1 inhibition. *J. Exp. Clin. Cancer Res.* **2019**, *38*, 51. [CrossRef]
96. Chaudhry, M.A. Radiation-induced microRNA: Discovery, functional analysis, and cancer radiotherapy. *J. Cell. Biochem.* **2014**, *115*, 436–449. [CrossRef] [PubMed]
97. Cellini, F.; Morganti, A.G.; Genovesi, D.; Silvestris, N.; Valentini, V. Role of microRNA in response to ionizing radiations: Evidences and potential impact on clinical practice for radiotherapy. *Molecules* **2014**, *19*, 5379–5401. [CrossRef] [PubMed]
98. Marta, G.N.; Garicochea, B.; Carvalho, A.L.; Real, J.M.; Kowalski, L.P. MicroRNAs, cancer and ionizing radiation: Where are we? *Rev. Assoc. Med. Bras.* **2015**, *61*, 275–281. [CrossRef] [PubMed]
99. Lima, F.; Ding, D.; Goetz, W.; Yang, A.J.; Baulch, J. High LET 56Fe ion irradiation induces tissue-specific changes in DNA methylation in the mouse. *Environ. Mol. Mutagen.* **2014**, *55*, 266–277. [CrossRef]
100. Nzabarushimana, E.; Miousse, I.R.; Shao, L.; Chang, J.; Allen, A.R.; Turner, J.; Stewart, B.; Raber, J.; Koturbash, I. Long-term epigenetic effects of exposure to low doses of 56Fe in the mouse lung. *J. Radiat. Res.* **2014**, *55*, 823–828. [CrossRef]
101. Miousse, I.R.; Shao, L.J.; Chang, J.H.; Feng, W.; Wang, Y.Y.; Allen, A.R. Exposure to low-dose Fe-56-ion radiation induces long-term epigenetic alterations in mouse bone marrow hematopoietic progenitor and stem cells. *Radiat. Res.* **2014**, *182*, 92–101. [CrossRef]
102. Kennedy, E.M.; Conneely, K.N.; Vertino, P.M. Epigenetic Memory of Space Radiation Exposure. Available online: https://three-jsc.nasa.gov/articles/Vertino.pdf (accessed on 30 July 2014).
103. Durante, M.; Cucinotta, F.A. Heavy ion carcinogenesis and human space exploration. *Nat. Rev. Cancer* **2008**, *8*, 465–472. [CrossRef]
104. Cucinotta, F.A. Space Radiation Risks for Astronauts on Multiple International Space Station Missions. *PLoS ONE* **2014**, *9*. [CrossRef]
105. Morano, A.; Angrisano, T.; Russo, G.; Landi, R.; Pezone, A.; Bartollino, S.; Zuchegna, C.; Babbio, F.; Bonapace, I.M.; Allen, B.; et al. Targeted DNA methylation by homology-directed repair in mammalian cells. Transcription reshapes methylation on the repaired gene. *Nucleic Acids Res.* **2014**, *42*, 804–821. [CrossRef]
106. Prior, S.; Miousse, I.R.; Nzabarushimana, E.; Pathak, R.; Skinner, C.; Kutanzi, K.R.; Allen, A.R.; Raber, J.; Tackett, A.J.; Hauer-Jensen, M.; et al. Densely ionizing radiation affects DNA methylation of selective LINE-1 elements. *Environ. Res.* **2016**, *150*, 470–481. [CrossRef]
107. Belinsky, S.A.; Klinge, D.M.; Liechty, K.C.; March, T.H.; Kang, T.; Gilliland, F.D.; Sotnic, N.; Adamova, G.; Rusinova, G.; Telnov, V. Plutonium targets the p16 gene for inactivation by promoter hypermethylation in human lung adenocarcinoma. *Carcinogenesis* **2004**, *25*, 1063–1067. [CrossRef]
108. Templin, T.; Amundson, S.A.; Brenner, D.J.; Smilenov, L.B. Whole mouse blood microRNA as biomarkers for exposure to γ-rays and 56Fe ions. *Int. J. Radiat. Biol.* **2011**, *87*, 653–662. [CrossRef] [PubMed]
109. Templin, T.; Young, E.F.; Smilenov, L.B. Proton radiation-induced miRNA signatures in mouse blood: Characterization and comparison with 56Fe-ion and gamma radiation. *Int. J. Radiat. Biol.* **2012**, *88*, 531–539. [CrossRef] [PubMed]
110. Ward, J.F. DNA damage produced by ionizing radiation in mammalian cells: Identities, mechanisms of formation, and reparability. *Prog. Nucleic Acid Res. Mol. Biol.* **1988**, *35*, 95–125. [CrossRef] [PubMed]
111. O'Neill, P.; Fielden, E.M. Primary Free Radical Processes in DNA. *Adv. Radiat. Biol.* **1993**, *17*, 53–120. [CrossRef]
112. De Lara, C.M.; Jenner, T.J.; Townsend, K.M.S.; Marsden, S.J.; O'Neill, P. The Effect of Dimethyl Sulfoxide on the Induction of DNA Double-Strand Breaks in V79-4 Mammalian Cells by Alpha Particles. *Radiat. Res.* **1995**, *144*, 43–49. [CrossRef]
113. Azzam, E.I.; Jay-Gerin, J.P.; Pain, D. Ionizing radiation-induced metabolic oxidative stress and prolonged cell injury. *Cancer Lett.* **2012**, *327*, 48–60. [CrossRef]
114. Shrishrimal, S.; Kosmacek, E.A.; Oberley-Deegan, R.E. Reactive Oxygen Species Drive Epigenetic Changes in Radiation-Induced Fibrosis. *Oxid. Med. Cell. Longev.* **2019**, *2019*. [CrossRef]

115. Cerda, S.; Weitzman, S.A. Influence of oxygen radical injury on DNA methylation. *Mutat. Res.* **1997**, *386*, 141–152. [CrossRef]
116. Franco, R.; Schoneveld, O.; Georgakilas, A.G.; Panayiotidis, M.I. Oxidative stress, DNA methylation and carcinogenesis. *Cancer Lett.* **2008**, *266*, 6–11. [CrossRef]
117. Ziech, D.; Franco, R.; Pappa, A.; Panayiotidis, M.I. Reactive Oxygen Species (ROS)-Induced genetic and epigenetic alterations in human carcinogenesis. *Mutat. Res.* **2011**, *711*, 167–173. [CrossRef] [PubMed]
118. Branco, M.R.; Ficz, G.; Reik, W. Uncovering the role of 5- hydroxymethylcytosine in the epigenome. *Nat. Rev. Genet.* **2012**, *13*, 7–13. [CrossRef] [PubMed]
119. Efimova, O.A.; Koltsova, A.S.; Krapivin, M.I.; Tikhonov, A.V.; Pendina, A.A. Environmental Epigenetics and Genome Flexibility: Focus on 5-Hydroxymethylcytosine. *Int. J. Mol. Sci.* **2020**, *21*, 3223. [CrossRef] [PubMed]
120. Maltseva, D.V.; Baykov, A.A.; Jeltsch, A.; Gromova, E.S. Impact of 7,8-dihydro-8-oxoguanine on methylation of the CpG site by Dnmt3a. *Biochemistry* **2009**, *48*, 1361–1368. [CrossRef] [PubMed]
121. Giorgio, M.; Dellino, G.I.; Gambino, V.; Roda, N.; Pelicci, P.G. On the epigenetic role of guanosine oxidation. *Redox Biol.* **2020**, *29*, 101398. [CrossRef] [PubMed]
122. Shaughnessy, D.T.; McAllister, K.; Worth, L.; Haugen, A.C.; Meyer, J.N.; Domann, F.E.; Houten, B.V.; Mostoslavsky, R.; Bultman, S.J.; Baccarelli, A.A.; et al. Mitochondria, energetics, epigenetics, and cellular responses to stress. *Environ. Health Perspect.* **2014**, *122*, 1271–1278. [CrossRef]
123. Balaban, R.S.; Nemoto, S.; Finkel, T. Mitochondria, oxidants, and aging. *Cell* **2005**, *120*, 483–495. [CrossRef]
124. Szumiel, I. Ionizing radiation-induced oxidative stress, epigenetic changes and genomic instability: The pivotal role of mitochondria. *Int. J. Radiat. Biol.* **2015**, *91*, 1–12. [CrossRef]
125. Baulch, J.E. Radiation-induced genomic instability, epigenetic mechanisms and the mitochondria: A dysfunctional ménage a trois? *Int. J. Radiat. Biol.* **2019**, *95*, 516–525. [CrossRef]
126. Wu, Q.; Ni, X. ROS-Mediated DNA Methylation Pattern Alterations in Carcinogenesis. *Curr. Drug Targets* **2015**, *16*, 13–19. [CrossRef]
127. Kietzmann, T.; Petry, A.; Shvetsova, A.; Gerhold, J.M.; Görlach, A. The epigenetic landscape related to reactive oxygen species formation in the cardiovascular system. *Br. J. Pharmacol.* **2017**, *174*, 1533–1554. [CrossRef]
128. Qian, W.; Miki, D.; Zhang, H.; Liu, Y.; Zhang, X.; Tang, K.; Kan, Y.; La, H.; Li, X.; Li, S.; et al. A histone acetyltransferase regulates active DNA demethylation in Arabidopsis. *Science* **2012**, *336*, 1445–1448. [CrossRef] [PubMed]
129. Nishiyama, A.; Yamaguchi, L.; Nakanishi, M. Regulation of maintenance DNA methylation via histone ubiquitylation. *J. Biochem.* **2016**, *159*, 9–15. [CrossRef] [PubMed]
130. Fuks, F. DNA methylation and histone modifications: Teaming up to silence genes. *Curr. Opin. Genet. Dev.* **2005**, *15*, 490–495. [CrossRef] [PubMed]
131. Robertson, K. DNA methylation and chromatin–unraveling the tangled web. *Oncogene* **2002**, *21*, 5361–5379. [CrossRef] [PubMed]
132. Cameron, E.; Bachman, K.; Myöhänen, S.; Herman, J.G.; Baylin, S.B. Synergy of demethylation and histone deacetylase inhibition in the re-expression of genes silenced in cancer. *Nat. Genet.* **1999**, *21*, 103–107. [CrossRef] [PubMed]
133. Cedar, H.; Bergman, Y. Linking DNA methylation and histone modification: Patterns and paradigms. *Nat. Rev. Genet.* **2009**, *10*, 295–304. [CrossRef]
134. Rose, N.R.; Klose, R.J. Understanding the relationship between DNA methylation and histone lysine methylation. *Biochim. Biophys. Acta* **2014**, *1839*, 1362–1372. [CrossRef]
135. Du, J.; Johnson, L.M.; Jacobsen, S.E.; Patel, D.J. DNA methylation pathways and their crosstalk with histone methylation. *Nat. Rev. Mol. Cell Biol.* **2015**, *16*, 519–532. [CrossRef]
136. Sun, X.; He, Y.; Huang, C.; Ma, T.-T.; Li, J. The epigenetic feedback loop between DNA methylation and microRNAs in fibrotic disease with an emphasis on DNA methyltransferases. *Cell. Signal.* **2013**, *25*, 1870–1876. [CrossRef]
137. Wang, S.; Wu, W.; Claret, F.X. Mutual regulation of microRNAs and DNA methylation in human cancers. *Epigenetics* **2017**, *12*, 187–197. [CrossRef] [PubMed]
138. Huan, T.; Mendelson, M.; Joehanes, R.; Yao, C.; Liu, C.; Song, C.; Bhattacharya, A.; Rong, J.; Tanriverdi, K.; Keefe, J.; et al. Epigenome-wide association study of DNA methylation and microRNA expression highlights novel pathways for human complex traits. *Epigenetics* **2020**, *15*, 183–198. [CrossRef] [PubMed]

139. Chen, D.; Jin, C. Histone variants in environmental-stress-induced DNA damage repair. *Mutat. Res.* **2019**, *780*, 55–60. [CrossRef] [PubMed]
140. Seymour, C.B.; Mothersill, C.E.; Alper, T. High yields of lethal mutations in somatic mammalian cells that survive ionizing radiation. *Int. J. Radiat. Biol.* **1986**, *50*, 167–179. [CrossRef]
141. Nagasawa, H.; Little, J.B. Induction of sister chromatid exchanges by extremely low doses of α-particles. *Cancer Res.* **1992**, *52*, 6394–6396.
142. Kadhim, M.A.; Macdonald, D.A.; Goodhead, D.T.; Lorimore, S.A.; Marsden, S.J.; Wright, E.G. Transmission of chromosomal instability after plutonium alpha-particle irradiation. *Nature* **1992**, *355*, 738–740. [CrossRef]
143. Marder, B.A.; Morgan, W.F. Delayed chromosomal instability induced by DNA damage. *Mol. Cell. Biol.* **1993**, *13*, 6667–6677. [CrossRef]
144. Mothersill, C.; Seymour, C. Radiation-induced bystander effects: Past history and future directions. *Radiat. Res.* **2001**, *155*, 759–767. [CrossRef]
145. Morgan, W.F. Non-targeted and delayed effects of exposure to ionizing radiation: I. Radiation-induced genomic instability and bystander effects in vitro. *Radiat. Res.* **2003**, *159*, 567–580. [CrossRef]
146. Morgan, W.F. Non-targeted and delayed effects of exposure to ionizing radiation: II. Radiation-induced genomic instability and bystander effects in vivo, clastogenic factors and transgenerational effects. *Radiat. Res.* **2003**, *159*, 581–596. [CrossRef]
147. Olivieri, G.; Bodycote, J.; Wolff, S. Adaptive response of human lymphocytes to low concentrations of radioactive thymidine. *Science* **1984**, *223*, 594–597. [CrossRef] [PubMed]
148. Nagasawa, H.; Little, J.B. Unexpected sensitivity to the induction of mutations by very low doses of alpha particle radiation: Evidence for a bystander effect. *Radiat. Res.* **1999**, *152*, 552–557. [CrossRef] [PubMed]
149. Blyth, B.J.; Sykes, P.J. Radiation-Induced Bystander Effects: What Are They, and How Relevant Are They to Human Radiation Exposures? *Radiat. Res.* **2011**, *176*, 139–157. [CrossRef] [PubMed]
150. Koturbash, I.; Boyko, A.; Rodriguez-Juarez, R.; McDonald, R.J.; Tryndyak, V.P.; Kovalchuk, I.; Pogribny, I.P.; Kovalchuk, O. Role of epigenetic effectors in maintenance of the long-term persistent bystander effect in spleen in vivo. *Carcinogenesis* **2007**, *28*, 1831–1838. [CrossRef]
151. Mancuso, M.; Pasquali, E.; Leonardi, S.; Tanori, M.; Rebessi, S.; Di Majo, V.; Pazzaglia, S.; Toni, M.P.; Pimpinella, M.; Covelli, V.; et al. Oncogenic bystander radiation effects in Patched heterozygous mouse cerebellum. *Proc. Natl. Acad. Sci. USA* **2008**, *105*, 12445–12450. [CrossRef]
152. Kadim, M.; Salomaa, S.; Wright, E.; Hildebrandt, G.; Belyakov, O.V.; Prise, K.M.; Little, M.P. Non-targeted effects of ionising radiation-implications for low dose risk. *Mutat. Res.* **2013**, *752*, 84–98. [CrossRef]
153. Campa, A.; Balduzzi, M.; Dini, V.; Esposito, G.; Tabocchini, M.A. The complex interactions between radiation induced non-targeted effects and cancer. *Cancer Lett.* **2015**, *356*, 126–136. [CrossRef]
154. Lorimore, S.A.; Coates, P.J.; Wright, E.G. Radiation-induced genomic instability and bystander effects: Inter-related nontargeted effects of exposure to ionizing radiation. *Oncogene* **2003**, *22*, 7058–7069. [CrossRef]
155. Huang, L.; Kim, P.M.; Nickoloff, J.A.; Morgan, W.F. Targeted and nontargeted effects of low-dose ionizing radiation on delayed genomic instability in human cells. *Cancer Res.* **2007**, *67*, 1099–1104. [CrossRef]
156. Lorimore, S.A.; Kadhim, M.A.; Pocock, D.A.; Papworth, D.; Stevens, D.L.; Goodhead, D.T.; Wright, E.G. Chromosomal instability in the descendants of unirradiated surviving cells after alpha-particle irradiation. *Proc. Natl. Acad. Sci. USA* **1998**, *95*, 5730–5733. [CrossRef]
157. Schwartz, J.L. Variability: The common factor linking low dose-induced genomic instability, adaptation and bystander effects. *Mutat. Res.* **2007**, *616*, 196–200. [CrossRef] [PubMed]
158. Mothersill, C.; Seymour, C. Changing paradigms in radiobiology. *Mutat. Res. Rev. Mutat. Res.* **2012**, *750*, 85–95. [CrossRef] [PubMed]
159. Morgan, W.F.; Sowa, M.B. Non-targeted effects induced by ionizing radiation: Mechanisms and potential impact on radiation induced health effects. *Cancer Lett.* **2015**, *356*, 17–21. [CrossRef] [PubMed]
160. Hall, E.J. The bystander effect. *Health Phys.* **2003**, *85*, 31–35. [CrossRef] [PubMed]
161. Sowa, M.B.; Goetz, W.; Baulch, J.E.; Pyles, D.N.; Dziegielewski, J.; Yovino, S.; Snyder, A.R.; De Toledo, S.M.; Azzam, E.I.; Morgan, W.F. Lack of evidence for low-LET radiation induced bystander response in normal human fibroblasts and colon carcinoma cells. *Int. J. Radiat. Biol.* **2010**, *86*, 102–113. [CrossRef]
162. Sowa, M.B.; Goetz, W.; Baulch, J.E.; Lewis, A.J.; Morgan, W.F. No evidence for a low linear energy transfer adaptive response in irradiated RKO cells. *Radiat. Prot. Dos.* **2011**, *143*, 311–314. [CrossRef]

163. Ilnytskyy, Y.; Kovalchuk, O. Non-targeted radiation effects-An epigenetic connection. *Mutat. Res.* **2011**, *714*, 113–125. [CrossRef]
164. Manti, L.; Jamali, M.; Prise, K.M.; Michael, B.D.; Trott, K.R. Genomic Instability in Chinese Hamster Cells After Exposure to X Rays or Alpha Particles of Different Mean Linear Energy Transfer. *Radiat. Res.* **1997**, *147*, 22–28. [CrossRef]
165. Morgan, W.F.; Corcoran, J.; Hartmann, A.; Kaplan, M.I.; Limoli, C.L.; Ponnaiya, B. DNA double-strand breaks, chromosomal rearrangements, and genomic instability. *Mutat. Res.* **1998**, *404*, 125–128. [CrossRef]
166. Snyder, A.R.; Morgan, W.F. Differential induction and activation of NF-κB transcription complexes in radiation-induced chromosomally unstable cell lines. *Environ. Mol. Mut.* **2005**, *45*, 177–187. [CrossRef]
167. Bright, S.; Kadhim, M. The future impacts of non-targeted effects. *Int. J. Radiat. Biol.* **2018**, *94*, 727–736. [CrossRef] [PubMed]
168. Merrifield, M.; Kovalchuk, O. Epigenetics in radiation biology: A new research frontier. *Front. Genet.* **2013**, *4*. [CrossRef] [PubMed]
169. Rugo, R.E.; Mutamba, J.T.; Mohan, K.; Yee, T.; Chaillet, J.R.; Greenberger, J.S.; Engelward, B.P. Methyltransferases mediate cell memory of a genotoxic insult. *Oncogene* **2011**, *30*, 751–756. [CrossRef] [PubMed]
170. Sisakht, M.; Darabian, M.; Mahmoodzadeh, A.; Bazi, A.; Shafiee, S.M.; Mokarram, P.; Khoshdel, Z. The role of radiation induced oxidative stress as a regulator of radio-adaptive responses. *Int. J. Radiat. Biol.* **2020**. [CrossRef] [PubMed]
171. Filkowski, J.N.; Ilnytskyy, Y.; Tamminga, J.; Koturbash, I.; Golubov, A.; Bagnyukova, T.; Pogribny, I.P.; Kovalchuk, O. Hypomethylation and genome instability in the germline of exposed parents and their progeny is associated with altered miRNA expression. *Carcinogenesis* **2010**, *1*, 1110–1115. [CrossRef]
172. Guéguen, Y.; Bontemps, A.; Ebrahimian, T.G. Adaptive responses to low doses of radiation or chemicals: Their cellular and molecular mechanisms. *Cell. Mol. Life Sci.* **2019**, *76*, 1255–1273. [CrossRef]
173. Esposito, G.; Campa, A.; Pinto, M.; Simone, G.; Tabocchini, M.A.; Belli, M. Adaptive response: Modelling and experimental studies. *Radiat. Prot. Dosim.* **2011**, *143*, 320–324. [CrossRef]
174. Timp, W.; Feinberg, A.P. Cancer as a dysregulated epigenome allowing cellular growth advantage at the expense of the host. *Nat. Rev. Cancer* **2013**, *13*, 497–510. [CrossRef]
175. Kim, J.G.; Park, M.T.; Heo, K.; Yang, K.M.; Yi, J.M. Epigenetics Meets Radiation Biology as a New Approach in Cancer Treatment. *Int. J. Mol. Sci.* **2013**, *14*, 15059–15073. [CrossRef]
176. Altucci, L.; Clarke, N.; Nebbioso, A.; Scognamiglio, A.; Gronemeyer, H. Acute myeloid leukemia: Therapeutic impact of epigenetic drugs. *Int. J. Biochem. Cell Biol.* **2005**, *37*, 1752–1762. [CrossRef]
177. Jones, P.A.; Gonzalgo, M.L. Altered DNA methylation and genome instability: A new pathway to cancer? *Proc. Natl. Acad. Sci. USA* **1997**, *94*, 2103–2105. [CrossRef] [PubMed]
178. Lengauer, C.; Kinzler, K.W.; Vogelstein, B. Genetic instability in colorectal cancers. *Nature* **1997**, *386*, 623–627. [CrossRef] [PubMed]
179. Zingg, J.M.; Jones, P.A. Genetic and epigenetic aspects of DNA methylation on genome expression, evolution, mutation and carcinogenesis. *Carcinogenesis* **1997**, *18*, 869–882. [CrossRef] [PubMed]
180. Zhao, Y.; Epstein, R.J. Programmed genetic instability: A tumor-permissive mechanism for maintaining the evolvability of higher species through methylation-dependent mutation of DNA repair genes in the male germ line. *Mol. Biol. Evol.* **2008**, *25*, 1737–1749. [CrossRef]
181. Feinberg, A.P.; Tycko, B. The history of cancer epigenetics. *Nat. Rev. Cancer* **2004**, *4*, 143–153. [CrossRef]
182. Rodriguez, J.; Frigola, J.; Vendrell, E.; Risques, R.A.; Fraga, M.F.; Morales, C.; Moreno, V.; Esteller, M.; Capellà, G.; Ribas, M.; et al. Chromosomal instability correlates with genome-wide DNA demethylation in human primary colorectal cancers. *Cancer Res.* **2006**, *66*, 8462–9468. [CrossRef]
183. Negrini, S.; Gorgoulis, V.G.; Halazonetis, T.D. Genomic instability-an evolving hallmark of cancer. *Nat. Rev. Mol. Cell Biol.* **2010**, *11*, 220–228. [CrossRef]
184. Esteller, M.; Silva, J.M.; Dominguez, G.; Bonilla, F.; Matias-Guiu, X.; Lerma, E.; Bussaglia, E.; Prat, J.; Harkes, I.C.; Repasky, E.A.; et al. Promoter hypermethylation and BRCA1 inactivation in sporadic breast and ovarian tumors. *J. Natl. Canc. Inst.* **2000**, *92*, 564–569. [CrossRef]
185. Kanwal, R.; Gupta, S. Epigenetics and cancer. *J. Appl. Physiol.* **2010**, *109*, 598–605. [CrossRef]

186. Durso, D.F.; Bacalini, M.G.; Faria do Valle, I.; Pirazzini, C.; Bonafé, M.; Castellani, G.; Caetano Faria, A.M.; Franceschi, C.; Garagnani, P.; Nardini, C. Aberrant methylation patterns in colorectal cancer: A meta-analysis. *Oncotarget* **2017**, *8*, 12820–12830. [CrossRef]
187. Romanenko, A.; Morell-Quadreny, L.; Lopez-Guerrero, J.A.; Pellin, A.; Nepomnyaschy, V.; Vozianov, A.; Llombart-Bosch, A. The INK4a/ARF locus: Role in cell cycle control for renal cell epithelial tumor growth after the Chernobyl accident. *Virchows Arch.* **2004**, *445*, 298–304. [CrossRef] [PubMed]
188. Little, J.B. Radiation carcinogenesis. *Carcinogenesis* **2000**, *21*, 397–404. [CrossRef] [PubMed]
189. Jones, P.A.; Baylin, S.B. The fundamental role of epigenetic events in cancer. *Nat. Rev. Genet.* **2002**, *3*, 415–428. [CrossRef] [PubMed]
190. Takai, D.; Jones, P.A. Comprehensive analysis of CpG islands in human chromosomes 21 and 22. *Proc. Natl. Acad. Sci. USA* **2002**, *99*, 3740–3745. [CrossRef]
191. Fraga, M.F.; Ballestar, E.; Villar-Garea, A.; Boix-Chornet, M.; Espada, J.; Schotta, G.; Bonaldi, T.; Haydon, C.; Ropero, S.; Petrie, K.; et al. Loss of acetylation at Lys16 and trimethylation at Lys20 of histone H4 is a common hallmark of human cancer. *Nat. Genet.* **2005**, *37*, 391–400. [CrossRef]
192. Seligson, D.B.; Horvath, S.; Shi, T.; Yu, H.; Tze, S.; Grunstein, M.; Kurdistani, S.K. Global histone modification patterns predict risk of prostate cancer recurrence. *Nature* **2005**, *435*, 1262–1266. [CrossRef]
193. Jones, P.; Issa, J.; Baylin, S. Targeting the cancer epigenome for therapy. *Nat. Rev. Genet.* **2016**, *17*, 630–641. [CrossRef]
194. Zhao, Z.; Shilatifard, A. Epigenetic modifications of histones in cancer. *Genome Biol.* **2019**, *20*, 245. [CrossRef]
195. Dhall, A.; Zee, B.M.; Yan, F.; Blanco, M.A. Intersection of Epigenetic and Metabolic Regulation of Histone Modifications in Acute Myeloid Leukemia. *Front. Oncol.* **2019**, *9*, 432. [CrossRef]
196. Reddy, K.B. MicroRNA (miRNA) in cancer. *Cancer Cell Int.* **2015**, *15*, 38. [CrossRef]
197. Peng, Y.; Croce, C.M. The role of MicroRNAs in human cancer. *Signal Transduct. Target. Ther.* **2016**, *1*, 15004. [CrossRef]
198. Chen, X.; Zhao, C.; Zhao, Z.; Wang, H.; Fang, Z. Specific Glioma Prognostic Subtype Distinctions Based on DNA Methylation Patterns. *Front. Genet.* **2019**, *10*, 786. [CrossRef] [PubMed]
199. Tanno, B.; Babini, G.; Leonardi, S.; Giardullo, P.; De Stefano, I.; Pasquali, E.; Ottolenghi, A.; Atkinson, M.J.; Saran, A.; Mancuso, M. Ex vivo miRNome analysis in Ptch1+/− cerebellum granule cells reveals a subset of miRNAs involved in radiation-induced medulloblastoma. *Oncotarget* **2006**, *7*, 68253–68269. [CrossRef]
200. Liu, C.; Li, B.; Cheng, Y.; Lin, J.; Hao, J.; Zhang, S.; Mitchel, R.E.J.; Sun, D.; Ni, J.; Zhao, L.; et al. MiR-21 plays an important role in radiation induced carcinogenesis in BALB/c mice by directly targeting the tumor suppressor gene Big-h3. *Int. J. Biol. Sci.* **2011**, *7*, 347–363. [CrossRef] [PubMed]
201. Wilke, C.M.; Hess, J.; Klymenko, S.V.; Chumak, V.V.; Zakhartseva, L.M.; Bakhanova, E.V.; Feuchtinger, A.; Walch, A.K.; Selmansberger, M.; Braselmann, H.; et al. Expression of miRNA-26b-5p and its target TRPS1 is associated with radiation exposure in post-Chernobyl breast cancer. *Int. J. Cancer* **2018**, *142*, 573–583. [CrossRef] [PubMed]
202. Sharma, S.; Kelly, T.K.; Jones, P.A. Epigenetics in cancer. *Carcinogenesis* **2010**, *31*, 27–36. [CrossRef]
203. Kamiya, K.; Yasukawa-Barnes, J.; Mitchen, J.M.; Gould, M.N.; Clifton, K.H. Evidence that carcinogenesis involves an imbalance between epigenetic high-frequency initiation and suppression of promotion. *Proc. Natl. Acad. Sci. USA* **1995**, *92*, 1332–1336. [CrossRef]
204. Clifton, K.H. Comments on the evidence in support of the epigenetic nature of radiogenic initiation. *Mutat. Res.* **1996**, *350*, 77–80. [CrossRef]
205. Feinberg, A.; Ohlsson, R.; Henikoff, S. The epigenetic progenitor origin of human cancer. *Nat. Rev. Genet.* **2006**, *7*, 21–33. [CrossRef]
206. Jordan, C.T.; Guzman, M.L.; Noble, M. Cancer stem cells. *N. Engl. J. Med.* **2006**, *355*, 1253–1261. [CrossRef]
207. Seibold, P.; Auvinen, A.; Averbeck, D.; Bourguignon, M.; Hartikainen, J.M.; Hoeschen, C.; Laurent, O.; Noël, G.; Sabatier, L.; Salomaa, S.; et al. Clinical and epidemiological observations on individual radiation sensitivity and susceptibility. *Int. J. Radiat. Biol.* **2020**, *96*, 324–339. [CrossRef] [PubMed]
208. D'Urso, A.; Brickner, J.H. Mechanisms of epigenetic memory. *Trends Genet.* **2014**, *30*, 230–236. [CrossRef] [PubMed]
209. Little, M.P.; Goodhead, D.T.; Bridges, B.A.; Bouffler, S.D. Evidence relevant to untargeted and transgenerational effects in the offspring of irradiated parents. *Mutat. Res. Rev. Mutat. Res.* **2013**, *753*, 50–67. [CrossRef] [PubMed]

210. Heard, E.; Martienssen, R.A. Transgenerational Epigenetic Inheritance: Myths and mechanisms. *Cell* **2014**, *157*, 95–109. [CrossRef] [PubMed]
211. Nelson, V.R.; Nadeau, J.H. Transgenerational genetic effects. *Epigenomics* **2010**, *2*, 797–806. [CrossRef]
212. Reik, W.; Dean, W.; Walter, J. Epigenetic Reprogramming in Mammalian Development. *Science* **2001**, *293*, 1089. [CrossRef]
213. Zeng, Y.; Chen, T. DNA Methylation Reprogramming during Mammalian Development. *Genes* **2019**, *10*, 257. [CrossRef]
214. Luning, K.G.; Frolen, H.; Nilsson, A. Genetic effects of 239Pu salt injections in male mice. *Mutat. Res.* **1976**, *34*, 539–542. [CrossRef]
215. Dubrova, Y.E. Radiation-induced transgenerational instability. *Oncogene* **2003**, *22*, 7087–7709. [CrossRef]
216. Dubrova, Y.E.; Plumb, M.A. Ionising radiation and mutation induction at mouse minisatellite loci. The story of the two generations. *Mutat. Res.* **2002**, *499*, 143–150. [CrossRef]
217. Kamstra, J.H.; Hurem, S.; Martin, L.M.; Lindeman, L.C.; Legler, J.; Oughton, D.; Salbu, B.; Brede, D.A.; Lyche, J.L.; Aleström, P. Ionizing radiation induces transgenerational effects of DNA methylation in zebrafish. *Sci. Rep.* **2018**, *8*, 15373. [CrossRef] [PubMed]
218. Gardner, M.J.; Snee, M.P.; Hall, A.J.; Powell, C.A.; Downes, S.; Terrell, J.D. Results of case-control study of leukaemia and lymphoma among young people near Sellafield nuclear plant in West Cumbria. *BMJ* **1990**, *300*, 423–429. [CrossRef]
219. Dubrova, Y.E.; Grant, G.; Chumak, A.A.; Stezhka, V.A.; Karakasian, A.N. Elevated minisatellite mutation rate in the post-Chernobyl families from Ukraine. *Am. J. Hum. Genet.* **2002**, *71*, 801–809. [CrossRef] [PubMed]
220. Dubrova, Y.E.; Bersimbaev, R.I.; Djansugurova, L.B.; Tankimanova, M.K.; Mamyrbaeva, Z.Z.; Mustonen, R.; Lindholm, C.; Hulten, M.; Salomaa, S. Nuclear weapons tests and human germline mutation rate. *Science* **2002**, *295*, 1037. [CrossRef]
221. Dubrova, Y.E.; Ploshchanskaya, O.G.; Kozionova, O.S.; Akleyev, A.V. Minisatellite germline mutation rate in the Techa River population. *Mutat. Res.* **2006**, *602*, 74–82. [CrossRef] [PubMed]
222. Ozasa, K.; Cullings, K.M.; Ohishi, W.; Hida, A.; Grant, E.J. Epidemiological studies of atomic bomb radiation at the Radiation Effects Research Foundation. *Int. J. Radiat. Biol.* **2019**, *95*, 879–891. [CrossRef]
223. Nomura, T.; Baleva, L.S.; Ryo, H.; Adachi, S.; Sipyagina, A.E.; Karakhan, N.M. Transgenerational effects of radiation on cancer and other disorders in mice and humans. *J. Radiat. Cancer Res.* **2017**, *8*, 123–134. [CrossRef]
224. Beresford, N.A.; Horemans, N.; Copplestone, D.; Raines, K.E.; Orizaola, G.; Wood, M.D.; Laanen, P.; Whitehead, H.C.; Burrows, J.E.; Tinsley, M.C.; et al. Towards solving a scientific controversy—The effects of ionising radiation on the environment. *J. Environ. Radioact.* **2020**, *211*. [CrossRef]
225. Horemans, N.; Spurgeon, D.J.; Lecomte-Pradines, C.; Saenen, E.; Bradshaw, C.; Oughton, D.; Rasnaca, I.; Kamstra, J.H.; Adam-Guillermin, C. Current evidence for a role of epigenetic mechanisms in response to ionizing radiation in an ecotoxicological context. *Environ. Pollut.* **2019**, *251*, 469–483. [CrossRef]
226. United Nations Scientific Committee on the Effects of Atomic Radiation (UNSCEAR). *UNSCEAR 2010 Report*; United Nations: New York, NY, USA, 2011; pp. 1–14.
227. United Nations Scientific Committee on the Effects of Atomic Radiation (UNSCEAR). Non-Targeted and Delayed Effects of Exposure to Ionizing Radiation. In *UNSCEAR 2006 Report*; United Nations: New York, NY, USA, 2006; Volume 2, pp. 1–79.
228. ICRP. ICRP Publication 118: ICRP Statement on Tissue Reactions/Early and Late Effects of Radiation in Normal Tissues and Organs-Threshold Doses for Tissue Reactions in a Radiation Protection Context. *Ann. ICRP* **2012**, *41*, 1–322. [CrossRef]
229. Meyers, C.A. Neurocognitive dysfunction in cancer patients. *Oncology* **2000**, *14*, 75–79. Available online: https://pubmed.ncbi.nlm.nih.gov/10680150/ (accessed on 18 August 2020).
230. Syndikus, I.; Tait, D.; Ashley, S.; Jannoun, L. Long-term follow-up of young children with brain tumors after irradiation. *Int. J. Radiat. Oncol. Biol. Phys.* **1994**, *30*, 781–787. [CrossRef]
231. Hall, P.; Adami, H.O.; Trichopoulos, D.; Pedersen, N.L.; Lagiou, P.; Ekbom, A.; Ingvar, M.; Lundell, M.; Granath, F. Effect of low doses of ionising radiation in infancy on cognitive function in adulthood: Swedish population based cohort study. *BMJ* **2004**, *328*, 19. [CrossRef]
232. Otake, M.; Schull, W.J. Radiation-related brain damage and growth retardation among the prenatally exposed atomic bomb survivors. *Int. J. Radia. T Biol.* **1998**, *74*, 159–171. [CrossRef]

233. Schull, W.J.; Otake, M. Cognitive function and prenatal exposure to ionizing radiation. *Teratology* **1999**, *59*, 222–226. [CrossRef]
234. Little, M.P. A review of non-cancer effects, especially circulatory and ocular diseases. *Radiat. Environ. Biophys.* **2013**, *52*, 435–449. [CrossRef]
235. Cucinotta, F.A.; Alp, M.; Sulzman, F.M.; Wang, M. Space radiation risks to the central nervous system. *Life Sci. Space Res.* **2014**, *2*, 54–69. [CrossRef]
236. Parihar, V.K.; Allen, B.; Tran, K.K.; Macaraeg, T.G.; Chu, E.M.; Kwok, S.F.; Chmielewski, N.N.; Craver, B.M.; Baulch, J.E.; Acharya, M.M.; et al. What happens to your brain on the way to Mars. *Sci. Adv.* **2015**, *1*, e1400256. [CrossRef]
237. Impey, S.; Jopson, T.; Pelz, C.; Tafessu, A.; Fareh, F.; Zuloaga, D.; Marzulla, T.; Riparip, L.K.; Stewart, B.; Rosi, S.; et al. Short- and long-term effects of 56Fe irradiation on cognition and hippocampal DNA methylation and gene expression. *BMC Genom.* **2016**, *17*, 825. [CrossRef]
238. Acharya, M.M.; Baddour, A.A.D.; Kawashita, T.; Allen, B.D.; Syage, A.R.; Nguyen, T.H.; Yoon, N.; Giedzinski, E.; Yu, L.; Parihar, V.K.; et al. Epigenetic determinants of space radiation-induced cognitive dysfunction. *Sci. Rep.* **2017**, *7*, 42885. [CrossRef]
239. Schultz-Hector, S.; Trott, K.R. Radiation-induced cardiovascular diseases: Is the epidemiologic evidence compatible with the radiobiologic data? *Int. J. Radiat. Oncol. Biol. Phys.* **2007**, *67*, 10–18. [CrossRef]
240. Little, M.P.; Tawn, E.J.; Tzoulaki, I.; Wakeford, R.; Hildebrandt, G. A systematic review of epidemiological associations between low and moderate doses of ionizing radiation and late cardiovascular effects, and their possible mechanisms. *Radiat. Res.* **2008**, *169*, 99–109. [CrossRef]
241. Mitchel, R.E.; Hasu, M.; Bugden, M.; Wyatt, H.; Little, M.P.; Gola, A.; Hildebrandt, G.; Priest, N.D.; Whitman, S.C. Low-dose radiation exposure and atherosclerosis in ApoE−/− mice. *Radiat. Res.* **2011**, *175*, 665–676. [CrossRef]
242. Lowe, D.; Raj, K. Premature aging induced by radiation exhibits pro-atherosclerotic effects mediated by epigenetic activation of CD44 expression. *Aging Cell* **2014**, *13*, 900–910. [CrossRef]
243. Koturbash, I.; Miousse, I.R.; Sridharan, V.; Nzabarushimana, E.; Skinner, C.M.; Melnyk, S.B.; Pavliv, O.; Hauer-Jensen, M.; Nelson, G.; Boerma, M. Radiation-induced changes in DNA methylation of repetitive elements in the mouse heart. *Mutat. Res.* **2016**, *787*, 43–53. [CrossRef]
244. Liu, Y.C.; Wilkins, M.; Kim, T.; Malyugin, B.; Mehta, J.S. Cataracts. *Lancet* **2017**, *390*, 600–612. [CrossRef]
245. Truscott, R.J.W. Age-related nuclear cataract—Oxidation is the key. *Exp. Eye Res.* **2005**, *80*, 709–725. [CrossRef]
246. Blakely, E.A.; Kleiman, N.J.; Neriishi, K.; Chodick, G.; Chylack, L.T.; Cucinotta, F.A.; Minamoto, A.; Nakashima, E.; Kumagami, T.; Kitaoka, T.; et al. Radiation Cataractogenesis: Epidemiology and Biology. *Radiat. Res.* **2010**, *173*, 709–717. [CrossRef]
247. Ainsbury, E.A.; Barnard, S.; Bright, S.; Dalke, C.; Jarrin, M.; Kunze, S.; Tanner, R.; Dynlacht, J.R.; Quinlan, R.; Graw, J.; et al. Ionizing radiation induced cataracts: Recent biological and mechanistic developments and perspectives for future research. *Mutat Res.* **2016**, *770*, 238–261. [CrossRef]
248. Lanza, M.; Benincasa, G.; Costa, D.; Napoli, C. Clinical Role of Epigenetics and Network Analysis in Eye Diseases: A Translational Science Review. *J. Ophthalmol.* **2019**, *2019*, 2424956. [CrossRef]
249. Alkozi, H.A.; Franco, R.; Pintor, J.J. Epigenetics in the Eye: An Overview of the Most Relevant Ocular Diseases. *Front. Genet.* **2017**, *8*, 144. [CrossRef] [PubMed]
250. HLEG. High Level and Expert Group Report on European Low Dose Risk Research-Radiation Protection. European Commission EUR 23884; Luxembourg: Office for Official Publications of the European Communities. 2009. Available online: https://cordis.europa.eu/docs/publications/1070/107087891-6_en.pdf (accessed on 18 August 2020).
251. Morgan, W.F.; Bair, W.J. Issues in Low Dose Radiation Biology: The Controversy Continues. A Perspective. *Radiat. Res.* **2013**, *179*, 501–510. [CrossRef] [PubMed]
252. Belli, M.; Tabocchini, M.A.; Jourdain, J.R.; Salomaa, S.; Repussard, J. The European initiative on low-dose risk research: From the HLEG to MELODI. *Radiat. Prot. Dosimetry.* **2015**, *16*, 178–181. [CrossRef] [PubMed]
253. MELODI SRA Working Group. Strategic Research Agenda of the Multidisciplinary European Low Dose Initiative (MELODI)-2019. Available online: http://www.melodi-online.eu/m_docs_sra.html (accessed on 18 August 2020).
254. Goodhead, D.T. Spatial and temporal distribution of energy. *Health Phys.* **1988**, *55*, 231–240. [CrossRef] [PubMed]

255. United Nations Scientific Committee on the Effects of Atomic Radiation (UNSCEAR). Sources and Effects of Ionizing Radiation. In *1993 Report to the General Assembly, with Scientific Annexes*; United Nations: New York, NY, USA, 1993.
256. Wakeford, R.; Tawn, E.J. The meaning of low dose and low dose-rate. *J. Radiol. Prot.* **2010**, *30*, 1–3. [CrossRef] [PubMed]
257. Rothkamm, K.; Löbrich, M. Evidence for a lack of DNA double-strand break repair in human cells exposed to very low x-ray doses. *Proc. Natl. Acad. Sci. USA* **2003**, *100*, 5057–5062. [CrossRef]
258. Grudzenski, S.; Raths, A.; Conrad, S.; Rübe, C.E.; Löbrich, M. Inducible response required for repair of low-dose radiation damage in human fibroblasts. *Proc. Natl. Acad. Sci. USA* **2010**, *107*, 14205–14210. [CrossRef]
259. Amundson, S.A.; Do, K.T.; Fornace, A.J. Induction of Stress Genes by Low Doses of Gamma Rays. *Radiat. Res.* **1999**, *152*, 225. [CrossRef]
260. Amundson, S.A.; Lee, R.A.; Koch-Paiz, C.A.; Bittner, M.L.; Meltzer, P.; Trent, J.M.; Fornace, A.J., Jr. Differential Responses of Stress Genes to Low Dose-Rate γ Irradiation. *Mol. Cancer Res.* **2003**, *1*, 445–452.
261. Ding, L.H.; Shingyoji, M.; Chen, F.; Hwang, J.J.; Burma, S.; Lee, C.; Cheng, J.F.; Chen, D.J. Gene Expression Profiles of Normal Human Fibroblasts after Exposure to Ionizing Radiation: A Comparative Study of Low and High Doses. *Radiat. Res.* **2005**, *164*, 17–26. [CrossRef]
262. Sokolov, M.; Neumann, R. Global Gene Expression Alterations as a Crucial Constituent of Human Cell Response to Low Doses of Ionizing Radiation Exposure. *Int. J. Mol. Sci.* **2016**, *17*, 55. [CrossRef] [PubMed]
263. Mezentsev, A.; Amundson, S.A. Global Gene Expression Responses to Low- or High-Dose Radiation in a Human Three-Dimensional Tissue Model. *Radiat. Res.* **2011**, *175*, 677–688. [CrossRef] [PubMed]
264. Goldberg, Z.; Rocke, D.M.; Schwietert, C.; Berglund, S.R.; Santana, A.; Jones, A.; Lehmann, J.; Stern, R.; Lu, R.; Siantar, C.H. Human in vivo dose-response to controlled, low-dose low linear energy transfer ionizing radiation exposure. *Clin. Cancer Res.* **2006**, *12*, 3723–3729. [CrossRef] [PubMed]
265. Mitchel, R.E.; Jackson, J.S.; Morrison, D.P.; Carlisle, S.M. Low doses of radiation increase the latency of spontaneous lymphomas and spinal osteosarcomas in cancer-prone, radiation-sensitive Trp53 heterozygous mice. *Radiat. Res.* **2003**, *159*, 320–327. [CrossRef]
266. Bernal, A.J.; Dolinoy, D.C.; Huang, D.; Skaar, D.A.; Weinhouse, C.; Jirtle, R.L. Adaptive radiation-induced epigenetic alterations mitigated by antioxidants. *FASEB J.* **2013**, *27*, 665–671. [CrossRef]
267. Vaiserman, A.M. Hormesis and epigenetics: Is there a link? *Ageing Res. Rev.* **2011**, *10*, 413–421. [CrossRef]
268. Kovalchuk, O.; Burke, P.; Besplug, J.; Slovack, M.; Filkowski, J.; Pogribny, I. Methylation changes in muscle and liver tissues of male and female mice exposed to acute and chronic low-dose X-ray-irradiation. *Mutat. Res.* **2004**, *548*, 75–84. [CrossRef]
269. Taki, K.; Wang, B.; Nakajima, T.; Wu, J.; Ono, T.; Uehara, Y.; Matsumoto, T.; Oghiso, Y.; Tanaka, K.; Ichinohe, K.; et al. Microarray Analysis of Differentially Expressed Genes in the Kidneys and Testes of Mice after Long-term Irradiation with Low-dose-rate γ-rays. *J. Radiat. Res.* **2009**, *50*, 241–252. [CrossRef]
270. Ye, S.; Yuan, D.; Xie, Y.; Pan, Y.; Shao, C. Role of DNA methylation in long-term low-dose γ-rays induced adaptive response in human B lymphoblast cells. *Int. J. Radiat. Biol.* **2013**, *89*, 898–906. [CrossRef]
271. United Nations Scientific Committee on the Effects of Atomic Radiation (UNSCEAR). Sources and Effects of Ionizing Radiation. In *2008 Report to the General Assembly*; United Nations: New York, NY, USA, 2010; Volume 1.
272. Planel, H.; Soleilhavoup, J.P.; Tixador, R.; Richoilley, G.; Conter, A.; Croute, F.; Caratero, C.; Gaubin, Y. Influence on cell proliferation of background radiation or exposure to very low, chronic gamma radiation. *Health Phys.* **1987**, *52*, 571–578. [CrossRef]
273. Smith, G.B.; Grof, Y.; Navarrette, A.; Guilmette, R.A. Exploring biological effects of low background radiation from the other side of the background. *Health Phys.* **2011**, *100*, 263–265. [CrossRef] [PubMed]
274. Castillo, H.; Schoderbek, D.; Dulal, S.; Escobar, G.; Wood, J.; Nelson, R.; Smith, G. Stress induction in the bacteria Shewanella Oneidensis and Deinococcus Radiodurans in response to below-background ionizing radiation. *Int. J. Radiat. Biol.* **2015**, *3002*, 1–33. [CrossRef]
275. Satta, L.; Augusti-Tocco, G.; Ceccarelli, R.; Esposito, A.; Fiore, M.; Paggi, P.; Poggesi, I.; Ricordy, R.; Scarsella, G.; Cundari, E. Low environmental radiation background impairs biological defence of the yeast Saccharomyces cerevisiae to chemical radiomimetic agents. *Mutat. Res.* **1995**, *347*, 129–133. [CrossRef]

276. Satta, L.; Antonelli, F.; Belli, M.; Sapora, O.; Simone, G.; Sorrentino, E.; Tabocchini, M.A.; Amicarelli, F.; Ara, C.; Nisi, S. Influence of a low background radiation environment on biochemical and biological responses in V79 cells. *Radiat. Environ. Biophys.* **2002**, *41*, 217–224. [CrossRef]
277. Carbone, M.C.; Pinto, M.; Antonelli, F.; Amicarelli, F.; Balata, M.; Belli, M.; Conti Devirgiliis, L.; Sapora, O.; Simone, G.; Sorrentino, E.; et al. Effects of deprivation of background environmental radiation on cultured human cells. *Il Nuovo Cim.* **2010**, *4*, 469–477. [CrossRef]
278. Fratini, E.; Carbone, C.; Capece, D.; Esposito, G.; Simone, G.; Tabocchini, M.A.; Tomasi, M.; Belli, M.; Satta, L. Low-radiation environment affects the development of protection mechanisms in V79 cells. *Radiat. Environ. Biophys.* **2015**. [CrossRef]
279. Morciano, P.; Iorio, R.; Iovino, D.; Cipressa, F.; Esposito, G.; Porrazzo, A.; Satta, L.; Alesse, E.; Tabocchini, M.A.; Cenci, G. Effects of reduced natural background radiation on Drosophila melanogaster growth and development as revealed by the FLYINGLOW program. *J. Cell Physiol.* **2018**, *233*, 23–29. [CrossRef]
280. Morciano, P.; Cipressa, F.; Porrazzo, A.; Esposito, G.; Tabocchini, M.A.; Cenci, G. Fruit flies provide new insights in low radiation background biology at the INFN underground Gran Sasso National Laboratory (LNGS). *Radiat. Res.* **2018**, *190*, 217–225. [CrossRef]
281. Little, M.P.; Tawn, E.J.; Tzoulaki, I.; Hildebrandt, G.; Paris, F.; Tapio, S.; Elliott, P. Review and meta-analysis of epidemiological associations between low/moderate doses of ionizing radiation and circulatory disease risks, and their possible mechanisms. *Radiat. Environ. Biophys.* **2010**, *49*, 139–153. [CrossRef]
282. Vaiserman, A.M. Hormesis, adaptive epigenetic reorganization, and implications for human health and longevity. *Dose Response* **2010**, *8*, 16–21. [CrossRef]
283. Azzam, E.I.; Colangelo, N.W.; Domogauer, J.D.; Sharma, N.; de Toledo, S.M. Is Ionizing Radiation Harmful at any Exposure? An Echo That Continues to Vibrate. *Health Phys.* **2016**, *110*, 249–251. [CrossRef] [PubMed]
284. Vaiserman, A.; Koliada, A.; Zabuga, O.; Socol, Y. Health Impacts of Low-Dose Ionizing Radiation: Current Scientific Debates and Regulatory Issues. *Dose Response* **2018**, *16*, 1559325818796331. [CrossRef] [PubMed]
285. Feinendegen, L.E. Evidence for beneficial low level radiation effects and radiation hormesis. *Br. J. Radiol.* **2005**, *78*, 3–7. [CrossRef] [PubMed]
286. Calabrese, E.J.; O'Connor, M.K. Estimating risk of low radiation doses—A critical review of the BEIR VII report and its use of the linear no-threshold (LNT) hypothesis. *Radiat. Res.* **2014**, *182*, 463–474. [CrossRef]

International Journal of
Molecular Sciences

MDPI

Review

The Effect of Low Temperatures on Environmental Radiation Damage in Living Systems: Does Hypothermia Show Promise for Space Travel?

Hisanori Fukunaga [1,2]

[1] Patrick G Johnston Centre for Cancer Research, Queen's University Belfast, 97 Lisburn Road, Belfast BT9 7AE, UK; hfukunaga01@qub.ac.uk

[2] Tohoku Medical Megabank Organization, Tohoku University, 2-1 Seiryo-machi, Aoba-ku, Sendai 980-8573, Japan

Received: 29 July 2020; Accepted: 31 August 2020; Published: 1 September 2020

Abstract: Low-temperature treatments (i.e., hypothermia) may be one way of regulating environmental radiation damage in living systems. With this in mind, hibernation under hypothermic conditions has been proposed as a useful approach for long-term human space flight. However, the underlying mechanisms of hypothermia-induced radioresistance are as yet undetermined, and the conventional risk assessment of radiation exposure during hibernation remains insufficient for estimating the effects of chronic exposure to galactic cosmic rays (GCRs). To promote scientific discussions on the application of hibernation in space travel, this literature review provides an overview of the progress to date in the interdisciplinary research field of radiation biology and hypothermia and addresses possible issues related to hypothermic treatments as countermeasures against GCRs. At present, there are concerns about the potential effects of chronic radiation exposure on neurological disorders, carcinogenesis, ischemia heat failures, and infertility in astronauts; these require further study. These concerns may be resolved by comparing and integrating data gleaned from experimental and epidemiological studies.

Keywords: environmental radiation; hibernation; HIF-1; hypothermia; hypoxia; radioresistance; ROS; space travel

1. Introduction

Since Wilhelm Röntgen's discovery of X-rays in 1895 and Henri Becquerel's discovery of radioactivity in 1896, countless studies have aimed to understand the characteristics of ionizing radiation (IR) and its effects on biological systems. Generally speaking, radiation is a double-edged sword: On the one hand, it has become an essential diagnostic and treatment tool in modern medicine; on the other hand, its carcinogenic properties are well-known. Radiation can induce a broad spectrum of DNA lesions, including single- and double-strand DNA breaks, damage to nucleotide bases, and cross-linking; further, exposure to radiation can seriously damage biological systems by triggering cell death or inducing mutations that lead to radiation-induced cancers [1]. The first radiation-related solid cancer was reported in 1902; it appeared in an ulcerated area of the skin. The first case of leukemia was documented in radiation workers in 1911 [2]. Almost immediately, animal model systems were applied to study the effects of radiation on what we now know as DNA. In 1927, Hermann Muller demonstrated that X-rays can cause mutations in fruit flies at a frequency with an approximately linear correlation with its dose [3]. According to monographs by the International Agency for Research on Cancer (IARC) [4], all types of IR have been classified as 'Group 1'. This category is used to describe situations where there is sufficient evidence to expect carcinogenicity in humans. Thus, proper protection from IR is of importance for both human and non-human species.

Jacobi [5] was the first to develop the "effective dose" concept in 1975. Since then, it has since been used by the International Commission on Radiological Protection (ICRP) as a key measurement when assessing the risks of radiation exposure's stochastic effects (e.g., hereditary effects and carcinogenesis) and dose exposure limits [6]. IR deposits energy directly into any matter being irradiated. The quantity used to express this energy is the absorbed dose; this is a quantified physical dose that is dependent on both the level of incidental radiation and the irradiated object's absorption properties. The International System of Units (SI) has adopted the gray (Gy) as the unit for measuring absorbed doses. A 'gray' is defined as one joule of energy absorbed per kilogram of matter. As a physical quantity, an absorbed dose alone is not a satisfactory indicator of biological response, as responses may be driven by many accompanying factors. To accommodate the consideration of stochastic radiological risks, the ICRP and the International Commission on Radiation Units and Measurements devised the dose quantity's equivalent dose and the effective dose. These are used to estimate the biological effectiveness of a given absorbed dose. The SI unit for an effective dose is the sievert (Sv), which currently represents, among the entire population, a 5.5% probability of developing cancer [6]. An 'effective dose' refers to the type of radiation and the characteristics of each organ or tissue that has been irradiated, since different organs have different levels of radiation sensitivity [7]. We know that the average annual effective dose from background radiation is around 3 mSv, while the typical effective doses of nuclear medical and radiology examinations are as follows: Standard radiographic examinations (approximately 0.01–10 mSv), most nuclear medicine procedures (0.3–20 mSv), computed tomographic examinations (2–20 mSv), and interventional radiological procedures (5–70 mSv) [8].

Low-temperature treatments (i.e., hypothermia) seem to be a promising means of regulating environmental radiation biological effects, although the underlying mechanisms of the approach remain unclear. It is well known that the hibernation of heterothermic, and the cooling of poikilothermic, animals in cold environments provides them with temporary protection against the acute effects of radiation [9]. Several studies over the past few decades have sought to explain the interactions of IR and temperature in animals, and there is evidence that hypothermia does provide at least some radioprotective effect. Some early studies showed that hypothermia in living organisms, such as fish (*Carassius carassius*) [10], frogs (*Rana pipiens*) [11], marmots (*Marmota monax*) [12], CF1 strain mice [13], T strain mice [14,15], rats [16], and ground squirrels (*Citellus tridecemlineatus*) [17,18], following total-body irradiation prolongs their survival. Furthermore, the radiosensitivity of certain organs in vivo and cells in vitro has been studied, such as the testes [19], spleen [20], ovaries [21], and the hemopoietic system [22]. All of these works were related to the acute effects of radiation.

In 1961, Bloch and colleagues were the first to prove that the irradiation of high-grade cerebral astrocytoma (glioblastoma multiforme) under conditions of mild whole-body hypothermia (rectal temperature maintained at 31–32 °C) led to an increase in radiosensitivity [23,24]. That same year, Joseph Weiss indicated the possibility that hypothermic treatments could increase a tumor's radiosensitivity while decreasing the sensitivity of the normal tissues surrounding it [25]. The development of a system for evaluating localized hypothermia's radioprotective effects is important for potential clinical applications [26]. A ground-breaking clinical study from 2017 showed the significant radioprotective effect associated with the use of localized hypothermia (15 °C) when accompanied by a single large dose of radiation aimed at mitigating a rectal obstruction and/or bleeding [27]. When used in tandem, whole-body and local hypothermic treatments have shown great promise for protecting normal tissue functions.

While it is known that hypothermia can have a protective effect against acute radiation injuries in living systems, it could also be useful for radiological protection in the future. This article aims to review recent progress in studies on the effect of hypothermia on environmental radiation damage in biomolecules, cells, and living systems. In the coming era of space travel, if hypothermia is shown to be capable of providing a radioprotective effect, hibernation could be an effective solution for those engaged in long-term missions. Here, from the point of view of radiological protection, we also discuss the possible impact of hypothermia on the future of space exploration.

2. Environmental Radiation Damage in Living Systems

The quality and quantity of DNA damage is assessed in terms of the radiation type and dose [28]. For example, high and low linear energy transfer (LET) radiation induce different spectra and qualities/complexity of DNA lesions, because of the differences in radiation track structures [29]. This also has an effect on the dose delivered to each cell by individual tracks at low doses, which is radiation quality dependent. Higher LET radiation delivers an increased dose per track. More specifically, in cases of low-dose exposure, such as environmental radiation, radiation's energy deposition is localized along its track, which leads to a non-uniform distribution of exposed or unexposed cells in irradiated tissues [30,31]. For this reason, there are possible interactions between the irradiated and non-irradiated cells and the dynamics of those cells in the tissues that are involved in environmental radiation-induced biological responses at the whole-tissue level [32].

The "radiation-induced bystander effects (RIBEs)" refer to radiation-induced responses that are observed in cells that did not directly receive a radiation dose but did receive signals from nearby or neighboring irradiated cells. They behave as though they have been exposed, showing sister chromatid exchanges (SCEs) [33], chromosomal instability [34], micronuclei formations [35], gene mutations [36], and apoptosis [35,37]. In 1992, Nagasawa and Little first reported on RIBEs. They observed SCEs in ~30% of immortalized Chinese hamster ovary cells when only 1% of the population were calculated to be traversed through the nucleus by an a-particle following irradiation. Although most studies on bystander responses have reported cell damage in the non-irradiated cells, there are some reports on bystander-mediated adaptive responses [38]. Due to the complexity of these responses and the variety of positive and negative cellular endpoints, there is still controversy [39]. These responses are mediated either through gap junctions or via soluble factors released by irradiated cells.

In 1909, Köhler was the first to report on clinical observations of a tissue-sparing response during grid radiotherapy, wherein spatially fractionated radiation was delivered using a grid-like pattern of beams [40]. In 1995, a notable "tissue-sparing effect (TSE)" was reported in rat brain tissues during a study of microbeam radiotherapy (MRT) [41] performed at the National Synchrotron Light Source, at Brookhaven National Laboratory in Upton, New York. Since then, an MRT-related TSE has been confirmed in a variety of species and tissue types [40–47]. The TSE is the phenomenon by which normal tissues tolerate single exposures to narrow planes of synchrotron-generated X-rays up to several hundred Gy [48]. The TSE of spatial-fractionated radiation indicates significant implications for both clinical applications and the improvement of risk assessments for exposure to non-uniform radiation, such as environmental radiation.

Intercellular responses could be involved in non-targeted effects, including RIBEs and TSEs, in response to spatially fractionated radiation fields [49]. Furthermore, to achieve tissue homeostasis, cell competition is essential as a cell fitness-sensing mechanism. This is seen across an array of species, from insects to mammals. The process eliminates cells that, while viable, are less fit than their neighbors [50]. Damaged cells induced by spatially fractionated radiation are removed by the neighboring cells through cell competition, resulting in the prevention of a pathological state, such as carcinogenesis, at the tissue level [51]. After the complete clearance of radiation-induced damaged cells, tissue regeneration generally occurs for the purpose of maintaining normal tissue functions, namely homeostasis. Somatic stem cells migrate from the intact to the defective parts and regenerate the structure and function of the tissue via their proliferation and differentiation [52,53]. Such tissue homeostasis mechanisms could be involved in radiation-induced biological responses at the tissue level. However, there is little knowledge of this to date.

3. Mechanisms of the Effect of Hypothermia-Induced Radioprotection

Hypothermia is a condition that results from the drop in the core body temperature to a level at or below 35 °C. It develops when a body's rate of heat production is exceeded by its rate of heat loss [54]. Physiological and environmental stresses that induce behavioral hypothermia include dehydration, hypercapnia, and anemia [55]. Given the marked effect that body temperature has on oxygen uptake in resting animals, hypothermia is beneficial in these conditions, as it reduces the demand for oxygen, decreases evaporative water loss (in amphibians), protects the brain's metabolic status and function, and decreases xenobiotic compounds' toxicity. In short, hypothermia elicits hypoxia.

It has long been known that the amount of available oxygen in tissues plays an important role in determining these tissues' IR sensitivity; this is referred to as the "oxygen effect" [56,57] In 1953, Louis Gray and his colleagues hypothesized that tumors are generally more anoxic than their surrounding normal tissues. As a result, since they are anoxic, tumors are found at a low position on the oxygen-radiosensitivity curve, when compared with normal well-vascularized tissues [56]. In fact, hypoxia can lead to an up to three-fold increase in radioresistance [58]. Furthermore, cellular antioxidants have shown a significant decrease of radiosensitivity, such as glutathione [59]. Thus, when discussing the effects of hypothermia on radiosensitivity, the degree of tissue oxygenation must be considered. In 1960, Joseph Weiss used aerobic cultures from HeLa cells, the first immortal human cell line, to demonstrate that there is no difference in radiosensitivity when cultures are irradiated at 1 and 37 °C [60]. He also revealed that no histological differences were detected between the spleens of mice irradiated under hypoxic hypothermic conditions [20] and those irradiated while hypoxic at a normal body temperature [61]. This indicates that hypothermia's effect on irradiated mammalian cells and tissues is dependent on oxygen levels. From this, we can infer that the observed increase in radiosensitivity is mainly the result of concomitant hypoxia.

In clinical practice, hypoxia is a hallmark of solid tumors and a major obstacle to the effectiveness of radiotherapies, which kill cancer cells through the generation of reactive oxygen species (ROS) [62]. Several radiobiological studies have shown that hypoxic cells are resistant to ROS insults because of the shortage of ROS substrate oxygen. Further, and paradoxically, there is evidence that ROS are produced more in hypoxic than in normoxic cells and serve as signaling molecules that render cells adaptive to hypoxia. In 1995, Gregg Semenza's research group discovered hypoxia-inducible factor (HIF-1) [63], which is involved in these hypoxic responses and regulates several genes in ROS homeostasis [64]. HIF-1 consists of an inducible alpha subunit (HIF-1α) and a constitutively expressed beta subunit (HIF-1β) [65]. Under normoxic conditions, the lysine and proline residues on HIF-1α's oxygen-dependent degradation domain are hydroxylated, and the modified HIF-1α interacts with the Von Hippel-Lindau E3 ubiquitin ligase complex. This is followed by degradation via the ubiquitin proteasome pathway [66]. However, as previously described by Ridder and colleagues [67], hypoxia induces the accumulation of HIF-1α through the prevention of protein degradation, or by the upregulation of gene expression via ROS-mediated pathways. Then, as a result of increased HIF-1α, HIF-1 is activated and regulates more than 100 genes, conferring radioresistance by acting on multiple mechanisms at different levels. In addition, hypoxia- and radiation-induced ROS could trigger a feedback loop that favors the generation of antioxidants. In this way, the combination of hypoxia, ROS, and HIF-1 signaling demonstrates its important role in hypothermia-induced radioresistance for both tumors and normal tissues.

In non-uniform radiation fields, such as environmental radiation exposures, radiation-hit cells and non-hit cells co-exist. Regarding this, in vitro experimental configurations containing in-field and out-of-field cells have been established, showing that intercellular communications from cells in-field to cells out-of-field reduces the survival of out-of-field cells using a 50% in-field and 50% out-of-field (half-field) irradiation [68]. Under hypoxic conditions, while in-field responses were oxygen dependent, out-of-field effects were observed to be independent of oxygen, with similar or greater cell killing [69]. This highlighted the need for further understanding of intercellular signaling under hypoxic conditions.

There are other possible mechanisms for eliciting hypothermia-induced radioprotection. For example, when enzymes are exposed to IR at low temperatures, there is a progressive decrease in radiation sensitivity; namely, greater levels of enzymatic activity after the same dose of radiation delivered at a low temperature compared to room temperature [70]. Furthermore, cold-inducible RNA binding protein (CIRP), which responds to mild cold shock, assists cells in adapting to hypothermic conditions by stabilizing specific mRNAs and facilitating their translation. CIRP protects cells from ultraviolet radiation and hypoxia-induced senescence processes [71]; however, these radioprotective effects for maintaining cellular activity require further study and investigation.

4. Impairment of Radiation Damage Repair under Hypothermic Conditions

In 1966, Egami and Etoh applied lethal acute or fractionated exposures of radiation to fish (*Oryzias latipes*) at 23 and 11 °C [72]. At a water temperature of 23 °C, the delivery of a lethal dose of 40 Gy into two fractions, separated by a three-day interval, helped many fish to survive the 30-day experimental period; by contrast, at a lower water temperature, 11 °C, the overall response to the fractionated dose was the same as to the whole dose delivered at once. This indicates that, at low temperatures, the efforts to repair radiation damage injuries are ineffective; thus, a fractional dose does not help to reduce the radiation damage [73].

Poikilothermic animals are good experimental subjects for studies on the influence of low temperature on radiation sensitivity. According to a 1982 report by the United Nations Scientific Committee on the Effects of Atomic Radiation (UNSCEAR), tissue repair/regeneration and recovery from radiation injury in self-renewing tissues in fish are considerably inhibited when kept at suboptimal temperatures [74]. In cases of chronic radiation exposure, radiation damage accumulates without repair in fish in cold environments, such as Arctic winters. In contrast, those fish in warmer environments, which are continuously able to repair some of their radiation-induced damage, demonstrate lower levels of radiation damage from chronic exposure [9]. The impairment of radiation damage repair capacities may be involved in other living systems, including those of homoiothermic animals. In addition, other endpoints, specifically non-cancer outcomes, such as neurological disorders and ischemic heart disease, could benefit from further investigation.

The radiation-induced effects in genotoxicity under hypothermic conditions are yet to be determined. In recent years, Baird and colleagues showed that hypothermia (13 °C) postpones DNA damage repair in irradiated BJ-hTERT cells and protects against cell death [75]. By contrast, a series of studies investigating hypothermia-induced DNA damage in human peripheral blood lymphocytes [76–80] has suggested the possibility of DNA repair promotion that leads to a reduced transformation of DNA damage to chromosomal aberrations. Therefore, to our knowledge, hypothermia-induced modulations of DNA damage repair in vitro remain controversial.

5. Possible Issues Related to Human Space Exploration and the Induction of Hibernation

In 1961, Yuri Gagarin was the first human to journey to outer space. Since then, the technology associated with space travel and exploration has advanced significantly. As future missions explore realms beyond low-Earth orbit (LEO) and outside the protection zone of the Earth's magnetic shielding, the nature of the radiation exposure that astronauts encounter will include higher levels of exposure [81]. While travelling outside LEO, every cell in an astronaut's body will be traversed by proton or electron rays every few days, and by high atomic number and energy (HZE) ion-charged particles every few months [82]. Shielding may be able to reduce their exposure to galactic cosmic rays (GCRs), including HZE particles, but the technique is not likely to resolve the problem entirely [83]. From the viewpoint of radiation micro-dosimetry, for GCR exposures, the energy deposition of radiation is localized along its track. This results in a spatially fractionated distribution of exposed or unexposed cells in the irradiated tissue. For this reason, the detection of the temporospatial dose's distribution could be of scientific importance, allowing for more accurate individual risk assessments of exposure associated with environmental radiation.

The concept of the effective dose does not yield an individual-specific dose but instead uses a reference person for a given exposure situation. Furthermore, effective doses do not take into account the genomic diversity of individuals' radiosensitivity, so it is not an appropriate means for estimating individuals' radiation-induced health risks among the general population. In fact, as explained in *ICRP Publication 103*, the effective dose is a risk-adjusted quantity for the control of exposure; it was never intended to serve as a measure of risk [6]. This dose is calculated using reference phantoms for the purpose of enabling the summation of doses from all radiation exposures for comparisons with limits, constraints, and reference levels (set at the same quantity), and for the optimization of protection. Implicit in its use is the central assumption of a linear non-threshold dose–response relationship between dose and risk; it is a reasonable assumption for protection purposes but has not been proven for low doses [84]. A single set of tissue-weighting factors is used in the calculation of the effective dose, despite previously recognized differences in the age and sex dependence of the relative contributions of cancer types to the overall detriment and, crucially, in the overall magnitude of cancer detriment.

Risk assessment and proper GCR protection are essential for humans' long-term activities in space [85]. In the US, the National Aeronautics and Space Administration (NASA) bases its safety standards on the acute exposure levels recorded among Japanese atomic bomb survivors [86], but this ultimately seems to be an insufficient approach, as the survivors' acute irradiation scenario differs from the chronic exposure to GCRs that astronauts on a two- or three-year mission to Mars would encounter [87]. In addition, physiological factors, such as age, sex, and DNA repair-deficiency, are clearly important in estimating the biological effects induced by exposure to radiation. In fact, NASA has developed a risk-based approach to radiation exposure limits that accounts for individual factors (e.g., age, gender, and smoking history) and uses them when assessing the uncertainties related to risk estimates [88]. However, many more epidemiological studies are still needed; the Million Person Study of Low-Dose Health Effects (MPS) [89], for example, can contribute to our understanding of the health effects of chronic exposure. The MPS' large study size, approximately one million subjects, makes it capable of providing more precise estimates of lifetime risks of radiation and may indicate reasonable approaches for addressing specific issues of interest to not only NASA but also ICRP, UNSCEAR, and other organizations concerned with radiological protection.

The idea of hibernation has been proposed as a possible approach for use in human space travel. Recently, a procedure to induce a metabolic state known as "synthetic torpor" in non-hibernating mammals was successfully developed; this could be an efficient means of conserving resources, reducing the incidence of mental disorders related with long-term missions and mitigating or preventing radiation-induced acute effects [90]. However, GCR-induced long-term or chronic health risks for humans, such as cancer, heart failure, and dementia, during hibernation or synthetic torpor are still unclear. In addition, environmental radiation damage on reproductive potential remains to be determined. An epidemiological study from 2013 examined 83 healthcare workers who had been exposed to radiation at work. The study shed light on the effects of this exposure on spermatozoa, noting changes in their motility characteristics, global hypermethylation, increased incidences of morphological abnormalities in sperm, and sperm DNA fragmentation [91]. Such epidemiological investigation is of importance to provide new insights into environmental radiation risks on human fertility [92]. If a hypothermic state can induce an impairment of the radiation-induced damage repair response, hibernating astronauts might be more vulnerable to the chronic effects of radiation, compared to those allowed to travel under optimum-temperature sleeping conditions. At present, there is a lack of experimental studies on the combination of chronic radiation exposure and hypothermia. Taken together, current concerns about the possible effects of chronic radiation exposure on carcinogenesis, neurological disorders, ischemic heat failures, and infertility for astronauts require further elucidation that may be resolved by comparing and integrating observed data from epidemiological and experimental studies, as well as biophysical models and computational approaches.

6. Conclusions

This literature review offers a summary of the historical progress in the interdisciplinary research field of radiobiology and hypothermia. As shown in Figure 1, conventional hypothermia appears to be a double-edged sword: On the one hand, it protects organisms against the acute effects of radiation via a hypoxic response; on the other hand, it can lead to impaired radiation damage repair capabilities. Long-term or chronic radiation-induced health risks, including carcinogenesis, neurological disorders, ischemic heart disease, and sterility under hypothermic conditions, are still controversial; nevertheless, it is important for humans' long-term activities in space.

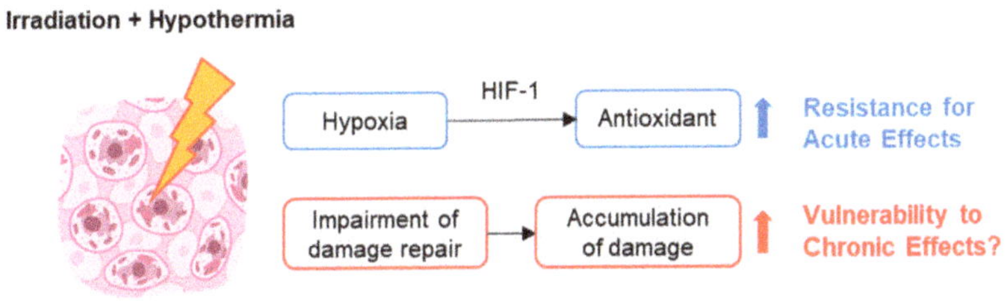

Figure 1. Overview of radiation-induced effects in cells in hypothermic conditions. Hypothermia treatments protect organisms from the effects of acute radiation via a hypoxic response. Additionally, it can induce the impairment of radiation damage repair responses.

To establish a personalized risk assessment for environmental radiation exposures, we should consider the tissue-level responses induced by heterogenous tempo-spatial radiation exposure as well as individual differences in radiation sensitivity. In addition, further investigations are required to provide novel insights into our understanding of the temperature regulation of environmental radiation damage in living systems, as this will be of increasing importance in the coming era of space exploration. Consequently, it will be of importance to monitor not only the total dose but also the tempo-spatial distribution as well as the temperature in the circumstances of the irradiation for risk assessment.

Funding: This research received no external funding.

Conflicts of Interest: The author declares no conflict of interest.

Abbreviations

IR	Ionizing radiation
IARC	International Agency for Research on Cancer
UV	Ultraviolet radiation
ICRP	International Commission on Radiological Protection
SI	International System of Units
Gy	Gray
Sv	Sievert
LET	Linear energy transfer
RIBE	Radiation induced bystander effect
SCE	Sister chromatid exchanges
TSE	Tissue-sparing effect
MRT	Microbeam radiotherapy
ROS	Reactive oxygen species
HIF	Hypoxia inducible factor
CIRP	Cold-inducible RNA binding protein
UNSCEAR	United Nations Scientific Committee on the Effects of Atomic Radiation
LEO	Low-Earth orbit
HZE	High atomic number and energy
GCR	Galactic cosmic ray
NASA	National Aeronautics and Space Administration
MPS	Million Person Study of Low-Dose Health Effects

References

1. Little, J.B. Radiation carcinogenesis. *Carcinogenesis* **2000**, *21*, 397–404.
2. Shah, D.J.; Sachs, R.K.; Wilson, D.J. Radiation-induced cancer: A modern view. *Br. J. Radiol.* **2012**, *85*, e1166–e1173.
3. Muller, H.J. Artificial Transumutation of the Gene. *Science* **1927**, *66*, 84–87.
4. El Ghissassi, F.; Baan, R.; Straif, K.; Grosse, Y.; Secretan, B.; Bouvard, V.; Benbrahim-Tallaa, L.; Guha, N.; Freeman, C.; Galichet, L.; et al. A review of human carcinogens—Part D: Radiation. *Lancet Oncol.* **2009**, *10*, 751–752.
5. Jacobi, W. The concept of the effective dose—A proposal for the combination of organ doses. *Radiat. Environ. Biophys.* **1975**, *12*, 101–109.
6. International Commission on Radiological Protection. The 2007 Recommendations of the International Commission on Radiological Protection. ICRP publication 103. *Ann. ICRP* **2007**, *37*, 1–332.
7. Barnett, G.C.; West, C.M.L.; Dunning, A.M.; Elliott, R.M.; Coles, C.E.; Pharoah, P.D.P.; Burnet, N.G. Normal tissue reactions to radiotherapy: Towards tailoring treatment dose by genotype. *Nat. Rev. Cancer* **2009**, *9*, 134–142.
8. Mettler, F.A.; Huda, W.; Yoshizumi, T.T.; Mahesh, M. Effective Doses in Radiology and Diagnostic Nuclear Medicine: A Catalog. *Radiology* **2008**, *248*, 254–263.
9. Sazykina, T.G.; Kryshev, A.I. Manifestation of radiation effects in cold environment: Data review and modeling. *Radiat. Environ. Biophys.* **2011**, *50*, 105–114.
10. Gros, C.M.; Keiling, R.; Bloch, J. Radiation protection effect of cold in fish. *Strahlentherapie* **1959**, *109*, 241–245.
11. Patt, H.M.; Swift, M.N. Influence of temperature on the response of frogs to X irradiation. *Am. J. Physiol.* **1948**, *155*, 388–393.
12. Smith, F.; Grenan, M.M. Effect of hibernation upon survival time following whole-body irradiation in the marmot (Marmota monax). *Science* **1951**, *113*, 686–688.
13. Storer, J.B.; Hempelmann, L.H. Hypothermia and increased survival rate of infant mice irradiated with X-rays. *Am. J. Physiol.* **1952**, *171*, 341–348.
14. Hornsey, S. Protection from whole-body x-irradiation afforded to adult mice by reducing the body temperature. *Nature* **1956**, *178*, 87.
15. Hornsey, S. Discussion -The effect of hypothermia on the radiosensitivity of mice to whole-body X-irradiation. *Proc. R. Soc. London Ser. B Biol. Sci.* **1957**, *147*, 547–549.
16. Hajduković, S.; Hervé, A.; Vidović, V. Diminution de radiosensibilité du rat adulte en hypothermie profonde. *Experientia* **1954**, *10*, 343–344.
17. Musacchia, X.J.; Barr, R.E. Survival of whole-body-irradiated hibernating and active ground squirrels; Citellus tridecemlineatus. *Radiat. Res.* **1968**, *33*, 348–356.
18. Barr, R.E.; Musacchia, X.J. The Effect of Body Temperature and Postirradiation Cold Exposure on the Radiation Response of the Hibernator Citellus tridecemlineatus. *Radiat. Res.* **1969**, *38*, 448.
19. Weiss, L. Alteration of radio-sensitivity of the testis by extreme hypothermia. *J. Endocrinol.* **1959**, *19*, 22–28.
20. Weiss, L. The alteration in radiosensitivity of the intact mouse spleen by extreme hypothermia. *Br. J. Radiol.* **1960**, *33*, 32–35.
21. Hornsey, S. Fertility and lifespan of mice protected by hypothermia against total-body irradiation. *Gerontology* **1959**, *3*, 128–136.
22. Weiss, L. Alterations in radiosensitivity of the haemopoietic system of the mouse produced by extreme hypothermia. *Int. J. Radiat. Biol.* **1960**, *2*, 409–423.
23. Bloch, M.; Bloom, H.J.; Penman, J.; Walsh, L. Irradiation of cerebral astrocytomata under whole-body hypothermia. *Lancet* **1961**, *2*, 906–909.
24. Bloch, M.; Bloom, H.J.; Penman, J.; Walsh, L. Observations on patients with cerebral astrocytoma (Glioblastoma multiforme) treated by irradiation under whole-body hypothermia. *Br. J. Cancer* **1966**, *20*, 722–728.
25. Weiss, L. Sensitivity of Hypothermic Mammals to X Irradiation. *Br. Med. Bull.* **1961**, *17*, 70–73.
26. Hrycushko, B.A.; Bing, C.; Futch, C.; Wodzak, M.; Stojadinovic, S.; Medin, P.M.; Chopra, R. Technical Note: System for evaluating local hypothermia as a radioprotector of the rectum in a small animal model. *Med. Phys.* **2017**, *44*, 3932–3938.

27. Hrycushko, B.A.; Chopra, R.; Sayre, J.W.; Richardson, J.A.; Folkert, M.R.; Timmerman, R.D.; Medin, P.M. Local Hypothermia as a Radioprotector of the Rectal Wall During Prostate Stereotactic Body Radiation Therapy. *Int. J. Radiat. Oncol. Biol. Phys.* **2017**, *98*, 75–82.
28. Averbeck, D.; Salomaa, S.; Bouffler, S.; Ottolenghi, A.; Smyth, V.; Sabatier, L. Progress in low dose health risk research. *Mutat. Res. Mutat. Res.* **2018**, *776*, 46–69.
29. Goodhead, D.T. Initial events in the cellular effects of ionizing radiations: Clustered damage in DNA. *Int. J. Radiat. Biol.* **1994**, *65*, 7–17.
30. Booz, J.; Feinendegen, L.E. A microdosimetric understanding of low-dose radiation effects. *Int. J. Radiat. Biol. Relat. Stud. Phys. Chem. Med.* **1988**, *53*, 13–21.
31. International Commission on Radiation Units and Measurements. *Microdosimetry. ICRU Report 36*; International Commission on Radiation Units and Measurements: Bethesda, MD, USA, 1983.
32. Prise, K.M.; Folkard, M.; Michael, B.D. Bystander responses induced by low LET radiation. *Oncogene* **2003**, *22*, 7043–7049.
33. Nagasawa, H.; Little, J.B. Induction of sister chromatid exchanges by extremely low doses of alpha-particles. *Cancer Res.* **1992**, *52*, 6394–6396.
34. Lorimore, S.A.; Kadhim, M.A.; Pocock, D.A.; Papworth, D.; Stevens, D.L.; Goodhead, D.T.; Wright, E.G. Chromosomal instability in the descendants of unirradiated surviving cells after alpha-particle irradiation. *Proc. Natl. Acad. Sci. USA* **1998**, *95*, 5730–5733.
35. Belyakov, O.V.; Prise, K.M.; Trott, K.R.; Michael, B.D. Delayed lethality, apoptosis and micronucleus formation in human fibroblasts irradiated with X-rays or alpha-particles. *Int. J. Radiat. Biol.* **1999**, *75*, 985–993.
36. Nagasawa, H.; Little, J.B. Unexpected sensitivity to the induction of mutations by very low doses of alpha-particle radiation: Evidence for a bystander effect. *Radiat. Res.* **1999**, *152*, 552–557.
37. Belyakov, O.V.; Folkard, M.; Mothersill, C.; Prise, K.M.; Michael, B.D. Bystander-induced apoptosis and premature differentiation in primary urothelial explants after charged particle microbeam irradiation. *Radiat. Prot. Dosim.* **2002**, *99*, 249–251.
38. Iyer, R.; Lehnert, B.E. Low dose, low-LET ionizing radiation-induced radioadaptation and associated early responses in unirradiated cells. *Mutat. Res.* **2002**, *503*, 1–9.
39. Goldberg, Z.; Lehnert, B.E. Radiation-induced effects in unirradiated cells: A review and implications in cancer. *Int. J. Oncol.* **2002**, *21*, 337–349.
40. Schültke, E.; Balosso, J.; Breslin, T.; Cavaletti, G.; Djonov, V.; Esteve, F.; Grotzer, M.; Hildebrandt, G.; Valdman, A.; Laissue, J. Microbeam radiation therapy—grid therapy and beyond: A clinical perspective. *Br. J. Radiol.* **2017**, *90*, 20170073.
41. Slatkin, D.N.; Spanne, P.; Dilmanian, F.A.; Gebbers, J.O.; Laissue, J.A. Subacute neuropathological effects of microplanar beams of x-rays from a synchrotron wiggler. *Proc. Natl. Acad. Sci. USA* **1995**, *92*, 8783–8787.
42. Laissue, J.A.; Geiser, G.; Spanne, P.O.; Dilmanian, F.A.; Gebbers, J.O.; Geiser, M.; Wu, X.Y.; Makar, M.S.; Micca, P.L.; Nawrocky, M.M.; et al. Neuropathology of ablation of rat gliosarcomas and contiguous brain tissues using a microplanar beam of synchrotron-wiggler-generated X rays. *Int. J. Cancer* **1998**, *78*, 654–660.
43. Dilmanian, F.A.; Zhong, Z.; Bacarian, T.; Benveniste, H.; Romanelli, P.; Wang, R.; Welwart, J.; Yuasa, T.; Rosen, E.M.; Anschel, D.J. Interlaced x-ray microplanar beams: A radiosurgery approach with clinical potential. *Proc. Natl. Acad. Sci. USA* **2006**, *103*, 9709–9714.
44. van der Sanden, B.; Bräuer-Krisch, E.; Siegbahn, E.A.; Ricard, C.; Vial, J.-C.; Laissue, J. Tolerance of arteries to microplanar X-ray beams. *Int. J. Radiat. Oncol. Biol. Phys.* **2010**, *77*, 1545–1552.
45. Bouchet, A.; Lemasson, B.; Christen, T.; Potez, M.; Rome, C.; Coquery, N.; Le Clec'h, C.; Moisan, A.; Bräuer-Krisch, E.; Leduc, G.; et al. Synchrotron microbeam radiation therapy induces hypoxia in intracerebral gliosarcoma but not in the normal brain. *Radiother. Oncol.* **2013**, *108*, 143–148.
46. Grotzer, M.A.; Schültke, E.; Bräuer-Krisch, E.; Laissue, J.A. Microbeam radiation therapy: Clinical perspectives. *Phys. Med.* **2015**, *31*, 564–567.
47. Fukunaga, H.; Kaminaga, K.; Sato, T.; Butterworth, K.T.; Watanabe, R.; Usami, N.; Ogawa, T.; Yokoya, A.; Prise, K.M. High-precision microbeam radiotherapy reveals testicular tissue-sparing effects for male fertility preservation. *Sci. Rep.* **2019**, *9*, 12618.
48. Dilmanian, F.A.; Qu, Y.; Feinendegen, L.E.; Peña, L.A.; Bacarian, T.; Henn, F.A.; Kalef-Ezra, J.; Liu, S.; Zhong, Z.; McDonald, J.W. Tissue-sparing effect of x-ray microplanar beams particularly in the CNS: Is a bystander effect involved? *Exp. Hematol.* **2007**, *35*, 69–77.

49. Fukunaga, H.; Prise, K.M. Non-uniform radiation-induced biological responses at the tissue level involved in the health risk of environmental radiation: A radiobiological hypothesis. *Environ. Health* **2018**, *17*, 93.
50. Di Gregorio, A.; Bowling, S.; Rodriguez, T.A. Cell Competition and Its Role in the Regulation of Cell Fitness from Development to Cancer. *Dev. Cell* **2016**, *38*, 621–634.
51. Maruyama, T.; Fujita, Y. Cell competition in mammals—Novel homeostatic machinery for embryonic development and cancer prevention. *Curr. Opin. Cell Biol.* **2017**, *48*, 106–112.
52. Biteau, B.; Hochmuth, C.E.; Jasper, H. Maintaining Tissue Homeostasis: Dynamic Control of Somatic Stem Cell Activity. *Cell Stem Cell* **2011**, *9*, 402–411.
53. Eming, S.A.; Martin, P.; Tomic-Canic, M. Wound repair and regeneration: Mechanisms, signaling, and translation. *Sci. Transl. Med.* **2014**, *6*, 265sr6.
54. Turk, E.E. Hypothermia. *Forensic Sci. Med. Pathol.* **2010**, *6*, 106–115.
55. Wood, S.C. Interactions Between Hypoxia and Hypothermia. *Annu. Rev. Physiol.* **1991**, *53*, 71–85.
56. Gray, L.H.; Conger, A.D.; Ebert, M.; Hornsey, S.; Scott, O.C. The concentration of oxygen dissolved in tissues at the time of irradiation as a factor in radiotherapy. *Br. J. Radiol.* **1953**, *26*, 638–648.
57. Conger, A.D. The effect of oxygen on the radiosensitivity of mammalian cells. *Radiology* **1956**, *66*, 63–69.
58. Hall, E.J.; Giaccia, A.J. *Radiobiology for the Radiologist*; Lippincott Williams & Wilkins: Philadelphia, PA, USA, 2006; ISBN 0781741513.
59. Morse, M.L.; Dahl, R.H. Cellular glutathione is a key to the oxygen effect in radiation damage. *Nature* **1978**, *271*, 660–662.
60. Weiss, L. Some effects of hypothermia and hypoxia on the sensitivity of hela cells to x-rays. *Int. J. Radiat. Biol.* **1960**, *2*, 20–27.
61. Weiss, L. Decrease in radiosensitivity of the Intact Mouse Spleen produced by Hypoxia. *Nature* **1959**, *184*, 1156–1157.
62. Harada, H. How Can We Overcome Tumor Hypoxia in Radiation Therapy? *J. Radiat. Res.* **2011**, *52*, 545–556.
63. Wang, G.L.; Semenza, G.L. Purification and characterization of hypoxia-inducible factor. *J. Biol. Chem.* **1995**, *270*, 1230–1237.
64. Fukuda, R.; Zhang, H.; Kim, J.W.; Shimoda, L.; Dang, C.V.; Semenza, G.L.L. HIF-1 Regulates Cytochrome Oxidase Subunits to Optimize Efficiency of Respiration in Hypoxic Cells. *Cell* **2007**, *129*, 111–122.
65. Movafagh, S.; Crook, S.; Vo, K. Regulation of hypoxia-inducible Factor-1a by reactive oxygen species: New developments in an old debate. *J. Cell. Biochem.* **2015**, *116*, 696–703.
66. Pereira, T.; Zheng, X.; Poellinger, L. Degradation of the hypoxia-inducible factor 1α: Where does it happen? *Cell Cycle* **2006**, *5*, 2720–2722.
67. Wang, H.; Jiang, H.; Van De Gucht, M.; De Ridder, M. Hypoxic radioresistance: Can ROS be the key to overcome it? *Cancers* **2019**, *11*, 112.
68. Butterworth, K.T.; McGarry, C.K.; Trainor, C.; O'Sullivan, J.M.; Hounsell, A.R.; Prise, K.M. Out-of-Field Cell Survival Following Exposure to Intensity-Modulated Radiation Fields. *Int. J. Radiat. Oncol. Biol. Phys.* **2011**, *79*, 1516–1522.
69. Thompson, H.F.; Butterworth, K.T.; McMahon, S.J.; Ghita, M.; Hounsell, A.R.; Prise, K.M. The Impact of Hypoxia on Out-of-Field Cell Survival after Exposure to Modulated Radiation Fields. *Radiat. Res.* **2017**, *188*, 636–644.
70. Kempner, E.S.; Haigler, H.T. The influence of low temperature on the radiation sensitivity of enzymes. *J. Biol. Chem.* **1982**, *257*, 13297–13299.
71. Liao, Y.; Tong, L.; Tang, L.; Wu, S. The role of cold-inducible RNA binding protein in cell stress response. *Int. J. Cancer* **2017**, *141*, 2164–2173.
72. Egami, N.; Etoh, H. Effect of Temperature on the Rate of Recovery from Radiation-Induced Damage in the Fish Oryzias latipes. *Radiat. Res.* **1966**, *27*, 637.
73. Egami, N. Kinetics of recovery from injury after whole-body x-irradiation of the fish Oryzias latipes at different temperatures. *Radiat. Res.* **1969**, *37*, 192–201.
74. United Nations Scientific Committee on the Effects of Atomic Radiation. *Ionizing Radiation: Sources and Biological Effects: 1982 Report to the General Assembly, with Annexes*; United Nations: New York, NY, USA, 1982.
75. Baird, B.J.; Dickey, J.S.; Nakamura, A.J.; Redon, C.E.; Parekh, P.; Griko, Y.V.; Aziz, K.; Georgakilas, A.G.; Bonner, W.M.; Martin, O.A. Hypothermia postpones DNA damage repair in irradiated cells and protects against cell killing. *Mutat. Res. Fundam. Mol. Mech. Mutagen.* **2011**, *711*, 142–149.

76. Dang, L.; Lisowska, H.; Manesh, S.S.; Sollazzo, A.; Deperas-Kaminska, M.; Staaf, E.; Haghdoost, S.; Brehwens, K.; Wojcik, A. Radioprotective effect of hypothermia on cells a multiparametric approach to delineate the mechanisms. *Int. J. Radiat. Biol.* **2012**, *88*, 507–514.
77. Lisowska, H.; Wegierek-Ciuk, A.; Banasik-Nowak, A.; Braziewicz, J.; Wojewodzka, M.; Wojcik, A.; Lankoff, A. The dose-response relationship for dicentric chromosomes and γ-H2AX foci in human peripheral blood lymphocytes: Influence of temperature during exposure and intra- and inter-individual variability of donors. *Int. J. Radiat. Biol.* **2013**, *89*, 191–199.
78. Lisowska, H.; Brehwens, K.; Zölzer, F.; Wegierek-Ciuk, A.; Czub, J.; Lankoff, A.; Haghdoost, S.; Wojcik, A. Effect of hypothermia on radiation-induced micronuclei and delay of cell cycle progression in TK6 cells. *Int. J. Radiat. Biol.* **2014**, *90*, 318–324.
79. Cheng, L.; Lisowska, H.; Sollazzo, A.; Wegierek-Ciuk, A.; Stepień, K.; Kuszewski, T.; Lankoff, A.; Haghdoost, S.; Wojcik, A. Modulation of radiation-induced cytogenetic damage in human peripheral blood lymphocytes by hypothermia. *Mutat. Res. Genet. Toxicol. Environ. Mutagen.* **2015**, *793*, 96–100.
80. Lisowska, H.; Cheng, L.; Sollazzo, A.; Lundholm, L.; Wegierek-Ciuk, A.; Sommer, S.; Lankoff, A.; Wojcik, A. Hypothermia modulates the DNA damage response to ionizing radiation in human peripheral blood lymphocytes. *Int. J. Radiat. Biol.* **2018**, *94*, 551–557.
81. Chancellor, J.C.; Blue, R.S.; Cengel, K.A.; Auñón-Chancellor, S.M.; Rubins, K.H.; Katzgraber, H.G.; Kennedy, A.R. Limitations in predicting the space radiation health risk for exploration astronauts. *npj Microgravity* **2018**, *4*, 8.
82. Cucinotta, F.A.; Durante, M. Cancer risk from exposure to galactic cosmic rays: Implications for space exploration by human beings. *Lancet Oncol.* **2006**, *7*, 431–435.
83. Durante, M.; Cucinotta, F.A. Heavy ion carcinogenesis and human space exploration. *Nat. Rev. Cancer* **2008**, *8*, 465–472.
84. Fukunaga, H.; Yokoya, A.; Taki, Y.; Butterworth, K.T.; Prise, K.M. Precision Radiotherapy and Radiation Risk Assessment: How Do We Overcome Radiogenomic Diversity? *Tohoku J. Exp. Med.* **2019**, *247*, 223–235.
85. Furukawa, S.; Nagamatsu, A.; Nenoi, M.; Fujimori, A.; Kakinuma, S.; Katsube, T.; Wang, B.; Tsuruoka, C.; Shirai, T.; Nakamura, A.J.; et al. Space Radiation Biology for "Living in Space". *Biomed Res. Int.* **2020**, *2020*, 4703286.
86. United Nations Scientific Committee on the Effects of Atomic Radiation. *Effects of Ionizing Radiation: United Nations Scientific Committee on the Effects of Atomic Radiation: UNSCEAR 2006*; United Nations: New York, NY, USA, 2008.
87. Boice, J.D. The Million Person Study relevance to space exploration and mars. *Int. J. Radiat. Biol.* **2019**, 1–9.
88. Cucinotta, F.A. Review of NASA approach to space radiation risk assessments for mars exploration. *Health Phys.* **2015**, *108*, 131–142.
89. Boice, J.D.; Cohen, S.S.; Mumma, M.T.; Ellis, E.D. The Million Person Study, whence it came and why. *Int. J. Radiat. Biol.* **2019**, 1–14.
90. Tinganelli, W.; Hitrec, T.; Romani, F.; Simoniello, P.; Squarcio, F.; Stanzani, A.; Piscitiello, E.; Marchesano, V.; Luppi, M.; Sioli, M.; et al. Hibernation and radioprotection: Gene expression in the liver and testicle of rats irradiated under synthetic torpor. *Int. J. Mol. Sci.* **2019**, *20*, 352.
91. Kumar, D.; Salian, S.R.; Kalthur, G.; Uppangala, S.; Kumari, S.; Challapalli, S.; Chandraguthi, S.G.; Krishnamurthy, H.; Jain, N.; Kumar, P.; et al. Semen abnormalities, sperm DNA damage and global hypermethylation in health workers occupationally exposed to ionizing radiation. *PLoS ONE* **2013**, *8*, e69927.
92. Fukunaga, H.; Butterworth, K.T.; Yokoya, A.; Ogawa, T.; Prise, K.M. Low-dose radiation-induced risk in spermatogenesis. *Int. J. Radiat. Biol.* **2017**, *93*, 1291–1298.

MDPI
St. Alban-Anlage 66
4052 Basel
Switzerland
Tel. +41 61 683 77 34
Fax +41 61 302 89 18
www.mdpi.com

International Journal of Molecular Sciences Editorial Office
E-mail: ijms@mdpi.com
www.mdpi.com/journal/ijms

www.ingramcontent.com/pod-product-compliance
Lightning Source LLC
Chambersburg PA
CBHW041939260926
51850CB00018B/295
9783036574875